Therapeutic Potential of Cell Cycle Kinases in Breast Cancer

Manzoor Ahmad Mir
Editor

Therapeutic Potential of Cell Cycle Kinases in Breast Cancer

Springer

Editor
Manzoor Ahmad Mir
Department of Bioresources
University of Kashmir
Srinagar, Jammu and Kashmir, India

ISBN 978-981-19-8910-0 ISBN 978-981-19-8911-7 (eBook)
https://doi.org/10.1007/978-981-19-8911-7

This Springer imprint is published by the registered company Springer Nature Singapore Pte Ltd.
The registered company address is: 152 Beach Road, #21-01/04 Gateway East, Singapore 189721, Singapore

This Book is Dedicated to my Family
Ayzel Manzoor Mir
Aariz Manzoor Mir
Sumaira Manzoor Mir

Foreword

I am pleased to provide the foreword to Dr. Manzoor Ahmad Mir's valuable book, *Therapeutic Potential of Cell Cycle Kinases in Breast Cancer*. Breast cancer is a dreadful disease that causes physical and mental suffering to individuals who are diagnosed with it. Despite huge investments in breast cancer treatment, the number of new cases and deaths continues to rise. As per the GLOBOCON 2021 report, breast cancer is the second most prevalent cancer diagnosed in women after skin cancer. Breast cancer can strike both men and women, and 1 among 8 in females and 1 among 1000 in males can get affected. The rate of incidence of breast cancer is more than 1.3 million on a yearly basis. BC can be cured in almost 70–80% of patients having early-stage, non-metastatic BC.

The cell cycle is controlled by various cyclins and Cyclin-Dependent Kinases (CDKs). The importance of cyclins and CDKs in the cell cycle was revealed in a study of fission yeast and cell division, in which the significance of complex Cdc2 (CDK1) was genetically studied and exhibited an effective role in mitosis. Dr. Mir discusses critical issues about the incidence, dysregulation of CDKs in breast cancer, treatment, and prevention of breast cancer using CDK inhibitors. The book has particularly highlighted the conventional as well as the newly developed treatment approaches including various CDK inhibitors in breast cancer. In this regard, the new innovative treatment method, especially the targeted therapies, CDK inhibitors, and the nanotechnology intervention approaches, has revolutionized the field of breast cancer.

Thus, targeting CDKs with their specific inhibitors in BC is considered to be very useful. In this book, Dr. Mir has shed light on the role of cell cycle in cancer progression, role of CDKs and their dysregulation in breast cancer, their importance in BC progression and metastasis, their prognostic significance, and the specific CDK inhibitors used to overcome the BC progression.

Raid Saleem Albaradie
Department of Medical Laboratory
Majmaah University
Al Majma'ah, Saudi Arabia

Preface

As per the GLOBOCON 2021 report breast cancer is the second most prevalent cancer diagnosed in women after skin cancer. Breast cancer can strike both men and women, and 1 among 8 in females and 1 among 1000 in males can get affected. The rate of incidence of breast cancer is more than 1.3 million on a yearly basis. BC can be cured in almost 70–80% of patients having early-stage, non-metastatic BC. In this book, we will shed light on the role of cell cycle in cancer progression, role of cyclin-dependent kinases (CDKs) and their dysregulation in breast cancer, their importance in BC progression and metastasis, their prognostic significance, and the specific CDK inhibitors used to overcome the BC progression. Among the various hallmarks of cancer, increased cell proliferation is one of the most important aspects that need to be taken into consideration. The proliferation of cells is synchronized by the cell cycle, and there are well-defined regulatory mechanisms that regulate the cell cycle. The cell cycle is controlled by various cyclins and CDKs. The importance of cyclins and CDKs in the cell cycle was revealed in a study of fission yeast and cell division, in which the significance of complex Cdc2 (CDK1) was genetically studied and exhibited an effective role in mitosis. CDKs belong to the serine-threonine kinases that can form an association with cyclins, phosphorylate them and thus activate them at specific positions during the cell cycle progression. The cell division during an individual's life span as well as during development takes place only at specific places as well as a specific time and divides the content of the cell in a very accurate way. The coherence, integrity as well as maintenance of every step in the cell cycle are well maintained by the cell cycle checkpoints. Breast cancer like many other cancers involves increased proliferation of cells, which results from the disruption in the cell cycle regulation by dysregulated CDKs. The dysregulation in the CDKs during BC leads to the uncontrolled proliferation of cancer cells, thus maintaining the progression of BC along with other factors. The studies have revealed that BC is associated with dysregulation of various cyclin/CDKs and their dysregulation does have a role in developing different phenotypes of BC.

In this book, Dr. Mir has discussed critical issues about the incidence, dysregulation of CDKs in breast cancer, treatment, and prevention of breast cancer using CDK inhibitors. The book has particularly highlighted the conventional as well as the newly developed treatment approaches including various CDK inhibitors in breast cancer. In this regard, the new innovative treatment method, especially the targeted therapies, CDK inhibitors, and the nanotechnology intervention approaches, has revolutionized the field of breast cancer. In this book, we will shed light on the role of cell cycle in cancer progression, role of CDKs and their dysregulation in breast cancer, their importance in BC progression and metastasis, their prognostic significance, and the specific CDK inhibitors used to overcome the BC progression.

Srinagar, Jammu and Kashmir, India

Manzoor Ahmad Mir

Acknowledgments

"First and foremost, I would like to thank the Prime Mover Almighty in whom I have great faith."

Although it is difficult to thank everybody and interweave in words the genuine efforts made by the people directly or indirectly to make this book possible, I would like to take this opportunity to pen down my appreciation for a great number of people whose contribution in numerous ways immensely helped me to move towards my destination. It gives me immense pleasure to express my sense of gratitude and respect to my mentor Dr. Javed Naim Agrewala, whose constant support, utmost patience, invaluable advice, and guidance rekindled my interest in science from time to time. I will benefit for a long time to come from his sincerity, originality, and truthfulness, which has nourished my intellectual maturity. A special vote of thanks goes to Prof. Talat Ahmad and Prof. Neolofar Khan for their efficient guidance to work hard with sincerity, dedication, and devotion both in professional and personal life. The staff, scholars Mr. Umar Mehraj, Mr. Basharat, Mr. Bashir, Mr. Wajahat, Miss Hina Qayoom, Miss Hafsa Qadri, Miss Shazia Sofi, Miss Shariqa Jan, and the students at the Department of Bioresources deserve a special mention for all the help rendered at the hour of need.

I am also thankful to my soulmate Miss Sumaira Manzoor and my parents Mohammad Abdullah Mir, Zoona Begum and my in laws Mohd Ashraf Malik, and Posha Begum for their constant support, love, and care throughout the whole journey. I would like to mention here the affection and love showered by my angels Ayzel Manzoor and Aariz Manzoor throughout the writing process of the book. I am short of words to justify their care, affection, and untiring support showered by these very good friends.

Finally, the author would like to thank and acknowledge the *Jammu and Kashmir Science Technology and Innovation Council (JKST&IC)* for providing the financial support vide *Project No. JKST&IC/SRE/885-87, dated 27-10-2021* sanctioned to Dr. Manzoor Ahmad Mir (Principal Investigator), and this book is part of the same project entitled *"Targeting Drug Resistant Cancer Stem Cells by Combinational Delivery of Paclitaxel and Quercetin Zerumbone in Breast Cancer."*

Dr. Manzoor A Mir

Contents

1 **Introduction to Breast Cancer** 1
Manzoor Ahmad Mir and Hina Qayoom

2 **Current Treatment Approaches to Breast Cancer** 23
Manzoor Ahmad Mir and Abrar Yousuf Mir

3 **Introduction to Cell Cycle and Its Regulators** 53
Manzoor Ahmad Mir and Asma Jan

4 **Cell Cycle and Cancer** 83
Manzoor Ahmad Mir and Shazia Sofi

5 **Cell Cycle Dysregulation in Breast Cancer** 103
Manzoor Ahmad Mir, Sameer Ullah Khan, and Shariqa Aisha

6 **Molecular Subtypes of Breast Cancer and CDk Dysregulation** 133
Manzoor Ahmad Mir and Ifshana Mohi Ud Din

7 **Breast Tumor Microenvironment and CDKs** 149
Manzoor Ahmad Mir and Abrar Yousuf Mir

8 **CDK Dysregulation in Breast Cancer: A Bioinformatics Analysis** . . 175
Manzoor Ahmad Mir, Shazia Sofi, and Pir M. Ishfaq

9 **CDK1 Dysregulation in Breast Cancer** 195
Manzoor Ahmad Mir and Burhan Ul Haq

10 **Cdk4/Cdk6 Dysregulation in Estrogen-Positive Receptor Breast Cancers** 211
Manzoor Ahmad Mir and Ulfat Jan

11 **Therapeutic Implications of CDKs in Breast Cancer** 233
Manzoor Ahmad Mir and Burhan Ul Haq

12 **Novel CDK Inhibitors in Breast Cancer** 253
Manzoor Ahmad Mir and Tabish Javeed

13 Targeting CDKs with Other Chemotherapeutic Drugs: A Combinatorial Approach 269
Manzoor Ahmad Mir and Umar Y. Mir

14 CDKs in Cell-Cycle Progression and Therapeutic Strategies in Human Breast Cancer 291
Manzoor Ahmad Mir and Ulfat Jan

15 CDk Inhibitor for Treatment of Breast Cancer 313
Manzoor Ahmad Mir, Aabida Gul, Shazia Sofi, and M. Sultan Khan

16 Response of Therapy in Cell-Cycle Regulatory Genes in Breast Cancer 335
Manzoor Ahmad Mir, Shariqa Aisha, Kaneez Fatima, and Fayaz Ahmad Malik

17 Different Cyclins and Their Significance in Breast Cancer 355
Manzoor Ahmad Mir and Hina Qayoom

Editor and Contributors

About the Editor

Manzoor Ahmad Mir M.Sc., Ph.D., FRSB Dr. Manzoor Ahmad Mir holds a master's degree in Zoology with Gold Medal from HNBG Central University and after qualifying prestigious national level CSIR-JRF-NET examination he worked jointly for his Ph.D. at Jawaharlal Nehru University, New Delhi, and CSIR-Institute of Microbial Technology Chandigarh in the field of Immunopathology. His basic research interests include molecular immunology and immunobiology of microbes (MTB). His Ph.D. research title was "Role of co-stimulation in the survival of intracellular pathogens." He has published more than 50 high-impact research papers and book chapters, in recognition of which he has received several awards and royalties from international publishing houses. Dr. Manzoor has authored more than 15 books with international publishers. He is on the editorial board and reviewer of some prestigious journals and has been an invited speaker at various scientific meetings/conferences within India and abroad. He is member of many scientific organizations and societies like American Association of Cancer Research, Fellow of Royal British Society, International Immunology Association, Indian Cancer Society, Indian Immunology Society, Indian National Science Association, IMMUNOCON, etc. Dr. Manzoor was awarded Teachers Associateship for Research Excellence (TARE) Fellowship by DST Govt. of India. He has been awarded Summer Research Fellowship Programme (SRFP-2019) by the Indian Academy of Sciences and the National Science Academy. He was awarded a research project on combination therapy in breast cancer by J&K Science Technology and Innovation. Dr. Manzoor has developed Massive Open Online Courses (MOOCs) in Immunology and Endocrinology for UG students sanctioned by UGC-Consortium for Educational Communication (CEC) SWAYAM Ministry of HRD, Govt of India. He has research and teaching experience of 15 years. He currently teaches Cancer Biology and Immunology at the Department of Bioresources, School of Biological Sciences, University of Kashmir. He is presently heading the Department of Bioresources, University of Kashmir.

Department of Bioresources, School of Biological Sciences, University of Kashmir, Srinagar, India

Contributors

Shariqa Aisha Department of Bioresources, School of Biological Sciences, University of Kashmir, Srinagar, J&K, India

Ifshana Mohi Ud Din Department of Bioresources, School of Biological Sciences, University of Kashmir, Srinagar, J&K, India

Kaneez Fatima Pharmacology Division, CSIR-Indian Institute of Integrative Medicine, Sanat Nagar, Srinagar, J&K, India

Aabida Gul Department of Bioresources, School of Biological Sciences, University of Kashmir, Srinagar, J&K, India

Burhan Ul Haq Department of Biotechnology, Central University of Kashmir, Ganderbal, J&K, India

Asma Jan Department of Bioresources, School of Biological Sciences, University of Kashmir, Srinagar, J&K, India

Ulfat Jan Department of Bioresources, School of Biological Sciences, University of Kashmir, Srinagar, J&K, India

Tabish Javeed Department of Bioresources, School of Biological Sciences, University of Kashmir, Srinagar, J&K, India

M. Sultan Khan Neurobiology and Molecular Chronobiology Laboratory, Department of Animal Biology, School of Life Sciences, University of Hyderabad, Hyderabad, India

Sameer Ullah Khan Pharmacology Division, CSIR-Indian Institute of Integrative Medicine, Srinagar, J&K, India

Fayaz Ahmad Malik Pharmacology Division, CSIR-Indian Institute of Integrative Medicine, Sanat Nagar, Srinagar, J&K, India

Pir Ishfaq Department of Bioresources, School of Biological Sciences, University of Kashmir, Srinagar, J&K, India

Abrar Yousuf Mir Department of Bioresources, School of Biological Sciences, University of Kashmir, Srinagar, J&K, India

Manzoor Ahmad Mir Department of Bioresources, School of Biological Sciences, University of Kashmir, Srinagar, J&K, India

Hina Qayoom Department of Bioresources, School of Biological Sciences, University of Kashmir, Srinagar, J&K, India

Shazia Sofi Department of Bioresources, School of Biological Sciences, University of Kashmir, Srinagar, J&K, India

Umar Yousuf Department of Bioresources, School of Biological Sciences, University of Kashmir, Srinagar, J&K, India

Introduction to Breast Cancer 1

Manzoor Ahmad Mir and Hina Qayoom

1.1 Introduction

Breast cancer is the leading cause of death among women worldwide (Fig. 1.1) (Spitale et al. 2009). It is a heterogenous disease that comprises many different subgroups that differ in their distinct pathological features and clinical significance (Tang et al. 2008; Desmedt et al. 2009; Sotiriou and Pusztai 2009; Spitale et al. 2009; Iwamoto and Pusztai 2010; Reis-Filho et al. 2010; Weigelt et al. 2010). These different subgroups with distinct histopathological and biological features lead to different responses to treatments and hence require different therapeutic strategies. Therefore, specific grouping of breast cancer into its subtypes is imperative for therapeutic decision-making and personalized treatment (Blows et al. 2010). The high-throughput screening of gene expression analysis like microarrays has revealed that the response of tumor cells to treatment is determined by intrinsic molecular characteristics than anatomical prognostic factors (Sotiriou and Pusztai 2009; Iwamoto and Pusztai 2010; Reis-Filho et al. 2010; Weigelt et al. 2010). This stratified and personalized approach accordingly will prove beneficial to increase the accuracy and reproducibility of diagnosing disease with better efficacy (Pusztai et al. 2008). A study was conducted using 456 cDNA clones, in accordance to which breast cancer is classified into five intrinsic subtypes such as Luminal A, Luminal B, HER2 overexpression, basal, and normal-like tumors (Perou et al. 2000; Sørlie et al. 2001). The fundamental rationale behind this classification is the underlying differential expression patterns among various breast cancer subtypes at the molecular level (Sørlie et al. 2003). These subtypes have been characterized by their immunohistochemistry status (IHC) as shown in Table 1.1.

M. A. Mir (✉) · H. Qayoom
Department of Bioresources, School of Biological Sciences, University of Kashmir, Srinagar, J&K, India
e-mail: drmanzoor@kashmiruniversity.ac.in

M. A. Mir (ed.), *Therapeutic Potential of Cell Cycle Kinases in Breast Cancer*,
https://doi.org/10.1007/978-981-19-8911-7_1

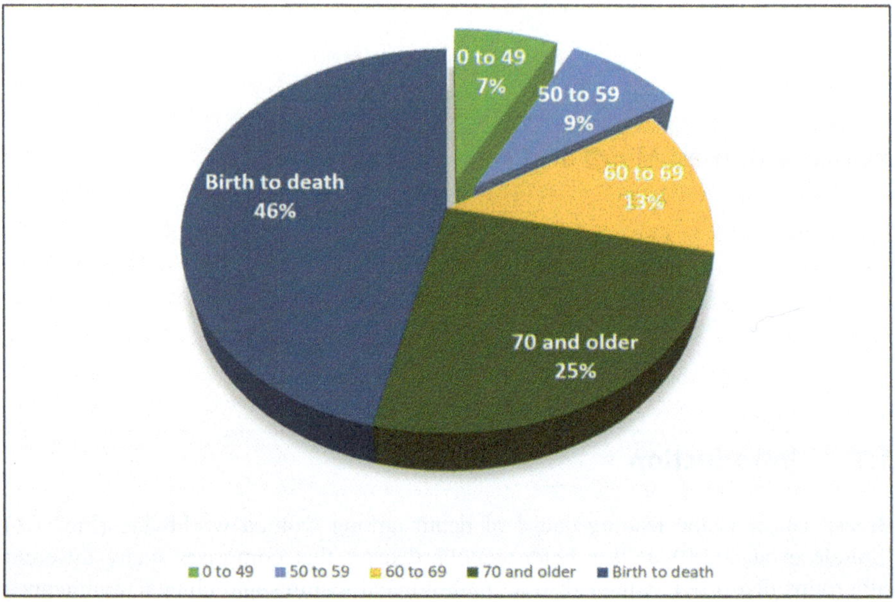

Fig. 1.1 Probability (%) of developing invasive breast cancer selected by age intervals (American cancer society, 2022)

Table 1.1 Breast cancer subtypes according to IHC status

Intrinsic subtype	Grade	IHC status	Outcome	Incidence
Luminal A	1/2	ER+, PR+, HER2−, Ki67−	Good	23.7% [p1] Cheang et al. (2009)
Luminal B	2/3	ER+, PR+, HER2−, Ki67+ ER+, PR+, HER2+, Ki67+	Intermediate Poor	38.8% [p1] Cheang et al. (2009) 14% [p1] Cheang et al. (2009)
HER2 overexpression	2/3	[ER−PR−] HER2+	Poor	11.2% [p1] Cheang et al. (2009)
Basal	3	[ER−PR−] HER2−, basal marker +	Poor	12.3% [p1] Cheang et al. (2009)
Normal-like	1/2/3	ER−PR−] HER2−, Ki67−	Intermediate	7.8% [p2] Smid et al. (2008)

World Health Organization (WHO) has defined breast cancer into 21 types based on histomorphology and growth patterns (Dieci et al. 2014). Breast cancer has been characterized into two broad categories such as in situ carcinoma and invasive carcinoma. In situ carcinoma include: Ductal carcinoma in situ (DCIS) and lobular carcinoma in situ (LCIS); both of these are differentiated by cytological features and growth patterns with DCIS further classified by tumor architecture (Gautam and Malhotra 2010). Whereas, the invasive carcinoma subtypes are characterized on the basis of their structural form (tubular, medullary, and papillary), by their secretion (mucinous/colloid), and architecture.

Breast cancer has been characterized into two broad categories such as in situ carcinoma and invasive carcinoma. In situ carcinoma include: Ductal carcinoma in situ (DCIS) and lobular carcinoma in situ (LCIS); both of these are differentiated by cytological features and growth patterns with DCIS further classified by tumor architecture (Gautam and Malhotra 2010). Whereas, the invasive carcinoma subtypes are characterized on the basis of their structural form (tubular, medullary, and papillary), by their secretion (mucinous/colloid), and architecture.

1.2 History of Breast Cancer

Breast cancer as a disease has been studied since time immemorial. The history of this disease dates back to more than 3500 years ago in Egyptian history (Hellman 1993). Hippocrates (Father of Western medicine) described breast cancer as a humoral disease in 460 BC. Galen (131–203) described the secretion of black bile as the cause of breast cancer in the body. After the onset of the sixteenth century, new technologies were invented in surgery. Various experts suggested the surgical removal of pectoral muscle along with breasts (Guillemeau, 1550–1601), Vesalius (1514–64) used ligatures, Severinus (1580–1659) supported the removal of axillary nodes along with breasts as he was of the opinion that axillary nodes are part of the malignant process. A revolution came when Descartes (1596–1650) gave a lymph theory of origin that was later on perpetuated by John Hunter (1728–93), their theory was conceptually better than black bile theory and encouraged more and more surgeons to remove the already affected lymph nodes. In 1757, a French surgeon Henry LeDran made progressions to the theory that breast cancer at its earliest stage is a local disease, which then spreads to lymph nodes and then enters the circulation. This theory offered the hope that surgery might cure the disease if performed early. Other surgeons also embraced this concept. In 1871, Henry Arnott also reiterated the local origin of breast cancer and advocated the principle of curative surgery with en bloc operation at the earliest moment (Fig. 1.2) (Hellman 1993).

1.3 Surgery

The radical mastectomy approach was begun at the Johns Hopkins Hospital, Baltimore in 1882. The operation was performed by removing the skin and breast tissues beneath the pectoralis muscles with axillary lymph nodes. This surgery proved successful as it improved the survival of patients by reducing the relapse of the disease by 6%, previously known to range between 51–82% earlier on (Halsted

Fig. 1.2 Pioneers in the history of breast cancer (Lukong 2017)

1907). However, this radical surgery was associated with several problems like deformity in shape and sensory abnormalities of the arm and chest with lymphedema. Therefore, in this view by the onset of 1912, surgeons led by J.B. Murphy discontinued the resection of the pectoralis muscles (Murphy 1912). In 1948, in place of surgical removal of breast tissues, focus was shifted toward breast conservation with the advent of adjuvant radiation therapy (McWhirter 1948).

Later on, several studies were published in 1989 that approved the trend to move toward minimum surgery with more thrust on preservation of an essentially normal breast. Nowadays, surgical management of patients with breast cancer involves both primary tumors and region-specific lymphatics. The primary abnormal tissues are managed either by mastectomy (surgical removal of complete one or both breasts) or lumpectomy (removal of only abnormal tumors) with removal of lymph nodes or their biopsy. For treating locally advanced breast cancer neoadjuvant chemotherapy is the first step followed by modified radical mastectomy or lumpectomy (breast conservation therapy) according to the treatment requirement (McWhirter 1948) Fig. 1.3.

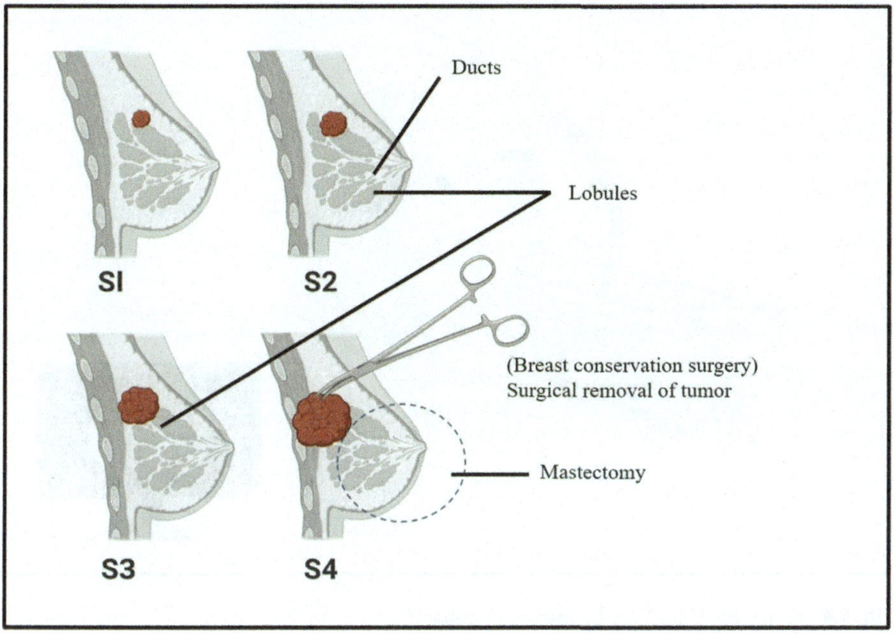

Fig. 1.3 Surgical resection of breast cancer tissue by both: Mastectomy (complete removal of breasts) and Lumpectomy (removal of abnormal tissues from breast)

1.4 Radiation Therapy

With the onset of the twentieth century, radiotherapy came into existence and was reported to be effective in treating breast cancer. In 1932, Pfahler reported that on application of radiotherapy 53 out of 1022 patients had early disease recovered from breast cancer. Therefore, it was concluded that radiotherapy was significant for patients with weak response for surgery or who refused to go under surgery (Pfahler 1932). In the mid of the twentieth century, Robert McWhirter was the first proponent of radiotherapy as he achieved a 5-year survival rate of 62% from treating nearly 759 patients with radiation therapy that was comparable to standard radical mastectomy. He also concluded that radiation therapy will prove beneficial in treating nodal disease (McWhirter 1948). Baclesse from France clearly concluded radiation therapy was most effective in treating breast cancers with keeping tumor size into consideration the likelihood of local control (Baclesse 1984). In support of it, a first randomized controlled trial of conservative surgery and radiotherapy versus radical mastectomy was performed at Guy's Hospital London by Atkins et al. in 1972 (Atkins et al. 1972). The three major studies such as the Danish study, the Canadian study, and the Danish study on the significance of local control and its effect on survival of patients have highlighted the importance of micro metastases in locoregional lymphatics are a potent source of systemic metastases, suggesting that

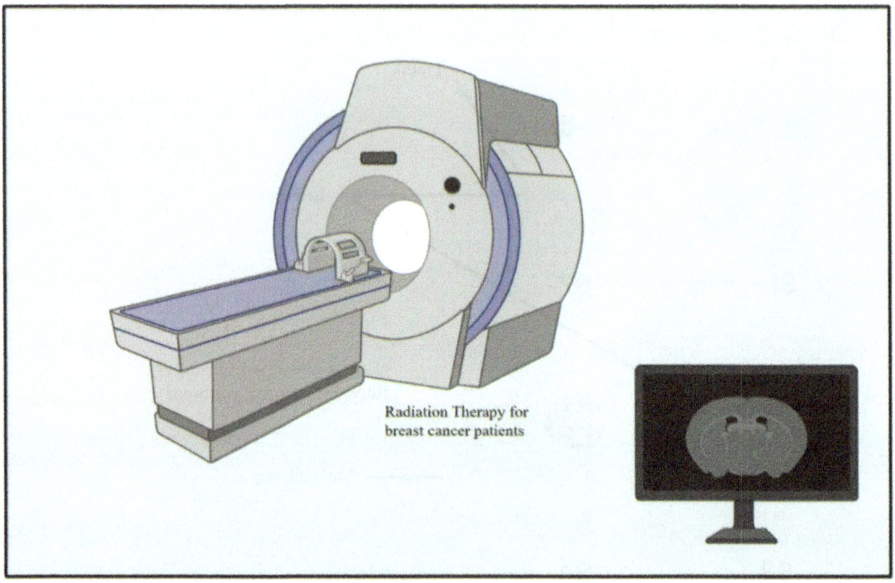

Fig. 1.4 Radiation therapy for breast cancer patients

locoregional metastatic eradication enhances the survival. In today's world, the radiation treatment of breast cancer has evolved from 2D to 3D Conformal Radiotherapy (3DCRT) and to accelerated partial breast irradiation (APBI) aiming to reduce normal tissue toxicity and overall treatment (Overgaard 1999; Overgaard et al. 1997; Ragaz et al. 1997) Fig. 1.4.

1.5 Systemic Therapy

The systemic therapy includes three main therapies for breast cancer treatment. The oldest being endocrine therapy (ET) for patient's expressing the three hormonal receptors, the second being chemotherapy (CT) and targeted therapy (TT) that include an increased list of new agents targeted against molecules responsible for tumor progression.

1.6 Hormone Therapy

By the end of the nineteenth century, Thomas Beatson had demonstrated the hormonal dependency of breast cancer in patients after their surgical oophorectomy. After which the importance of hormones was also confirmed by the use of hypophysectomy and adrenalectomy (Huggins and Bergenstal 1952; Luft and Olivecrona 1955; Beatson 1989). With the discovery of estrogen receptor (ER) in breast tumor

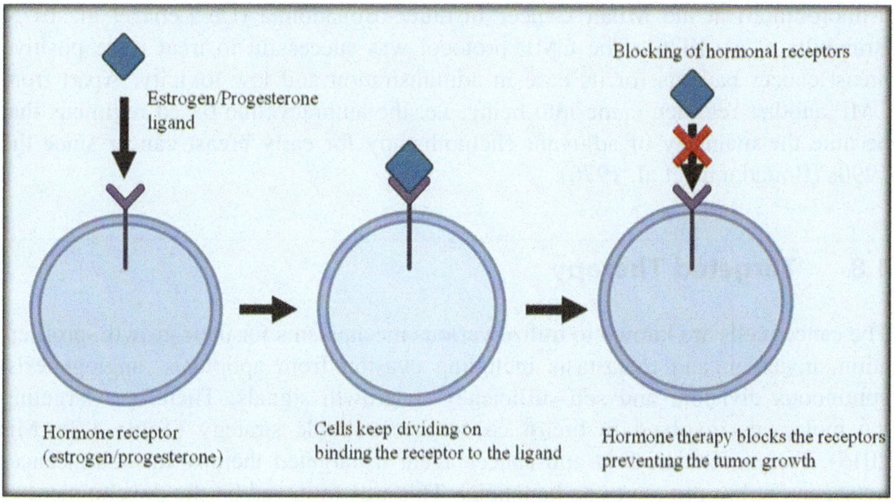

Fig. 1.5 Hormonal therapy to block and prevent tumor growth

the endocrine surgery has largely been replaced by endocrine therapies. The antagonist of estrogen receptor tamoxifen has replaced surgical oophorectomy (Mir et al. 2020). Similarly, hypophysectomy has been replaced by luteinizing hormone-releasing hormone (LHRH) agonists and adrenalectomy by aromatase inhibitors. Quite recently, the hormone receptors (estrogen and progesterone) and human epidermal growth factor receptor 2 (HER2) status identify key molecular subtypes of human breast tumors and nowadays guide choice of therapy. For patients with hormone-receptor positive (ER, PR) breast cancer, hormone therapy with tamoxifen and aromatase inhibitors (anastrozole, letrozole, exemestane) or ovarian suppression (in premenopausal patients) represent the backbone of treatment along with chemotherapy (Mir et al. 2020). For patients with hormone-receptor positive breast cancer the introduction of hormone receptor inhibitors has improved the prognosis of breast cancer patients (Dawood et al. 2008) Fig. 1.5.

1.7 Chemotherapy

During the fifth and sixth decade of the twentieth century, chemotherapy came into existence for the treatment of patients with several types of advanced solid tumors and hematologic neoplasms. Earlier breast cancer was treated using single-agent chemotherapy in hormone-resistant metastatic setting (Qayoom and Bhat 2020; Mehraj et al. 2021a). The chemotherapeutic agents were cyclophosphamide, phenylalanine mustard, vincristine, vinblastine, methotrexate and 5-fluorouracil. However, later on combination chemotherapy was approached, the first being Cooper's regimen (Akram and Siddiqui 2012; Qayoom et al. 2021a). In view of this, later on a protocol was introduced known as CMF protocol (cyclophosphamide-Methotrexate-

5-fluorouracil) at the Milan Cancer Institute, Bonadonna (De Lena et al. 1975; Brambilla et al. 1976). The CMF protocol was successful to treat node positive breast cancer patients for its ease in administration and low toxicity. Apart from CMF another regimen came into being, i.e. the anthracycline-based regimens that became the mainstay of adjuvant chemotherapy for early breast cancer since the 1990s (Bonadonna et al. 1976).

1.8 Targeted Therapy

The cancer cells are known to utilize various mechanisms for their growth, proliferation, invasion, and metastasis including evasion from apoptosis, angiogenesis, continuous division, and self-sufficiency in growth signals. Therefore, targeting the molecules involved in breast cancer is a reliable strategy (Table 1.2) (Mir 2015). Tamoxifen, the first anti-cancer agent in targeted therapy in breast cancer management has proven to be beneficial. This was followed by the development of Trastuzumab a recombinant humanized monoclonal antibody that inhibited HER-2/neu protein that also proved to be a significant cornerstone. As it was previously reported that 20–25% of breast cancer patients have HER2 overexpression that is directly involved in the dysregulation of the intracellular mitogenic signaling that is known to lead to aggressive tumor behavior (Slamon et al. 1987). Similarly, Bevacizumab was approved in combination with paclitaxel as a first-line treatment against vascular endothelial growth factor (VEGF) receptor family in HER2-negative metastatic breast cancer (Folkman 1995; Miller et al. 2007). Another, HER1/2 inhibitor Lapatinib in combination with capecitabine was approved to be beneficial in trastuzumab-pretreated metastatic breast cancer patients (Cetin et al. 2014). Recently published data indicate that a combination of two biological agents such as lapatinib and trastuzumab can be effective as a treatment beyond trastuzumab-related progression (Burris Iii 2004; Konecny et al. 2006). PARP inhibitors (poly ADP-ribose polymerase inhibition), mediated by a new class of small molecules, are an interesting area of investigation (Muñoz-Gámez et al. 2005; Drew and Calvert 2008). Future directions of research in HER2-positive breast cancer should focus on the evaluation of novel antibodies (pertuzumab, T-DM1), and irreversible TKIs (neratinib, BIBW 2992) and inhibitors of HER2-related downstream signaling (mTOR, TORC 1/2, PI3K/Akt), and of receptor cross-talk between estrogen receptor and insulin-like growth factor (IGFR) (Mir and Agrewala 2008; Mehraj et al. 2022b; Mehraj et al. 2022c) Fig. 1.6.

1.9 Intrinsic Molecular Sub-Typing of Breast Cancer

Since the last fifteen years, breast cancer has been subdivided into four main molecular subtypes: Luminal A, Luminal B, HER2-enriched, and Basal-like (Fig. 1.7).

Table 1.2 Additional genes associated with breast cancer

Gene	Abnormality in breast cancer	Location	Function	Reference
P53	Mutations in 30% of breast cancers	17p13.1	Tumor suppressor gene	Varna et al. (2011) and Hientz et al. (2017)
NME1	SNP of NME1 gene associated with higher breast cancer-specific mortality and patients with an early-stage cancer	17q21.3	Metastasis suppressor gene	Qu et al. (2008) and Roberts et al. (2017)
RB1	Rb1 inactivation in 20–35% of breast cancers	13q14.2	Tumor suppressor gene	Cheng et al. (2010)
PTEN	Loss of PTEN protein expression in up to 33% of breast cancers	10q23.3	Tumor suppressor gene	Loibl et al. (2016)
ATM	Mutation of ATM increases the risk of two–threefold in general and five–ninefold in women under age 50	11q22. q23	Tumor suppressor gene	Choi et al. (2016)
CDH1 (E-cadherin)	Inactivation of CDH1 in 85% of lobular breast carcinomas	16q22.1	Tumor suppressor gene	Desmedt et al. (2016)
FHIT	The rate of FHIT hypermethylation in breast cancer was 8.4 fold higher than that in normal breast tissues	3p14.2	Putative tumor suppressor gene	Su et al. (2015)
Maspin	Expression of maspin in 20–80% invasive breast cancer	18q21.33	Tumor suppressor genes	Berardi et al. (2013)
PIK3CA	Mutations in 37% of the HR^{+}/ HER2− metastatic and 40% of early breast cancer	3q26.3	Oncogene	Lefebvre et al. (2016)
CCND1 (cyclin D1)	Overexpression in 50% of breast tumors	11q13	Oncogene	Montalto and De Amicis (2020)

1.9.1 Luminal A and Luminal B Subtypes

Luminal A and Luminal B subtypes are mainly classified at the RNA and protein levels on the basis of their involvement in the overexpression of cell cycle/proliferation-related and luminal/hormone-regulated pathways. In comparison to Luminal A tumors, Luminal B tumors are distinguished by their higher expression of proliferation/cell cycle related genes (Qayoom et al. 2022; Sofi et al. 2022a, 2022b) or proteins; for instance, AURKA and MK167 and lower expression of several luminal-related genes or proteins such as the progesterone receptor (PR) and FOXA1 but not the estrogen receptor (Prat et al. 2013), which is found similarly expressed between the two luminal subtypes and can only help distinguish luminal

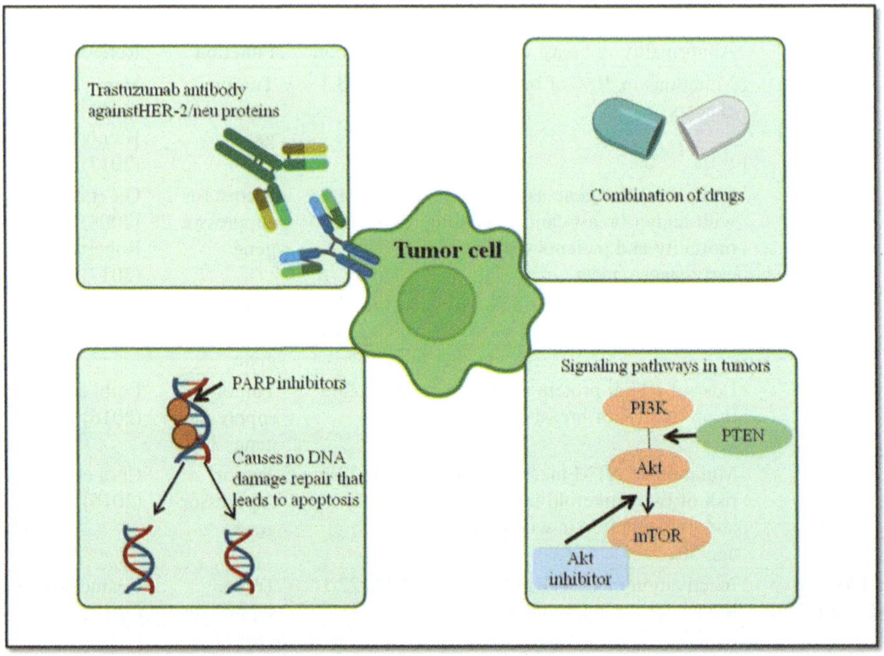

Fig. 1.6 Different forms of targeted therapy to prevent tumor growth

from non-luminal disease. Moreover, at the DNA level Luminal A tumors show a lower number of mutations across the genome, lower number of chromosomal copy-number changes (e.g., lower rates of CCND1 amplification), less TP53, GATA3 and more PIK3CA and MAP3K1 mutations compared to Luminal B tumors (Cancer Genome Atlas Network et al. 2012).

1.9.2 The HER2 Enriched Subtype

The HER2-enriched subtype is mainly categorized for the upregulation of HER2-related and proliferation-related genes and proteins (e.g., ERBB2/HER2 and GRB7), intermediate expression of luminal-related genes and proteins (e.g., ESR1 and PGR), and low expression of basal-related genes and proteins (e.g., keratin 5 and FOXC1). This subgroup of breast cancer is known to feature majority of mutations at the DNA level, for instance TP53 and PIK3CA mutations, respectively. However, the HER2-enriched tumors are known to have high frequency of APOBEC3B-associated mutations (Roberts et al. 2013). APOBEC3B is a subclass of APOBEC cytidine deaminases, which converts cytosine to uracil and has been implicated as a source of mutations in many cancer types (Kuong and Loeb 2013).

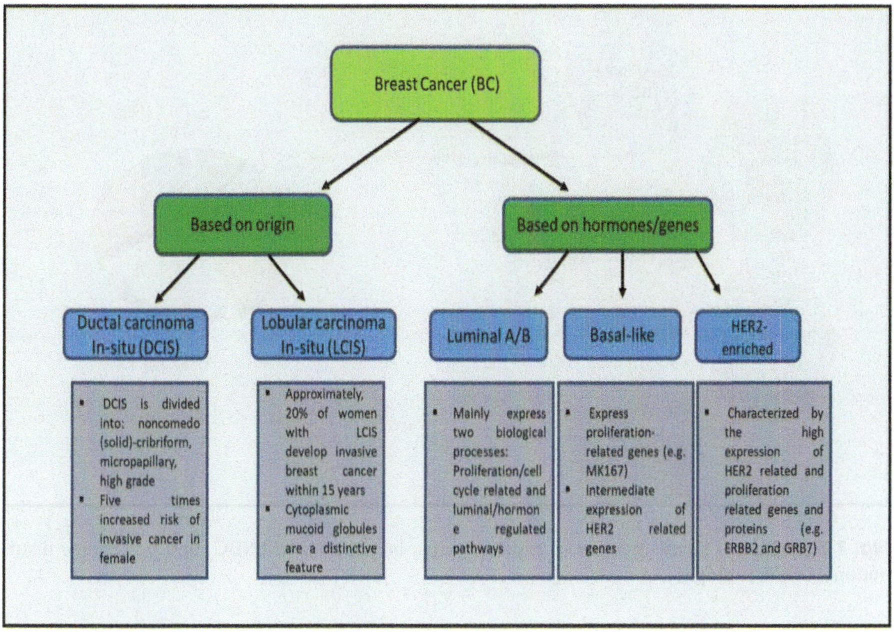

Fig. 1.7 Classification of breast cancer into various subtypes

1.9.3 The Basal-Like Subtype

The basal-like breast cancer subtype at the RNA and protein level is characterized by the overexpression of certain proliferation-related genes (e.g., MK167) and keratins typically expressed by the basal layer of the skin (e.g., keratins 5, 14, and 17), intermediate expression of HER2− related genes, and very low expression of luminal-related genes. Basal-like tumors constitute to be the subtype with second highest mutations across the genome with mostly hypomethylated, TP53, BRCA1, and PIK3CA mutated (Foulkes et al. 2003; Prat et al. 2014). In this subtype, ERBB2/HER2 overexpression/amplification is found to be associated.

1.9.4 Triple-Negative Breast Cancer

Triple-negative breast cancer (TNBC) is another subtype defined by the lack of/or negative expression of hormone receptors (ER, PR, and HER2) (Wolff et al. 2014). The basal-like subtype is often interchangeably used with TNBC. That is because of nearly 56% of gene expression of TNBC subtype overlaps with the basal-like breast cancer ranging from 60 to 90% (Perou et al. 2000; Prat et al. 2015). This breast cancer subtype is the most complex, lethal, and aggressive of all subtypes. TNBC is highly invasive with reduced survival rates among the patients and higher mortality rates (Dent et al. 2007). The aggressiveness of TNBC is attributed to their

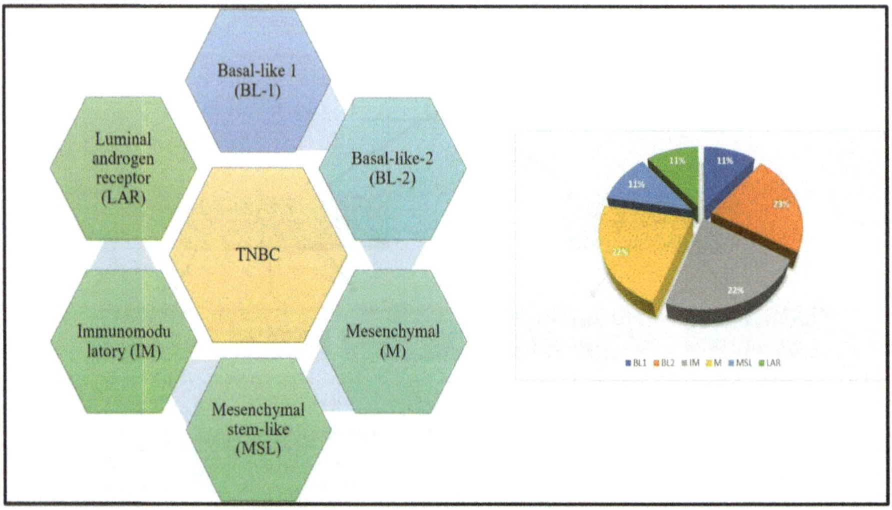

Fig. 1.8 Lehmann's classification of triple-negative breast cancer (TNBC) and percentage distribution of each subtype

non-responsiveness toward endocrine therapy or molecular targeted therapy. Therefore, chemotherapy is the main systemic treatment against TNBC, however, the higher metastatic load results in tumor recurrence (Fig. 1.8) (Chaudhary et al. 2018).

1.10 TNBC Sub-Typing and Clinical Implications

In 2011, Lehmann et al. performed gene expression profiling of tumor samples and divided TNBC into six subtypes: basal-like 1 (BL-1), basal-like 2 (BL-2), mesenchymal (M), mesenchymal stem-like (MSL), immunomodulatory (IM), and luminal androgen receptor (LAR) (Table 1.3) (Lehmann et al. 2011).

When comparing the Lehmann's classification with the previous intrinsic subtype BL1, BL2, IM, and M types correspond to basal-like types, MSL type is similar to normal-like type, and LAR shares characteristics with a part of the luminal or HER2 subtype (Fig. 1.9) (Saha and Nanda 2016).

1.10.1 Basal-Like Subtype

Due to the overexpression of genes related to the cell cycle and genes that respond to DNA damage, the basal-like subtype is known to have a strong proliferative ability (Sofi et al. 2022a, 2022b). This subtype can proliferate more readily than other subtypes, which is mostly related to the overexpression of the Ki-67 gene. According to reports, the majority of individuals with the basal-like subtype had germline BRCA gene mutations. As abnormalities in the homologous recombination repair

Table 1.3 Genomic TNBC subtypes and assignment of TNBC cell lines to subtypes

TNBC subtype	Genetic abnormalities	Mutations	Cell line
Basal-like 1	Cell cycle gene expression DNA repair gene (ATR-BRCA pathway) Proliferation genes	BRCA1; STAT4; UTX BRCA2; CTNND1; TOP2B; CAMK1G BRCA1; MAPK13; MDC1 TP53 BRCA1 PTEN; RB1; SMAD4; TP53 CDKN2A; TP53	HCC2157 HCC1599 HCC1937 HCC1143 HCC3153 MDA-MB-468
Basal-like 2	Growth factor-signaling pathways (EGFR, MET, NGF, Wnt/β-catenin, IGF-1R) Glycolysis, gluconeogenesis Expression of myoepithelial markers	BRCA1 RB1; TP53 PTEN; TP53 CDKN2A; TP53; UTX TP53	SUM149PT CAL-851 HCC70 HCC1806 HDQ-P1
Immunomodulatory	Immune cell processes (CTLA4, IL2, IL7 pathways, antigen processing/presentation) Gene signature for medullary BC (rare TNBC with a favorable prognosis)	TP53; CTNNA1; DDX18; HUWE1; NFKBIA APC; BRAF; MAP 2 K4; RB1	HCC1187 DU4475
Mesenchymal-like	Cell motility Cell differentiation Growth factor signaling EMT	PTEN; RB1; TP53 PIK3CA TP53	BT-549 CAL-51 CAL-120
Mesenchymal stem-like	Similar to M+ Low proliferation Angiogenesis genes	CDKN2A; HRAS; TP53 NF1; TP53 PIK3CA; TP53 HRAS BRCA1; TP53 BRAF; CDKN2A; KRAS; NF2; TP53; PDGFRA	HS578T MDA-MB-157 SUM159PT MDA-MB-436 MDA-MB-231
Luminal androgen receptor	Androgen receptor gene Luminal gene expression pattern Molecular apocrine subtype	PIK3CA; CDH1; PTEN PIK3CA PIK3CA PIK3CA; RB1; TP53; PTEN	MDA-MB-453 SUM185PE HCC2185 CAL-148

Lehmann et al. (2011)

pathway and genomic instability are thought to be caused by BRCA gene mutations. Regardless of the BRCA mutation in the basal-like subtype, the resulting genomic instability is caused (Lehmann et al. 2011). It can be deduced that individuals with

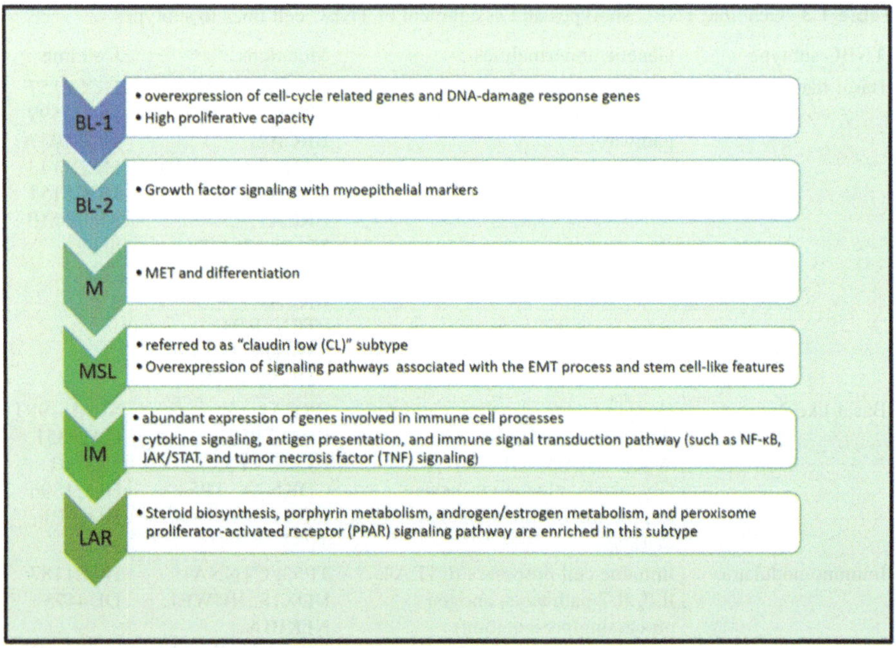

Fig. 1.9 Main characteristics of different TNBC subtypes

the basal-like subtype may benefit clinically from treatments that target highly proliferative tumors based on the distinctive molecular characteristics of the basal-like type. Patients with the basal-like subtype showed a greater pCR rate than those with another subtype, supporting this idea (Yin et al. 2020). Additionally, therapeutic effects may result from anti-cancer medications that target DNA damage response pathways (such as platinum-based chemotherapy and PARP inhibitors).

1.10.2 Immunomodulatory (IM) Subtype

The IM subtype of TNBC is a different subtype of TNBC that exhibits elevated expression of several genes involved in immune cell functions such as cytokine signaling, immune signal transduction pathway (including NF-B, JAK/STAT, and tumor necrosis factor (TNF) signaling), and cell signaling. Low M2-like macrophages are a feature of this subtype (Mehraj et al. 2022d; Lehmann et al. 2011; Burstein et al. 2015; Jézéquel et al. 2015). Because the increase in tumor-infiltrating lymphocytes (TILs) has been linked to the IM subtype of TNBC having a better prognosis than other TNBC subtypes (Jézéquel et al. 2015; Ray et al. 2021), these subtypes are significant targets for tailored therapy (Burstein et al. 2015; Mir and Mehraj 2019; Thomas et al. 2021).

However, it is well known that cancer cells elude immune response primarily by MYC amplification or PI3K pathway activation and upregulation of immune checkpoint markers in the presence of scarce immune cell infiltration. The immune checkpoint inhibitors, such as programmed death ligand 1 (PD-L1) and cytotoxic T lymphocyte-associated antigen 4 (CTLA-4), have recently attracted a lot of attention for the treatment of cancer (Zhao et al. 2020; Mehraj et al. 2021c). Further research is therefore important given the improved prognosis of this subtype and the immune checkpoint inhibition (Lee et al. 2020).

1.10.3 Mesenchymal-Like Subtype

A new molecular subtype of breast cancer known as the "claudin low (CL)" subtype was described by Herschkowitz et al. in 2007 (Herschkowitz et al. 2007). This subtype was linked to claudin cluster downregulation (genes involved in tight junctions and cell–cell adhesion, including Claudins 3, 4, 7, Occludin, and E-cadherin). This subtype also includes genes related to immune system functions, metastases (EMT), and mammary stem cells (Mir and Agrewala 2007; Hennessy et al. 2009; Prat et al. 2010; Sabatier et al. 2014; Mehraj et al. 2021b; Mir et al. 2022). However, this particular subtype had similarities to the basal-like subtype but differed in that it did not have highly expressed genes for proliferation (Prat et al. 2010; Mehraj et al. 2022a). Although this specific subtype displayed a CD44+/CD24/low expression profile and high expression of ALDH1A1 stem cell-like characteristics (Prat et al. 2010; Qayoom et al. 2021b).

Compared to luminal A and other breast cancer subtypes, this CL subtype has showed poor clinical outcomes and poor reactivity to chemotherapy (Prat et al. 2010; Yersal and Barutca 2014; Dias et al. 2017). Numerous genes involved in DNA repair, cell cycle regulation, immunological signaling, and abnormal expression of genes related to osteocytes (OGN) and adipocytes have been reported to be underexpressed in the MSL subtype (ADIPOQ, PLIN1). These results also demonstrate the CL subtype's commonality (Burstein et al. 2015). Furthermore, the MSL subtype's histological findings showed discohesive tumor cell growth in addition to metaplastic characteristics (Ray et al. 2021).

1.10.4 Luminal Androgen Receptors (LAR) Subtype

Similar to the luminal subtype, the LAR subtype is estrogen receptor negative and abundant in hormone-regulated pathways (Lehmann et al. 2011). This subtype is also abundant in the metabolism of porphyrins, steroid biosynthesis, androgen/estrogen metabolism, as well as the peroxisome proliferator-activated receptor (PPAR) signaling pathway (Liu et al. 2016; Mehraj et al. 2022a). A molecular apocrine subtype with a strong apocrine characteristic and positive AR was previously described by Farmer et al. (2005). This collection of activated genes is involved in metabolism, particularly the generation of fatty acids and lipids. The

LAR subtype is connected to a low Ki-67 proliferation index, apocrine differentiation, and a good histologic grade (Ray et al. 2021).

The previously discovered molecular apocrine subtype is part of the LAR subtype (Lehmann et al. 2011). Additionally, this subtype is more prevalent in tumors that are HER2-positive (Farmer et al. 2005), and a recent study found that 75% of HER2-enriched breast cancers belong to the LAR subtype (Bareche et al. 2018). Despite being a subtype of TNBC, it demonstrates its involvement in the activation of the estrogen signaling system, supporting the idea that treating this subtype with anti-estrogen, anti-androgen, and anti-HER2 therapy may be helpful (Sanga et al. 2009; Mir 2015; Liu et al. 2016). Since the projected values of AR expression are still unknown, it has not yet been possible to identify and categorize patients who will benefit from AR targeted therapy (Anestis et al. 2020). Two recent phase II clinical trials showed that AR blocking therapy benefited patients with AR-positive TNBC clinically (Gucalp et al. 2013; Traina et al. 2018).

1.11 Brand-New Drugs

1.11.1 Fulvestrant

For patients with locally advanced or metastatic breast cancer who test positive for ER, tamoxifen medication is typically the first line of treatment. Tamoxifen has some estrogenic activity, which can lead to endometrial hyperplasia or cancer and put patients at risk for thrombosis, as was previously mentioned. However, unlike tamoxifen, which is a partial agonist, fulvestrant is fundamentally a pharmacological antagonist. The term "pure" anti-estrogen is used to describe it. The medication, known as a selective estrogen receptor down-regulator (SERD), binds directly to the ER, blocking ER dimerization and accelerating the receptor's fast degeneration (Wakeling 2000).

AstraZeneca's Fulvestrant was given FDA approval in 2002 as a second-line endocrine therapy for postmenopausal women with metastatic hormone-receptor positive breast cancer whose disease had progressed after receiving anti-estrogen therapy.

Aromatase inhibitors: In premenopausal women, the main source of estradiol (estrogen) is the ovaries. However, in postmenopausal women, the ovaries stop producing estrogen, and estrogen is instead synthesized in a number of extragonadal locations, such as the breast adipose tissue (Samavat and Kurzer 2015). In postmenopausal women, estrogens are primarily produced through the conversion of adrenal androgens into estrone and estradiol, with estradiol serving as the major physiological hormone. The rate-limiting and last stage of this estrogen production is catalyzed by the enzyme aromatase.

By inhibiting the activity of this enzyme, aromatase inhibitors (AIs) reduce the amount of estrogen that some breast tumors can use as fuel (Chumsri et al. 2011). AIs are well tolerated and have been demonstrated to be more effective than the anti-estrogen tamoxifen in lowering the risk of breast cancer recurrence and spread. The first of a new generation of AI medications, letrozole (commonly known as Femara), was authorized by the FDA in 2005. The Novartis-produced letrozole was initially authorized for long-term usage in postmenopausal women who had finished five years of tamoxifen therapy. Clinical trials showed that the medication lowers the chance of breast cancer spread and recurrence even more than tamoxifen does by itself.

Another FDA-approved AI, anastrozole (made by AstraZeneca and sold under the brand name Arimidex), has been demonstrated to prolong survival and lessen cancer symptoms in postmenopausal women with advanced breast cancer (Clarke and Khosla 2009). Raloxifene, which had been prescribed to postmenopausal women since 1997 to prevent and treat osteoporosis, was later discovered to be equally as effective as tamoxifen in lowering the chance of developing invasive breast cancer. Although it can be used to treat non-invasive breast cancer and lower the chance of developing it, raloxifene also works by limiting the effects of estrogen on breast tissue (Clarke and Khosla 2009).

1.11.2 HER2 Blockers

Lapatinib, also marketed under the name Tykerb by its maker GlaxoSmithKline, received FDA approval in 2007 for the treatment of HER2-positive patients whose conditions had stopped responding to Herceptin. HER2 is overexpressed in around 25% of breast tumors, and as was already established, this overexpression results in a more aggressive phenotype and a poor prognosis. A receptor tyrosine kinase called HER2 sends signals to support various cellular functions, including cell division. Lapatinib blocks the tyrosine kinase activity of HER2 and HER1, which reduces these receptors' signaling.

The medication was also authorized for use in patients with advanced breast cancer whose tumors overexpressed the HER2 protein in conjunction with the medication capecitabine. For individuals with HER-2 positive breast cancer, lapatinib and letrozole were also approved as an initial treatment in 2010 (Rimawi et al. 2015).

1.12 Conclusion

Breast cancer is women the leading cause of mortality among women worldwide. Being a heterogenous disease, it comprises several subgroups that differ in their characteristic features and their importance in the disease. The complexity of the disease varies according to the subtypes such as: Luminal A, Luminal B, HER2 overexpression, Basal, and normal-like. Among the varied subtypes triple-negative breast cancer (TNBC) is the most complex and lethal subtype of breast cancer. TNBC itself consists of several subtypes such as: Basal-like 1/2, immunomodulatory, mesenchymal-like, mesenchymal stem-like, and luminal androgen receptor.

1.13 Further Readings

For more insights about the topic we would suggest detailed findings from the books of (Mir MA, 2022) https://doi.org/10.1016/C2021-0-02565-7, https://doi.org/10.1016/C2022-0-00074-X and (Mir MA, 2021) https://doi.org/10.52305/WXJL6770, and from the cancer.net website on the following mentioned below links,

https://www.cancer.net/cancer-types/breast-cancer/types-treatment

https://discovery.ucl.ac.uk/id/eprint/1472740/

https://www.jmedsciences.com/doi/JMEDS/pdf/10.5005/jp-journals-10045-00138

For diagrammatic illustrations, descriptive tables (Lazzeroni, 2012) http://www.eurekaselect.com/article/49928

See video links on over all status of Cancer, its various types, current new treatment possible options available

https://www.sciencedirect.com/science/article/pii/S205970292032278X

The readers can have a look upon the following video YouTube links for the better understanding of the chapter:

https://youtu.be/wIsdjfwPUxY

https://youtu.be/SVjJt984PlU

https://www.youtube.com/watch?v=9hgrfXleNsM

https://www.youtube.com/watch?v=ZWqfoBj2bsA

https://www.sciencedirect.com/science/article/pii/S221464741630054X

https://www.ncbi.nlm.nih.gov/pmc/articles/PMC5715522/

References

Akram M, Siddiqui SA (2012) Breast cancer management: past, present and evolving. Indian J Cancer 49(3):277

Anestis A et al (2020) Androgen receptor in breast cancer—clinical and preclinical research insights. Molecules 25(2):358

Atkins H et al (1972) Treatment of early breast cancer: a report after ten years of a clinical trial. Br Med J 2(5811):423–429

Baclesse F (1984) Roentgen therapy as the sole method of treatment of cancer of the breast. In: Modern radiation oncology, vol 2. Harper and Row Publishers, Inc, New York, pp 115–129

Bareche Y et al (2018) Unravelling triple-negative breast cancer molecular heterogeneity using an integrative multiomic analysis. Ann Oncol 29(4):895–902

Beatson GT (1989) On the treatment of inoperable cases of carcinoma of the mamma: suggestions for a new method of treatment, with illustrative cases. Cancer J (Villejuif) 2(10):347–350

Berardi R et al (2013) Role of maspin in cancer. Clin Transl Med 2(1):1–19

Blows FM et al (2010) Subtyping of breast cancer by immunohistochemistry to investigate a relationship between subtype and short and long term survival: a collaborative analysis of data for 10,159 cases from 12 studies. PLoS Med 7(5):e1000279

Bonadonna G et al (1976) Combination chemotherapy as an adjuvant treatment in operable breast cancer. N Engl J Med 294(8):405–410

Brambilla C et al (1976) Response and survival in advanced breast cancer after two non-cross-resistant combinations. Br Med J 1(6013):801–804

Cancer Genome Atlas Network et al (2012) Comprehensive molecular portraits of human breast tumors. Nature 490(7418):61–70
Burris Iii HA (2004) Dual kinase inhibition in the treatment of breast cancer: initial experience with the EGFR/ErbB-2 inhibitor lapatinib. Oncologist 9(S3):10–15
Burstein MD et al (2015) Comprehensive genomic analysis identifies novel subtypes and targets of triple-negative breast cancer identification of four unique subtypes of TNBCs. Clin Cancer Res 21(7):1688–1698
Cetin B et al (2014) Lapatinib plus capecitabine for HER2-positive advanced breast cancer: a multicentre study of Anatolian Society of Medical Oncology (ASMO). J Chemother 26(5): 300–305
Chaudhary LN et al (2018) Triple-negative breast cancer: who should receive neoadjuvant chemotherapy? Surg Oncol Clin 27(1):141–153
Cheang MCU et al (2009) Ki67 index, HER2 status, and prognosis of patients with luminal B breast cancer. J Natl Cancer Inst 101(10):736–750
Cheng L et al (2010) Rb inactivation accelerates neoplastic growth and substitutes for recurrent amplification of cIAP1, cIAP2 and Yap1 in sporadic mammary carcinoma associated with p53 deficiency. Oncogene 29(42):5700–5711
Choi M et al (2016) ATM mutations in cancer: therapeutic implications. Mol Cancer Ther 15(8): 1781–1791
Chumsri S et al (2011) Aromatase, aromatase inhibitors, and breast cancer. J Steroid Biochem Mol Biol 125(1–2):13–22
Clarke BL, Khosla S (2009) New selective estrogen and androgen receptor modulators. Curr Opin Rheumatol 21(4):374
Dawood SS et al (2008) Prognosis of women with stage IV breast cancer by HER2 status and trastuzumab treatment: an institutional based review. J Clin Oncol 26(15_suppl):1018–1018
De Lena M et al (1975) Adriamycin plus vincristine compared to and combined with cyclophosphamide, methotrexate, and 5-fluorouracil for advanced breast cancer. Cancer 35(4):1108–1115
Dent R et al (2007) Triple-negative breast cancer: clinical features and patterns of recurrence. Clin Cancer Res 13(15):4429–4434
Desmedt C et al (2009) Development and validation of gene expression profile signatures in early-stage breast cancer. Cancer Investig 27(1):1–10
Desmedt C et al (2016) Genomic characterization of primary invasive lobular breast cancer. J Clin Oncol 34(16):1872–1881
Dias K et al (2017) Claudin-low breast cancer; clinical & pathological characteristics. PLoS One 12(1):e0168669
Dieci MV et al (2014) Rare breast cancer subtypes: histological, molecular, and clinical peculiarities. Oncologist 19(8):805–813
Drew Y, Calvert H (2008) The potential of PARP inhibitors in genetic breast and ovarian cancers. Ann N Y Acad Sci 1138(1):136–145
Farmer P et al (2005) Identification of molecular apocrine breast tumors by microarray analysis. Breast Cancer Res 7(2):1–1
Folkman J (1995) Angiogenesis in cancer, vascular, rheumatoid and other disease. Nat Med 1(1): 27–30
Foulkes WD et al (2003) Germline BRCA1 mutations and a basal epithelial phenotype in breast cancer. J Natl Cancer Inst 95(19):1482–1485
Gautam M, Malhotra XZ (2010) Histological, molecular and functional subtypes of breast cancers. Cancer Biol Ther 10:955–960
Gucalp A et al (2013) Phase II trial of bicalutamide in patients with androgen receptor–positive, estrogen receptor–negative metastatic breast cancer. Clin Cancer Res 19(19):5505–5512
Halsted WS (1907) I. The results of radical operations for the cure of carcinoma of the breast. Ann Surg 46(1):1
Hellman S (1993) Dogma and inquisition in medicine. Breast cancer as a case study. Cancer 71(7): 2430–2433

Hennessy BT et al (2009) Characterization of a naturally occurring breast cancer subset enriched in epithelial-to-mesenchymal transition and stem cell characteristics. Cancer Res 69(10): 4116–4124

Herschkowitz JI et al (2007) Identification of conserved gene expression features between murine mammary carcinoma models and human breast tumors. Genome Biol 8(5):1–17

Hientz K et al (2017) The role of p53 in cancer drug resistance and targeted chemotherapy. Oncotarget 8(5):8921

Huggins C, Bergenstal DM (1952) Inhibition of human mammary and prostatic cancers by adrenalectomy. Cancer Res 12(2):134–141

Iwamoto T, Pusztai L (2010) Predicting prognosis of breast cancer with gene signatures: are we lost in a sea of data? Genome Med 2(11):1–4

Jézéquel P et al (2015) Gene-expression molecular subtyping of triple-negative breast cancer tumors: importance of immune response. Breast Cancer Res 17(1):1–16

Konecny GE et al (2006) Activity of the dual kinase inhibitor lapatinib (GW572016) against HER-2-overexpressing and trastuzumab-treated breast cancer cells. Cancer Res 66(3): 1630–1639

Kuong KJ, Loeb LA (2013) APOBEC3B mutagenesis in cancer. Nat Genet 45(9):964–965

Lee Y-M et al (2020) Molecular subtypes of triple-negative breast cancer: understanding of subtype categories and clinical implication. Genes Genom 42(12):1381–1387

Lefebvre C et al (2016) Mutational profile of metastatic breast cancers: a retrospective analysis. PLoS Med 13(12):e1002201

Lehmann BD et al (2011) Identification of human triple-negative breast cancer subtypes and preclinical models for selection of targeted therapies. J Clin Invest 121(7):2750–2767

Liu Y-R et al (2016) Comprehensive transcriptome analysis identifies novel molecular subtypes and subtype-specific RNAs of triple-negative breast cancer. Breast Cancer Res 18(1):1–10

Loibl S et al (2016) Integrated analysis of PTEN and p4EBP1 protein expression as predictors for pCR in HER2-positive breast cancer PTEN and p4EBP for pCR prediction. Clin Cancer Res 22(11):2675–2683

Luft R, Olivecrona H (1955) Hypophysectomy in man. Experiences in metastatic cancer of the breast. Cancer 8(2):261–270

Lukong KE (2017) Understanding breast cancer–the long and winding road. BBA Clin 7:64–77

McWhirter R (1948) V. The value of simple mastectomy and radiotherapy in the treatment of cancer of the breast. Br J Radiol 21:583

Mehraj U et al (2021a) Tumor microenvironment promotes breast cancer chemoresistance. Cancer Chemother Pharmacol 87(2):147–158

Mehraj U et al (2021b) The tumor microenvironment as driver of stemness and therapeutic resistance in breast cancer: new challenges and therapeutic opportunities. Cell Oncol 44: 1209–1229

Mehraj U et al (2021c) Prognostic significance and targeting tumor-associated macrophages in cancer: new insights and future perspectives. Breast Cancer 28(3):539–555

Mehraj U et al (2022a) Expression pattern and prognostic significance of baculoviral inhibitor of apoptosis repeat-containing 5 (BIRC5) in breast cancer: a comprehensive analysis. Adv Cancer Biol-Metastasis 4:100037

Mehraj U et al (2022b) Cryptolepine targets TOP2A and inhibits tumor cell proliferation in breast cancer cells-an in vitro and in silico study. Anticancer Agents Med Chem 22:3025

Mehraj U et al (2022c) Adapalene inhibits the growth of triple-negative breast cancer cells by S-phase arrest and potentiates the antitumor efficacy of GDC-0941. Front Pharmacol 13:958443

Mehraj U et al (2022d) Chemokines in triple-negative breast cancer heterogeneity: new challenges for clinical implications. Semin Cancer Biol 86(Pt 2):769

Miller K et al (2007) Paclitaxel plus bevacizumab versus paclitaxel alone for metastatic breast cancer. N Engl J Med 357(26):2666–2676

Mir MA (2015) Developing costimulatory molecules for immunotherapy of diseases. Academic Press, London. https://doi.org/10.1016/C2014-0-02898-5. ISBN: 9780128025857

Mir MA, Agrewala JN (2007) Influence of CD80 and CD86 co-stimulation in the modulation of the activation of antigen presenting cells. Curr Immunol Rev 3(3):160–169
Mir MA, Agrewala JN (2008) Signaling through CD80: an approach for treating lymphomas. Expert Opin Ther Targets 12(8):969–979
Mir MA et al (2022) Immuno-onco-metabolism and therapeutic resistance. In: Immuno-oncology crosstalk and metabolism. Springer, Singapore, pp 45–89. https://link.springer.com/chapter/10.1007/978-981-16-6226-3_3
Mir MA, Mehraj U (2019) Double-crosser of the immune system: macrophages in tumor progression and metastasis. Curr Immunol Rev 15(2):172–184
Mir MA et al (2020) Targeting different pathways using novel combination therapy in triple negative breast cancer. Curr Cancer Drug Targets 20(8):586–602
Montalto FI, De Amicis F (2020) Cyclin D1 in cancer: a molecular connection for cell cycle control, adhesion and invasion in tumor and stroma. Cells 9(12):2648
Muñoz-Gámez JA et al (2005) PARP inhibition sensitizes p53-deficient breast cancer cells to doxorubicin-induced apoptosis. Biochem J 386(1):119–125
Murphy JB (1912) Surgical clinics of John B. Murphy. University of Chicago, Chicago. Cholelithiasis **1**: 417
Overgaard M (1999) Overview of randomized trials in high risk breast cancer patients treated with adjuvant systemic therapy with or without postmastectomy irradiation. Semin Radiat Oncol 9(3):292
Overgaard M et al (1997) Postoperative radiotherapy in high-risk premenopausal women with breast cancer who receive adjuvant chemotherapy. N Engl J Med 337(14):949–955
Perou CM et al (2000) Molecular portraits of human breast tumors. Nature 406(6797):747–752
Pfahler GE (1932) Results of radiation therapy in 1022 private cases of carcinoma of the breast from 1902 to 1928. Am J Roentgenol Rad Ther 27:497–408
Prat A et al (2010) Phenotypic and molecular characterization of the claudin-low intrinsic subtype of breast cancer. Breast Cancer Res 12(5):1–18
Prat A et al (2013) Prognostic significance of progesterone receptor–positive tumor cells within immunohistochemically defined luminal a breast cancer. J Clin Oncol 31(2):203
Prat A et al (2014) Molecular features of the basal-like breast cancer subtype based on BRCA1 mutation status. Breast Cancer Res Treat 147(1):185–191
Prat A et al (2015) Clinical implications of the intrinsic molecular subtypes of breast cancer. Breast 24:S26–S35
Pusztai L et al (2008) Effect of molecular disease subsets on disease-free survival in randomized adjuvant chemotherapy trials for estrogen receptor–positive breast cancer. J Clin Oncol 26(28): 4679–4683
Qayoom H, Bhat BA (2020) U Mehraj U, Mir MA (2020) Rising trends of cancers in Kashmir valley: distribution pattern, incidence and causes. J Oncol Res Treat 5(150):2
Qayoom H et al (2021a) Integrating immunotherapy with chemotherapy: a new approach to drug repurposing. In: Drug repurposing-molecular aspects and therapeutic applications. IntechOpen, London. https://doi.org/10.5772/intechopen.100183
Qayoom H et al (2021b) An insight into the cancer stem cell survival pathways involved in chemoresistance in triple-negative breast cancer. Future Oncol 17(31):4185–4206
Qayoom H et al (2022) Expression patterns and therapeutic implications of CDK4 across multiple carcinomas: a molecular docking and MD simulation study. Med Oncol 39(10):1–13
Qu S et al (2008) Genetic polymorphisms of metastasis suppressor gene NME1 and breast cancer survival. Clin Cancer Res 14(15):4787–4793
Ragaz J et al (1997) Adjuvant radiotherapy and chemotherapy in node-positive premenopausal women with breast cancer. N Engl J Med 337(14):956–962
Ray T et al (2021) Therapeutically targeting cancers that overexpress FOXC1: a transcriptional driver of cell plasticity, partial EMT, and cancer metastasis. Front Oncol 11:721959
Reis-Filho JS et al (2010) Molecular profiling: moving away from tumor philately. Sci Transl Med 2(47):47ps43

Rimawi MF et al (2015) Targeting HER2 for the treatment of breast cancer. Annu Rev Med 66:111
Roberts SA et al (2013) An APOBEC cytidine deaminase mutagenesis pattern is widespread in human cancers. Nat Genet 45(9):970–976
Roberts MR et al (2017) Single nucleotide variants in metastasis-related genes are associated with breast cancer risk, by lymph node involvement and estrogen receptor status, in women with European and African ancestry. Mol Carcinog 56(3):1000–1009
Sabatier R et al (2014) Claudin-low breast cancers: clinical, pathological, molecular and prognostic characterization. Mol Cancer 13(1):1–14
Saha P, Nanda R (2016) Concepts and targets in triple-negative breast cancer: recent results and clinical implications. Ther Adv Med Oncol 8(5):351–359
Samavat H, Kurzer MS (2015) Estrogen metabolism and breast cancer. Cancer Lett 356(2):231–243
Sanga S et al (2009) Gene expression meta-analysis supports existence of molecular apocrine breast cancer with a role for androgen receptor and implies interactions with ErbB family. BMC Med Genet 2(1):1–16
Slamon DJ et al (1987) Human breast cancer: correlation of relapse and survival with amplification of the HER-2/neu oncogene. Science 235(4785):177–182
Smid M et al (2008) Subtypes of breast cancer show preferential site of relapse. Cancer Res 68(9): 3108–3114
Sofi S et al (2022a) Targeting cyclin-dependent kinase 1 (CDK1) in cancer: molecular docking and dynamic simulations of potential CDK1 inhibitors. Med Oncol 39(9):1–15
Sofi S et al (2022b) Cyclin-dependent kinases in breast cancer: expression pattern and therapeutic implications. Med Oncol 39(6):1–16
Sørlie T et al (2001) Gene expression patterns of breast carcinomas distinguish tumor subclasses with clinical implications. Proc Natl Acad Sci 98(19):10869–10874
Sørlie T et al (2003) Repeated observation of breast tumor subtypes in independent gene expression data sets. Proc Natl Acad Sci 100(14):8418–8423
Sotiriou C, Pusztai L (2009) Gene-expression signatures in breast cancer. N Engl J Med 360(8): 790–800
Spitale A et al (2009) Breast cancer classification according to immunohistochemical markers: clinicopathologic features and short-term survival analysis in a population-based study from the south of Switzerland. Ann Oncol 20(4):628–635
Su Y et al (2015) The clinicopathological significance and drug target potential of FHIT in breast cancer, a meta-analysis and literature review. Drug Des Devel Ther 9:5439
Tang P et al (2008) Molecular classifications of breast carcinoma with similar terminology and different definitions: are they the same? Hum Pathol 39(4):506–513
Thomas R et al (2021) Immune checkpoint inhibitors in triple negative breast cancer treatment: promising future prospects. Front Oncol 10:600573
Traina TA et al (2018) Enzalutamide for the treatment of androgen receptor–expressing triple-negative breast cancer. J Clin Oncol 36(9):884
Varna M et al (2011) TP53 status and response to treatment in breast cancers. J Biomed Biotechnol 2011:284584
Wakeling AE (2000) Similarities and distinctions in the mode of action of different classes of antioestrogens. Endocr Relat Cancer 7(1):17–28
Weigelt B et al (2010) The contribution of gene expression profiling to breast cancer classification, prognostication and prediction: a retrospective of the last decade. J Pathol 220(2):263–280
Wolff AC et al (2014) Recommendations for human epidermal growth factor receptor 2 testing in breast cancer: American Society of Clinical Oncology/College of American Pathologists clinical practice guideline update. Arch Pathol Lab Med 138(2):241–256
Yersal O, Barutca S (2014) Biological subtypes of breast cancer: prognostic and therapeutic implications. World J Clin Oncol 5(3):412
Yin L et al (2020) Triple-negative breast cancer molecular subtyping and treatment progress. Breast Cancer Res 22(1):1–13
Zhao S et al (2020) Molecular subtypes and precision treatment of triple-negative breast cancer. Ann Transl Med 8(7):499

Current Treatment Approaches to Breast Cancer

2

Manzoor Ahmad Mir and Abrar Yousuf Mir

2.1 Introduction

The treatment of breast carcinoma is comprehensive. Early detection of breast cancer has a lesser risk of progression, a minor rate of reappearance, and a high probability of survival (McPhail et al. 2015). Timely detected breast carcinoma may be treated with local and systemic approaches (Anampa et al. 2015). However, invading, aggressive breast tumor survival rates often remain less (García Rodríguez et al. 2010; Mehraj et al. 2021a, 2021b). Treatment for malignancy necessitates a planned, holistic approach in which competent onco-surgeons and professional workers coordinate within a functional infrastructure to run the important therapies while decreasing the patient's economic and logistical constraints. These days, various options are available for the treatment of breast tumors, including radiotherapy, chemical therapy, surgical methods, hormonal therapy, immunotherapy, targeted therapy, and, novel, genetic therapy. The mortality graph due to breast cancer is flattening due to progression in screening, diagnosis, and treatment. Most female patients receive adjuvant systemic therapy because it has been shown to improve survival rates and provide better results, and molecular profiling to personalize cure based on the threat is now a medical reality for patients with HR-positive carcinoma. Breast-conserving surgeries with radiation therapy or mastectomy are options for most females with low-grade breast tumors. Fractional mastectomy, lumpectomy, quadrantectomy, and local excision are some of the options for breast-conserving surgery. Breast conservation therapy has been practiced for more than two decades, with these techniques followed by five to seven weeks of phototherapy. Breast cancer is a divergent disease at the molecular level, molecular trait includes

M. A. Mir (✉) · A. Y. Mir
Department of Bioresources, School of Biological Sciences, University of Kashmir, Srinagar, J&K, India
e-mail: drmanzoor@kashmiruniversity.ac.in

M. A. Mir (ed.), *Therapeutic Potential of Cell Cycle Kinases in Breast Cancer*,
https://doi.org/10.1007/978-981-19-8911-7_2

activation of HER2 which is encoded by ERBB2, activation of hormone receptors (ER and PR), and mutations in BRCA genes. So, breast cancer diagnosis and treatment necessitate a multidisciplinary approach involving several subspecialties. Diagnostic imaging work-up and biopsy are critical in initiating a diagnosis and influencing surgical choices on primary tumor management, axillary staging, and the treatment sequence (Qayoom et al. 2022, p. 1122). After a breast tumor diagnosis is made, the degree of the illness is accessed, which directs whether systemic neoadjuvant therapy is necessary or not. Breast cancer in advanced stage IV is considered incurable unless there is a reason for palliative resection of the primary tumor, stage four is treated solely with systemic therapy. The death rate of breast tumors has lowered in the last few decades due to progression in chemotherapy for handling breast cancer (Osborne 1998; Sofi et al. 2022). In addition, the efficiency of biological therapies like as anti-HER2 monoclonal antibodies established the opportunity and significance of the molecularly targeted method in BC therapy (Gianni et al. 2011). TNBC which can invade other parts of the body is still a grave illness with few therapeutic options. The biological processes underlying the diverse treatment response in BC have been better understood in recent years. Tamoxifen and other endocrine medications have transformed breast cancer management, resulting in significant reductions in cancer-related deaths.

2.2 Breast Cancer Radiation Therapy

Radiation therapy is a cancer-killing and tumor-shrinking treatment that employs high-energy radiation. Elevated energy radiation destroys cellular genetic pools, preventing them from further growth and expansion (Jackson and Bartek 2009; Mehraj et al. 2021a, 2021b). Even though radiation damages both cancerous and non-cancerous cells, the main objective of radiotherapy is to significantly augment the radiation amount to tumor cells while dropping risks to healthy cells located close to carcinoma cells or in their radiation path. Cancer cells generally are not as effective as normal cells in repairing radiation damage, which leads to the death of differing cancer cells (Begg et al. 2011). Radiotherapy plays a key function in the management of both "non-invasive and invasive BC". In both therapeutic and palliative care settings, radiation therapy is beneficial in the management of breast malignancy (Bese et al. 2008). Following "breast-conserving surgery" radiation of the breast yields oncologic results comparable to those of a mastectomy in early illness (Fisher et al. 2002; Veronesi et al. 2002). In individuals with advanced-stage cancer, adjuvant locoregional radiation to the breast wall and nearby lymphatic nodes lowers the risk of local and distant recurrence with a benefit to disease-specific overall survival (Braunstein et al. 2017). This advantage is enhanced by risk factors including node positive, adolescence, triple-negative molecular markers, high-grade cancer as well as lymph vascular invasion are present (Vrieling et al. 2017; Mir et al. 2020). Once radiation exposure levels are raised to include regional disease, the danger of toxicity, particularly to the "lungs and heart" rises. Providing radiation to patients with MBC is beneficial in several cases (McGale et al. 2014).

When surgical resection is not an option, palliative radiation can help with local control in cases of dermal recurrence or localized disease. More exact and homogeneous radiation therapy can be provided with advances in CT image analysis, simulation, patient monitoring, and delivery methods. After BCT, radiotherapy options include "hypo-fractionated, entire breast radiation, accelerated restricted breast radiation with external beam treatment beam therapy." Radiation can be delivered to the tumor spot mainly in two ways. External beam radiation from outside the body is transmitted to the tumor site by targeting high-energy rays (protons or particle radiation and photons). This is the most commonly used approach in clinical settings.

2.2.1 External Beam Radiation Therapy

Unless systemic chemotherapy is administered, teletherapy usually starts three to six weeks after surgical treatment. The simulation process is the first step in cancer treatment. Each patient receives a unique "breast boards, wing boards" or set of cradles or molds that can be customized. This ensures that the sufferer is in a consistent state for each treatment. Patients are usually positioned supine, with their torsos inclined 10–15 degrees. The shoulder is outwardly rotated and the ipsilateral arm is grabbed at about 100–120 degrees. At this moment, radio opaque wires are positioned as well as managed to secure beside the operational scars. The treatment field borders are then defined by the radiation oncologist to include breast target and if necessary, nearby lymph nodes, the stimulated CT is carried out. The iso-center is selected then patient's skin is marked with daily set-up marks. Treatment planning in three dimensions is carried out. The volumes of treatment and key structures are determined and defined. The best beam configurations are selected.

Tangential fields encompassing the most anterior thorax are commonly used for early-stage BC. "Level I and II lymph node chains" are examples of these fields. Though most of the levels I and II axillary nodes are to be included, although special attention must be paid to the tangent field margins particularly the contact between the cranial and posterior chest wall (Schlembach et al. 2001). Radio-therapy to the supraclavicular fossa with or without posterior axillary boost may be beneficial for patients who have un-dissected nodes, or one to more lymph node metastases, or one to three positive nodes. 4–6 MV-photon energy is typically employed to treat the breast and lymph nodes. Whole-breast radiation therapies are given five days a week, with a total dose of about "50 Gy" administrated in 25–28 fragments. Electrons are commonly used in the boost treatment. At 1.8–2 Gy per fraction, the "lumpectomy cavity" is strengthened in favor of additional 10–16 Gy. "Intensity-modulated radiation therapy (IMRT) or planning IMRT" for treatment of breast has been developed due to discoveries in radiation therapy design and delivery systems. With IMRT, the dosage of the "contralateral breast" is lowered (Borghero et al. 2007). In comparison to standard radiotherapy, it has been shown that planning IMRT can produce more uniform plans with very few hot spots (Bamett et al. 2009) and (Herrick et al. 2008).

2.3 Breast Brachytherapy

"Breast brachy-therapy" was once used as "boost" after outside entire breast radiation therapy for treating the lumpectomy cavity. Utilization of "accelerated partial breast irradiation (APBI)" has now been adopted by various clinics either with various applications such as Contura multi-lumen balloon, SAVI, MammoSite balloon or with interstitial needle implants, or 3D conformal external "radio-therapy after breast-conserving surgery" as the sole mode for radiation treatment. Higher dosages of radiation per fraction deposited on the tumor bed by irradiating smaller volume. When particularly in comparison to daily whole breast external beam radiotherapy, this dramatically reduces treatment times and the patient's travel time. Even after breast-conserving surgery, "interstitial breast brachytherapy" was used efficiently for ten years. Proxima Therapeutics' MammoSite* balloon catheter has been sanctioned by the FDA in 2002 for intra-cavitary sufficient dosage rate of breast brachytherapy. A potential multi-center study assesses efficacy of the "MammoSite® balloon catheter" registered 70 patients in the beginning. Following that, only light to medium self-limited side effects were found in 43 patients who were eligible for the therapy (Keisch et al. 2003; Mehraj et al. 2021a, 2021b). Recently, the "American Society of Breast Surgeon" released the outcome of a registration study consisting of 1440-women who were allowed to treat with the "MammoSite® catheter after breast-conserving surgery." Frequency of "Ipsilateral BC as well as axillary rates were 2.15 percent and 0.36 percent" including both, after three years (Nelson et al. 2009). The balloon catheter has the benefits of being simpler to insert into the cavity, quite repeatable placement, and enhanced care comfort. It has the most widespread use as well as a prolonged track record.

One of the most recent brachytherapy devices in the market is the ClearPath™ multi-catheter device. The catheter is inserted via a single access point, but there is no need for a separate radiation source. In comparison to a single catheter system, using a multi-catheter hybrid can decrease dosage of normal breast cells and skin (Dickler 2009). The "SAVI device" such as single-entry, multi-catheter applicator that enables "radiation oncologist" to preferentially head-on radiation through up to "11 catheter channels" enabling additional customized manipulation of the iso-dose partitions is another latest change to the internal radiotherapy option. The equipment consists of collection of extensible catheters which are arranged throughout the central-lumen. This device attempts to combine the benefits of "interstitial brachytherapy" consisting the convenience of a single-entry device (Scanderbeg et al. 2009).

2.4 Breast Cancer Chemotherapies

Chemotherapy is the use of anti-cancer medications to treat tumor cells (Fig. 2.1). Breast cancer treatment will be depending on a number of variables including overall health, medical history, age (including whether or not menstruation is present), type and stage of cancer, and sensitivity to prescribed drugs and protocols. Chemotherapy choices are typically given in cycles: a treatment for a pre-determined amount of

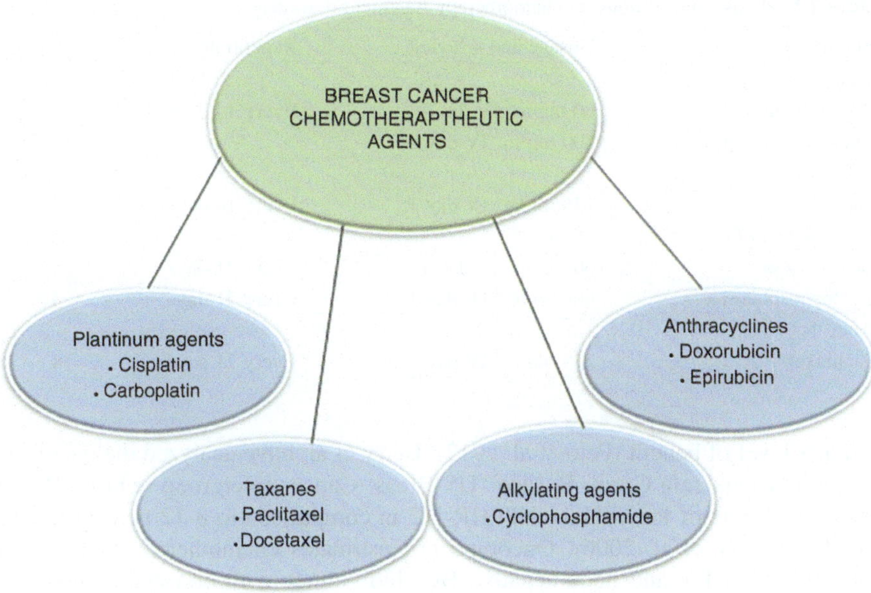

Fig. 2.1 Different kinds of chemo-therapeutic factors are used in the diagnosis of breast carcinoma

time followed by recovery phase and then another treatment. Before surgery, chemotherapy can be used to reduce the size of the tumor and, in some situations, allow for breast-conserving surgery rather than a mastectomy. It is routinely given after surgery and is sometimes given in a "dose-dense" manner.

2.4.1 Adjuvant Chemotherapy

For high-risk patients systematic chemotherapies are recommended. Innumerable number of chemotherapies are available such as anthracycline and a taxane. "Dox and cyclophosphamide" for 4-cycles accompanied by paclitaxel for 4-cycles (AC-T) is an effective treatment regimen in the United States. AC-T given in a constant dose (cd) manner after every two weeks with growth factor assistance after the chemotherapy cycle outperforms the previous schedule of every three weeks (Citron et al. 2003). Other optimum AC schedule after taxane involves weekly 12-week paclitaxel or 4-cycle docetaxel every three weeks (Sparano et al. 2008) (Table 2.1). DAC, docetaxel with AC, is also another common option, but not more advanced than the above treatments and docetaxel is more toxic than paclitaxel, and particularly has increased rates of febrile neutropenia (Swain et al. 2013; Mehraj et al. 2021a, 2021b).

The advantage of adjuvant chemotherapy was shown by meta-analyses to reduce recurring and breast cancer death rates in those suffering from HR negative BC with

Table 2.1 Breast cancer adjuvant chemotherapy treatment regimens

Regimen	Dosage and schedule	Repetition	Cycles
Dose-dense			
Doxorubicin	60 mg/m^2 IV day 1	Every 14 days	4
Cyclophosphamide	600 mg/m^2 IV day 1		
After that			
Paclitaxel	175 mg/m^2 IV day 1	Every 14 days	4
AC ⇒ Taxol (T)			
Doxorubicin	60 mg/m^2 IV day 1	Every 21 days	4
Cyclophosphamide	600 mg/m^2 IV day 1	Every 21 days	4
After that			
Paclitaxel	175 mg/m^2 IV day 1	Every 21 days	4

a higher level of benefit (Peto et al. 2012). Berry et al. have analyzed the data from cancer and leukemia Group B Also "US Breast Cancer Intergroup reduced 21–25 percent risk rate of individuals with HR-BC in comparison to 8-12 percent having HR+BC (Berry et al. 2006). Oncotype DX estimates chemotherapy benefits for patients with HR+ and node-negative BC. Individuals with elevated "Onco-type re-occurrence scores ≥31" leading to a significant decrease in recurrence risk with chemo-therapy (relative risk—0.26), on the other hand, those with poor scores receive little, whether any, advantage from chemotherapy (Paik et al. 2006). Although worries about anthracycline-associated cardiotoxicity or leukemogenic possibilities exist, anthracycline-carrying adjuvant therapy treatments had been utilized in the early-stage diagnosis of BC. "*Anthracycline*-based regimens" related to per annum risk of cardiopulmonary arrest of "0.08″ percent/year, in the 2000 "Early Breast Cancer Trialists' Collaborative Group (EBCTCG)" summary, compared to 0.06 percent a year in diagnosis of individuals with "non-anthracycline-based regimens." 9735 studies in US Oncology validated TC (docetaxel/cyclophosphamide) a feasible treatment for women with "early-stage breast cancer" particularly for those who are at a significant risk of cardiopathy or require treatment only for twelve-weeks (Jones et al. 2009). In this trial, 1016 females with treatable "breast cancer (stages I–III)" were randomly assigned phases of TC or 4-phases of "standard-dose AC (Adriamycin/cyclophosphamide)." Both DFS (81% vs 75%) and OS (87% vs 82%) were higher in the TC group after an average duration 7 years. Additionally, a meta-analysis comparing "anthracycline-based with non-anthracycline-based regimens" in 6564 females with primary-stage BC found that anthracycline treatment was only beneficial in individuals with HER2-positive cancer (Gennari et al. 2008). Topoisomerase IIa, the gene (TOP2A) which is situated after "HER2 gene on chromosome 17" is inhibited by anthracyclines in a biological sense. In around 35% of HER2 over-expressing breast tumors, TOP2A is co-amplified (Press et al. 2011).

TAILORx (Trial-Assigning-Individualized-Options-for-Treatment) randomly selected individuals suffering via Onco-type, incidence scores of 11–25 either with hormone therapy alone or hormone therapy in conjunction with chemotherapy.

Sufferers in this given cohort receive "anthracycline-containing or anthracycline-sparing" chemotherapy. Hormone therapy alone would be sufficient in sufferers with reduced Onco-type repetition scores, particularly those under 11.

2.4.2 Neo-Adjuvant Chemotherapy

Initially, neo-adjuvant chemotherapy was utilized to produce regionally advanced, uncurable resectable BC. NAC has lately been utilized to "downstage illness in the breast and axilla in treatable tumors" to allow breast conservation, and, in certain cases, to minimize axillary lymph node dissection. Various randomized trials have looked at the oncologic protection and survival rates of NAC (Van der Hage et al. 2001; Fisher et al. 1998). A systemic review of diagnosed individuals with "NAC vs surgery" after chemo-therapy found no significant distinction viability or loco-regional recurrence with "NAC," but 17% reduced rate of mastectomy in those receiving NAC (Mieog et al. 2007). Since a larger number of women involved in these trials were candidates for BCT at the time of diagnosis and therefore could not benefit from NAC, 17% is a conservative estimate. NAC is more likely to permit BCT in women with uni-centric malignity that are enormous in comparison to the breast size, as well as who have "HER2+" or triple-negative breast cancers. In clinically node-negative females, the use of NAC dramatically lowers the risk of axillary metastases. Higher rates of PCR in the breast and axilla have been observed after NAC in response to more effective systemic therapies. In three prospective randomized clinical studies, the effectiveness of sentinel node biopsy performed after NAC in patients with nodal metastases was examined. According to the "ACOSOG Z1071 and SENTINA studies, false-negative rates" are less than 10%, when is equal to what is acceptable for sentinel nodal biopsy in the fundamental surgical context. In a randomized trial done by "Memorial Sloan Kettering Cancer Center" 288 patients with nodal metastases who were clinically node-negative after NAC had a nodal pCR in 48% of the cases (Mamtani et al. 2016).

2.5 Breast Cancer Endocrine Therapy

The principal regulators of breast tissue development and differentiation are estrogen and progesterone. The ovaries are the primary source of both steroid hormones. They act on cells by attaching to "activating nuclear receptors" called "estrogen-receptors (ERs)" and "progesterone receptors (PRs)." These "receptors" demonstrate membrane-localized as well as transcriptional signaling activity once activated. Two of the most common ERs are ERα and ERβ. ERα is expressed in 70% of breast tumors; however, ERβ is poorly identified (Bland and Copeland 2009).

George T. Beatson was the first to notice the potential significance of estrogen in breast tissue, seeing that oophorectomy in rabbits led to lactation loss. Based on this finding, Beatson conducted an "oophorectomy" on a premenopausal woman with an unresectable breast carcinoma on June 15, 1895. She was cured and lived for another four years. Beatson's early work established the groundwork for hormone therapy. Most cancers that screened positive for estrogen or progesterone receptors respond well to endocrine therapy. Endocrine therapy could be used for 5–10 years, and probably much longer, in some cases. The main types of endocrine therapies include—tamoxifen and AIs.

2.5.1 Tamoxifen

There is a strong evidence to support the use of Tamoxifen as an adjuvant endocrine treatment for early-stages (Jankowitz and Davidson 2013). A nonsteroidal antiestrogen Tamoxifen was licensed by the US FDA in the 1970s as a hormonal therapy for postmenopausal women with metastatic breast cancer (mBC) (Fig. 2.2). According to guidelines from the "National Institutes of Health Consensus Conference on Breast Cancer Chemotherapy 1985" tamoxifen's use was extended to the "adjuvant setting with the treatment of postmenopausal women with node+ and ER+ tumors." Since 1958, randomized controlled studies have been carried out by the "National Surgical Adjuvant Breast and Bowel Project (NSABP)" to examine various aspects of adjuvant and surgical therapy (Band 2010). 2644 patients with node-negative, receptor-positive cancer underwent surgery and they are randomly assigned to receive either 5-years of Tamoxifen or *Placebo* in NSABP B-14 trail (Mamounas 2003). When tamoxifen was used instead of a placebo, the trial proved a

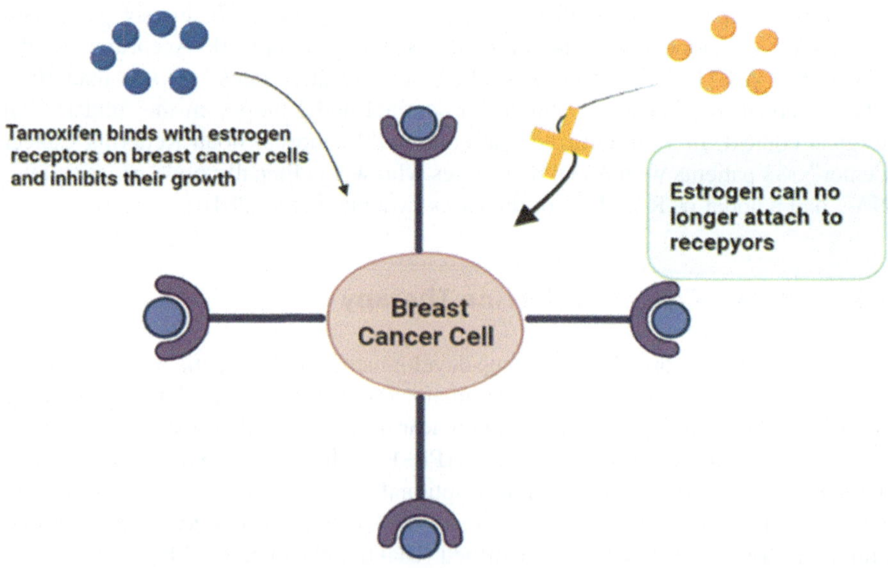

Fig. 2.2 Mechanism of action of the Tamoxifen: Tamoxifen blocks the action of estrogens binding to the "hormonal receptors" on the cancer cells; thereby estrogen is no longer able to bind to these receptors, due to which the mammary epithelium no longer functions as an estrogen receptor and proliferation of cancer cells is gradually slowed or stopped due to a lack of estrogen

remarkable improvement in "disease-free survival (DFS)." In addition, 194 RCT (randomized clinical trials) evaluated by "Early Breast Cancer Trialists' Collaborative Group (EBCTCG)" also discovered that adjuvant-tamoxifen therapy for 5-years in individuals with ER+ BC lowered breast tumor death rates by 31% and was considered highly efficient than 1 or 2 years of tamoxifen therapy. The EBCTCG found that 5-years of "adjuvant tamoxifen" in women with ER+ BC remarkably decreased relapse by one-third over the first ten years and decreased BC death rate by one-third over the first fifteen years in a follow-up meta-analysis. Tamoxifen treatment for five years possesses backbone of "adjuvant-hormone therapy" particularly for pre-menopausal women having BC, due to a slew of good research (Jankowitz and Davidson 2013).

2.5.2 Aromatase Inhibitors

Because ovarian tissue no longer produces estrogen, post-menopausal women instead use the enzyme aromatase to manufacture it predominantly from non-glandular sources. In addition to being separated from BC cells, aromatase has also been found in a number of other organs such as muscle, sub-cutaneous fat, and the liver (Bland and Copeland 2009). Because estrogen suppression has been so successful in the past, "aromatase inhibition" has received a lot of attention as a BC treatment option (Table 2.2).

Aromatase inhibitors (AIs) were efficient in the treatment of BC for the first two generations, but they had substantial adverse effects since they blocked other androgenic hormones like "cortisol and aldosterone." AIs third generation possesses greater aromatase selectivity so it is classified either as "steroidal (type I) or nonsteroidal (type II)." Non-steroidal impediment are reversible competitive inhibitors, whereas steroidal inhibitors irreversibly decrease enzyme activity. Women with metastatic BC were the first to be examined using third-generation AIs. When the efficacy of AI therapy in metastatic BC was discovered, the focus switched to its application in adjuvant settings. Several tamoxifen-treated patients had recurrences due to medication resistance or experienced detrimental effect including endometrial malignancy as well as "venous thromboembolic disorder." Alone in the *Arimidex, Tamoxifen*, or in Combination (ATAC) studies, 9366 post-menopausal women are localized with BC, were given *anastrozole* or *tamoxifen* for 5 years (Howell et al. 2005). *Anastrozole* substantially improved DFS (575 versus 651 occurrences; "hazard ratio = 0.87; confidence interval of 95 percent [CI] 0.78–0.97; $P = 0.01$)" following an average duration of 68 months. Patients on *anastrozole* had decreased threat of "venous thromboembolic" occurrences, endometrial malignancy, and "hot flashes" (Howell et al. 2005). Letrozole, a second type II AI, and tamoxifen were compared against one another in the "Breast International Group (BIG) 1-98 Collaborative Group" investigation. 8010 participants in the double-blind phase 3 trial were randomly assigned to "*tamoxifen* upto 5 years, *letrozole* for 5 years, *tamoxifen* for 2 years followed by *letrozole*, or *letrozole* for 2 years followed by tamoxifen." The original trial compared the two

Table 2.2 Clinical trials of aromatase inhibitors

Trial id or name	Treatment plans	No. of individuals	Endpoints	Results
ATAC, 2005	Five years of 1-milligram anastrozole vs. 20-mg tamoxifen vs. combination therapy	9366	DFS, OS, safety, the incidence of contralateral breast cancer, and time to distant recurrence	Anastrozole significantly extended DFS.
BIG 1-98, 2006	Five years of 20-milligram tamoxifen vs. 5 years of 2.5-mg letrozole vs. 2 years of 20-mg tamoxifen following 3 years of 2.5-mg letrozole vs. 2 years of 2.5-mg letrozole following 3 years of 20-mg tamoxifen	8010	DFS, systemic DFS, OS, time to distant recurrence.	Letrozole improved DFS, decreased distant recurrences, and extended the time to distant metastasis.
FACE, 2017	5 years of 2.5-mg letrozole vs. 5 years of 1-mg anastrozole	4136	Safety and effectiveness	5 year DFS rate was 84.9% for letrozole vs. 82.9% for anastrozole.
EBCTCG, 2015	5 years of an AI (group 1) vs. 5 years of tamoxifen (group 2) vs. 2–3 years of tamoxifen following AI to year 5(group 3) vs. 2–3 years of an AI following tamoxifen to year 5 (group 4)	31,920	Recurrence, BC mortality, death without recurrence, and all-cause mortality	Compared to tamoxifen, AIs lowered recurrence rates by about 30%.

groups assigned to receive tamoxifen, letrozole has a greater 5-year survival rate compared to the tamoxifen arm"84.0 percent and 81.4 percent, respectively." According to this study "Thromboembolism, endometrial cancer, and vaginal bleeding were all common in the *tamoxifen* arm (Thürlimann 2006).

Based on such trials, as well as on the 51-month to check out "BIG 1-98 trial" FDA recommended "*anastrozole* and *letrozole*" for hormone-sensitive early-stage BC as initial adjuvant therapy. "*Femara* Versus *Anastrozole* Clinical Evaluation (FACE)" a recently randomized phase-3-trial, examined the effectiveness and safety of *anastrozole* plus *letrozole* (Smith et al. 2017). Letrozole has been shown in earlier pharmacotherapeutic studies to block estradiol more effectively, but the FACE study that letrozole was not significantly safer or more effective than anastrozole in 4136 postmenopausal women with hormone receptor-positive and node-positive BC (Smith et al. 2017). Meta-analysis of randomized trials contrasting ALs and tamoxifen in early BC was also carried out by EBCTCG, and results were compared after 31,920 post-menopausal women with early BC and ER+ were divided into several therapeutic groups. The different sub-groups were; group 1: 5 years of AI, group 2: 5 years of *tamoxifen*, group 4: 2–3 years of an AI followed by *tamoxifen* for the total of 5 years. A five-year comparison between AI and changeover 2–3 years of *tamoxifen strategy subsequently* followed by AI for five years, indicated a reduction in recurrence in the "first year" of the AI-group, however, this advantage lost, when simultaneously taking both groups as AI.

2.5.3 Switching Trails in Endocrine Therapy

After two-three years of *tamoxifen*, AIs were also examined as a follow-up therapy. In conjunction with the *Arimidex-Nolvadex* 95, The (ABCSG) group "Austrian-Breast and Colorectal Cancer Study Group' trial 8" analyzed effectiveness of converting "*anastrozole*" for three years besides the adjuvant *tamoxifen* treatment for 2-years (Jakesz et al. 2005). 3224 post-menopausal women having HR+ BC, who had been taking tamoxifen for 2 years are randomly assigned to have "1 mg *Anastrozole*, 20 mg *Tamoxifen*, or 30 mg *Tamoxifen*" have been set up by investigators in the course of study. 40% depletion was observed in average duration of 28-months "67 *anastrozole* versus 110 *tamoxifen* events; 0.60; 95% CI 0.44–0.81; P =0.0009" in threat for an event with anastrozole (Jakesz 2005). The "Intergroup *Exemestane* Study (IES) examined-*Exemestane*" which is a type I-AI, in the adjuvant setting after *tamoxifen* therapy (Coombes 2004). The researchers evaluated 4742 "post-menopausal women with ER+ or ER unknown" BC, which after 2–3 years of adjuvant-*tamoxifen*, had no disease signs and had been randomly assigned for the entire five years for either exemestane or *tamoxifen* therapy. They discovered a 32% risk decline in the *exemestane* group after an average duration of 30.6-months, leading to a remarkable benefit, "DFS of 4.7 percent at 3-years" following randomization. These switching studies demonstrated that using AIs and *tamoxifen* sequentially offered significant benefits.

The Netherlands-based "analysis on the Duration of Extended Adjuvant *Letrozole* (IDEAL)" experiments randomized participants to 2.5–5 years of "*letrozole*" following 5 years of endocrine therapy. However, about 74% participants have completed "letrozole" for 2.5 years and just 57% have completed *letrozole* for 5 years (Blok et al. 2016).

2.6 Targeted Therapies in the Treatment of Breast Carcinoma

"Targeted therapies" for BC are utilized to diagnose individuals to whom which "BC cells" have an aberrant growth pattern due to the overexpression of particular distinctive proteins on their cell surface. Antibodies, which act identically to the human immune system, are primarily utilized as BC targeted therapy. Targeting up-regulation of the "HER2 protein" on the periphery of the breast cancerous cells is the most productive "breast cancer targeted therapy" available today (Fig. 2.3). There are currently some extensively used BC targeted therapies that are successful in blocking many molecular pathways (Table 2.3) such as *Trastuzumab*, also known as Herceptin, which inhibits cancer cells from receiving signals that tell them to grow (Giordano et al. 2014). *Pertuzumab*, also known as perjeta, works by blocking the signals that cancer cells use to proliferate (Giordano et al. 2014; Baselga et al. 2012). *Bevacizumab,* also known as *Avastin*, stops cancer cells from growing new blood capillaries that supply them oxygen, as well as nutrients (Gianni et al. 2013) and so on.

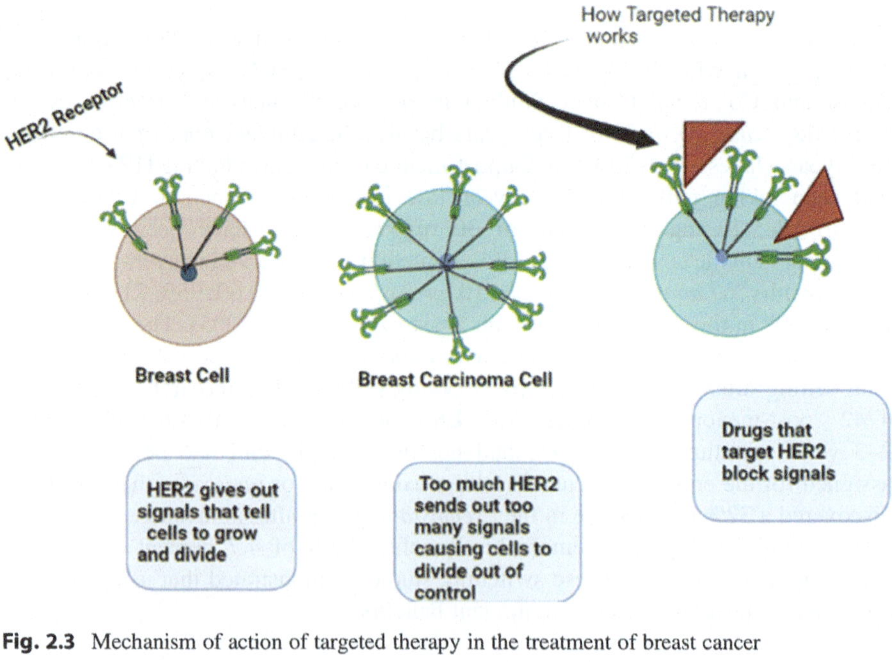

Fig. 2.3 Mechanism of action of targeted therapy in the treatment of breast cancer

Table 2.3 Targeted therapies clinical trails

Agent	Clinical trial id	Phase	Recruitment status
Trastuzumab	NCT00004067	III	Complete
	NCT00045032	III	Complete
Pertuzumab	NCT00567190	III	Complete
	NCT00545688	II	Complete
T-DM1	NCT00829166	III	Complete
	NCT01772472	III	Active, not recruiting

2.6.1 HER2+ Malignant Tumor of Breast

Human-epidermal-growth-receptor positive is an ErbB family "tyrosine kinase receptor" that is involved in pathways of cell signaling that promote cell proliferation, differentiation, and apoptotic suppression (García-Aranda and Redondo 2017). HER2, the most often differentially expressed "receptor in breast cancer" is highly expressed in around 15% cases of BCs, resulting in dysregulated cell proliferation (García-Aranda and Redondo 2017; Masoud and Pagès 2017). Because HER2 powerfully triggers tumor-growth-related downstream pathways, its expression levels are closely related to breast cancer metastasis and prognosis. "HER2" has been used as a biological target for discovery of novel therapeutics for these reasons throughout the last 20 years. Anti-HER2 therapy has enhanced survival in HER2 + breast cancers, even though they develop quickly and are more aggressive than other forms (Nami and Wang 2017) and (Mir et al. 2020).

2.6.2 Trastuzumab

Trastuzumab or Herceptin is the cornerstone of HER2+ BC treatment. It is the first FDA-approved humanized immunoglobulin G1kappa monoclonal antibody (mAb) which has been found to significantly increase DFS and OS in numerous clinical studies (Masoud and Pagès 2017). The "B-31 National Surgical Adjuvant Breast and Bowel" (NCT00004067) study contrasted a 4-cycle AC chemotherapy strategy, followed by weekly 12-week *paclitaxel*, against the same chemo-therapy strategy "plus 1 year of *trastuzumab*" beginning on the Ist-day of the therapy with "*paclitaxel*" (Romond et al. 2005). Participants present in "*trastuzumab* group had an elevated DFS rate than those found in the control after an average duration of two years. Furthermore, the possibility of mortality was lowered by nearly a third (Romond et al. 2005).

The trial HERA (NCT00045032) enrolled women who had undergone surgery and had finished 04 rounds of "adjuvant chemotherapy." The trial compared was *trastuzumab* ($n = 1693$) with an additional year or two years of *trastuzumab* ($n = 1694$) once this regimen was completed. In combined 1–2 year *trastuzumab* cohort exhibited greater value of "DFS" than the examination groups after an average duration of 1 year, with "127 DFS" occurrences in the "*trastuzumb*" cohort in comparison with 220 in the "observation cohort" (Piccart-Gebhart et al. 2005).

In these trails, the use of "adjuvant *trastuzumab plus anthracycline*-based regimens" (such as *doxorubicin*) was linked to high risk of cardiopulmonary arrest in these trials (Romond et al. 2005) and (Piccart-Gebhart et al. 2005). Therefore, the BCIRG 006 research looked into a trastuzumab-based non-anthracycline regimen (group 3) and compared three groups: Every three weeks, four rounds of "doxorubicin and cyclophosphamide were subsequently followed by four doses of docetaxel (AC-T) (group 1)" "AC-T followed by one year of trastuzumab beginning with the first-docetaxel dose (group 2), as well as 6 docetaxel rounds in combination with carboplatin and "contemporaneous *trastuzumab* (TCH)" followed-by 34 weeks of "*trastuzumab.*" When compared to normal AC-T therapy, *trastuzumab*-groups demonstrated improvements in DFS as well as in OS. Additionally, AC-T plus *trastuzumab* group had a "5-year DFS" value of 84% and 92% OS value, both with P-value <0.001.TCH had a "five-year DFS" rate of 81% as well as OS rate of 91%, with a P-value of 0.04 (Slamon 2011). Regardless of ER status, clinical studies with adjuvant trastuzumab have shown clinically substantial progression in DFS. In addition, clinically substantial improvements in the OS were shown in the B31 and HERA studies.

2.6.3 Pertuzumab

The monoclonal antibody (mAb), *Pertuzumab, which is recombinantly humanized,* binds to HER2 and inhibits its dimerizing with other HER-receptors (Franklin et al. 2004). Since "*pertuzumab*" interacts with HER2 at a distinct-epitope than "*trastuzumab*" and stimulates "antibody-dependent cell-mediated cytotoxicity" via complementary pathways, when used jointly, *pertuzumab* and *trastuzumab* exhibit stronger antitumor efficacy than either therapy alone (Franklin et al. 2004). To treat individuals with HER2+ mBC, pertuzumab is used in conjunction with *trastuzumab* and *docetaxel.* It is also utilized as a neo-adjuvant treatment for early HER2 +,"advanced, or inflammatory BC," in addition being used in conjunction with *trastuzumab plus* and chemotherapy for adjuvant therapy of HER2+ early BC with an extreme chance of re-occurrence. *Pertuzumab's* effectiveness and safety have been studied in some clinical trials. *Pertuzumab* with "*trastuzumab* plus *docetaxel*" as contrast to a *placebo* and "*trastuzumab* with *docetaxel*" in the CLEOPATRA phase 3 trial (NCT00567190) to treat mBC (Baselga et al. 2012). The *pertuzumab* group showed a clinically meaningful improvement in PFS. The phase 2 trial of Neosphere (NCT00545688) compared four groups to check the efficacy of "*pertuzumab*: *trastuzumab* with *docetaxel, pertuzumab* with *trastuzumab* plus *docetaxel, pertuzumab* with *trastuzumab*(no chemotherapy), and *pertuzumab* with docetaxel" (Gianni et al. 2012). The treatment combining two mAbs was highly effective than chemotherapy using only one mAb. Treatment with "*pertuzumab* in association with trastuzumab plus docetaxel" demonstrated a considerable enhancement in PCR when compared with trastuzumab plus docetaxel alone.

2.6.4 Conjugates of Antibodies and Drugs

"Antibody-drug conjugates (ADCs)" are made up of synthetic connection between recombinant mAb and a cytotoxic drug (Beck et al. 2017). Internalization of the antibody occurs after it binds to a specific antigen on the surface of the cancerous cell. The drug is subsequently delivered into the cell, where it causes cytotoxicity (Fig. 2.4).

T-DM1 is an "ADC" used to cure individuals with "HER2 (+) mBC" which have already been mediated with "*trastuzumab*" plus taxane (individually or in conjunction). EMILIA (NCT00829166) and KATHERINE (NCT01772472) were two clinical trials that looked at their safety and effectiveness (Verma 2012). In the phase 3 EMILIA study, T-DM1 and lapatinib with capecitabine were compared in individuals with "HER2 + mBC" who had earlier been mediated with "*trastuzumab*" and taxane chemotherapeutic agent. T-DM1 therapy substantially increased the PFS up to 10 months compared to *lapatinib* plus capecitabine PFS (6 months), (Verma 2012). In the KATHERINE study, the protection and effectiveness of T-DM1 were investigated in individuals with "HER2+ early BC" who had previously provided with neo-adjuvant treatment with "taxane and *trastuzumab*." Individuals mediated with "T-DM1" had a incomparably better invasive DFS than those treated with *trastuzumab* after an average duration of forty-months.

Another ADC trastuzumab deruxtecan which is also recognized as Ds-8201 is made up of "anti-HER2" immunoglobin and a "topoisomerase inhibitor." It is for adults who had two or more "anti-HER2" therapies in metastatic-context and have advanced or unresectable BC (Beck et al. 2017).

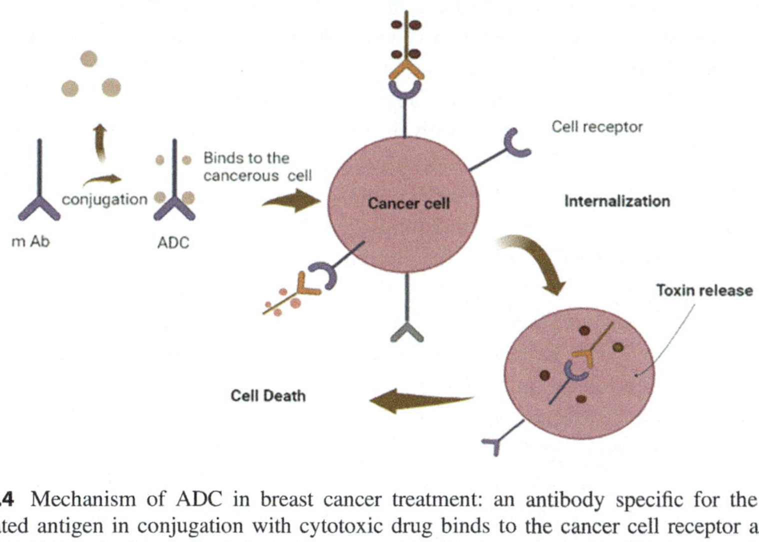

Fig. 2.4 Mechanism of ADC in breast cancer treatment: an antibody specific for the tumor-associated antigen in conjugation with cytotoxic drug binds to the cancer cell receptor and then internalization and release of cytotoxic drugs occurs inside the cancer cell which leads to the death of cancer cell

2.6.5 mTOR Pathways

The mTOR pathway appears to play a considerable position in targeted tumor therapy. In 70% of breast tumors, the PI3K/AKT/mTOR pathway is highly activated, and the kinase proteins discovered along these pathways could be prospective therapeutic targets for BC treatment. the mTOR pathway with the drug "everolimus" in conjunction with "HER-2 or ER inhibitors" is perhaps a hopeful perspective approach to make use of, and restore the "sensitivity of BC cells" to conventional therapies as well as overcome resistance processes that appear to evolve, "when the mTOR pathway" is overactive (Grunt and Mariani 2013). Glaysher et al. discovered that utilizing EGFR and mTOR inhibitors to target breast epithelial cell lines with knocked-in mutations resulted in improved responsiveness to therapeutic drugs. The study investigated the impacts of impeding "mTOR and EGFR" both with the associated drug activity of "sirolimus /ZSTK474 and gefitinib / erlotinib" on the parental cell line, finding a more efficient signaling blockade than with single agents regardless of the "knocked-in mutations in EGF, PI3K, KRAS, BRAF, or AKT" (Glaysher et al. 2014).

2.6.6 Receptor Tyrosine Kinase Inhibitors

Basically "Kinase-inhibitors" are tiny molecular compounds that interact with the "ATP-binding region of a tyrosine-kinase-receptor, such as HER2" and inhibit its activity (García-Aranda and Redondo 2017). Lapatinib, also known as Tykerb, is a kinase inhibitor that suppresses growth induced by the EGFR ErbB1 and HER2 pathways. Individuals with LA or mBC who progressed following a standard therapy with taxanes, anthracyclines, or trastuzumab were enrolled in "phase-3 trial" and assess the safety as well as effectiveness of lapatinib in conjunction with "capecitabine. Lapatinib + capecitabine" reduced probability of recurrence by 43%, in comparison to capecitabine alone (Geyer et al. 2006).

A second kinase inhibitor, neratinib or Nerlynx, is approved as a single drug for adjuvant therapy of HER2+ EBC after adjuvant trastuzumab therapy. It is also used with "capecitabine to treat HER2+ mBC" in people who had two or even more anti-HER2 therapies before (Segovia-Mendoza et al. 2015). The "phase-3-trial" ExteNET "NCT00878709" found that individuals which got "*neratinib*" instead of "*Placebo*" following conventional trastuzumab-based adjuvant treatment had substantially lower invasive DFS survival results after an average duration of roughly five years (Martin 2017). Recently tucatinib, also known as Tukysa—another kinase inhibitor recommended by the "FdA" in 2020 for the diagnosis of "HER2$^+$ mBC" in conjunction with trastuzumab and capecitabine. The "HER2CLIMB trial (NCT02614794)" investigated "*tucatinib*" in conjunction with *trastuzumab* as well as *capecitabine* in individuals having "HER2 + MBC" who had earlier been diagnosed with—*pertuzumab, trastuzumab*, and T-DM1. At one year, the *tucatinib* examined group ($n = 410$) had 46% reduced possibility of illness recurrence and mortality than the one administered with *capecitabine* plus *trastuzumab* alone ($n = 202$) (Murthy et al. 2020).

2.7 Function of Immunotherapies in the Diagnosis of Mammary Cancer

Immune system has a multifaceted function to play in cancer detection/prevention, early elimination, and progression. Elimination, equilibrium, and escape are three separate processes of host immune-surveillance and responses described by the "immunoediting" theory (Schreiber et al. 2011). Both innate and acquired immune systems work simultaneously to identify as well as remove cancer during the elimination phase. Few cancerous cells that survived the phase of elimination are thought to gain entry into the phase of equilibrium, in which the adaptive immune responses stop tumor expansion and keep them in a dormant state without eradicating the tumor. Tumor cells develop the capability to avoid immune detection and destruction during the escape phase in many ways (Schreiber et al. 2011). As a result, tumor-immune system evasion is a defining feature of malignancy (Hanahan and Weinberg 2011). Enhancing innate immune systems for cancer has been begun with high-dose interleukin-2 several decades ago, which has shown a sustainable therapeutic advantage for metastatic cancer patients, especially melanoma and renal cell carcinoma (Klapper et al. 2008; Qayoom et al. 2021). Following the discovery of therapeutically meaningful inhibitory routes of T cell activation: the "PD-1/PD-L1 and CTLA-4" there has been substantial upsurge in utilization of immuno-oncology methods (Keir et al. 2007). Immune checkpoint MAB inhibitors that block these pathways have shown to be effective in a range of malignancies, including melanoma, renal cell carcinoma, and NSC lung carcinoma. Furthermore, checkpoint inhibitors have been shown to produce long-lasting therapeutic improvements as well as increased percentage of "patients" living far away from the historic median, as earlier described in standard chemotherapeutic agents.

2.7.1 Role of Checkpoint Inhibitors in Immunotherapy

They are regarded as the major effectors of cell-mediated immunity because activated CD8+ (cytotoxic T lymphocytes) identify and kill pathogen-infected or abnormal cells such as tumor cells (Fig. 2.5). T cells, on the other hand, augment immunoglobin reactions via the activity of "Cd4+ [T-helper cells]" as well as augmentation of the anti-body synthesis by B-cells, therefore their stimulation is an important stage in the commencement and control of the immunological responses.

Both antigen-specific-stimulatory signals directing the interaction of "TCR to HLA-II" displayed on cell surface of APCs play a functional part in the activation and maturation of T-cells, according to lymphocyte-activated two-signal model (Fontana and Vance 2011; Mir 2015). At several moments throughout the immune response, stability of co-stimulatory and suppressive signals, also known as immune checkpoints, regulates the ensuing response, which prevents tissue damage and maintains self-tolerance (Pardoll 2012), and (Buchbinder and Desai 2016). Dysregulated production of inhibiting signals indicates a considerable benefit in the tumor micro-environment, resulting in immune escape, due to their immunosuppressive activities. Combinations of immunological checkpoint receptors and ligands are currently being studied as methods for cancer therapy by reinstating immune systems functionality, either as monotherapies or in combo therapies due to their relationship with lymphocyte activity suppression and consequent energy (Pardoll 2012; Darvin et al. 2018). In particular, all the pathways, i.e., "CTLA4, CD152 and Pd-1, CD279 or Pd-L1, CD274" have illustrated their validity for novel cancer therapy advancement and permitted clinical approval of these two pathways due to their involvement in the immune responses and peripheral tolerance.

T cell stimulation is regulated by additional costimulatory signals including Cd28 and CTLA-4, aside from the T cell communication with HLA-II. In this regard, "CTLA-4 signaling" impedes activation of T-cells, in contrast to CD-28 signaling, that are needed for activation of "T-cell and cytokine production." This is particularly essential in lymphatic nodes where CTLA-4 counteracts possibly auto-reactive T-cells during early activation stages of Cd-4 and Cd-8 cells (Buchbinder and Desai 2016). CD80 and CD86 ligands displayed on activated APCs can stimulate CD28

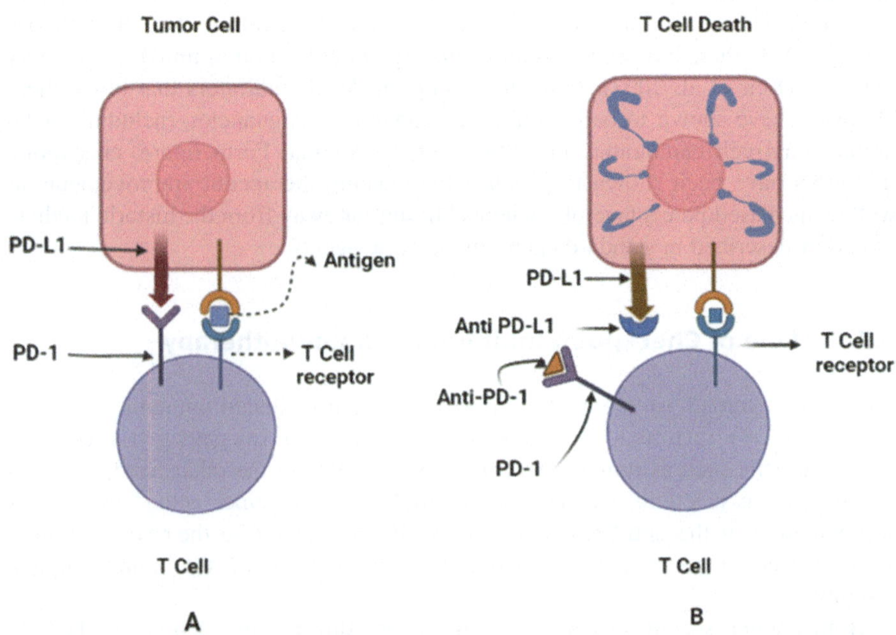

Fig. 2.5 (**a**) Attachment of "Pd-L1 to Pd-1" holds T cells from harming cancer cells inside our body. (**b**) Inhabiting the binding "PD-L1 to PD-1" with an "immune checkpoint inhibitor" (anti-PD-L1 or anti-PD-1) permits the T cells to cause death of cancerous cells

and CTLA-4, resulting in T-cell differentiation and proliferation via the formation of growth cytokines or by T-cell receptor signaling proteins dephosphorylation by tyrosine phosphatases when the CD28:CD80/CD86 ratio is high (Bell et al. 2018; Guntermann and Alexander 2002). Because "CTLA-4" has a strong affinity for CD80/86, it promotes immunosuppression by vying for Cd-28 and driving Cd80/86 removal from APCs surfaces (Seidel et al. 2018). As a result, "CTLA-4 inhibitors" can avoid T-cell depletion as well as increase the anti-tumor T-cell responses by limiting the association of "CTLA-4 and Cd80/86 ligands" (Fellner 2012). Although *ipilimumab* has been shown to improve lifespan in individuals with complex tumors, significant immuno-controlled complications, a hefty price tag, and limited response rates (varying from 4% to 16%) remained the main barriers to its usage (Fellner 2012).

"Pd-1" however, is primarily associated in the regulation of formerly "activated T-cells" in subsequent phases of an "immune response" primarily in tissues and cancers (Pardoll 2012; Buchbinder and Desai 2016). In T-cells with the "exhausted phenotype" membrane receptor expression can momentarily generated in "activated Cd8 T-cells" NK T-cells or myeloid-cells after T-cell receptor stimulation and activation by cytokines and interleukins is constitutive. When "PD-1" binds its ligand, PD-L1 enhances "T-cell-receptor proximal signal components dephosphorylation and inhibits signaling pathways controlled by protein kinases such as PTEN, CK2, PI3K/AKT, and RAS/MEK/ERK, T-cell generation, survival, chemokine synthesis, and rest repressor activities are-all reduced. Therefore, "checkpoint inhibitors" have ability to reinstate anticancer immune-response as well as increase "immune-mediated" tumor cell clearance by disturbing the link between "Pd-1 and Pd-L1"(Darvin et al. 2018). Even though *nivolumab* alone and in combination with *ipilimumab* enhances response rates completely in patients with metastatic-melanoma as compared to *ipilimumab* alone, response rates to "PD-1/PD-L1" inhabiting treatments hardly vary from 20–38% across tumor types, implying that these treatments are ineffective for the vast majority of advanced-stage patients (Garcia-Aranda and Redondo 2019). Although immune-related side effects and poor response rates in some patients with cancer, "CTLA4, and PD-1/PD-L1 inhibitors" have been shown to improve strong and long-lasting anti-tumor response and extend the average lifespan of patients with advanced cancer.

2.7.2 Breast Cancer Immunogenicity

Unlike nephroma and melanoma, which possess the major susceptivity for biological therapies, BC has not been generally thought to be immunogenic. Furthermore, immuno-suppressive substances are released by the tumor microenvironment in BC, which makes antigen presentation complex and harms the immune response (Mittendorf et al. 2007; Mehraj et al. 2021a, 2021b). It is also feasible to avoid immunological damage by suppressing autogenous immune-check-points which ordinarily end body's defense following the activation of antigen.

Despite having a minor impact on prime cancer growth, the defense system appears to be successful in avoiding BC metastases (Bidwell et al. 2012). With proper immune activation, it appears that every tumor can be immunogenic. The immune system appears to have played a key role in reaction to "monoclonal antibodies and tyrosine kinase inhibitors" and apart of data suggests it may also play a role in responses to endocrine therapy. Trastuzumab's mode of activity has traditionally been linked with antibody-dependent cellular cytotoxicity (Musolino et al. 2008; Tamura et al. 2011). As a result, a functional immune system is required for complete tumor response following molecularly targeted treatments, leading the way toward fundamentally novel regimens combining a targeted and immunological approach (Rakhra et al. 2010). To boost the anti-tumor responses, mAbs targeting "antigen tumor targets or immune-regulatory" substances, cell-mediated treatments such as "adoptive transfer of ex vivo-activated T and NK cells" or T-reg cell inhibition could be used.

2.8 Surgical Treatment of Breast Cancer

During surgery, tumor and some normal tissues in the nearby area are removed. Lumpectomy (excision of the lump solely) or mastectomy (surgical excision of the whole breast) is done based on the stage and type of malignancy. The surgeon must confirm that the margins of the tissues removed in the surgery are free of malignancy, showing that cancer has been entirely excised, according to standard procedure. Additional surgeries to excise more tissue might be required if the excised tissues do not provide clear margins. A portion of the pectoralis major, the front chest wall's principal muscle, may need to be removed in some cases. Breast cancer surgical treatment has evolved significantly over time. Previously, therapy required significant surgery and a lengthy hospital stay; however, it is now typically carried out an out-patient operation with rapid improvement (Lowery et al. 2012). Surgical treatment aims to improve local management, avoid locoregional relapse, and prolong survival. Simple mastectomy or reconstructed, with "primary or delayed" or "breast conservation treatment (BCT)" with or with-out the application of angioplasty procedures, are the various surgical methods for the treatment of breast tumors (Fig. 2.6) (Lowery et al. 2012).

2.8.1 Breast-Conserving-Therapy (BCT)

Breast-conserving-therapy may be diagnosed with a basic "broad excision" or varying levels of onco-plastic. BCT is regarded as an absolute surgical treatment for subset of BC patients and is nosologically acceptable (Hartmann-Johnsen et al. 2015; Mansell et al. 2017). BCT comprises lumpectomy which is followed by adjuvant radiation treatment to the whole breast. To perform BCT, tumor is removed with excellent cosmetic results and negative-margins, the individual should allow to receive radiation, and the breast should be suitable for the follow-up imaging to

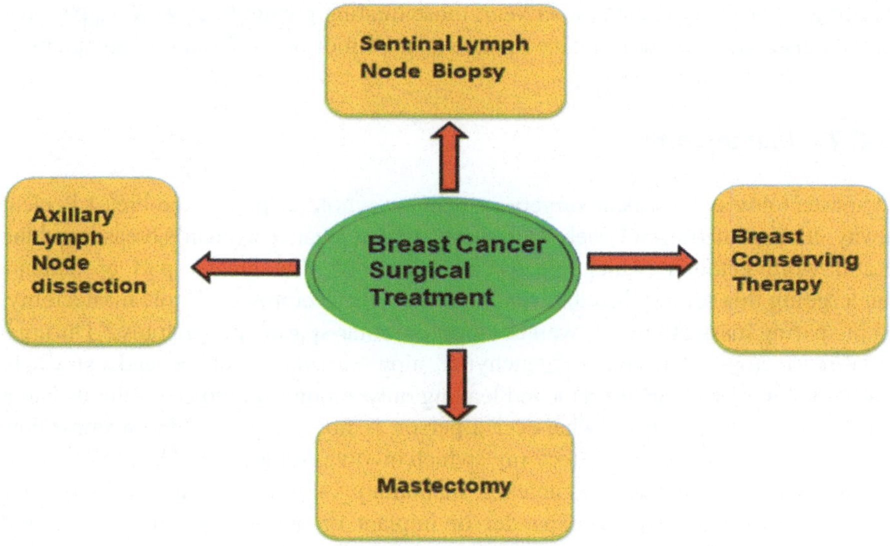

Fig. 2.6 Different types of surgical treatments in breast cancer treatment

swiftly identify local recurrence, a negative margin signifies "NO ink on the tumor" (Moran et al. 2014). Greater margin clearance does not enhance local-control in metastatic mammary cancer which is not necessary for "BC." If negative margins can be achieved with acceptable cosmetic results, lumpectomies can be performed regardless of the size of the tumor (Houssami et al. 2014). Females with large tumors compared to their breast size may benefit from neo-adjuvant their malignancies. "Bilateral-mastectomy" is examined in individuals with "BRCA1/2 mutations" as the chance of developing a new primary BC in the 20 years after diagnosis might vary from 26 to 40 percent, based on the age of initiation of first cancer, oophorectomy, and administration of endocrine therapy (Kuchenbaecker et al. 2017).

The imaging techniques commonly used to evaluate patients for BCT are clinical examination, mammograms, and diagnostic ultrasonography. According to "population-based research" of 1984, females with "DCIS and stages I and II" aggressive malignancies found that 88% of those who attempted BCT were effective. Because many females were switched to mastectomy without attempting re-excision, this is likely an underestimation of the proportion of females suitable for BCT (Morrow et al. 2009). The number of individuals who receive BCT and the number of individuals who only require one surgical treatment varies depending on the surgeon's skill. A study of surgical margins and treatment success rates in the United States demonstrated this (Morrow et al. 2017). It was a population-based study with roughly 7000 potential individuals, which was later decreased to 3279 in the analytical cohort. A total of 342 of the 488 surgeons who treated these individuals completed a survey on margins following lumpectomy. When compared to those treating 20 or fewer instances per year, those treating greater than 50 cases per year were substantially more probable to state a "no tumor on ink" sides as acceptable.

2.8.2 Mastectomy

A mastectomy is a medical surgery in which a whole or part of the breast is clear away. It originated from Greek word masts, which means "woman's breast" and the Latin word ectomia, which means "excision of." The greater part of patients undergoing this surgery has the choice of "total mastectomy or simple mastectomy, skin-sparing mastectomy" as well as "nipple areolar-sparing mastectomy." During a total-mastectomy, the breast parenchyma, nipple-areolar complex, and extra skin from the breast wall are all removed leaving only enough skin to cover the incision. This is frequently utilized whenever a person is not going through an immediate repair. The "skin-sparing mastectomy" which involves removing both "breast parenchyma and nipple-areolar" complexes, maintaining skin as a suitable-envelope for implantation of the "tissue expander or implant or donor flap, allows for rapid regeneration." The oncological efficacy of "skin-sparing-mastectomy" has been shown to be more effective against cancer than with basic mastectomy, with local recurrence of rate of 6% (Meretoja et al. 2007; Carlson et al. 2003; Lanitis et al. 2010). "The skin envelope and nipple-areolar" complex is preserved after a "nipple-areolar-sparing mastectomy." The first employed as a preventative measure, but it is currently being utilized more frequently in individuals having aggressive cancer. Recurrence rates of 2–5% have been observed locally, having an average duration of "2–5 years" (Moo et al. 2016; De Alcantara Filho et al. 2011). Patients must be cautiously chosen for this operation till long-term oncological safety has been proven, as a majority of the data come from single-institution retrospective studies with minimal follow-up.

2.8.3 Dissection of the Axillary Lymph Nodes

The cancer-specialist eliminates a clump of lymph nodes from underneath the armpit in an "axillary lymph node dissection." Axillary lymph node dissection (ALND) is frequently utilized in breast tumor surgery as a diagnostic and predictive marker. ALND is suggested for women with invasive BC as a component of their primary surgical therapy. It is a critical component in the surgical treatment of BC. Pathologic evaluation of the excised lymph nodes provides data that aids in determining the disease's pathological stage and is an important aspect of BC therapy (Le et al. 2016). ALND may improve the whole survival of women with BC by controlling regional node disease.

2.8.4 Sentinel Node Biopsy

Considering the first stage BC, SLNB is a comfortable and reliable procedure. The SLNB has turned into the standard process for assessing metastatic progression to the lymph node area (Lyman et al. 2005; McMasters et al. 2000). The "sentinel lymph node" is considered as the primary lymph gland in the "lymphatic basin" to acquire outflow from an anatomic area and is immunologically accountable for that area (Faries et al. 2000). Because of its reduced "false-negative rate" between 5% and 10% and remarkable sensitivity-value of around 90–95 percent in cancer observation in the lymphatic node region, this least invasive procedure has become the gold standard (Veronesi et al. 2010). At present, the SLNB has surpassed the ALND in axilla staging due to the exactness and low morbidity and invasion for the clinical node-negative disease.

2.9 Summary

Breast carcinoma is considered one of the most challenging diseases with a significant death rate. On the basis of hormonal expression and growth factor receptors, it is divided into a few primary molecular subtypes. Significant progress has been achieved in the discovery of novel treatments to treat BC over the past few years. With a comprehensive understanding of physiological variability of BC, a more successful and customized therapeutic strategy has been developed. Improvements in cancer risk prediction, precise prognosis evaluation, and innovative treatments in the adjuvant and neo-adjuvant settings, including immune-therapies or antibody-drug conjugates (ADCs), have all contributed to a steady advancement in therapy. Chemotherapeutic combinations comprising *anthracyclines* and *taxanes* have shown effective results in the treatment of BC. Moreover, the efficacy of hormonal therapies like "anti-HER2 MABs" possesses verified possibility along with a great significance of "molecular targeting in breast cancer" management. Checkpoint blockade has developed as an exciting treatment option for BC, particularly TNBC. Following BCT, radiotherapy reduces death rate and relapse. Most breast tumors with physiologically negative ALNs are candidates for SLNB, which eliminates the arm inflammation and discomfort involved in ALND. Individuals having BC are surviving much more than ever, thanks to the deployment of such treatments that leads to lower mortality. To conclude, it is evident that BC treatment is a subject that is continually evolving, with advancements being made all the time.

2.10 Further Readings

For more insights about the topic we would suggest detailed findings from the books of (Mir 2022) https://doi.org/10.1016/C2021-0-02565-7, https://doi.org/10.1016/C2022-0-00074-X (Mir 2021) https://doi.org/10.52305/WXJL6770, from cancer.net website, https://www.cancer.net/cancer-types/breast-cancer/types-treatment

For diagrammatic illustrations, descriptive tables (lazzeroni, 2012) http://www.eurekaselect.com/article/49928

https://www.researchgate.net/publication/342829505_Molecular_Complexity_of_Lymphovascular_Invasion_The_Role_of_Cell_Migration_in_Breast_Cancer_as_a_Prototype/figures?lo=1

See video links on over all status of Cancer, its various types, current new treatment possible options available

https://www.sciencedirect.com/science/article/pii/S205970292032278X; https://youtu.be/wIsdjfwPUxY
https://youtu.be/SVjJt984PlU

References

Anampa J, Makower D, Sparano JA (2015) Progress in adjuvant chemotherapy for breast cancer: an overview. Br J Cancer 13:195

Bamett G, Wilkinson J, Moody AM et al (2009) A randomized controlled trial of forward-planned radiotherapy (IMRT) for early breast cancer: baseline characteristics and dosimetry results. Radiother Oncol 92:34

Begg AC, Stewart F, Vens C (2011) Strategies to improve radiotherapy with targeted drugs. Nat Rev Cancer 11:239–253

Blok EJ, Kroep JR, Meershoek-Klein Kranenbarg E et al (2016) Optimal duration of extended letrozole treatment after 5 years of adjuvant endocrine therapy; results of the randomized phase III ideal trial *breast cancer symposium.* San Antonio, TX, S-04: p2-09-01

Baselga J, Cortés J, Kim SB, Im SA, Hegg R, Im YH, Roman L, Pedrini JL, Pienkowski T, Knott A et al (2012) Pertuzumab plus trastuzumab plus docetaxel for metastatic breast cancer. N Engl J Med 366:109–119

Bidwell BN, Slaney C, Withana NP, Forster S, Cao Y, Loi S et al (2012) Silencing of irf7 pathways in breast cancer cells promotes bone metastasis through immune escape. Nat Med 18:1224–1231

Bell RB, Fernandes RF, Bifulco CB, Leidner R, Weinberg A, Fox BA (2018) 15 immunotherapy. In: Oral, head and neck oncology and reconstructive surgery. Elsevier, St. Louis, pp 314–340

Buchbinder EI, Desai A (2016) Ctla-4 and pd-1 pathways: similarities, differences, and implications of their inhibition. Am J Clin Oncol 39:98–106

Darvin PT, Toor SM, Sasidharan Nair V, Elkord E (2018) Immune checkpoint inhibitors: recent progress and potential biomarkers. Exp Mol Med 50:1

Bese NS, Munshi A, Budrukkar A, Elzawawy A, Perez CA (2008) Breast radiation therapy guideline implementation in low- and middle-income countries. Cancer 113:2305–2314

Beck A, Goetsch L, Dumontet C, Corvaïa N (2017) Strategies and challenges for the next generation of antibody-drug conjugates. Nat Rev Drug Discov 16:315–337

Band PR (2010) The birth of the subspecialty of medical oncology and examples of its early scientific foundations. J Clin Oncol 1:3653–3658

Bland KI, Copeland E (2009) Comprehensive management of benign and malignant diseases. Elsevier, Philadelphia

Borghero YO, Salehpour M, McNeese MD et al (2007) Multileaf field-in-field forward-planned intensity-modulated dose compensation for whole-breast irradiation is associated with reduced contralateral breast dose: a phantom model comparison. Radiother Oncol 83(3):324–328
Braunstein LZ, Taghian A, Niemierko A et al (2017) Breast-cancer subtype, age, and lymph node status as predictors of local recurrence following breast-conserving therapy. Breast Cancer Res Treat 161(1):173–179
Citron ML, Berry D, Cirrincione C, Hudis C, Winer EP, Gradishar WJ et al (2003) Randomized trial of dose-dense versus conventionally scheduled and sequential versus concurrent combination chemotherapy as postoperative adjuvant treatment of node-positive primary breast cancer: first report of Intergroup Trial c9741/Cancer and Leukemia Group B Trial 9741. J Clin Oncol 21: 1431–1439
Coombes RC, Hall E, Gibson LJ, Intergroup Exemestane Study et al (2004) A randomized trial of exemestane after two to three years of tamoxifen therapy in postmenopausal women with primary breast cancer. N Engl J Med 11:1081–1092
Carlson GW, Styblo T, Lyles RH, Jones G, Murray DR, Staley CA et al (2003) The use of skin sparing mastectomy in the treatment of breast cancer: the Emory experience. Surg Oncol 12: 265–269
Dickler A (2009) Axxent electronic brachytherapy: a new device for delivering brachytherapy to the breast. Nat Clin Pract Oncol 6:138–142
De Alcantara Filho P, Capko D, Barry JM, Morrow M, Pusic A, Sacchini VS (2011) Nipple-sparing mastectomy for breast cancer and risk-reducing surgery: the memorial Sloan-Kettering cancer center experience. Ann Surg Oncol 18:3117–3122
Fisher B, Anderson S, Bryant J et al (2002) Twenty-year follow-up of a randomized trial comparing total mastectomy, lumpectomy, and lumpectomy plus irradiation for the treatment of invasive breast cancer. N Engl J Med 347:1233–1241
Faries MB, Bedrosian I, Reynolds C et al (2000) Active macromolecule uptake by lymph node antigen-presenting cells: a novel mechanism in determining sentinel lymph node status. Ann Surg Oncol 7:98–105
Fellner C (2012) Ipilimumab (yervoy) prolongs survival in advanced melanoma: serious side effects and a hefty price tag may limit its use. Pharm Ther 37(9):503–530
Fontana MF, Vance RE (2011) Two signal models in innate immunity. Immunol Rev 243:26–39
Fisher B, Bryant J, Wolmark N, Mamounas E, Brown A, Fisher ER et al (1998) Effect of preoperative chemotherapy on the outcome of women with operable breast cancer. J Clin Oncol 16:2672–2685
Franklin MC, Carey K, Vajdos FF et al (2004) Insights into ErbB signaling from the structure of the erbb2-pertuzumab complex. Cancer Cell 5:317–328
García-Aranda M, Redondo M (2017) Protein kinase targets in breast cancer. Int J Mol Sci 18:2543
García Rodríguez J, Colmenero CG, Soler RC, Sánchez IO (2010) Five years survival of women diagnosed with breast cancer during the period 1997-1999 in Toledo-Centro and Mancha Area, Spain. Rev Esp Salud Publica 86:843
Garcia-Aranda M, Redondo M (2019) Targeting protein kinases to enhance the response to anti-pd-1/pd-l1 immunotherapy. Int J Mol Sci 20(9):2296
Geyer CE, Forster J, Lindquist D et al (2006) Lapatinib plus capecitabine for HER2-positive advanced breast cancer. N Engl J Med 355:2733–2743
Guntermann C, Alexander DR (2002) Ctla-4 suppresses proximal tcr signaling in resting human cd4+ T cells by inhibiting zap-70 tyr319 phosphorylation: a potential role for tyrosine phosphatases. J Immunol 168:4420–4429
Gennari A, Sormani M, Pronzato P, Puntoni M, Colozza M, Pfeffer U, Bruzzi P et al (2008) Her2 status and efficacy of adjuvant anthracyclines in early breast cancer: a pooled analysis of randomized trials. J Natl Cancer Inst 100:14–20
Gianni L, Dafni U, Gelber RD, Azambuja E, Muehlbauer S, Goldhirsch A et al (2011) Treatment with trastuzumab for 1 year after adjuvant chemotherapy in patients with her2-positive early breast cancer: a 4-year follow-up of a randomized controlled trial. Lancet Oncol 236:1016

Glaysher S, Bolton L, Johnson P, Torrance C, Cree IA (2014) The activity of EGFR, mTOR, and PI3k inhibitors in an isogenic breast cell line model. BMC Res Notes 7:397
Grunt TW, Mariani G (2013) Novel approaches for molecular targeted therapy of breast cancer: interfering with PI3k/AKT/mTOR signaling. Curr Cancer Drug Targets 13:188–204
Gianni L, Pienkowski T, Im Y-H et al (2012) Efficacy and safety of neoadjuvant pertuzumab and trastuzumab in women with locally advanced, inflammatory, or early HER2-positive breast cancer (neosphere): a randomised multicentre, open-label, phase 2 trial. Lancet Oncol 13:25–32
Gianni L, Romieu G, Lichinitser M, Serrano SV, Mansutti M, Pivot X, Mariani P, Andre F, Chan A, Lipatov O et al (2013) AVEREL: a randomized phase III trial evaluating bevacizumab in combination with docetaxel and trastuzumab as first-line therapy for HER2-positive locally recurrent/metastatic breast cancer. J Clin Oncol 31:1719–1725
Giordano SH, Temin S, Kirshner JJ, Chandarlapaty S, Crews JR, Davidson NE, Esteva FJ, Gonzalez-Angulo AM, Krop I, Levinson J et al (2014) Systemic therapy for patients with advanced human epidermal growth factor receptor 2-positive breast cancer: American Society of Clinical Oncology Clinical Practice Guideline. J Clin Oncol 32:2078–2099
Peto R, Davies C, Godwin J, Gray R, Pan HC et al (2012) Comparisons between different polychemotherapy regimens for early breast cancer: meta-analyses of long-term outcome among 100,000 women in 123 randomised trials. Lancet 379:432–444
Howell A, Cuzick J, Baum M et al (2005) Results of the atac (arimidex, tamoxifen, alone or in combination) trial after completion of 5 years & 39;djuvant treatment for breast cancer. Lancet 1:60–62
Hanahan D, Weinberg R (2011) Hallmarks of cancer: the next generation. Cell 144:646–674
Hartmann-Johnsen OJ, Kåresen R, Schlichting E, Nygard JF (2015) Survival is better after breast conserving therapy than mastectomy for early stage breast cancer: a registry-based follow-up study of Norwegian women primary operated between 1998 and 2008. Ann Surg Oncol 22: 3836–3845
Herrick JS, Neill CJ, Rosser PF (2008) A comprehensive clinical 3-dimensional dosimetric analysis of forward planned IMRT and conventional wedge planned wedged techniques for intact breast radiotherapy. Med Dosim 33:62–70
Houssami N, Macaskill P, Marinovich ML, Morrow M (2014) The association of surgical margins and local recurrence in women with early-stage invasive breast cancer treated with breast-conserving therapy: a meta-analysis. Ann Surg Oncol 21:717–730
Jackson SP, Bartek J (2009) The DNA-damage response in human biology and disease. Nature 461: 1071–1078
Jakesz R, Jonat W, Gnant M, ABCSG and the GABG et al (2005) Switching of postmenopausal women with endocrine-responsive early breast cancer to anastrozole after 2 years & 39; adjuvant tamoxifen: combined results of abcsg trial 8 and arno 95 trial. Lancet 6:455–462
Jankowitz RC, Davidson N (2013) Adjuvant endocrine therapy for breast cancer: how long is long enough. Oncology 27:1210–1216
Jones S, Holmes F, Shaughnessy J, Blum JL, Vukelja SJ, McIntyre KJ et al (2009) Docetaxel with cyclophosphamide is associated with an overall survival benefit compared with doxorubicin and cyclophosphamide: 7-year follow-up of us oncology research trial 9735. J Clin Oncol 27:1177–1183
Keir ME, Butte MJ, Freeman GJ, Sharpe AH (2007) Pd-1 and its ligands in t-cell immunity. Curr Opin Immunol 19:309–314
Klapper JA, Downey S, Smith FO et al (2008) High-dose interleukin-2 for the treatment of metastatic renal cell carcinoma: a retrospective analysis of response and survival in patients treated in the surgery branch at the national cancer institute between 1986 and 2006. Cancer 113: 293–301
Keisch M, Vicini F, Kuske R et al (2003) Initial clinical experience with the mammo site breast brachytherapy applicator in women with early-stage breast cancer treated with breast-conserving therapy. Int J Radiat Oncol Biol Phys 55(2):289–293

Kuchenbaecker KB, Hopper J, Barnes DR, Phillips KA, Mooij TM, Roos-Blom MJ et al (2017) Risks of breast, ovarian, and contralateral breast cancer for BRCA1 and BRCA2 mutation carriers. JAMA 317:2402–2416

Lowery AJ, Kell M, Glynn RW, Kerin MJ, Sweeney KJ (2012) Locoregional recurrence after breast cancer surgery: a systematic review by receptor phenotype. Breast Cancer Res Treat 133:831–841

Lyman GH, Giuliano A, Somerfield MR et al (2005) American society of clinical oncology guideline recommendations for sentinel lymph node biopsy in early-stage breast cancer. J Clin Oncol 23:7703–7720

Lanitis S, Tekkis P, Sgourakis G, Dimopoulos N, Al Mufti R, Hadjiminas DJ (2010) Comparison of skin-sparing mastectomy versus non-skin-sparing mastectomy for breast cancer: a meta-analysis of observational studies. Ann Surg 251:632–639

Mamounas EP (2003) Neoadjuvant chemotherapy for operable breast cancer: is this the future? Clin Breast Cancer Suppl 1:S10–S19. https://doi.org/10.3816/cbc.2003.s.010

McGale P, Taylor C, Correa C et al (2014) Effect of radiotherapy after mastectomy and axillary surgery on 10-year recurrence and 20-year breast cancer mortality: meta-analysis of individual patient data for 8135 women in 22 randomised trials. Lancet 383:2127–2135

Mehraj U, Dar A, Wani NA, Mir MA (2021a) Tumor microenvironment promotes breast cancer chemoresistance. Cancer Chemother Pharmacol 9:1–2

Mehraj U, Qayoom H, Mir MA (2021b) Prognostic significance and targeting tumor-associated macrophages in cancer: new insights and future perspectives. Breast Cancer 28:539–555

Mir MA, Qayoom H, Mehraj U, Nisar S, Bhat B, Wani NA (2020) Targeting different pathways using novel combination therapy in triple negative breast cancer. Curr Cancer Drug Targets 20: 586–602

Mittendorf EA, Peoples G, Singletary SE (2007) Breast cancer vaccines: promise for the future or pipe dream. Cancer 110:1677–1686

Musolino A, Naldi N, Bortesi B, Pezzuolo D, Capelletti M, Missale G et al (2008) Immunoglobulin G fragment C receptor polymorphisms and clinical efficacy of trastuzumab-based therapy in patients with HER-2/neu-positive metastatic breast cancer. J Clin Oncol 26(11):1789–1796

Martin M, Holmes F, Ejlertsen B, ExteNET Study Group et al (2017) Neratinib after trastuzumab-based adjuvant therapy in HER2-positive breast cancer (ExteNET): 5- year analysis of a randomised, double-blind, placebo-controlled, phase 3 trial. Lancet Oncol 18(12):1688–1700

Mir MA (2015) Introduction to costimulation and costimulatory molecules. In: Developing costimulatory molecules for immunotherapy of diseases. Elsevier Academic Press, Cambridge, pp 1–299. https://doi.org/10.1016/c2014-0-02898-5

Murthy RK, Loi S, Okines A et al (2020) Tucatinib, trastuzumab, and capecitabine for HER2-positive metastatic breast cancer. N Engl J Med 382:597–609

Mansell J, Weiler-Mithoff E, Stallard S, Doughty JC, Mallon E, Romics L (2017) Oncoplastic breast conservation surgery is oncologically safe when compared to wide local excision and mastectomy. Breast 32:179–185

Moran MS, Schnitt S, Giuliano AE, Harris JR, Khan SA, Horton J et al (2014) Society of Surgical Oncology-American Society for Radiation Oncology consensus guideline on margins for breast-conserving surgery with whole-breast irradiation in stages I and II invasive breast cancer. J Clin Oncol 32:1507–1515

Masoud V, Pagès G (2017) New challenges to fight against resistance. World J Clin Oncol 8:120–134

McMasters KM, Tuttle T, Carlson DJ et al (2000) Sentinel lymph node biopsy for breast cancer: a suitable alternative to routine axillary dissection in multi-institutional practice when optimal technique is used. J Clin Oncol 18:2560–2566

McPhail S, Johnson S, Greenberg D, Peake M, Rous B (2015) Stage at diagnosis and early mortality from cancer in England. Br J Cancer 112:s108

Mamtani A, Barrio A, King TA, Van Zee KJ, Plitas G, Pilewskie M et al (2016) How often does neoadjuvant chemotherapy avoid axillary dissection in patients with histologically confirmed nodal metastases? Results of a prospective study. Ann Surg Oncol 23:3467–3474

Meretoja TJ, Rasia S, Von Smitten KA, Asko-Seljavaara SL, Kuokkanen HO, Jahkola TA (2007) Late results of skin-sparing mastectomy followed by immediate breast reconstruction. Br J Surg 94:1120–1125

Morrow M, Abrahamse P, Hofer TP, Ward KC, Hamilton AS, Kurian AW (2017) Trends in reoperation after initial lumpectomy for breast cancer: addressing overtreatment in surgical management. JAMA Oncol 3:1352–1375

Morrow M, Jagsi R, Alderman AK, Griggs JJ, Hawley ST, Hamilton AS et al (2009) Surgeon recommendations and receipt of mastectomy for treatment of breast cancer. JAMA 302:1551–1561

Moo TA, Pinchinat T, Mays S, Landers A, Christos P, Alabdulkareem H et al (2016) Oncologic outcomes after nipple-sparing mastectomy. Ann Surg Oncol 23:3221–3225

Mieog JS, van der Hage JA, Van de Velde CJ (2007) Neoadjuvant chemotherapy for operable breast cancer. Br J Surg 94:1189–1200

Nami B, Wang Z (2017) Her2 in breast cancer stemness: a negative feedback loop towards trastuzumab resistance. Cancers 9:40

Nelson J, Beitsch P, Vicini F et al (2009) Four year clinical update from the American Society of Breast Surgeons MammoSite brachytherapy trial. Am J Surg 198(1):83–91

Pardoll DM (2012) The blockade of immune checkpoints in cancer immunotherapy. Nat Rev Cancer 12:252–264

Paik S, Tang G, Shak S, Kim C, Baker J, Kim W et al (2006) Gene expression and benefit of chemotherapy in women with node-negative, estrogen receptor-positive breast cancer. J Clin Oncol 24:3726–3734

Piccart-Gebhart MJ, Procter M, Leyland-Jones B et al (2005) Herceptin adjuvant (hera)trial study team. Trastuzumab after adjuvant chemotherapy in her2-positive breast cancer. N Engl J Med 353:1659–1672

Press MF, Sauter G, Buyse M, Bernstein L, Guzman R et al (2011) Alteration of topoisomerase II-alpha gene in human breast cancer: association with responsiveness to anthracycline-based chemotherapy. J Clin Oncol 29:859–867

Qayoom H, Wani N, Alshehri B, Mir MA (2021) An insight into the cancer stem cell survival pathways involved in chemoresistance in triple-negative breast cancer. Future Oncol 17:4185–4206

Qayoom H et al (2022) Expression patterns and therapeutic implications of CDK4 across multiple carcinomas: a molecular docking and MD simulation study. Med Oncol 39:158. https://doi.org/10.1007/s12032-022-01779-9

Romond E, Perez E, Bryant J et al (2005) Trastuzumab plus adjuvant chemotherapy for operable her2-positive breast cancer. N Engl J Med 353:1673–1684

Rakhra K, Bachireddy P, Zabuawala T, Zeiser R, Xu L, Kopelman A et al (2010) Cd4(+) t cells contribute to the remodeling of the microenvironment required for sustained tumor regression upon oncogene inactivation. Cancer Cell 18:485–498

Schlembach PJ, Buchholz TA, Ross MI et al (2001) Relationship of sentinel and axillary level I and II lymph nodes to tangential fields used in breast irradiation. Int J Radiat Oncol Biol Phys 51(3): 671–678

Scanderbeg J, Yashar C, Rice R et al (2009) Clinical implementation of a new brachytherapy device for partial breast irradiation. Radiother Oncol 90:3642

Sparano JA, Wang M, Martino S, Jones V, Perez EA, Saphner T et al (2008) Weekly paclitaxel in the adjuvant treatment of breast cancer. N Engl J Med 358:1663–1671

Schreiber RD, Old L, Smyth MJ (2011) Cancer immunoediting: integrating immunity's roles in cancer suppression and promotion. Science 331:1565–1570

Segovia-Mendoza M, González-González M, Barrera D, Díaz L, García-Becerra R (2015) Efficacy and mechanism of action of the tyrosine kinase inhibitors gefitinib, lapatinib and neratinib in the treatment of her2-positive breast cancer: preclinical and clinical evidence. Am J Cancer Res 5: 2531–2561

Seidel JA, Otsuka A et al (2018) Anti-pd-1 and anti-ctla-4 therapies in cancer: mechanisms of action, efficacy, and limitations. Front Oncol 8:86

Slamon D, Eiermann W, Robert N, Breast Cancer International Research Group et al (2011) Adjuvant trastuzumab in her2-positive breast cancer. N Engl J Med 365:1273–1228

Smith I, Yardley D, Burris H et al (2017) Comparative efficacy and safety of adjuvant letrozole versus anastrozole in postmenopausal patients with hormone receptor-positive, node-positive early breast cancer: final results of the randomized phase III femara versus anastrozole clinical evaluation (face) trial. J Clin Oncol 1:1041–1048

Sofi S, Mehraj U, Qayoom H, Aisha S, Almilaibary A, Alkhanani M et al (2022) Targeting cyclin-dependent kinase 1 (cdk1) in cancer: molecular docking and dynamic simulations of potential cdk1 inhibitors. Med Oncol 39:1–15

Swain SM, Tang G, Geyer CE Jr, Rastogi P, Atkins JN, Donnellan PP et al (2013) Definitive results of a phase III adjuvant trial comparing three chemotherapy regimens in women with operable, node-positive breast cancer: the NSABP B-38 trial. J Clin Oncol 31:3197–3204

Thürlimann B, Keshaviah A, Coates AS, Breast International Group (BIG) 1-98 Collaborative Group et al (2006) A comparison of letrozole and tamoxifen in postmenopausal women with early breast cancer. N Engl J Med 18:2200

Tamura K, Shimizu C, Hojo T, Akashi-Tanaka S, Kinoshita T, Yonemori K et al (2011) Fcγr2a and 3a polymorphisms predict clinical outcome of trastuzumab in both neoadjuvant and metastatic settings in patients with her2-positive breast cancer. Ann Oncol 22:1302–1307

Veronesi U, Cascinelli N, Mariani L et al (2002) Twenty-year follow-up of a randomized study comparing breast-conserving surgery with radical mastectomy for early breast cancer. N Engl J Med 347:1227–1232

Vrieling C, Werkhoven EV, Maingon P et al (2017) Prognostic factors for local control in breast cancer after long-term follow-up in the eortc boost vs no boost trial: a randomized clinical trial. JAMA Oncol 3:42–48

Van der Hage JA, van de Velde CJ, Julien JP, Tubiana-Hulin M, Vandervelden C, Duchateau L (2001) Preoperative chemotherapy in primary operable breast cancer: results from the European organization for research and treatment of cancer trial 10902. J Clin Oncol 19:4224–4237

Verma S, Miles D, Gianni L, EMILIA Study Group et al (2012) Trastuzumab emtansine for her2-positive advanced breast cancer. N Engl J Med 367:1783–1791

Berry DA, Cirrincione C, Henderson IC, Citron ML, Budman DR, Goldstein LJ et al (2006) Estrogen-receptor status and outcomes of modern chemotherapy for patients with node-positive breast cancer. JAMA 295(14):1658–1657

Osborne CK (1998) Tamoxifen in the treatment of breast cancer. N Engl J Med 339(18):1056

Le VH, Brant K, Blackhurst DW et al (2016) The impact of the American College of Surgeons Oncology Group (ACOSOG) z0011 trial: an institutional review. Breast 29:117–119

Veronesi U, Viale G, Paganelli G et al (2010) Sentinel lymph node biopsy in breast cancer: ten-year results of a randomized controlled study. Ann Surg Oncol 251(4):595–600

Introduction to Cell Cycle and Its Regulators 3

Manzoor Ahmad Mir and Asma Jan

3.1 Introduction

The ability to divide is a trait that all cells possess. Rudolf Virchow, a German scientist, came to conclusion in 1855 that "omnis cellula e cellula" (every cell is made up of pre-existing cells), that became the third postulate of contemporary Cell Theory. Cell division produces new daughter cells. From a single-celled zygote (fertilized egg), succession of cell divisions creates all multicellular creatures with sexual cycles. As a result, cell proliferation acts as heart for cellular development and growth and it happens at all stages of its life. For example, man's blood contains 2.5 × 1013 RBCs (5 liters of blood with 5,00,000 RBCs/mm^3) and an RBC's average life span is 120 days, 2.5 × 1013 cells must have to be produced every 107 s to maintain a constant blood supply. Nerve cells and skeletal muscle cells, on the other hand, do not divide after differentiating.

Prokaryotes divide their cells quickly and easily (Bacteria). In contrast to prokaryotes, eukaryotic cells contain a variety of cell organelles. Binary fission is the most common way that prokaryotes divide their cells. Each copy of the prokaryotic chromosome is attached to a separate area of the cell membrane when it replicates. The replicated chromosomes get separated from the parent chromosome, when the cells start to pull apart. When a cell divides, it produces two identical cells with the same genetic makeup (Sofi et al. 2022a, 2022b). All life forms in a colony are genetically equivalent as a result of asexual reproduction in prokaryotes.

Eukaryote cell division is more sophisticated because of their greater chromosome number, complexity, and organelles; however, the same replication, segregation, and cytokinesis processes continue to take place. Throughout the eukaryotic

M. A. Mir (✉) · A. Jan
Department of Bioresources, School of Biological Sciences, University of Kashmir, Srinagar, J&K, India
e-mail: drmanzoor@kashmiruniversity.ac.in

M. A. Mir (ed.), *Therapeutic Potential of Cell Cycle Kinases in Breast Cancer*, https://doi.org/10.1007/978-981-19-8911-7_3

cell cycle, various cytoplasmic and nuclear activities must be synchronized (Qayoom et al. 2022).

3.2 Cell Cycle

The term "cell cycle" relates to the cellular and molecular activities that take place between one cell division and the next. The specifics of the occurrences may differ from one organism to the next, as well as throughout various stages of the organism's life cycle. Certain traits, although, are shared, since the cell cycle requires only a few processes which a cell must go through in order to complete its most basic task: copying and passing its genetic information to subsequent generations of cells. To do so, DNA needs to be accurately replicated, and the chromosome that was copied should be precisely split into two new cells, ensuring that all cells get a complete copy of the genome (Mehraj et al. 2022a, 2022b).

The parent cell (mother cell) splits and develops to produce a new cell (daughter cell) that contains all of the parent cell's genetic material. As a result, during the usual process of cell division, all of the parent cell's DNA must be replicated and properly distributed to the daughter cells to ensure genetic consistency. A cell goes through a number of sequential steps as part of this process, which is referred to as the cell cycle as shown in Fig. 3.1. There are two phases to the cell cycle and that is shown as a flowchart in Fig. 3.2.

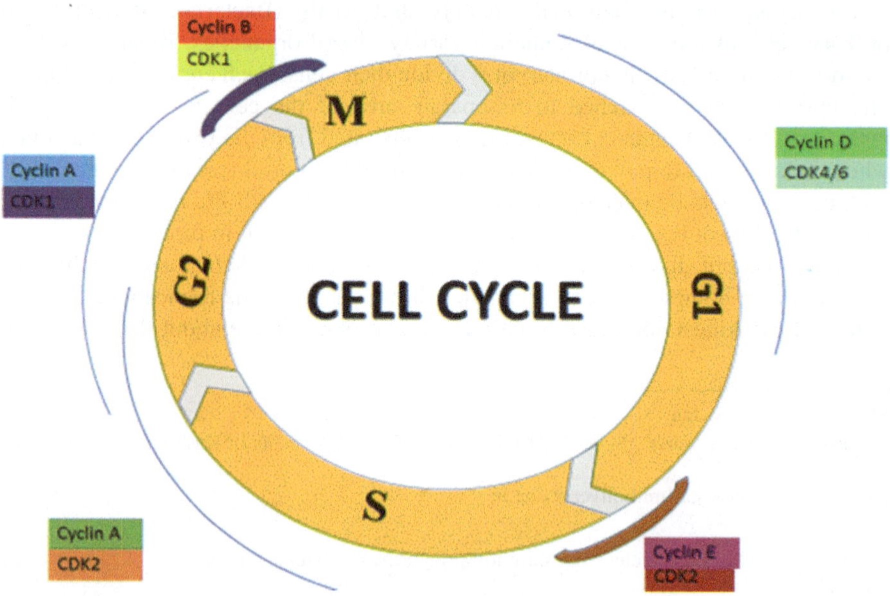

Fig. 3.1 Cell cycle and their regulation by cyclins and CDKs

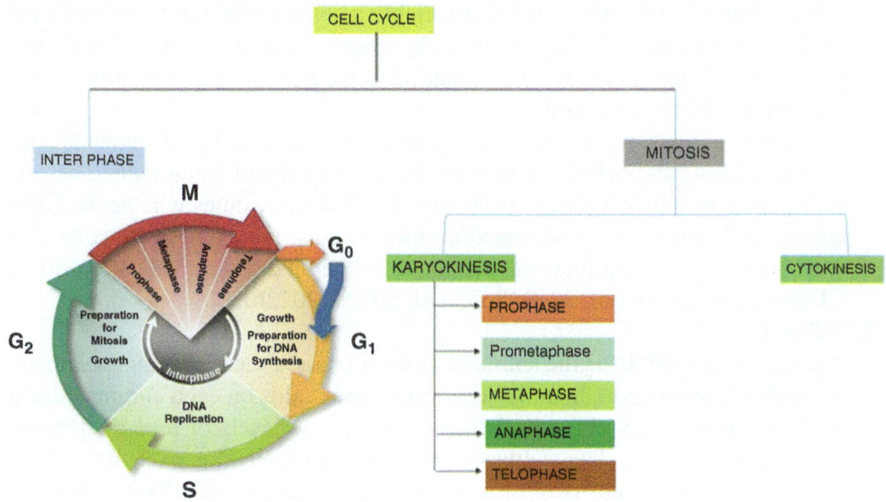

Fig. 3.2 Flowchart of cell cycle

3.2.1 Interphase

Interphase looks to be a resting state cytologically and helps to prepare the cells to reach M phase. Interphase is classified into two phases—

G1 (Gap Period 1) = Growth and chromosomal preparation for replication.

Synthesis period (S) = DNA synthesis (and centrosome).

G2 (Gap Period 2) = Mitosis Preparation.

During G2, a cell possesses twice as much DNA (4C) as it had during the diploid stage (2C). After mitosis, the daughter cells reach the G1 phase and have DNA content equal to 2C once more.

3.2.2 M Phase: The M Phase Is Classified into Two Phases

I. Mitosis, which involves the separation of duplicated chromosomes into two nuclei.

II. The process of a cell dividing into two daughter cells occurs during the cytokinesis.

I. **G1 Phase, Gap 1 Phase, or Growth 1 Phase**

The cells reach to the G1 phase from either the G0 or M phases. Extracellular mitogenic signals control the cellular transition from G0 to G1 phase (Limas and Cook 2019). The growth phase is known as G1. In M phase, the cell's biosynthetic activities are significantly inhibited; nevertheless, in G1 phase, they resume at a rapid rate. The cells amplify organelles such as mitochondria and ribosomes, synthesize numerous proteins, and expand their size within the G1 phase. Phases of the cell cycle last a variable amount of time in numerous kinds of cells. If we estimate that a normal developing human cell has a total cycle period of 24 h, the G1 phase lasts around 11 h, the S phase lasts 8 h, the G2 phase 4 h, and the M phase lasts about 1 h.

Various signals, including as stress, metabolic cues, and environmental cues interfere during the G1 phase to alter the cell's developmental path. The cells incorporate and interpret these messages. The cell determines whether to differentiate, self-renew, or perish based on these inputs; nonetheless, all cells must meet one crucial condition to enter S phase and begin their renewal: activation of CDKs (Limas and Cook 2019; Sofi et al. 2022a, 2022b).

II. **S Phase**

The synthesis of DNA is the hallmark of the S phase. After doubling the amount of DNA via replication, there are two sister chromatids on each chromosome in S phase. However, S phase is characterized by low levels of protein synthesis and gene expression. The synthesis of histone is an obvious exception. The S phase is when the majority of histones are made (Nelson et al. 2002). There is a checkpoint within the S phase which is thought to exist, to regulate the S phase progression. As a result of DNA damage and other replication-related stresses, checkpoint within the S phase shuts down Cdk2, which stops damaged DNA from being replicated by stopping origin firing (Ciardo et al. 2019). ATR (Ataxia-Telangiectasia and Rad3-related) an active Checkpoint Kinase that mediates cell cycle transition between S to G2 phase (Saldivar et al. 2018).

III. **G2 Phase**

After completing the S phase, the cell moves into the G2 phase. Mitosis signals the conclusion of the G2 phase. The main function of cells in the G2 phase is to make preparations for mitosis. Significant lipid/protein synthesis and cell proliferation are hallmarks of the G2 phase (Choudhuri et al. 2005). Protein synthesis inhibitors are well known to cause arrest of cells in the G2 phase. According to a new study, this is because p38 is inhibited, and protein synthesis is not required for mitotic entrance (Lockhead et al. 2020). Several cell types, including Xenopus embryos and cancer cells, lack the G2 phase. From the S phase to the M phase, the cell cycle moves quickly. In the G2 phase, cell size is thought to affect proliferation, but it is only been shown in fission yeast (Moseley et al. 2009). Repairing DNA double-strand breaks is another step that takes place during the G2 phase.

DNA double-strand breakage develops in the cells during or after DNA replication and must be repaired before the cell can progress to G2/M checkpoint (Glover et al. 2019).

3.2.3 Mitosis and Cytokinesis

Mitosis, in which nucleus of the cell divides and cytokinesis, in which cytoplasm of the cell divides and generates two new cells, are both part of the M phase. Mitosis has been further separated into—

I. *Prophase*
II. *Prometaphase*
III. *Metaphase*
IV. *Anaphase*
V. *Telophase*

I. *Prophase;* Prophase is marked by chromatin and chromosomal condensation, dissociation of centrosome, and disintegration of the nuclear envelope. The centrosome's transfer to two opposing poles is crucial for creation of the mitotic spindle machinery with bipolar polarity later on. The interphase organization gets quickly lost in prophase in such a condensin-dependent way, according to recent comprehensive research (Gibcus et al. 2018). In early prophase, chromosomes become visible as linearly ordered structures, according to microscopy observations (Liang et al. 2015). In early prophase, sister chromatids are intermingled, but during late prophase, they separate. Every chromatid is depicted like a series of rings extending from a central core containing Topoisomerase II Alpha as well as condensing complexes (Samejima et al. 2012). Rising Cdk1-cyclin B expression during prophase is a hallmark molecular event (Gavet and Pines 2010).

II. *Prometaphase* begins with the breaking of nuclear membrane, that occurs at the completion of prophase, and concludes with the completion of chromosomal aligning along the equator of the spindle, that indicates the start of metaphase. It is critical for accurate chromosomal separation to produce a metaphase plate with entire chromosomes aligned just at the equator of the cell adheres with the microtubules in the mitotic spindle. The perfect integration of numerous mitotic processes, including chromosomes' congressing to the spindle equator, nuclear envelope collapse and connection between microtubules of the mitotic spindle assembly and chromosome kinetochores are required to produce this configuration. In duplicated chromatids, kinetochore is a ring-shaped proteinaceous structure present in them (Santaguida and Musacchio 2009). The chromatids shorten and thicken during prometaphase (Liang et al. 2015), eventually forming fully condensed metaphase chromosomes (Paulson and Laemmli 1977).

III. *Metaphase;* Whenever replicated chromosomes are oriented within metaphase plate in the center of cell, the metaphase begins. The kinetochore microtubules drag the sister chromatids backward and forward till they have aligned themselves in the equatorial plane during the metaphase. The SAC pathway monitors the chromosome segregation process to ensure that all the kinetochores get linked to opposing poles of microtubules prior to segregation begins just after the shift to anaphase from metaphase.

The cohesiveness between sister chromatids is destroyed when each and every chromosome have been appropriately aligned as well as the kinetochores were also properly linked, causing the completely separated chromatids to move toward the opposing ends of a cell due to the pulling effect of the spindle microtubules. The cell is currently in anaphase stage (Dhatchinamoorthy et al. 2018).

IV. *Anaphase;* The kinetochore-microtubules shortening and elongation of spindle in mid-zone are two mechanistically separate stages in anaphase. Each chromatid migrates toward its respective pole as kinetochore microtubules shorten. The detached sister chromatids further separate in the mid-zone by spindle elongation. In certain creatures, these two processes may be separated in time, while in others, they may occur simultaneously. These two stages are referred to as anaphase A and B, respectively (Wordeman 2010). Anaphase B typically begins in human mitotic cells 30–50 s after anaphase A (Su et al. 2016). The spindle lengthens by 8 μm m during anaphase and by another 3 μm m during telophase (Wordeman 2010).

Telophase occurs after anaphase and begins with the recondensation of the chromosomes and the reconstruction of the nuclear membrane (Afonso et al. 2014). During telophase, a parent cell's duplicated chromosomes split into the two identical daughter cells which are identical in the nucleus. To segregate nuclear DNA from cytoplasm, the nuclear membrane arises which surrounds each set of chromosomes. At the same time, chromosomal decondensation starts (Güttinger et al. 2009).

Cytokinesis; The physical division of a mother cell's cytoplasm into two daughter cells is known as cytokinesis (Fededa and Gerlich 2012). To produce offspring with the correct complement of the chromosomes. To produce offspring with the correct complement of the chromosomes, cytoplasm and chromosome segregation must be precisely synchronized (Lens and Medema 2019). Reduced Cdk1 activity causes the mitotic spindle to reorganize and microtubules to stabilize during anaphase, resulting in cell cytokinesis. An essential early event is the building of the central spindle, which serves as a framework for midbody and aids in division plane specifications. Between two sets of chromosomes that are separated is the division plane. To avoid segregation errors, the plane's precise position is crucial. The constriction of actomyosin ring, that separates cytoplasm into two domains of developing new cells, triggers ingression of the cytokinetic furrow of connected plasma membrane. The final phase of cytokinesis is abscission (Mierzwa and Gerlich 2014). The plasma membranes of both the two new daughter cells are physically detached and it is known as abscission. The cytoskeletal structures of the intercellular bridge are removed by cells during abscission, which is following by contraction of cell cortex and plasma membrane separation (Schiel et al. 2012).

3.3 Time Determination for Cell Cycle

How can the duration of each cell cycle phase be determined? To begin, the duration of the whole cycle must be determined, which may be simply done in a homogeneous population of cultivated cells by estimating the number of cells present in a microscope and noting the quantity of time needed for overall cell number to get double. Alternately, after this period has been determined, the duration of S phase can be measured by adding 3H Thymidine to the culture for a short amount of time.

Because this is the period of DNA replication, 3H thymidine will only be absorbed into S phase cells. After that, the cells are autoradiographed, and the proportion of cells that have incorporated radioisotope is calculated by measuring the proportion of cells having exposed/reduced Ag grains.

The duration of each stage of the cell cycle is determined by multiplying the proportion of cells in that phase at any given time by the overall cell cycle time and a correction factor. In a population that is constantly dividing, a correction factor is required since there are always more young cells than elderly cells. For G1, a correction factor is 0.7, M have a correction factor of 1.4, and S-Phase cells have an intermediate value.

Light microscopy was used to scan the cell population and estimate the proportion of cells with condensed chromosomes at any given moment, the length of M phase can be estimated in a similar manner. The length of M phase is calculated by multiplying this by the entire cell cycle time and the correction factor.

As culturing cell samples are taken unless tagged mitotic chromosomes are seen, the length of the G2 phase can be determined. At the start of the incubation with 3H thymidine, the first cells where mitotic chromosomes were labeled had to be in the last stages of DNA synthesis. The duration of G2 refers to the time from the beginning of the labeling period and the appearance of cells with labeled mitotic (Toteja 2008).

Due to the lack of a G1 marker, the length of the G1 phase can be determined by adding G2 + S + M and subtracting this number from the overall cell cycle time. Cell's DNA amount at distinct cell cycle stages in a cell can also be used to differentiate them. G1 animal cells, for example, are diploid (meaning they have two copies of each chromosome), hence their DNA content is referred to as 2n (n designates the genome's haploid DNA content). Replication raises the cell's DNA content from 2n to 4n during S phase, hence S cells have DNA levels varying from 2n to 4n. In G2 and M Cells possesses a content of DNA of 4n, which drops to 2n after cytokinesis. Using a cell sorter with fluorescence activation or a flow cytometer, DNA content of the cell can be assessed by using a fluorescent dye that attaches to DNA while cells are being incubated, then analyzing the intensity of each individual cell's fluorescence, thereby identifying cells in the G1, S, and G2/M stages of the cell cycle.

3.4 G1 Is the Period of the Cell Cycle with the Most Variability

The S phase lasts 6–8 h in cultured mammalian cells, but the M phase only lasts a few minutes. G2 is often smaller than G1 and has a more constant length, lasting 4–6 h on average. Based on the type of cell, the length of G1 can vary significantly. A typical G1 cycle lasts 8–10 h; however, a few cells stay in G1 for minutes or even in hours, while others spend weeks, months, or years in G1. During G1, the cell must decide whether or not to divide again and when. A Go state is defined as cells that are still in the G1 phase for a prolonged amount of time. The amount of DNA present in G1 period is found in tissues that generally do not separate [like skeletal muscle or nerve cells] or that divide infrequently [like circulating lymphocytes]. Density-dependent growth inhibition (or contact inhibition) causes cultured cells to cease proliferating at G1 (Toteja 2008).

At cell division, eukaryotic chromosomes go through condensation-decondensation cycles, whereas prokaryotic DNA does not. Under a microscope, the chromosomes in the interphase state are decondensed and therefore cannot be identified. It is now feasible to see interphase chromosomes because of advances in techniques such as somatic cell hybridization. When a mitotic and interphase cell gets united, the interphase nuclei undergo chromosomal condensation. Premature chromosomal condensation (PCC) is a term used to describe this phenomenon. PCC from G1 nuclei only shows one chromatid, but PCC from G2 nuclei shows two. This shows that DNA replication begins following the G1 phase but prior to the G2 phase.

When a mitotic cell joins an S-phase cell, the S-phase chromatin condenses as well. Condensation in the S-phase nucleus, however, can result in the creation of "pulverized" chromosomal fragments rather than whole condensed chromosomes because replicating DNA is particularly vulnerable to damage.

3.5 Molecular Events During Cell Cycle

During the interphase, transcription occurs. All of the interphase nuclei become labeled once the cells are labeled with 3H uridine for a short length of time. This indicates that RNA production continues during the interphase period. RNA synthesis, on the other hand, decreases drastically in late prophase, and no transcription occurs during metaphase or anaphase. During metaphase, the chromosomes are severely condensed, and the DNA cannot be accessed by RNA polymerase, hence the chromosomes do not transcribe.

Replication of DNA is the most important molecular event which happens within the S phase of interphase. The amount of DNA in the cell becomes double at this moment, and sister chromatids develop. Factors that stimulate DNA synthesis are found in S phase cells. Experiments with cell fusion show that fusion with S phase cells helps to promote the replication initiation in G1 phase. This factor has no effect on G2 nuclei. This clearly demonstrates that some mechanism prevents DNA synthesis during the G2 phase. The chromosomes' more condensed, heterochromatin regions replicate late in S phase in all cells. The dormant X chromosome of female mammals, centromeric heterochromatin, is so late replicating.

Protein synthesis occurs during the interphase, although it slows down once the cell reaches mitosis. During the S phase of interphase, the key basic protein histones, which interact with DNA to form chromatin, are created. The cell cycle is connected to a number of other chemical events, including:

i). The drop in C-AMP levels that occurs during mitosis.
ii). During chromatin condensation, histones (particularly H1) are phosphorylated (Toteja 2008).

3.6 Cell Cycle Regulation and Its Regulators

The discovery of the processes at the molecular level which mediates the advancement of eukaryotic cells via the cycle of division is probably among the most fascinating achievements in contemporary cell biology. The regulation of cell cycle that we currently comprehend is the consequence of a convergence of findings on a wide range of organisms, including frogs, yeasts, sea urchins, and mammals. According to recent research, a common collection of protein kinases regulates the cell cycle of all eukaryotes that are in charge of initiating the main cell cycle transitions. CDKs, cyclins, cell-cycle checkpoints, and cell cycle signaling pathways are all part of cell cycle regulation.

3.6.1 Positive Regulators—Cyclins and CDKs

CDKs and cyclins are two types of proteins that work together. Pioneering research in yeasts have demonstrated the basic control of the cell cycle. A single CDK that binds to particular cyclins at various phases of the cycle regulates cell cycle progression in these organisms. It is known as Cdc2 in *Schizosaccharomyces pombe* and Cdc28 in *Saccharomyces cerevisiae.* During evolution, the number of cyclins and CDKs has greatly risen. A, B1, B2, C, D1, D2, D3, E, F, and H are the at least 11 distinct cyclins that have been identified. The cyclin frame is a conserved sequence of amino acids found in all types of cyclins that facilitates the interaction of cyclins with CDK. In CDK/cyclin complexes the regulating consequences of substrates such as localization, activation, and inactivation are determined by cyclins, which impart substrate specificity. Only a few cyclin–CDK complexes, however, are assumed to influence cell cycle advancement. Regulation of cell cycle according to the conventional model is constructed in eukaryotic cells based on this notion. Different CDKs and cyclins at different cell stages are shown in Table 3.1.

This model postulates that, unique Cyclin–CDK complexes are essential for initiating the numerous activities that occur during interphase in the mammalian cell cycle in a sequential and ordered manner. Cyclin D1, D2, and D3 are the first to detect mitogenic signals which particularly interact and stimulate CDK4 and CDK6 during G1, the cell cycle stage when cells get ready to begin synthesis of DNA (Malumbres and Barbacid 2001). When these complexes are activated, the pocket proteins rB, rB11 (or p107), and rB12 (or p130) are partially inactivated, allowing E-type cyclins (E1 and E2) to associate and stimulate CDK2 (Harbour et al. 1999). These pocket proteins are also phosphorylated by cyclin E–CDK2 complexes, resulting in their total inactivation (Harbour et al. 1999). Cyclins E are only available during the early stages of DNA synthesis and are strictly controlled during the cell cycle. This finding, combined with the capacity of dominant-negative mutants and CDK2 antibodies to stop the cell cycle in particular human cancer cell lines (van den Heuvel and Harlow 1993), led to the conclusion that cyclin E–CDK2 was needed for the shift from G1 phase to S phase to occur (Hochegger et al. 2008). In the final stage

Table 3.1 CDKs and cyclins at particular cell stage

CDKs	CDKs	Cell cycle stage
CDK1	Cyclin A	G2/M phase transition
CDK1	Cyclin B	Mitosis
CDK2	Cyclin A	S phase
CDK2	Cyclin E	G1/S phase transition
CDK4	Cyclin D1/D2/D3	G1 phase
CDK6	Cyclin D1/D2/D3	G1 phase

of DNA replication, cyclin A2 (in germ cells, cyclin A1) triggers CDK2 to accelerate the shift from the S phase to the mitotic phase, a stage referred as the G2 phase. Finally, A-type cyclins are hypothesized to stimulate CDK1 at the completion of interphase to speed up the commencement of mitosis. A-type cyclins are destroyed once the nuclear envelope is broken, allowing the CDK1–cyclin B complexes to form, which are crucial for pushing cells through mitosis5. Finally, in M phase, Cdk1/cyclin B complexes develop and drive mitotic completion (Riabowol et al. 1989). Throughout the mammalian cell cycle, other CDKs also gets involved.

Stimulation of cyclin D and cyclin E can assist cells in transitioning from G1 toward the S phase, whereas stimulation of CDC2 can help cells transition from S to G2/M phase (Lim et al. 2014). CDK activity is critical in controlling checkpoints of spindle polymerization and controlling cell cycle transcription. CDKs have the ability to start, promote, and finish cell cycle events (Angius et al. 2020). The activation of CDKs can help a cell progress from one stage to the next. A large number of CDK and cyclin complexes regulate the cell cycle. CDK1 that has been activated can phosphorylate target proteins, resulting in physiological impacts like condensation of chromosomes, phosphorylation of nuclear fibrin, which causes nuclear membrane disappearance and nuclear fibrin breakdown. The purpose of CDK-G1 cyclin dimer is to govern G1 and S phases, whereas CDC2-cyclin A and B mediate the process of mitosis, according to diverse forms of CDC2 and CDK cyclin activation duration (Orlando et al. 2008).

The production of activating cyclins and the phosphorylation of the cyclin-CDK complex are required for CDK activity to be strongly regulated. During cell cycle, CDK activity remains constant. Unlike CDK expression, cyclin levels fluctuate with the cell cycle, allowing CDKs to be activated by cyclins on a regular basis (Vermeulen et al. 2003).

The CDK7-cyclin H complex, often termed as the CAK (CDK-Activating Kinase), phosphorylates the CDK, resulting in its expression and interaction with the CDK. CDK-Activating Kinase is required for complete CDK activity. CDKs are phosphorylated by CAK on preserved threonine residues, resulting in structural alterations which could increase cyclin adherence and therefore affect CDK activities. Stimulation of CDK4 necessitates phosphorylation of CDK4 threonine 172, stimulation of CDK2 necessitates phosphorylation of CDK2 threonine 160, and stimulation of CDK1 necessitates phosphorylation of CDK1 threonine 161 (Vermeulen et al. 2003). By phosphorylating CDK1 on tyrosine 15 and/or threonine 14 by the kinases Wee1 and Myt1, the cyclin A-CDK1 complex can be inhibited. CDK expression can be inhibited by phosphorylation of the cyclin-CDK complexes. This inhibiting phosphate can be removed by the enzyme Cdc25 phosphatase, and CDK1 dephosphorylation is necessary for complete stimulation and subsequently cell cycle advancement (Howell and Lew 2012).

Target proteins which mediate cell cycle advancement are phosphorylated., and then active CDKs trigger downstream signaling events (Pines and Hunter 1991). The retinoblastoma tumor suppressive protein (pRB), is inactivated by phosphorylation by the cyclin D-CDK4/CDK6 complex to render it inactive, is among the most often investigated CDK substrates. pRb forms complexes with the transcription factors DP-1and E2F-1, as well as the histone deacetylase (HDAC) protein, when it is active.

pRb is phosphorylated during the G1 phase, resulting in its deactivation and the consequent releasing of DP-1 and E2F-1. Gene transcription is stimulated by E2F-1 essential in S-phase advancement like cyclin A, cyclin E, and Cdc25 (Buchkovich and Greider 1996). E2F also controls the gene expression which code for enzymes that synthesize nucleotides like thymidylate synthase, dihydrofolate reductase, and thymidine kinase (Bracken et al. 2004). During the remaining portion of the cell cycle, pRb remains hyperphosphorylated, and the CDK2-cyclin E complex maintains such a hyperphosphorylated condition.

Additionally phosphorylating p27 (an inhibitor of complexes containing CDK2) all through the G1/S phase is the CDK2-cyclin E complex, resulting in its destruction (Hinds et al. 1992). The phosphorylation of histone H1 by the CDK2-cyclin E complexes, that is necessary for chromosomal condensation during replication of DNA. Cyclin B-CDK1 complexes use Histone H1 as a substrate. Finally, via phosphorylating the DNA polymerase alpha primase, cyclin A-dependent kinases regulate DNA replication initiation (Vermeulen et al. 2003).

3.6.2 Negative Regulators

Negative regulators, which normally limit CDK activity, also govern the cell cycle. CDK activity can be inhibited by interacting with CDK inhibitors, or CKIs, which are cellular proteins. CKIs can block CDK activation by attaching to individual CDKs or the cyclin-CDK complex and their established and emerging functions are given in Table 3.2.

CKIs are classified into two groups: Cip (CDK-Interacting Protein)/Kip (Kinase Inhibitor Protein) and INK4 (CDK4 inhibitor). p21 (Waf1, Cip1), p27 (Kip1), and p57 (Kip1) are all members of the Cip/Kip family (Kip2) whereas p16 (INK4a); p15 (INK4b); p18 (INK4c); and p19 (INK4d) are all the INK4 family members (INK4d) as shown in Table 3.3. Prior to cyclin binding, INK4 family members of CKIs deactivate CDKs by building stabilized compounds from separate CDKs. INK4 family members bound with CDK4 and CDK6 prevent INK4 family members from interacting with cyclin D and entering the G1 phase (Harper and Brooks 2005).

A conserved area in these inhibitors is engaged in the attachment of cyclins and the inhibition of kinases (Harper and Brooks 2005). Cip/Kip family members are

Table 3.2 Functions of CDK inhibitors

Protein	Established function	Emerging function	Reference
p21	Cdk/cyclin complexes inhibition	Silencing of Sox2 expression leads to NSC differentiation.	Marqués-Torrejón et al. (2013)
p27	Cdk/cyclin complexes inhibition Ngn2 stabilization leads to neuron induction.	Transcriptional co-repressors are recruited Silencing Sox2 expression leads to ESC differentiation	Pippa et al. (2012) Maskey et al. (2015)
p57	Cdk/cyclin complexes inhibition. Stabilization of MyoD promotes myoblast differentiation		

NSC Neural Stem Cell
ESC Embryonic Stem Cell

Table 3.3 List of CDK inhibitors

CDK inhibitors	CDK and CDK/cyclin complex	Effected Phase of cell cycle
CIP/KIP FAMILY (p21, p27, p57	CDK1/Cyclin A CDK1/Cyclin B CDK2/Cyclin A CDK2/Cyclin E	G2 G2/M S G1
INK4 Family (p15, p16, p18, p19)	CDK4/CDK6	G1

more specific than INK4 family members and they adhere to and prevent the processes of CDK2-cyclin A, CDK2-cyclin E, and CDK1-cyclin B complexes (Vermeulen et al. 2003; Harper and Brooks 2005). CKIs from the Cip/Kip family helps to activate the G1 phase by helping for the formation of cyclin D-CDK4/6 complex early in the G1 phase and keeping such a complexes stable all across G1 phase (Sherr and Roberts 1999). The transcription factor and Tumor suppressor p53 is responsible for p21 expression. p53 is stimulated by a variety of biological stimuli, including Chk1, Chk2, Ataxia Telangiectasia Mutated (ATM), Ataxia Telangiectasia and Rad3-related protein (ATR), and members of a signaling cascade system which responses when damaged DNA is detected, and promotes the expression of p21 and which inhibits the advancement of the cell cycle beyond the G1 phase in response to cellular stresses (Albert et al. 2014). Ultimately, pRB can be dephosphorylated by PP2A phosphatases., allowing it to bind to E2F and decrease E2F activity and cell cycle advancement (Kolupaeva and Janssens 2013; Mehraj et al. 2021a, 2021b).

3.7 Degradation of Cyclins

Unlike CDKs, cyclin concentration varies across the cell cycle and also during cell cycle entrance and exit, as shown in Fig. 3.3, (Evans et al. 1983). The progression of the cell cycle is significantly regulated by cyclins, their concentration of production and degeneration are strictly regulated. All across the cell cycle, certain cyclin transcript levels change, and the E2F transcription factors frequently transcribe some cyclins on a regular basis, as explained below (DeGregori et al. 1995). Breakdown of cyclins through the proteasome degradation pathway is also important for regulating their levels (Glotzer et al. 1991). Following a sequence of enzymatic processes which attaches ubiquitins on the proteins, the proteasome targets them for degradation (Gerber et al. 1995). Enzyme E1 that activates ubiquitin triggers the ubiquitin before being passed to the enzyme E2 that conjugates ubiquitin in these reactions. An E3 ubiquitin ligase confers specificity of the substrate by binding to the substrate and then attaches with the substrate and either permits immediate ubiquitin transference to the substrate from E2 or first moves ubiquitin toward the E3 and from there to the substrate. Multiple ubiquitinated substrates will become targets for degradation of proteasome. The cell cycle's rhythmic character is determined by the timely production and breakdown of cyclins (Davis et al. 2017).

Cyclins become ubiquitinated within the cell cycle, at the suitable period (Glotzer et al. 1991). Multiple ubiquitins on cyclins signal the proteasome to degrade them. The breakdown of cyclin is necessary for progression of cell cycle during various stages within the cell cycle (Sullivan and Morgan 2007). Mitotic cyclins and proteins that regulate sister-chromatid cohesion must be destroyed in order to enter anaphase (Harper and Brooks 2005) Fig. 3.3 Following the removal of cyclins at the conclusion of the G1 phase, cells can enter the S phase (Clurman et al. 1996).

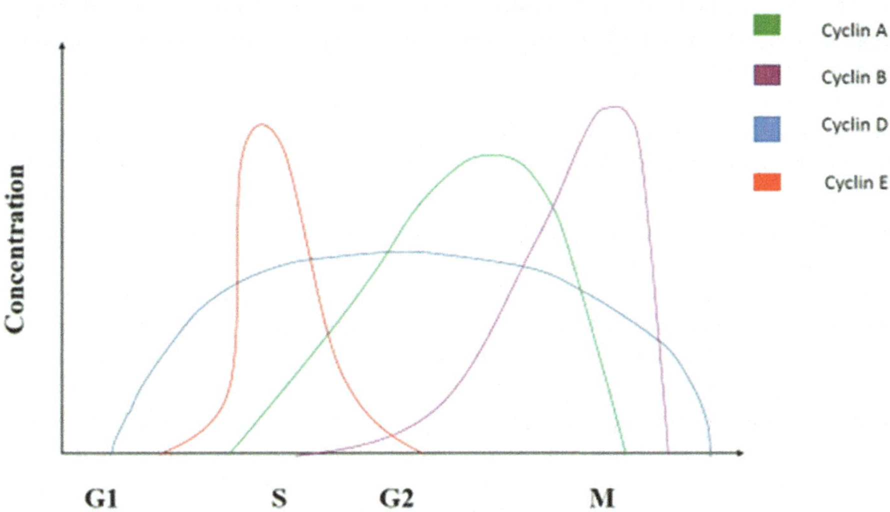

Fig. 3.3 Cyclins through different phases of cell cycle

The Anaphase-Promoting Complex (APC) (Sullivan and Morgan 2007) and the Skp1, Cullin, and F-box protein—named SCF complex—are three of its constituents which attaches to the target (Lisztwan et al. 1998) are two complexes which allow ubiquitination of proteins that depends on the cell cycle and subsequently being destroyed via the proteasome. Throughout the cell cycle, the SCF complex can be seen in about comparable levels. Proteins are normally targeted by the SCF if they are phosphorylated at single or more regions to accomplish cell-cycle-dependent activity regulation (Sullivan and Morgan 2007).

For its function in the breakdown of cyclin E, the SCF is vital for controlling of cell cycle. Concentration of cyclin E surges during G1/S transition and also decreases during replication of DNA (Dulić et al. 1992). For G1 progression, Cyclin E is both necessary and rate-limiting (Ohtsubo et al. 1995). The proteasome eagerly seeks for cyclin E and destroys it during S phase, its levels decrease (Hao et al. 2007). The SCF complex's F-box protein FBW7 detects phosphorylated cyclin E and directs it for degradation (Koepp et al. 2001). Following phosphorylation at conserved motifs, the Cdc4 phosphoserines, Cyclin E becomes a target for destruction (Koepp et al. 2001). FBW7 has a strong affinity for T380 and S384 phosphorylated cyclin E molecules (Welcker et al. 2003). Cyclin E must be phosphorylated, ubiquitinated, and then degraded in order for it to express itself periodically and for the cell cycle to progress. FBW7 phosphorylates the regions on cyclin E that are mutated or inactivation of FBW7, there is constant CDK2-cyclin E activity all across the cell cycle. Excessive proliferation, chromosomal instability, and aberrant S phase arise from uncontrolled cyclin E-CDK2 activity (Siu et al. 2012). The SCFFbw7's subunit known as Fbw7 is mutated in numerous tumors and tumor cell lines (Akhoondi et al. 2007) and is a haplo-insufficient tumor repressor in mice, which is uniform with Fbw7's role in inhibiting rapid proliferation of cells (Mao et al. 2004).

The APC, a huge multi-protein structure stimulates the ubiquitination and protein degradation, which orchestrates the metaphase to anaphase shift, is another complex which regulates protein breakdown in the cell cycle (Harper et al. 2002). The stimulation of the enzyme separase, which catalyzes the sister-chromatid segregation during anaphase, is triggered by the breakdown of securin by the APC (Irniger et al. 1995). During the completion of mitosis, by cytokinesis, as well as in G1 phase, the APC also is responsible for destroying S and M phase cyclins (Irniger et al. 1995). The APC ensures cell cycle progression by removing cyclins during parts of the cell cycle where they are not required.

Whereas phosphorylation of its targets triggers the SCF, activities of the APC are mostly controlled by activator subunit binding. Core APC subunits are phosphorylated by CDKs at cell cycle-specific sites which influences the binding of some activator proteins, results in various types of APC having variable target affinity during distinct phases of the cell cycle (Sullivan and Morgan 2007). The activator proteins are CDK substrates, which allows them to induce cell-cycle-specific APC activity. In mitosis, M-phase CDKs phosphorylate APC component CDC20 as they become activated (Sullivan and Morgan 2007) until the phosphorylation of CDC20 in mitosis, its association is prevented by an autoinhibitory loop

(Qiao et al. 2016). As a result, $APC^{Cdc20}$is formed, which is effective at mitosis once it degrades securin and releases separase. After that, the cleavage of cohesin by separase allows for the segregation of sister chromatids and the completion of mitosis in the cell. (Sullivan and Morgan 2007). APC^{Cdc20} similarly targets M-phase cyclins for degradation.˙ As a result, M-phase CDK activity activates APC which increases M-phase cyclin degradation, as a result, CDK activity in the M-phase and phosphorylation of APC gets reduced (Sullivan and Morgan 2007). This negatively feedback mechanism assures that cells experience a robust, quick increase in M-CDK activity followed by a rapid drop.

As the cell proceeds through late mitosis and early G1, the accessory component linked with the APC switches from CDC20 to Cdh1 and the major type of the APC is now the APC^{Cdh1}. Cdh1 binds better than Cdc20 when APC components are not phosphorylated (Sullivan and Morgan 2007). APC^{Cdh1} ubiquitinates and degrades Cdc20, allowing the APC to progress from its APC^{Cdc20} state in metaphase and anaphase to its APC^{Cdh1} state in G0/G1 (Huang et al. 2001). Until the cell commits toward the following cell cycle APC^{Cdh1} degrades S and M phase cyclins, during quiescence and G1, (Peters 2002). Cdh1 is phosphorylated in early mitosis, which inhibits it from associating with the APC (Hall et al. 2004). Cdh1 is phosphorylated at this moment in the cell cycle and APC^{Cdc20} accumulates again, allowing the cycle to continue (Peters 2002). Activation of phosphatases later in mitosis causes Cdh1 to be dephosphorylated and APCCdh1 activity to be restored (Jaspersen et al. 1999).

Cdh1 has been linked to quiescence and differentiation in a variety of species. Rapidly dividing embryos in Xenopus and Drosophila do not express Cdh1 (Peters 2002). Because of the differentiation and the beginning of the cell cycle's G1 phase, Cdh1 expression is induced (Peters 2002). Fission yeast cells with a Cdh1 deficiency when deprived of nutrients, they are able to multiply but do not cease in G1 phase (Peters 2002). Postmitotic cells, including those in the brain, express APCCdh1 (Gieffers et al. 1999).

3.8 Checkpoints of the Cell Cycle

The cell cycle checkpoint is a regulatory route that governs the order and duration of cell cycle transition and is important so that the cell maintains the allocation of chromosome and quality of DNA replication. Checkpoints give time for repairing by halting the cell-cycle and responding to DNA damage by activating genes or transcription that facilitate repairing (Elledge 1996). Some medications can disable cell cycle checkpoints at important stages in the cell cycle, permitting cells containing damaged DNA to bypass the repair process and proceed directly to the next stage, triggering the apoptosis pathway and inducing cell death.

The halting of the cell cycle, detection of DNA damage, and starting repairing of DNA are all done by the DDR (DNA Damage Response). Cell cycle checkpoints are being triggered during the G1, S, and the G2/M transition when DNA is damaged. DSBs activate Ataxia Telangiectasia Mutated (ATM) kinase, which initiates the G1 checkpoint by phosphorylating and activating checkpoint kinase 2 (CHK2). ATM has the ability to activate CHK2, which then activates p53. Cell cycle control, repairing of DNA, metabolism, senescence, autophagy, and cell death are just a few of the critical signaling pathways that are all controlled by activated p53 (Roos and Kaina 2006). For instance, during cell cycle advancement, p53 will effectively control the levels of expression of critical kinases CDK and p21, resulting in G1 blockage and cyclin E CDK2 complex inhibition (Khan et al. 2020). The p21 protein interferes by suppressing cyclin at the G1 checkpoint, the cell can move from the G1 phase to the S phase. The checkpoint within the S phase gets triggered to inhibit continued replication when DNA gets damaged during S phase, as a outcome of nucleotide excision/repair processes, DSB resolution intermediates, or as halted replication forks (Visconti et al. 2016). Ataxia telangiectasia and rad3 associated (ATR) kinase detects this damage and activates checkpoint kinase 1 (CHK1), which causes CDC25A protein body disintegration and stops S phase progression (Errico and Costanzo 2012). CHK1 and ATR also can be activated by G2/M checkpoints, which prohibit DNA-damaged cells to progress to mitosis. WEE1 is phosphorylated by CHK1, generating increased phosphorylation of phospho-Cdc2 (Tyr15) of CDK1and CDK2 and cell cycle halt during the G2 phase (Ronco et al. 2017). To enhance cell cycle arrest, ROS can activate CHK1 or directly influence the Cdc25 family of protein phosphatases (Cdc25A, B, and C) (Srinivas et al. 2019). WEE1, a bispecific kinase with a molecular weight of 96 kDa phosphorylates tyrosine 15 of CDK1 and is a major enzyme in blocking G2/M metastasis, has a significant impact on how the cell cycle progresses. WEE1 controls CDK1 activity, allowing the DDR process more time to repair DNA (Khan et al. 2020). Many cancers, including hepatocellular carcinoma, glioblastoma, and melanoma, overexpress WEE1 (Magnussen et al. 2012). The proper separation of the replicated genome is obtained in mitosis by a defensive process known as the SAC (Spindle Assembly Checkpoint) that delays entry into the later phases of mitosis to prevent errors in chromosomal separation (Masaki et al. 2003). In the final phases, SAC enhances complex/cyclic body (APC/C), prevents ubiquitin ligase and postponing the breakdown of cyclin B and the subsequent inhibitor securin unless all chromosome becomes linked bipolarly. The mitotic checkpoint complex (MCC) is used to apply SAC by attracting tension-free or detached centromeres (Visconti et al. 2016).

3.9 Signaling Pathways in the Cell Cycle

NF-κB signaling pathway, ATM CHK2/ATR-CHK1, p53 signaling pathway, PI3/AKT/mTOR signaling pathway or JAK-STAT signaling pathway are only a few of the signaling pathways that can control the cell cycle (Khan et al. 2020; Mir et al. 2020). The cell cycle can be controlled by altering the utterance of proteins, kinases, or genes across various signaling paths (Zhang et al. 2019). The DDR signal cascades ATR-CHK1 and ATM-CHK2 get activated by DSBs as well as SSBs, accordingly. These pathways must be activated in order for DNA repairing mechanism and checkpoints to work together. ATM is considered that it may reside as an dormant homodimer in healthy cells. ATM homodimer that is dormant was quickly driven into autophosphorylation at the intermolecular level in response to DSBs, resulting in dissociation and formation of part of the active monomer (Bakkenist and Kastan 2003). When ATM is active, CHK2 is activated, and downstream signaling pathways are activated as well. Base adducts, DSB, replication stress, and cross-linking are many kinds of DNA damage that activate ATR (Cimprich and Cortez 2008).

The phosphorylation of proteins is necessary for ATR-CHK1 and ATM-CHK2 stimulation and is shown in Fig. 3.4 (Chen et al. 2011). To activate ATR, it must bind to ssDNA (single-stranded DNA) bound with RPA (Replica Protein A) (Haahr et al. 2016). When ATR is active, p21CIP1 is upregulated, which subsequently activates CHK1. Cell cycle checkpoints get activated, the cells have entered the phase of DNA repairing, and the cell-cycle is stopped after ATM-CHK2 and ATR-CHK1 pathways get initiated (Mitri et al. 2015). DNA-damaged cells cannot enter the G1, S, or G2 phases when these two pathways are activated (Chen et al. 2011). Inhibition of the ATM-CHK2 and ATR-CHK1 pathways, on the other hand, inhibits checkpoints in the cell cycle, permitting damaged DNA-containing cells to begin the mitotic cycle without having to pass via repairing phase, and stimulates cycle advancement (Chen et al. 2011).

Tumor cell proliferation, survival, invasion, and immunosuppression are all attributed to the JAK-STAT3 signal (Yu et al. 2014; Mehraj et al. 2022a, 2022b). Multiple layers of regulation regulate the JAK-STAT pathway. JAKs can be inhibited by Protein Tyrosine Phosphatases (PTPs), cytokine signaling proteins (SOCS), and other proteins (Seif et al. 2017). STATs can be adversely regulated by PIAS protein, nuclear PTPs (like SHP2 and TCPTP), and intracellular PTPs (like

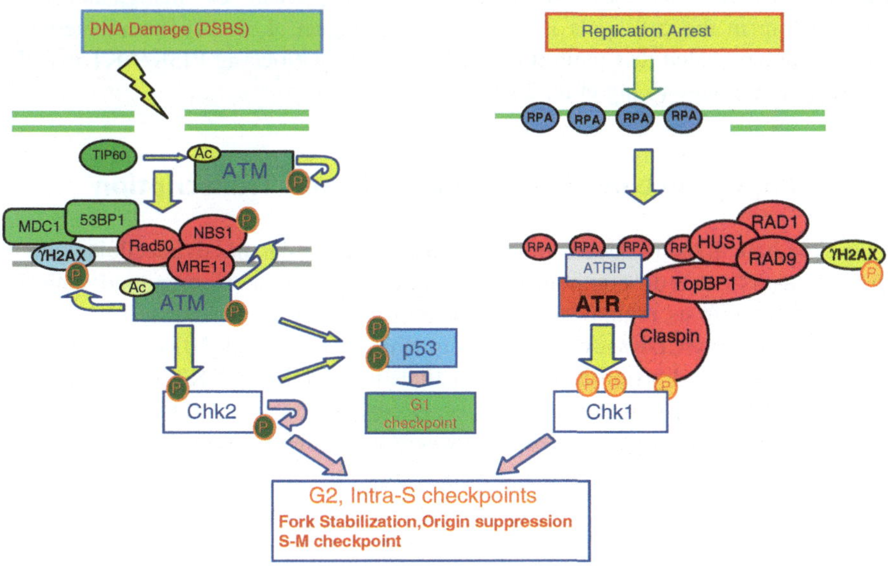

Fig. 3.4 ATM-Chk2 and ATR-Chk1 activation pathways

TCPTP and PTP1B). Activated STATs inhibitors bind with STATs in response to cytokine stimulation and reduce STATs transcriptional activity through various methods (Shuai and Liu 2003). A protein called p53 is the result of p53 gene mutations. In many forms of stress reactions, like as cell cycle halt and cell death, it has an anti-proliferative effect. The p53 gene, for example, is activated when cells are injured or their growth is aberrant, causing the cell cycle to be halted as well as even cell death (Vavrdová et al. 2019; Mehraj et al. 2021a, 2021b). Mutated genes and cellular proteins can easily influence p53 in tumor cells. The oncoprotein Mdm2 is a strong p53 inhibitor. Association of Mdm2 with the transcriptional activation domain of p53, preventing it from regulating target genes and acting as an anti-proliferator. In a controllable feedback mechanism loop of self-regulating feedback, p53 stimulates the expression of the Mdm2 gene (Haupt et al. 1997). Nutlin-1, a Mdm2 small molecule antagonist, will trigger the p53 pathway, prevent tumor development, and stop cell development in the G1 and G2 phases (Vassilev et al. 2004). Mammalian cells contain the transcription factor called Nuclear Factor-kappa B (NF-κB) that belongs to the Rel family of eukaryotic transcription factors. The NF-κB signaling system is a comprehensive mechanism that controls hundreds of genes' expression. Cell stress response, cell proliferation, cell survival, immune inhibition, and innate immunity are just a few of the activities that these genes are involved in a number of cells and organisms (Courtois and Gilmore 2006). NF-κB is strongly expressed in cancerous cells in comparison with normal cells and it promotes tumor metastasis significantly (Erez et al. 2010). Signal transmission, cell division, cell death, angiogenesis, and metabolism and other biological activities are all controlled by the signaling pathway of PI3/AKT/mTOR (Ying et al. 2020). The signaling pathway PI3/AKT/mTOR has a lot of potential in cancer treatment. PI3/AKT/mTOR increases the emergence and growth of malignancies by participating in the cell cycle in cancer cells (Ying et al. 2020). Simultaneously, studies have revealed that inhibiting tumor growth by blocking PI3K/AKT/mTOR-mediated autophagy (Ying et al. 2020).

3.10 Importance of CDKs in Cell Cycle and Transcription

CDKs mediate division of cells in responses to intracellular and extracellular stimuli by serving as catalytic components that form a heterodimer compound with the cyclins that serve as regulatory components (Canavese et al. 2012). Twenty-nine cyclins and twenty CDKs are present in human cells (Cao et al. 2014). Transition of cell-cycle and cell division is actively mediated by CDK1, CDK2, CDK3, CDK4, CDK6, and CDK7; however, CDK7–11 is engaged in gene transcription. CDK activity varies cyclically all across cell cycle (Cicenas and Valius 2011).

3.11 The Significance of CDKs in the Cell Cycle

The large proportion of cells having diploid DNA content in almost all of the adult tissues are halted in a quiescent G0 phase, which is either temporary (quiescence) or long lasting (senescence or terminal differentiation). By being activated by mitogenic chemicals (growth factor or hormone), the cell cycle is re-entered by quiescent cells following mitosis. Such components converge upon the cell cycle to trigger intracellular signaling network connections which interact with CDK4 as well as CDK6 to promote transition of cell cycle through G0 phase to G1 and then toward S phase. The Cyclin D1/D2/D3 positively control CDK6 and CDK4 activity, whereas inhibitors of CDK from the INK4 family negatively regulate it (p15INKB, p16INK4A, p18INK4C, and p19INK4D) (Malumbres and Barbacid 2001).

RB1 encodes the tumor-suppressing protein RB, and also highly similar proteins p107 (or RBL1) and p130 (or RBL2) which are phosphorylated by highly active cylcinD/CDK4 and CDK6. To restrict G1/S transition, the RB protein attaches co-repressors and suppresses the gene transcription mediated by Transcription Factors E2F. As a result, repeated phosphorylation of RB renders it inactive, permitting the cell cycle to progress through G1 phase to S phase. Furthermore, phosphorylated RB inhibits Transcription Factor E2F and promotes the gene transcription of G1/S like CCNA, CCNB, cyclin E (CCNE), RRM1 (ribonucleotide reductase M1), RRM2, dihydrofolate reductase (DHFR), BUB1 mitotic checkpoint serine/threonine kinase, spindle checkpoint protein MAD2 and Polo-like Kinase 1 (PLK1) are engaged in cell cycle advancement and G1–S phase transition (Asghar et al. 2015). The gene targets of E2Fs (cyclins E1 and E2) are stimulated in the late G1 phase, interacting and triggering of CDK2, which was previously isolated by 2 CDK inhibitors p27KIP1 and p21CIP1 and also p21CIP1 and p27KIP1 proteolysis by ubiquitin.

CDK2 is also triggered by CDC25A (cell division cycle 25A) that eliminates phosphorylation in CDK2 (ABE and WATANABE 1995). Besides that, CDK2 when activated has the ability of phosphorylating wide variety of substrate profile proteins necessary for cell advancement (like RB, E2F1, and p27KIP1), synthesis of histones (like coactivator of histone transcription (NPAT) and nuclear protein,), replication of DNA (including replication factors A and C), and duplication of centrosome like NPM (Nucleophosmin) (Deshpande et al. 2005). The active complex of CDK2 and cyclin E controls RB to nullify the G1/S phase point of restriction at the boundary, causing beginning of S phase and the formation of a positive feedback loop. CDK2 and CDK4/6 activity co-ordinate cell-cycle advancement toward S phase, known as the "restriction point," where the existing cell-cycle can be completed without the use of mitogens. As S phase comes to an end, cyclin E is removed by cyclin A and produces a novel structure, CDK2/cyclin A, where the ubiquitylation is caused by the FBXW7 protein (F-Box/WD repeat-containing protein 7) that promptly destroys cyclin E (Deshpande et al. 2005). By phosphorylation of E2F1 and CDC6, the CDK2 and cyclin A complexes conclude the S phase and push the shift toward the G2 from S phase. CDK1 is stimulated via cyclin A, allowing the cell to start the M phase transition. The complex cyclin B/CDK1 maintains CDK1 activity throughout mitosis. The phosphorylation of activated CDK1 causes the chromosome to condense, the nuclear envelope to break down, and the mitotic spindle to assemble. The SAC (Spindle Assembly Checkpoints) regulate the transitions from mitotic metaphase to anaphase, and by the breakdown of cyclin B by APC/C, CDK1 activity is suppressed and the anaphase is triggered (Gavet and Pines 2010). CDK1 expression is disrupted, allowing for chromosomal separation and mitotic and cytokinesis conclusion. CDK1 is almost the only CDK necessary for cell-cycle advancement, as this starts mitosis and helps to ensure major steps in cell replication happens in the correct sequence (Fisher 2011). Stimulation of CDK1 is mediated by such a balance between the phosphorylation of CDC25C phosphatases, a membrane-associated tyrosine- and threonine-specific cdc2-inhibitory kinase, MYT1 or PKMYT1 (Myelin Transcription Factor 1), and WEE1 G2 checkpoint kinase, in addition from its cyclin partners. CDK1 is phosphorylated at Tyr 15 by WEE1, whereas phosphorylation of MYT1 is done at Tyr 15 and Thr 14 to prevent CDK1 activity (Giannone et al. 2019) and CDC25C phosphatases relieve this phosphorylation. Cells are propelled from the G0 phase into the S phase by cyclin C/CDK3 phosphorylating RB. The cell leaves the cell cycle and enters the permanent or reversible G0 phase which is mediated via CDK3/cyclin C (Ren and Rollins 2004).

DNA damage arrests or stops the cell cycle, which is controlled through cell-cycle checkpoints, enabling for DNA repairing prior to cell cycle progresses into mitosis. Two important checkpoints in the cell cycle respond to DNA damage, which takes place before and after synthesis of DNA in the G1 and G2 phases, respectively, and have an impact on the activity of certain CDK complexes. The ATM (Ataxia Telangiectasia Mutated), ATR (Ataxia Telangiectasia and Rad3-related), PI3K (Phosphatidylinositol 3-Kinase)-like protein kinases, the CHEK1 gene and CHEK2 gene, which encodes the transducer Checkpoint Kinases CHK1 and CHK2, respectively, all are checkpoint kinases that are essential DNA damage signaling mediators (Zhao et al. 2001). ATM/ATR detect DNA damage signaling and subsequently phosphorylate and stimulate CHK1/CHK2 (Santo et al. 2015). The activation of p53 is due to the involvement of activated CHK2 which in response results in G1 halt during early stages which is p53-dependent, which allows chance for DNA repairing (Matsuoka et al. 1998).

The CKI, p21CIP1 gene is activated by p53 activation, which inhibits the cyclin E and CDK2 complex and upregulates the DNA repairing mechanism downstream. If the DNA repairing system fails or the cells are unable to program themselves to adapt to the rigorous cell-cycle halt that is alive, p53-induced apoptosis will cause the cells to die (Koniaras et al. 2001). Through phosphorylation of CDC25A, active CHK1 causes ubiquitination and proteolysis, producing brief S phase arrest.

Furthermore, CDC25C is phosphorylated and inactivated by active CHK1, resulting in cell cycle halt at G2 phase. Additionally, WEE1 is phosphorylated more frequently when CHK1 is active, which raises CDK1 and CDK2's prohibitory Tyr15 phosphorylation, as well as halt the cell cycle in the G2 phase (Yavuz et al. 2021). Reductions of CDK activities in the G2 phase also can boost WEE1 activity (Zheng et al. 2007). The Mitotic checkpoint or SAC serves as a moderator of chromosomal binding to the mitotic spindle in metaphase, that is also mediated through TTK protein kinase commonly referred as Monopolar Spindle 1 (MPS1). SAC stimulation causes a halt in the cell cycle in a temporary manner by suppressing APC/C activation.

The TTK binds several proteins of checkpoints to the kinetochores throughout the mitosis by phosphorylation of its substrates to assure proper chromosomal separation and integrity of the genome in order to create as well as manage the mitotic checkpoint (Lara-Gonzalez et al. 2019). SAC protects genomic instability caused by chromosomal segregation abnormalities in this way. Following SAC passage, the APC/C E3 ligase complex activates and labels cyclin B, preventing it from being broken down by ubiquitin and mitotic start (Thu et al. 2018). In short, checkpoints provide a failsafe mechanism for maintaining genetic integrity from parental to daughter cells. Checkpoint activation leads to CDK inhibition, indicating that CDK is a critical driver for progression of cell cycle.

3.12 CDKs and Transcription: What They Do

The largest component of RNAPII (RNA polymerase II) is Rpb1 has a repeat of the evolutionary preserved conserved heptapeptide (Tyr-Ser-Pro-Thr-Ser-Pro-Ser) in its (CTD) C-terminal domain (Whittaker et al. 2017), the synthesis of messenger RNAs (mRNAs) in mammals is highly controlled in mammals and therefore is split into different stages of initiation, pausing, elongation, and termination. Through changes in its phosphorylation level, the CTD plays an important function in the chromatin structuring and a transcriptional and co-transcriptional activity must be coordinated and in the processing of RNA (Jeronimo et al. 2015). Several CDK and Cyclin components, like CDK1 and CDK2, as well as the majority of transcriptional CDKs, like the subfamilies of CDK7, CDK8, and CDK9, phosphorylate, Ser2, Ser5, Ser7, Thr4, and Tyr1in the heptapeptide (Jeronimo et al. 2015). For the promoters' transcriptional activity to begin, the CTD-RNAPII must be phosphorylated at Ser5 and Ser7. Ser5 phosphorylation declines during starting transcription, while phosphorylation of Ser2 and Tyr1 rises, promoting elongation of transcription. First dephosphorylation of Tyr1 during termination of transcription, succeeded by Ser2, Ser5, and Ser7 allowing the transcriptional cycle to be restarted (Galbraith et al. 2019).

The Pre-Initiation Complex (PIC) is formed by DNA unwinding and promoter recognition, which initiates active transcription. RNAPII interacts with a large multi-subunit mediator complex as well as other general transcription factors. The core promoter is bound by the transcription factor II D's (TFIID) TATA binding protein, forming the PIC (Pre-Initiation Complex), to begin the process as shown in Fig. 3.5.

CDK8 and CDK19 bind to cyclin C, that are components of the MED (Mediator Complex Kinase Module), which serves as a molecular connection between the promoter's general RNAPII Pre-Initiation Complex transcriptional machinery and the signals specific to a gene from DNA linked Transcriptional Factors (Taylor-Harding et al. 2015). Med12 (Mediator Complex subunit 12), Med13, cyclin C, and

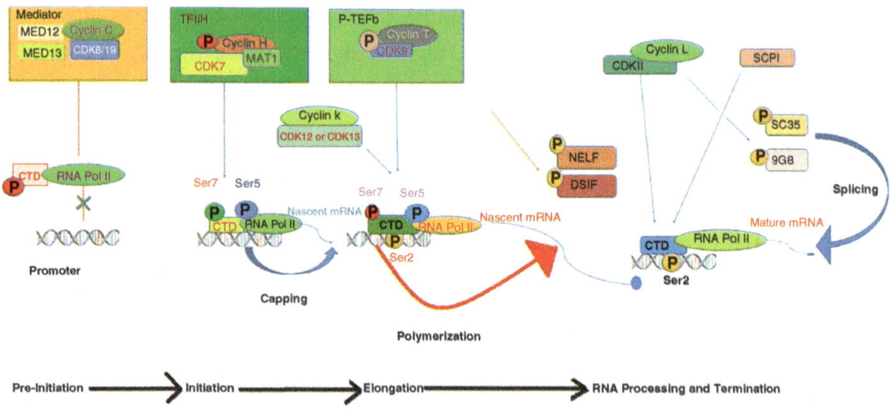

Fig. 3.5 The cycle of pre-initiation, elongation, and termination of RNAP II-based transcription is mediated by complexes of CDK and cyclin

CDK8 (or CDK19) make up the four components of kinase module of MED, which is usually related with transcription repression. Cyclin H is phosphorylated by MED to prevent formation of Pre-Initiation Complex which inhibits TFIIH's action on CTD, and it phosphorylates the CTD-RNAPII that prevents it from interacting to DNA promoter and obstruct the Pre-Initiating Complex(PIC) from forming (Lim and Kaldis 2013). The 10-subunit general transcription factor, which contains the TFIIH complex, is made up of the Mat1 (Menage a trois-1) a ring-finger protein, the catalytic subunit CDK7, and the regulatory component cyclin H that acts like an ATPase, protein kinase, and helicase, as well as being the last to be recruited. In the active site of RNAPII, expression of the helicase unwinds the DNA and forming single-stranded DNA at the transcription initiation site. The Ser5 and Ser7 of CTD-RNAPII are phosphorylated by the CDK7 component in the TFIIH complex that has a kinase activity, that aids transcription start and promoter clearing. The CTD that has been phosphorylated further aids the attachment of a capping enzyme, which catalyzes insertion of a methyl-guanosine cap to the nascent mRNA's 5'end (Whittaker et al. 2018). By discharging the promoter at proximate halt and encourage extension, CDK9 which is phosphorylated and stimulated by cyclin H/CDK7 that serves as a CAK (CDK-Activating Kinase) attaches to cyclin T1 and T2 as a component of P-TEFb (Positive Transcription Elongation Factor b). To generate elongation complex stalling, the pre-mRNA transcript is extended by the stimulated CDK9 and cyclin T by phosphorylating DSIF (5,6-dichloro-1-beta Dribofuranosyl-benzimidazole sensitivity-inducing factor) and NELF (Negative Elongation Factor) and CTD is also phosphorylated on serine 2 by it to activate its RNA polymerizing action (Wohlbold et al. 2012). P-TEFb binding requires CTD phosphorylation by CDK7, and CDK7 inhibition reduces Ser2 phosphorylation by CDK9 (Viladevall et al. 2009).

CDK12 and a closely related homolog CDK13, together as well as the related component cyclin K, are also involved in phosphorylating Ser 2 at CTD in recent research. Early in the transcription process, Ser 2 is phosphorylated by CDK9 and subsequently transfers its function to CDK12 for the greater part of the elongation process; whereas involvement of CDK12 in CTD phosphorylation is specific for a gene (Jeronimo et al. 2016). CDK12 is associated with alternative exon splicing, which is essential for the cells responses to DNA injury, forming a novel connection across the cell cycle regulation and transcriptional machinery (Blazek et al. 2011). To stop the transcription, CTD is phosphorylated by CDK, however its physiology is still unknown (Malumbres 2014). Cyclin L and CDK11 associates with a range of transcription elongation factors, including facilitates chromatin transcription (FACT), TFIIS (general Transcription Factor II S), TFIIF (general Transcription Factor II F), and ELL2 (RNA polymerase Elongation Factor 2). Furthermore, cyclin L/CDK11 regulates splicing of RNA by Phosphorylating Factors involved in splicing of pre-mRNA, like as 9G8 (Srfs7) and SC35 (Srfs2) (Lim et al., 2013). Termination of transcription is enhanced by Sarcoplasmic Calcium-Binding Protein 1 (SCP1) by dephosphorylating Ser 5 of CTD-RNAPII (Whittaker et al. 2017). The dephosphorylation mechanism needs to be investigated in more detail; however, Phosphatases that compete with some CDKs, like Cdc14, are thought to play a role (Clemente-Blanco et al. 2011).

CDK8 and CDK19, at the level of promoters, stimulate the transcription machinery. Cyclin H is phosphorylated by CDK8 and CDK19 as well to prevent the PIC from being assembled, which inhibits TFIIH activity, and to prevent RNAPII from attaching to promoter DNA, it phosphorylates its CTD and preventing the PIC from being assembled. CDK7 and CDK9 elongate mRNA by phosphorylating CTD-RNAPII in a stepwise manner. Ser2 phosphorylation at the CTD is likewise controlled by CDK12 and CDK13 and their co-factor cyclin K, facilitating mRNA elongation. The coordination of transcription and RNA splicing is facilitated by CDK11. SCP1 stimulates transcription termination whereas NELF and DSIF prevent elongation.

3.13 Cancer, CDKs, and CDK Inhibitors

Adult tissues achieve homeostasis when a dormant pool of stem cells produces daughter progenitor cells having a greater ability for proliferation on a regular basis. Quiescence is required to prevent adult stem cells capacity for repopulation from being exhausted prematurely during the course of a person's lifetime. Recent data suggests that the modulation of CDK-cyclin complexes by CDK inhibitors is necessary to maintain the state of dormancy in several stem cell types. Unlike CDK downregulation, which can lead to a loss of homeostasis in certain organs, CDK hyperactivation can promote tumor growth by causing unplanned division of cells in stem or progenitor cells.

3.14 Interphase CDKs Are Dysregulated in Tumor

Human cancer is largely caused by changes in CDKs and their regulators have been thoroughly explained. CDK6 and CDK4 activity deregulation has been linked to a range of tumors (Malumbres 2014). A mutation that causes miscoding (Arg24Cys) in CDK4 allows INK4 inhibitor binding to be prevented among a limited group of melanoma patients. As a result of adjacent translocations, CDK6 is highly expressed in some leukemia's. Cdk4 and Cdk6 are also overexpressed or amplified in numerous types of tumors (such as breast tumors, melanoma, glioma, sarcoma as well as lymphoma).

However, it seems difficult to determine the causal involvement of these changes in tumor growth; for instance, Cdk4 and MdM2 are amplified together in the majority of such tumors (Malumbres 2014). INK4 inhibitors and D-type cyclins are mis-regulated in almost all tumor types (Ortega et al. 2002) implying that CDK6 and CDK4 kinases are overactive during tumor in people, with CDK4 preferred in tumors of the epithelium (mucosae and endocrine tissues) and some sarcomas and CDK6 preferred in mesenchymal tumors (sarcomas and leukemias) and in human cancer, Cdk2 has not been identified to be mutated. On the other hand, E-type cyclins are strongly expressed in human cancers, and p21 as well as p27 inhibitors are typically repressed throughout tumor growth. These findings point to a possible role for CDK2 in human cancers.

In mice, experimental cell cycle-dysregulation causes the formation of cancers. As, CDK4 stimulation is fundamental in a knock-in strain carrying the Arg24Cys miscoding mutation observed in people with cancer leads to hyperplasia's of the epithelium (of the gut, breast, and liver), endocrine neoplasia's (Leydig cells, insulinomas, and pituitary tumors) and albeit with long latencies, sarcomas (Rane et al. 2002). Such mice do not acquire cancer until they are exposed to a skin cancer-causing substance (Sotillo et al. 2001). Concepts for CDK6- or CDK2-induced carcinogenesis are not known at this time. Tumor formation is caused by the deletion in the germ line of mice either of p21 or p27 (Martín et al. 2005). CDK2 somehow does not appear to perform a substantial function in cancers that lack these inhibitors, according to genetic research (Martín et al. 2005). Because p21 and p27 both suppress CDK1, it is plausible that deregulated CDK1 activity is to blame for tumor formation in cancers that lack p21 or p27 expression.

3.15 Summary

Cyclins, CDKs that binds with the CDKs, Inhibitors of CDK, as well as the substrates all regulate the cell cycle in some way. During quiescence, CDK activation piles up and regulates the transition of cell cycle. Cells are committed to cell cycle entrance after an abrupt change in this process. Complexes of CDK are well recognized. For their functions in progression of cell cycle control, cell proliferation, gene transcription and constituting a framework for regulating cell-cycle promoting activity in responses to numerous extracellular and intracellular signals. The discovery of the molecular mechanisms that control the passage of eukaryotic cells through the division cycle is one of the most exciting achievements in contemporary cell biology. Some features of checkpoint signaling, whether in the context of human growth and illness or as a basic principle, remain unclear. Undoubtedly, we will discover previously unknown features of checkpoint signaling, and the expanding arsenal of extremely advanced experimental methods and technologies will enable a more complete understanding of the cell cycle's extraordinary faithfulness.

3.16 Further Readings

For further understanding of cell cycle and its regulators; see Chapters 17 and 18 of; Molecular biology of the cell (4th edition) by Bruce Alberts et al. See also Chapter 18 of; The cell: A Molecular Approach (8th edition) by Geoffrey M. Cooper

For more insights about the topic we would suggest detailed findings from the books of (Mir 2022) https://doi.org/10.1016/C2021-0-02565-7, https://doi.org/10.1016/C2022-0-00074-X (Mir 2021) https://doi.org/10.52305/WXJL6770

The below links for video lectures may also be helpful:

https://youtu.be/foR2tZHj5Eo
https://youtu.be/nEMMKzYQf9A
https://youtu.be/EZTPHtwKB48
https://youtu.be/g7iAVCLZWuM

References

Abe K, Watanabe S (1995) Apoptosis of mouse pancreatic acinar cells after duct ligation. Arch Histol Cytol 58(2):221–229

Afonso O et al (2014) Feedback control of chromosome separation by a midzone Aurora B gradient. Science 345(6194):332–336

Akhoondi S et al (2007) FBXW7/hCDC4 is a general tumor suppressor in human cancer. Cancer Res 67(19):9006–9012

Albert T et al (2014) Characterization of molecular and cellular functions of the cyclin-dependent kinase CDK9 using a novel specific inhibitor. Br J Pharmacol 171(1):55–68

Angius G et al (2020) Prexasertib, a checkpoint kinase inhibitor: from preclinical data to clinical development. Cancer Chemother Pharmacol 85(1):9–20

Asghar U et al (2015) The history and future of targeting cyclin-dependent kinases in cancer therapy. Nat Rev Drug Discov 14(2):130–146

Bakkenist CJ, Kastan MB (2003) DNA damage activates ATM through intermolecular autophosphorylation and dimer dissociation. Nature 421(6922):499–506

Blazek D et al (2011) The Cyclin K/Cdk12 complex maintains genomic stability via regulation of expression of DNA damage response genes. Genes Dev 25(20):2158–2172

Bracken AP et al (2004) E2F target genes: unraveling the biology. Trends Biochem Sci 29(8):409–417

Buchkovich KJ, Greider CW (1996) Telomerase regulation during entry into the cell cycle in normal human T cells. Mol Biol Cell 7(9):1443–1454

Canavese M et al (2012) Cyclin dependent kinases in cancer: potential for therapeutic intervention. Cancer Biol Ther 13(7):451–457
Cao L et al (2014) Phylogenetic analysis of CDK and cyclin proteins in premetazoan lineages. BMC Evol Biol 14(1):1–16
Chen M et al (2011) Loss of the chromatin regulator MRG15 limits neural stem/progenitor cell proliferation via increased expression of the p21 Cdk inhibitor. Stem Cell Res 7(1):75–88
Choudhuri T et al (2005) Curcumin selectively induces apoptosis in deregulated cyclin D1-expressed cells at G2 phase of cell cycle in a p53-dependent manner. J Biol Chem 280(20):20059–20068
Ciardo D et al (2019) On the interplay of the DNA replication program and the intra-S phase checkpoint pathway. Genes 10(2):94
Cicenas J, Valius M (2011) The CDK inhibitors in cancer research and therapy. J Cancer Res Clin Oncol 137(10):1409–1418
Cimprich KA, Cortez D (2008) ATR: an essential regulator of genome integrity. Rev Mol Cell Biol 9(8):616–627
Clemente-Blanco A et al (2011) Cdc14 phosphatase promotes segregation of telomeres through repression of RNA polymerase II transcription. Nat Cell Biol 13(12):1450–1456
Clurman BE et al (1996) Turnover of cyclin E by the ubiquitin-proteasome pathway is regulated by cdk2 binding and cyclin phosphorylation. Genes Dev 10(16):1979–1990
Courtois G, Gilmore T (2006) Mutations in the NF-κB signaling pathway: implications for human disease. Oncogene 25(51):6831–6843
Davis RJ et al (2017) Anti-PD-L1 efficacy can be enhanced by inhibition of myeloid-derived suppressor cells with a selective inhibitor of PI3Kδ/γ. Cancer Res 77(10):2607–2619
DeGregori J et al (1995) Cellular targets for activation by the E2F1 transcription factor include DNA synthesis-and G1/S-regulatory genes. Mol Cell Biol 15(8):4215–4224
Deshpande A et al (2005) Cyclins and cdks in development and cancer: a perspective. Oncogene 24(17):2909–2915
Dhatchinamoorthy K et al (2018) Regulation of kinetochore configuration during mitosis. Curr Genet 64(6):1197–1203
Dulić V et al (1992) Association of human cyclin E with a periodic G1-S phase protein kinase. Science 257(5078):1958–1961
Elledge SJ (1996) Cell cycle checkpoints: preventing an identity crisis. Science 274(5293): 1664–1672
Erez N et al (2010) Cancer-associated fibroblasts are activated in incipient neoplasia to orchestrate tumor-promoting inflammation in an NF-κB-dependent manner. Cancer Cell 17(2):135–147
Errico A, Costanzo V (2012) Mechanisms of replication fork protection: a safeguard for genome stability. Crit Rev Biochem Mol Biol 47(3):222–235
Evans T et al (1983) Cyclin: a protein specified by maternal mRNA in sea urchin eggs that is destroyed at each cleavage division. Cell 33(2):389–396
Fededa JP, Gerlich DW (2012) Molecular control of animal cell cytokinesis. Nat Cell Biol 14(5): 440–447
Fisher D (2011) Control of DNA replication by cyclin-dependent kinases in development. In: Cell cycle in development. Springer, Berlin, pp 201–217
Galbraith MD et al (2019) Therapeutic targeting of transcriptional cyclin-dependent kinases. Transcription 10(2):118–136
Gavet O, Pines J (2010) Progressive activation of CyclinB1-Cdk1 coordinates entry to mitosis. Dev Cell 18(4):533–543
Gerber MR et al (1995) Cdc37 is required for association of the protein kinase Cdc28 with G1 and mitotic cyclins. Proc Natl Acad Sci 92(10):4651–4655
Giannone G et al (2019) Role of cyclin-dependent kinase inhibitors in endometrial cancer. Int J Mol Sci 20(9):2353
Gibcus JH et al (2018) A pathway for mitotic chromosome formation. Science 359(6376):eaao6135

Gieffers C et al (1999) Expression of the CDH1-associated form of the anaphase-promoting complex in postmitotic neurons. Proc Natl Acad Sci 96(20):11317–11322
Glotzer M et al (1991) Cyclin is degraded by the ubiquitin pathway. Nature 349(6305):132–138
Glover L et al (2019) Persistent DNA damage foci and DNA replication with a broken chromosome in the African trypanosome. mBio 10(4):e01252–e01219
Güttinger S et al (2009) Orchestrating nuclear envelope disassembly and reassembly during mitosis. Nat Rev Mol Cell Biol 10(3):178–191
Haahr P et al (2016) Activation of the ATR kinase by the RPA-binding protein ETAA1. Nat Cell Biol 18(11):1196–1207
Hall MC et al (2004) Multi-kinase phosphorylation of the APC/C activator Cdh1 revealed by mass spectrometry. Cell Cycle 3(10):1278–1284
Hao B et al (2007) Structure of a Fbw7-Skp1-cyclin E complex: multisite-phosphorylated substrate recognition by SCF ubiquitin ligases. Mol Cell 26(1):131–143
Harbour JW et al (1999) Cdk phosphorylation triggers sequential intramolecular interactions that progressively block Rb functions as cells move through G1. Cell 98(6):859–869
Harper JV, Brooks G (2005) The mammalian cell cycle. In: Cell cycle control. Humana Press, New York, pp 113–153
Harper JW et al (2002) The anaphase-promoting complex: it's not just for mitosis any more. Genes Dev 16(17):2179–2206
Haupt Y et al (1997) Mdm2 promotes the rapid degradation of p53. Nature 387(6630):296–299
Hinds PW et al (1992) Regulation of retinoblastoma protein functions by ectopic expression of human cyclins. Cell 70(6):993–1006
Hochegger H et al (2008) Cyclin-dependent kinases and cell-cycle transitions: does one fit all? Nat Rev Mol Cell Biol 9(11):910–916
Howell AS, Lew DJ (2012) Morphogenesis and the cell cycle. Genetics 190(1):51–77
Huang JN et al (2001) Activity of the APCCdh1 form of the anaphase-promoting complex persists until S phase and prevents the premature expression of Cdc20p. J Cell Biol 154(1):85–94
Irniger S et al (1995) Genes involved in sister chromatid separation are needed for B-type cyclin proteolysis in budding yeast. Cell 81(2):269–277
Jaspersen SL et al (1999) Inhibitory phosphorylation of the APC regulator Hct1 is controlled by the kinase Cdc28 and the phosphatase Cdc14. Curr Biol 9(5):227–236
Jeronimo C et al (2016) Tail and kinase modules differently regulate core mediator recruitment and function in vivo. Mol Cell 64(3):455–466
Jeronimo C et al (2015) The histone chaperones FACT and Spt6 restrict H2A. Z from intragenic locations. Mol Cell 58(6):1113–1123
Khan T et al (2020) Synergistic activity of agents targeting growth factor receptors, CDKs and downstream signaling molecules in a panel of pancreatic cancer cell lines and the identification of antagonistic combinations: implications for future clinical trials in pancreatic cancer. Oncol Rep 44(6):2581–2594
Koepp DM et al (2001) Phosphorylation-dependent ubiquitination of cyclin E by the SCFFbw7 ubiquitin ligase. Science 294(5540):173–177
Kolupaeva V, Janssens V (2013) PP1 and PP2A phosphatases–cooperating partners in modulating retinoblastoma protein activation. FEBS J 280(2):627–643
Koniaras K et al (2001) Inhibition of Chk1-dependent G2 DNA damage checkpoint radiosensitizes p53 mutant human cells. Oncogene 20(51):7453–7463
Lara-Gonzalez P et al (2019) The G2-to-M transition is ensured by a dual mechanism that protects cyclin B from degradation by Cdc20-activated APC/C. Dev Cell 51(3):313–325. e310
Lens S, Medema RH (2019) Cytokinesis defects and cancer. Nat Rev Cancer 19(1):32–45
Liang P et al (2015) CRISPR/Cas9-mediated gene editing in human tripronuclear zygotes. Protein Cell 6(5):363–372
Lim S, Kaldis P (2013) Cdks, cyclins and CKIs: roles beyond cell cycle regulation. Development 140(15):3079–3093

Lim T-G et al (2014) Curcumin suppresses proliferation of colon cancer cells by targeting CDK2 Curcumin inhibits CDK2 to suppress colon cancer cell growth. Cancer Prev Res 7(4):466–474
Limas JC, Cook JG (2019) Preparation for DNA replication: the key to a successful S phase. FEBS Lett 593(20):2853–2867
Lisztwan J et al (1998) Association of human CUL-1 and ubiquitin-conjugating enzyme CDC34 with the F-box protein p45SKP2: evidence for evolutionary conservation in the subunit composition of the CDC34–SCF pathway. EMBO J 17(2):368–383
Lockhead S et al (2020) The apparent requirement for protein synthesis during G2 phase is due to checkpoint activation. Cell Rep 32(2):107901
Magnussen GI et al (2012) High expression of Wee1 is associated with poor disease-free survival in malignant melanoma: potential for targeted therapy. PLoS One 7(6):e38254
Malumbres M (2014) Cyclin-dependent kinases. Genome Biol 15(6):1–10
Malumbres M, Barbacid M (2001) To cycle or not to cycle: a critical decision in cancer. Nat Rev Cancer 1(3):222–231
Mao J-H et al (2004) Fbxw7/Cdc4 is a p53-dependent, haplo insufficient tumour suppressor gene. Nature 432(7018):775–779
Marqués-Torrejón MÁ et al (2013) Cyclin-dependent kinase inhibitor p21 controls adult neural stem cell expansion by regulating Sox2 gene expression. Cell Stem Cell 12(1):88–100
Martín A et al (2005) Cdk2 is dispensable for cell cycle inhibition and tumor suppression mediated by p27Kip1 and p21Cip1. Cancer Cell 7(6):591–598
Masaki T et al (2003) Cyclins and cyclin-dependent kinases: comparative study of hepatocellular carcinoma versus cirrhosis. Hepatology 37(3):534–543
Maskey D et al (2015) Cell cycle-dependent ubiquitylation and destruction of NDE 1 by CDK 5-FBW 7 regulates ciliary length. EMBO J 34(19):2424–2440
Matsuoka M et al (1998) Interferon-α-induced G1 phase arrest through up-regulated expression of CDK inhibitors, p19Ink4D and p21Cip1 in mouse macrophages. Oncogene 16(16):2075–2086
Mehraj U et al (2022a) Expression pattern and prognostic significance of baculoviral inhibitor of apoptosis repeat-containing 5 (BIRC5) in breast cancer: a comprehensive analysis. Adv Cancer Biol Metastasis 4:100037
Mehraj U et al (2021a) Tumor microenvironment promotes breast cancer chemoresistance. Cancer Chemother Pharmacol 87:147–158
Mehraj U et al (2021b) Prognostic significance and targeting tumor-associated macrophages in cancer: new insights and future perspectives. Breast Cancer 28(3):539–555
Mehraj U et al (2022b) Expression pattern and prognostic significance of CDKs in breast cancer: an integrated bioinformatic study. Cancer Biomark 34:505–519
Mierzwa B, Gerlich DW (2014) Cytokinetic abscission: molecular mechanisms and temporal control. Dev Cell 31(5):525–538
Mir MA et al (2020) Targeting different pathways using novel combination therapy in triple negative breast cancer. Curr Cancer Drug Targets 20(8):586–602
Mitri Z et al (2015) A phase 1 study with dose expansion of the CDK inhibitor dinaciclib (SCH 727965) in combination with epirubicin in patients with metastatic triple negative breast cancer. Investig New Drugs 33(4):890–894
Moseley JB et al (2009) A spatial gradient coordinates cell size and mitotic entry in fission yeast. Nature 459(7248):857–860
Nelson DM et al (2002) Coupling of DNA synthesis and histone synthesis in S phase independent of cyclin/cdk2 activity. Mol Cell Biol 22(21):7459–7472
Ohtsubo M et al (1995) Human cyclin E, a nuclear protein essential for the G1-to-S phase transition. Mol Cell Biol 15(5):2612–2624
Orlando DA et al (2008) Global control of cell-cycle transcription by coupled CDK and network oscillators. Nature 453(7197):944–947
Ortega S et al (2002) Cyclin D-dependent kinases, INK4 inhibitors and cancer. Biochim Biophy Acta 1602(1):73–87

Paulson JR, Laemmli U (1977) The structure of histone-depleted metaphase chromosomes. Cell 12(3):817–828
Peters J-M (2002) The anaphase-promoting complex: proteolysis in mitosis and beyond. Mol Cell 9(5):931–943
Pines J, Hunter T (1991) Cyclin-dependent kinases: a new cell cycle motif? Trends Cell Biol 1(5): 117–121
Pippa R et al (2012) p27Kip1 represses transcription by direct interaction with p130/E2F4 at the promoters of target genes. Oncogene 31(38):4207–4220
Qayoom H et al (2022) Expression patterns and therapeutic implications of CDK4 across multiple carcinomas: a molecular docking and MD simulation study. Med Oncol 39(10):1–13
Qiao R et al (2016) Mechanism of APC/CCDC20 activation by mitotic phosphorylation. Proc Natl Acad Sci 113(19):E2570–E2578
Rane SG et al (2002) Germ line transmission of the Cdk4 R24C mutation facilitates tumorigenesis and escape from cellular senescence. Mol Cell Biol 22(2):644–656
Ren S, Rollins BJ (2004) Cyclin C/cdk3 promotes Rb-dependent G0 exit. Cell 117(2):239–251
Riabowol K et al (1989) The cdc2 kinase is a nuclear protein that is essential for mitosis in mammalian cells. Cell 57(3):393–401
Ronco C et al (2017) ATM, ATR, CHK1, CHK2 and WEE1 inhibitors in cancer and cancer stem cells. Medchemcomm 8(2):295–319
Roos WP, Kaina B (2006) DNA damage-induced cell death by apoptosis. Trends Mol Med 12(9): 440–450
Saldivar JC et al (2018) An intrinsic S/G2 checkpoint enforced by ATR. Science 361(6404): 806–810
Samejima K et al (2012) Mitotic chromosomes are compacted laterally by KIF4 and condensin and axially by topoisomerase IIα. J Cell Biol 199(5):755–770
Santaguida S, Musacchio A (2009) The life and miracles of kinetochores. EMBO J 28(17): 2511–2531
Santo L et al (2015) Targeting cyclin-dependent kinases and cell cycle progression in human cancers. Semin Oncol 42:788
Schiel JA et al (2012) FIP3-endosome-dependent formation of the secondary ingression mediates ESCRT-III recruitment during cytokinesis. Nat Cell Biol 14(10):1068–1078
Seif F et al (2017) The role of JAK-STAT signaling pathway and its regulators in the fate of T helper cells. Cell Commun Signal 15(1):1–13
Sherr CJ, Roberts JM (1999) CDK inhibitors: positive and negative regulators of G1-phase progression. Genes Dev 13(12):1501–1512
Shuai K, Liu B (2003) Regulation of JAK–STAT signalling in the immune system. Nat Rev Immunol 3(11):900–911
Siu KT et al (2012) An integrated view of cyclin E function and regulation. Cell Cycle 11(1):57–64
Sofi S et al (2022a) Targeting cyclin-dependent kinase 1 (CDK1) in cancer: molecular docking and dynamic simulations of potential CDK1 inhibitors. Med Oncol 39(9):1–15
Sofi S et al (2022b) Cyclin-dependent kinases in breast cancer: expression pattern and therapeutic implications. Med Oncol 39(6):1–16
Sotillo R et al (2001) Wide spectrum of tumors in knock-in mice carrying a Cdk4 protein insensitive to INK4 inhibitors. EMBO J 20(23):6637–6647
Srinivas US et al (2019) ROS and the DNA damage response in cancer. Redox Biol 25:101084
Su K-C et al (2016) A regulatory switch alters chromosome motions at the metaphase-to-anaphase transition. Cell Rep 17(7):1728–1738
Sullivan M, Morgan DO (2007) Finishing mitosis, one step at a time. Nat Rev Mol Cell Biol 8(11): 894–903
Taylor-Harding B et al (2015) Cyclin E1 and RTK/RAS signaling drive CDK inhibitor resistance via activation of E2F and ETS. Oncotarget 6(2):696
Thu K et al (2018) Targeting the cell cycle in breast cancer: towards the next phase. Cell Cycle 17(15):1871–1885

Toteja R (2008) Cell cycle and cell cycle regulation. National Institute of Science Communication and Information Resources (NISCAIR), New Delhi
van den Heuvel S, Harlow E (1993) Distinct roles for cyclin-dependent kinases in cell cycle control. Science 262(5142):2050–2054
Vassilev LT et al (2004) In vivo activation of the p53 pathway by small-molecule antagonists of MDM2. Science 303(5659):844–848
Vavrdová T et al (2019) Phosphorylation of plant microtubule-associated proteins during cell division. Front Plant Sci 10:238
Vermeulen K et al (2003) The cell cycle: a review of regulation, deregulation and therapeutic targets in cancer. Cell Prolif 36(3):131–149
Viladevall L et al (2009) TFIIH and P-TEFb coordinate transcription with capping enzyme recruitment at specific genes in fission yeast. Mol Cell 33(6):738–751
Visconti R et al (2016) Cell cycle checkpoint in cancer: a therapeutically targetable double-edged sword. J Exp Clin Cancer Res 35(1):1–8
Welcker M et al (2003) Multisite phosphorylation by Cdk2 and GSK3 controls cyclin E degradation. Mol Cell 12(2):381–392
Whittaker SR et al (2018) Molecular profiling and combinatorial activity of CCT 068127: a potent CDK 2 and CDK 9 inhibitor. Mol Oncol 12(3):287–304
Whittaker SR et al (2017) Inhibitors of cyclin-dependent kinases as cancer therapeutics. Pharmacol Ther 173:83–105
Wohlbold L et al (2012) Chemical genetics reveals a specific requirement for Cdk2 activity in the DNA damage response and identifies Nbs1 as a Cdk2 substrate in human cells. PLoS Genet 8(8):e1002935
Wordeman L (2010) How kinesin motor proteins drive mitotic spindle function: lessons from molecular assays. Semin Cell Dev Biol 21:260
Yavuz M, et al (2021) Condition medium of glioblastoma cell lines decreases the viability of glioblastoma cells by modulating gene expression profile. bioRxiv. https://doi.org/10.1101/2021.09.11.459916
Ying H et al (2020) TRIM59 promotes tumor growth in hepatocellular carcinoma and regulates the cell cycle by degradation of protein phosphatase 1B. Cancer Lett 473:13–24
Yu S-Y et al (2014) Induction of p21Waf1/Cip1 by garcinol via downregulation of p38-MAPK signaling in p53-independent H1299 lung cancer. J Agric Food Chem 62(9):2085–2095
Zhang J et al (2019) Targeting cyclin-dependent kinases in gastrointestinal cancer therapy. Discov Med 27(146):27–36
Zhao J et al (2001) Transcriptional activation of cyclin D1 promoter by FAK contributes to cell cycle progression. Mol Biol Cell 12(12):4066–4077
Zheng XD et al (2007) Phosphorylation of Rga2, a Cdc42 GAP, by CDK/Hgc1 is crucial for Candida albicans hyphal growth. EMBO J 26(16):3760–3769

Cell Cycle and Cancer

4

Manzoor Ahmad Mir and Shazia Sofi

4.1 Introduction

Cancer is the most lethal disease leading to a low life expectancy worldwide (Mir and Mir (2022); Mir et al. 2022a–e; Sung et al. 2021). According to World Health Organization (WHO) estimations for 2019, cancer is the third or fourth top cause of death before the age of 70 in 23 countries and the first or second leading cause in 112 of 183 nations (Bray et al. 2021). Hence, cancer is a major life-threatening disease that poses a great challenge to the present biomedical knowledge and treatments. Unfortunately, the complexity of the disease at the tissue level makes it difficult to accurately diagnose it and ensure that treatment is effective (Meacham and Morrison 2013; Fisher et al. 2013). Prostate, lung and bronchi, colon and rectum, and urine bladder are the main organs in men that are most severely impacted by cancer. Breast, lung, bronchus, colon, rectum, uterine corpus, and thyroid cancer prevalence in women have been found to be highest, correspondingly. This data estimates that prostate and breast cancer as the most prevalent type of cancer seen in men and women, respectively (Mir and Gul 2022; Siegel et al. 2020). While as blood cancer, and cancers related to the brain and lymph nodes, are the most common cancers found in children that account for about 28% of all cancers in children (Schottenfeld and Fraumeni 2006; Mehraj et al. 2022).

M. A. Mir (✉) · S. Sofi
Department of Bioresources, School of Biological Sciences, University of Kashmir, Srinagar, J&K, India
e-mail: drmanzoor@kashmiruniversity.ac.in

M. A. Mir (ed.), *Therapeutic Potential of Cell Cycle Kinases in Breast Cancer*,
https://doi.org/10.1007/978-981-19-8911-7_4

4.2 Cancer Development

A condition known as cancer causes some body cells to grow out of control and propagate to other parts of body to form new tumours (a process called metastasis). The progression of cancer, termed as carcinogenesis, can be best described by enlisting all the features of cancer cells that stir up the process and hence make them distinct characteristics of cancer cells. Cancer progression depends upon the procurement of several abnormal properties like: self-supporting proliferation, insensitivity to anti-proliferative signals, failure of cancer cells to undergo apoptosis, lower requirement of growth factors, angiogenesis and, for malignancy, tissue invasion and metastasis (Mir and Haq 2022; Hanahan and Weinberg 2000). The transition from normal cell to cancer cell that is termed as transformation is a multistep process and can be divided into three distinct stages: initiation, promotion, and progression (Kinzler and Vogelstein 1996). Initiation is a process in which genomic changes get accumulated in cells and they are able to form tumours. Promotion is associated with increased proliferation of initiated cells (Mir et al. 2022d). Progression is marked by acquiring additional genetic changes that lead to malignancy and metastasis. Progression encompasses a substantial growth in tumour size and either growth-related or mutually exclusive metastasis (Sherr 2000) Fig. 4.1.

4.3 Cancer: Cell Cycle Dysregulation

Cancer is being increasingly viewed as a malfunctioning cell cycle. It indicates that the most exhibiting cause of the tumorigenesis is the defective cell cycle machinery leading to unregulated cell proliferation. The main targets of the disease are either the components of the cell cycle itself or the upstream signalling events that ultimately trigger cell cycle events. Although the cancer development process suggests that every tumour is defective in one or more aspect of cell cycle control, but carcinogenesis implies that apart from inducing defects in cell cycle machinery. Cancer can be viewed as a stepwise process that eventually leads to a dysregulated cell cycle (Sherr and Roberts 2004). Human cancers have been linked to cell cycle dysregulation in the past two decades, supported by a vast body of literature (Malumbres and Barbacid 2001). Tumour cells acquire mutations that induce mitosis and create obstructions in responding to anti-mitogenic signalling that leads to abnormal proliferation (Malumbres and Barbacid 2001; Massagué 2004).

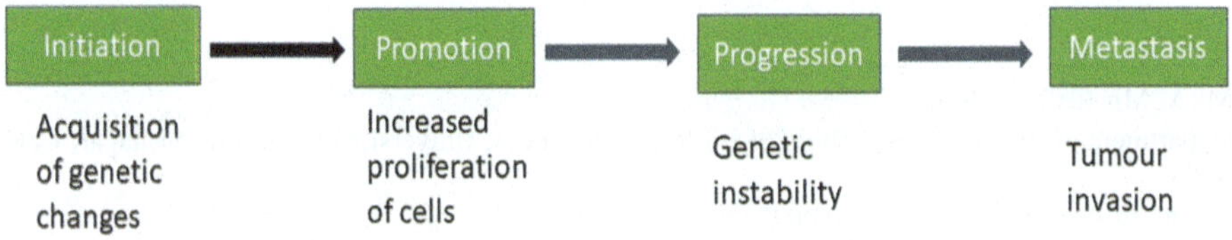

Fig. 4.1 Different stages of cancer development

Additionally, most tumours develop chromosomal instability (CIN), a malfunction that results in alterations to the number of chromosomes, and genomic instability (GIN), which causes additional mutations (Mir et al. 2022d; Kastan and Bartek 2004). Together, these changes lead to proliferative benefits as well as greater vulnerability to the accumulation of further genetic changes that aid in tumour development and the acquisition of more aggressive phenotypes. The three main cell cycle disorders that are either directly or indirectly brought on by insufficient cyclin-dependent kinase (CDK) regulation are unscheduled proliferation, GIN, and CIN (Kops et al. 2005).

4.4 Cell Cycle

The sequence of activities known as the cell cycle occurs when cellular components are duplicated and then properly divided into daughter cells. DNA replication in eukaryotes is restricted to a specific S-phase, also known as synthesis, and chromosomal segregation takes place during the M-phase of mitosis. S-phase and mitosis are separated by the two Gap phases, G1 and G2 (Malumbres and Barbacid 2005). Instead of being inactive, cells acquire mass during these times, as well as integrate growth signals, organize a replicated genome, and get ready for chromosome segregation. The cyclin-dependent kinases (CDKs) are the main enzymes that control how the cell cycle develops. These serine/threonine protein kinases phosphorylate important substrates to advance mitosis and boost DNA synthesis (Weinert and Hartwell 1988; Mehraj et al. 2021).

4.5 Cell Cycle Entry and Progression

The cells choose whether to start DNA replication and go through the cell cycle or to stay in the G1 phase, which is the pre-replicative phase, before going through the S-phase. During the G1 phase, cells can also enter the quiescent phase, also known as the G0 phase, which is a non-proliferative phase. Many of the cells in the adult body must enter the G1 phase in order to start DNA replication and the cell cycle (Pennycook and Barr 2020). Once DNA replication in S-phase is complete, cells can decide to enter M-phase by starting chromosomal condensation and central chromosome alliance. M-phase precisely separates the DNA which is duplicated (mitosis) and divides the whole cellular material into two new daughter cells. In M-phase, which also restarts the cell cycle so that interphase returns, cells commit to segregating the genetic material (Qayoom and Bhat 2020; Rubin et al. 2020).

4.6 Cell Cycle Checkpoints

Cell cycle checkpoints operate the cell cycle's integrity and appropriate advancement. Before moving on to the following phase of the cell cycle, these checkpoints ensure that the operations at each phase have been correctly completed. Cell cycle checkpoints are biochemical signalling pathways that can monitor and detect various kinds of structural DNA flaws or changes in how the DNA functions. They then trigger a cellular response that initiates DNA repair and slows the course of the cell cycle. Because the checkpoint pathways have not changed throughout time, checkpoint failure results in cancer cells continuing to develop (Nasmyth 1996). Checkpoint responses are an important factor in determining whether cells will survive or die. Seven checkpoints have been identified so far in the eukaryotic cell cycle: quiescent, G1/S, replicative S, and G2 checkpoints, the mitotic checkpoint, cytokinesis checkpoint, and the DNA damage checkpoints. Checkpoints remove those cells by causing permanent cell cycle arrest or cell death when DNA damage is irreparable (Mir et al. 2022d). Similar to this, cells fight off genotoxic stressors till the very end using a variety of strategies, including complex survival pathways between DNA synthesis and the divisional phase of the cell cycle, there are two gap periods (Pardee 1989). Eukaryotic cell cycle progression requires the coordinated activity of proteolytic enzymes and a number of kinase cascades (King et al. 1994, Malumbres and Barbacid 2005). Throughout the cell cycle, cyclins go through a continual cycle of synthesis and degradation, timely controlling kinase activity (Malumbres and Barbacid 2005). Three interphase CDKs (CDK2, CDK4, and CDK6), a mitotic CDK (CDK1, also known as cell division control protein 2 (CDC2), and ten cyclins from four different classes make up the CDK-cyclin that directly drives the cell cycle (the A-, B-, D-, and E-type cyclins) (Peng et al. 1998). The mammalian cyclins are broadly classified into A, B1, B2, C, D, E, H, T (Table 4.1). The cyclin box, a domain used to bind and activate Cdks, is a region of homology shared by all cyclins. However, not all cyclins and Cdks are involved in controlling the cell cycle. Apoptosis, DNA repair, differentiation, and transcription regulation are some of the additional roles that have been discovered (Roy et al. 1994; Rickert et al. 1996).

Table 4.1 Cyclins and cell cycle checkpoints

Cyclins	Associated CDKs	Function
Cyclin A	Cdk1, Cdk2	S-phase entry and G2/M transition
Cyclin B	Cdk1	Entry into mitosis and G2 exit
Cyclin C	Cdk8	Transcriptional regulation
Cyclin D (D1, D2, D3)	Cdk4, Cdk6	G0 to S-phase transition G1/S-phase transition, G2 to M-phase transition
Cyclin E	Cdk2	Entry into S-phase
Cyclin H	Cdk7	Transcriptional regulation, Cdk activation
Cyclin T	Cdk9	Transcriptional regulation

Checkpoints in the cell cycle let important cellular processes like DNA replication to stop. When complete cellular division might be harmful, such as in the presence of DNA damage, these checkpoints are used (Kim et al. 2005). Most DNA damage checkpoint signalling pathways culminate on the inactivation of either CDK1/cyclin or CDK2/cyclin complexes as the primary regulators of mammalian cellular progression (Richardson and Jasin 2000). The intra-S checkpoint in mammalian cells is crucial for stopping the advancement of the S-phase in the presence of DNA damage (Hartwell and Weinert 1989; Qayoom et al. 2021). The serine-threonine checkpoint kinases CHK1 and CHK2 are phosphorylated and activated upon the detection of a DSB by a variety of kinases, such as PI3 K's, ATM (Ataxia-Telangiectasia Mutated), and ATR (ATM and Rad3-related). CHK1 and CHK2 subsequently phosphorylate and stabilize TP53 (p53) (Sørensen and Syljuåsen 2012). Following p53 stabilization, the CDK inhibitory protein p21WAF1/CIP1 is transactivated by p53. Here, CDK2/cyclin E activity is efficiently suppressed by p21WAF1/CIP1, blocking the G1/S transition and the start of DNA synthesis (Nyberg et al. 2002; Shechter et al. 2004; Iyer and Rhind 2017; Chehab et al. 1999). One crucial step in maintaining the G1/S DNA damage checkpoint is the activation of p53 and p21WAF1/CIP1 through checkpoint-mediated activation, which inhibits CDK2. Loss of p53 or p21WAF1/CIP1 impairs the cellular response to DNA damage, and mice lacking these proteins are more prone to developing cancer (Chehab et al. 2000; Shieh et al. 2000; Bartek and Lukas 2003; Mehraj et al. 2022). By encouraging the degradation of CDC25 phosphatases, CHK1 and CHK2 can also have a secondary effect on CDK activity (Kastan et al. 1992; Gu et al. 1993; Harper et al. 1993; Mitra et al. 1999). The CDC25 phosphatases are strong CDK/cyclin complex activators that work in direct opposition to the WEE1/MYT1 phosphorylation-induced inhibition of the glycine-rich CDK inhibitory loop domain. These residues are threonine 14 (T14) and tyrosine 15 (Y15) in CDK1 and CDK2. Both CDK2/cyclin E and CDK1/cyclin B must have these residues dephosphorylated by the CDC25 dual-specificity phosphatases in order to fully activate their respective kinases (Donehower et al. 1992; Brugarolas et al. 1995; Mir and Mehraj 2019). Thus, DNA damage is a strong initiator of CDK inhibition that can be brought on by the stimulation of CDK inhibitory proteins as well as the destruction of CDK activators.

4.7 Regulation of Cyclin-CDK Complexes

Beyond the cell cycle, cyclins, Cdks, and CKIs can influence these cellular and developmental processes. Particular focus is placed on the possibility that kinase-dependent or -independent pathways may be used to carry out each of these procedures. Most cyclins enhance Cdk activity, but CKIs decrease it. CKIs are divided into two groups based on the structure and Cdk specificity of each group. The Ink4 family includes the genes p16INK4a, p15INK4b, p18INK4c, and p19INK4d. However, the Cip/Kip family members are more adaptable and characteristically prevent the actions of cyclin A-, B-, D-, and E-dependent kinase

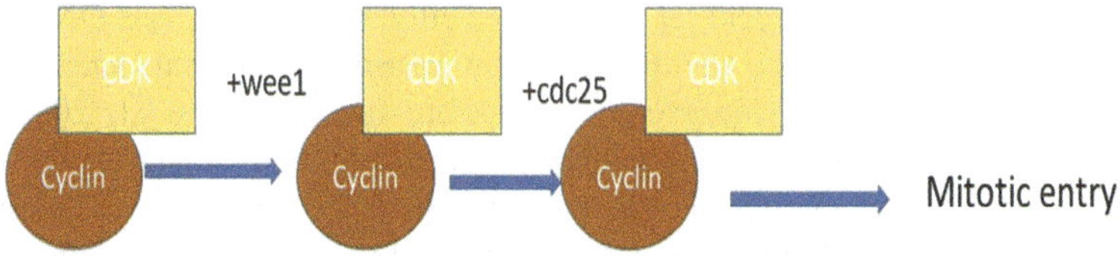

Fig. 4.2 Cell cycle regulation

complexes (Martín-Caballero et al. 2001). Based on sequence homology, more members have been added to the Cdk, cyclin, and CKI families., it has become evident that the original criteria used to classify the founding members are no longer applicable. For instance, it was originally believed that cyclins are solely Cdks' regulatory components, that Cdk/cyclin complexes are the only ones that CKIs can inhibit, and that Cdks and cyclins must interact for Cdks to become active. Despite this deviation from the usual cooperative behaviour, recent studies have amply shown the functions of separate subunits without complex formation, and as a result cyclins, cdks, and CKIs are now believed to have a diversity of cell cycle-independent functions in mammals. Cdk4 and Cdk6 are the primary targets of Cdkn2d (Mailand et al. 2000). Recent research has abundantly demonstrated the functions of individual subunits without complex formation. The Rb/E2F pathway, which is intimately tied to cell cycle control, is one of the most well-studied instances of how cell cycle regulators affect transcription (Mailand et al. 2002). Members of the E2F family of transcription factors are bound and sequestered by the retinoblastoma protein (Rb), p107 (Rbl1), and p130 (Rbl2) in the hypophosphorylated state (Busino et al. 2003). Cdk4/6 and Cdk2 are in charge of sequentially phosphorylating Rb, reducing its inhibition of E2F and enabling the activation of genes required for boosting S-phase entry and DNA synthesis. They do this in collaboration with their respective catalytic partners, D- and E-type cyclins (Fig. 4.2).

4.8 Activation by Phosphorylation

The protein kinase activity of Cyclin-cdk complexes depends on the phosphorylation state of CDK subunit. The activation is completed in two steps and involves binding of cyclins and subsequent phosphorylation by the CDK activating kinases (CAK). For efficient CAK phosphorylation, association of CDK with its cyclin subunit is required in human Cdc2 residue at 161 positions. This type of phosphorylation is activating in nature (Hoffmann et al. 1994; Sørensen et al. 2003; Jinno et al. 1994). Phosphorylation is enhanced by the binding of cyclins as it affects cyclin binding sites (Molinari et al. 2000). CDK activation is completed in two steps, first the binding of CDK2 with cyclin A brings a substantial conformational variation in the kinase activity and modulates the binding ability of ATP constituent of the substrate; second, the activation segment's threonine residue (Thr160 in the human CDK2 sequence) is phosphorylated by CAK to enhance protein substrate binding and align substrates for phosphoryl transfer (Malumbres et al. 2004; Malumbres and Barbacid 2009). In CDK7 phosphorylation occurs at activation site (threonine 170 in human sequence). But also has a second site of phosphorylation in the activation (Atherton-Fessler et al. 1993) segment (Ser 164) (Malumbres et al. 2009). When compared to the rest of CDKs, phosphorylation is not important for the CAK activity. CAKs actively phosphorylate CDKs that are bound to their relevant cyclins. They do not phosphorylate CDKs in monomeric form even if they do so they are phosphorylated very poorly. in monomeric state activation segment cannot be accessed by CAKs (Dyson 1998; Sherr and Roberts 1999).

4.9 CDK Inhibition by Phosphorylation

In contrast to the activation of CDK complexes by phosphorylation, cyclin-CDK complexes can also be inactivated by phosphorylation at the sites of inhibitory phosphorylation. In higher vertebrates the adjacent threonine residues at 14th position and Tyr at 15th position in CDC2 and CDK2 are the sites of inhibitory phosphorylation. The actual mechanism of inhibition is still not clear. Phosphorylation of CDK1 by wee1 at Thr 15 and Thr 14 is also inhibitory in nature that keeps kinase activity of CDk1 low and prevents cells from initiating mitosis until their size is adequate. During entry into M-phase the activity of wee1 is decreased by various regulators and hence activity of CDK1 is increased (Matsushime et al. 1992; Harbour and Dean 2000; Mir and Mehraj 2019).

4.9.1 CDK Inhibitors (CKI's)

Regulation of cyclin-CDK complexes is also contributed by CDK inhibitor proteins. These inhibitor proteins inhibit the kinase activity of CDKs by interfering with their binding with cyclins that is necessary for the activation of cyclin-CDK complexes. There are two types of CDK inhibitor proteins.

4.9.2 CDK Interacting Protein/Kinase Inhibitory Protein (CIP/KIP)

Family of CKIs are the negative regulators of G1 phase cell cycle progression [70]. CIP/KIP family includes P21, P27, P57 that inhibit a wide array of cyclin-CDK complexes. CIP/KIP proteins play many other important roles outside the nucleus. P27kip1 regulates actin dynamics and cell migration (Won et al. 1992). Another member of the family P21cip1 has an ability of inhibiting Rho-kinase (ROCK). P57kip2 regulates subcellular localization (Sherr and Roberts 1999). A cell cycle arrest occurs in G-1 phase in variety of cell types by forming complexes of cyclins D1-D3, CDK4 or CDK6 and cyclin E or cyclin A CDK2 (El-Deiry et al. 1993).

4.9.3 Inhibitors of Kinase (INK4)

INK-4 Family is another type of CKI's and include that contribute to cell cycle control in mammals. INK4 members include P15, P16, P18, and p19. These proteins inhibit the activity of CDK4 and CDK6 with D-type cyclins (Harper et al. 1993) (Fig. 4.3).

P16 has an important role in regulating the Rb. P16 is a tumour suppressor protein that plays a major role in slowing down the pace of Rb and hence deregulates the cell cycle. In human tumours P16 gene is mutated in a high proportion. Cells in which P16 is deleted, P15 also gets affected simultaneously. In such cells the levels of Rb do not influence P15 but in turn get incited by growth-inhibitory cytokine TGF-β (Polyak et al. 1994; Toyoshima and Hunter 1994; Tanaka et al. 2002) that binds to CDK4 and CDK6 and carries on the phosphorylation.

P18 and P19—regulate the activities of cyclin/CDK4 and cyclin/CDK6 complexes but exert no effect over cyclin E/CDK2. Cyclin A/CDK2 or cyclin B/CDK2. The net effect of the inhibition applied by P18 and P19 coordinates with inhibition of G1 phase progression in mammalian cells (Okamoto et al. 1994; Otterson et al. 1994; Koh et al. 1995). The inactivation of INK4 inhibitors or the overexpression of D-type cyclins, cdk4 and cdk6, are thought to be the causes of Rb′ s functional inactivation. Rb that has been hyperphosphorylated cannot bind to or inhibit E2F transcription factors, as was previously mentioned. The discovery that ectopic production of D-type cyclins in dormant cells increases the expression of at least some E2F-regulated genes supports this concept (Ouelle et al. 1995; Pomerantz et al. 1998; Zhang et al. 1998). Although E2F gene mutations in human malignancies have not yet been discovered, there is compelling circumstantial evidence that dysregulation of E2F transcriptional control is a critical step in carcinogenesis. In cell culture-based transformation tests, some E2F genes have been demonstrated to serve as oncogenes (Hunter and Pines 1994; Sherr 1996). Furthermore, it has recently been demonstrated that uncontrolled expression of E2F1 in a transgenic

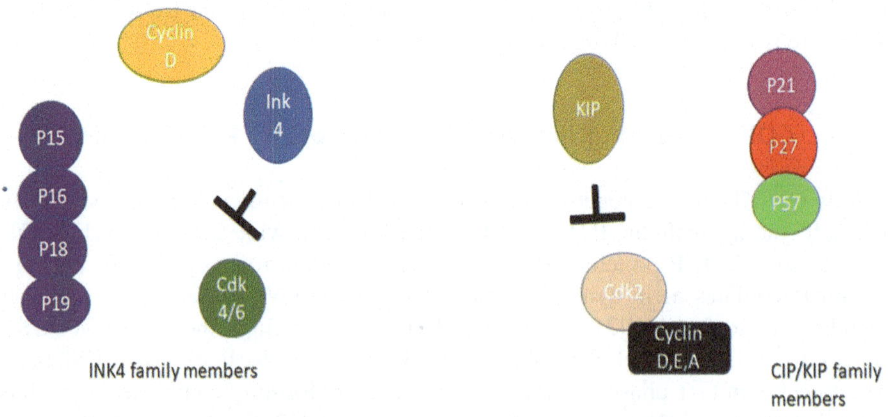

Fig. 4.3 Classes of CDK inhibitors

Table 4.2 Cyclin-dependent kinase inhibitors

CKI's	Different Types of CKI's	Role played
CIP/KIP	P21	Inhibition of Cdk2
	P27	Inhibition of CyclinE-Cdk2
	P57	Cdk4 and Cdk6
INK4	P15	
	P16	
	P18	
	P19	

mice model works in conjunction with either an active Ras gene or a p53 deficit to promote the growth of skin cancers (Reynisdóttir and Massagué 1997; Sangfelt et al. 1997) (Table 4.2).

4.10 Role of M-C

M-CDK commonly called as mitosis promoting factor or maturation promoting factor is the cyclin-CDK complex that is synthesized during the S and G-2 phase. M-CDK promotes the entry into mitosis (M-phase) and meiosis by causing phosphorylation of a wide variety of proteins. M-CDK activity is inhibited by wee1 protein kinase which phosphorylates a tyrosine residue at 15th position in the CDK subunit therefore inhibiting the premature entry of cells into mitosis. The inhibitory role played by wee1 is opposed by a protein phosphatase cdc25 that removes the inhibitory phosphate group and results in the activation of M-CDK and drives the G2/M transition. Yoshio Masui, a researcher in Toronto, identified MPF as a component that promotes egg maturation that involves the meiotic phase. After purifying MPF from the Xenopus frog, Jim Maller and Fred Lokha in Denver further refined Yoshio's cell free assays for monitoring MPF.

4.11 Role of APC/C Activators During Mitotic Division

APC/C is an E3 ubiquitin ligase that facilitates the metaphase to anaphase transition and exit from mitosis by targeting a set of regulatory proteins. APC/C activation requires association with two homologous activators cdc20 and cdh1 (cdc-homologue1). APC/C initiates metaphase-anaphase transition by mediating the degradation of anaphase inhibitor Pds1/securing ensuing separation of cohesion complex which holds the sister chromatids together. After anaphase, APC/CCdh1 mediates the final degradation of mitotic B-type cyclins and several other proteins (Motokura et al. 1991; Bodrug et al. 1994; Lovec et al. 1994; Wang et al. 1994; Morse et al. 1997) as the cell exits mitosis and enters G1. In S-phase and G2, the APC/C is inactive to allow accumulation of proteins required for building the mitotic spindle. APC/C mediated proteolysis of key regulatory proteins drives the cell from

G2 through M-phase into G1 (m. Accordingly, the APC/C is under a strict temporal control so these targets are destroyed in the correct order. APC/CCdc20 is controlled by at least four ways to achieve this. First, transient transcription from the S-phase through the G2 phase and proteolysis in the G1 phase both affect Cdc20 levels (Mir et al. 2022d). Once linked, Mad2p, a part of the spindle assembly checkpoint (SAC) pathway, inhibits APC/CCdc20 in G2 (Leach et al. 1993; Wölfel et al. 1995; Easton et al. 1998). Additionally, the Protein Kinase A (PKA) enzyme directly phosphorylates Cdc20 to block its function when the DNA damage checkpoint pathway is activated (Kamb 1998). The spindle checkpoint signal is silenced when bi-polar attachment of the chromosomes on the metaphase plate allowing securin (Pds1) ubiquitylation/destruction and anaphase to occur.

4.12 Spindle Assembly Checkpoint

Spindle assembly checkpoint ensures correct chromosomal alignment and microtubule attachment at the metaphasic plate. The spindle assembly checkpoint keeps track of the mitotic spindle's flaws and delays sister chromatid segregation until all flaws have been fixed (Mir et al. 2022d) (Fig. 4.4). APC/C is blocked by these spindle microtubules' improper kinetochore attachment, which sends out a negative signal: thereby Inhibiting the metaphase to anaphase transition as a result. The mitotic checkpoint pathway's best-studied components are Mad1, Mad2, Mad3, Bubr1, Bub3, and Mps1, which were first discovered in budding yeast (Lapointe

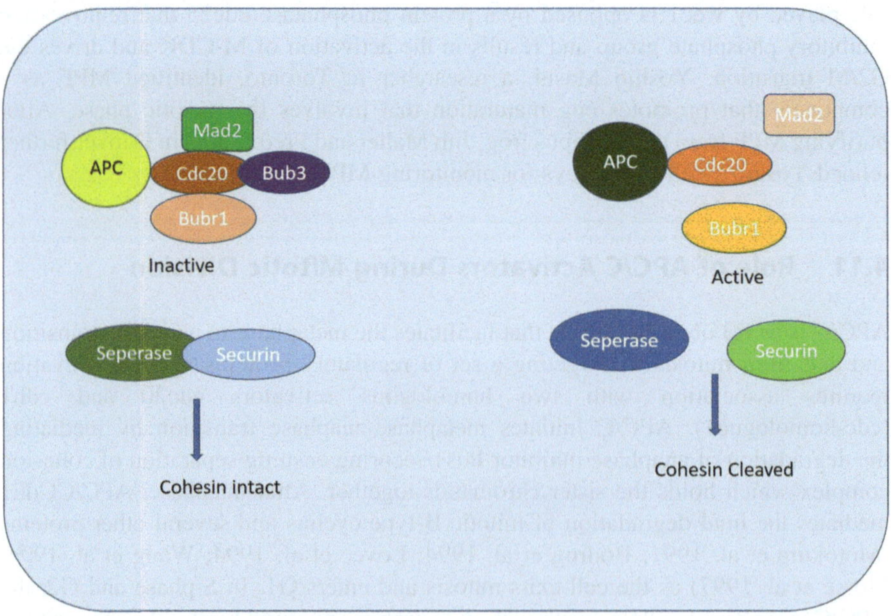

Fig. 4.4 Spindle assembly checkpoint signalling

Table 4.3 Spindle assembly checkpoint proteins and their functions

SAC proteins	Functions
Mad1	Inhibits the activity of APC/C and prevents anaphase onset before the spindle is built
Mad2	APC/C inhibitor
Bubr1	Inhibits Cdc-20-Apc activity
Bub3	Prevents early anaphase entry and mitotic exit

et al. 1996; Johnson 1995). The downstream target of the multi subunit machinery is APC/C complex that results in destruction of several proteins and mitotic cyclins (Shao and Robbins 1995). Mad2 is an essential APC/C inhibitor and prevents anaphase onset. Bubr1 works in harmony with Mad2 and inhibits cdc20-APC activity (Table 4.3). Only after the proper alignment of all the chromosomes at kinetochore correctly at the metaphasic plate spindle assembly checkpoint is finally turned off the localization of the Mad2 and Bubr1 to the kinetochore may be dependent on one or many proteins like Aurora B kinase (Mir et al. 2022b) (Fig. 4.4).

4.13 DNA Damage Checkpoints

Prevent the daughter cells from acquiring mutant DNA. A signal transduction mechanism set off by the damaged DNA prevents cell cycle advancement until the DNA is fully repaired. DNA double-strand breaks (DSBs) during interphase result in an immediate signalling response that is reliant on the checkpoint protein kinase mutant ataxia telangiectasia (ATM). The resultant alteration of ongoing transcription levels and patterns, activation of DNA repair machinery, and interaction with cell cycle regulators all result in a slowing or cessation of the cell cycle (Mir et al. 2022d). The primary mechanism for limiting the accumulation and spread of genetic errors during cell division is this biological response to DNA damage. Once activated by the DNA damage sensor complex MRN (MRE1, RAD50, and NBS1), ATM phosphorylates a wide range of substrates. The transcription factor p53 and the protein kinase CHK2 are significant targets for cell cycle regulation. The mutant checkpoint protein kinase ataxia telangiectasia is required for the fast-signalling response that DNA double-strand breaks (DSBs) during interphase cause (ATM). The response modifies ongoing transcription levels and patterns, activates DNA repair machinery, and interacts with cell cycle regulators, slowing or halting the advancement of the cell cycle (Hartwell and Weinert 1989). This biological response to DNA damage substantially prevents the accumulation and spread of genetic mistakes during cell division. Despite the fact that the protein kinase CHK2 and the transcription factor p53 are essential for cell cycle regulation, a variety of substrates are phosphorylated by ATM when the DNA damage sensor complex MRN (MRE1, RAD50, and NBS1) activates it. To prevent the commencement of the S-phase, P53 activates the CDK inhibitor p21, which significantly

Fig. 4.5 DNA damage signalling cascade

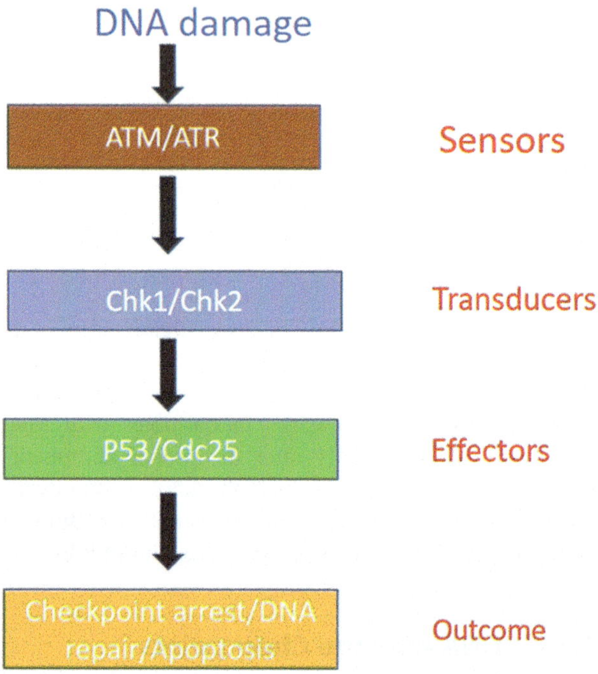

inhibits cyclin-CDK complexes in G1 (Fig. 4.5). When CDC2551 is degraded during the S and G2 phases, CDK1 is phosphorylated under the direction of WEE1 to delay the onset of mitosis. P53 and ATM are not as crucial for slowing or stopping cell cycle progression during tumour growth because of some protein redundancy with other proteins. Despite the fact that the protein kinase CHK2 and the transcription factor p53 are essential for cell cycle regulation, a variety of substrates are phosphorylated by ATM when the DNA damage sensor complex MRN (MRE1, RAD50, and NBS1) activates it. To prevent the commencement of the S-phase, P53 activates the CDK inhibitor p21, which significantly inhibits cyclin-CDK complexes in G1. When CDC2551 is degraded during the S and G2 phases, CDK1 is phosphorylated under the direction of WEE1 to delay the onset of mitosis. P53 and ATM are not as crucial for slowing or stopping cell cycle progression during S and G2 stages because of some redundancy with other proteins. DNA end resection at DSBs is regulated by the cell cycle, which has an impact on the repair method of choice (Mir et al. 2022d). Because the concept of "severe" varies depending on the environment and the type of cell, judgments about a cell's fate are not uniform or always easy to predict. Apoptosis, permanent cell cycle stoppage, and senescence are the three events that cells can experience. The cell cycle arrest is either reversible (quiescence) or irreversible (apoptosis) if the cell does not go through this process during the pre-replicative G1 phase (senescence) (Hayles et al. 1994). Long-term arrest during the S or G2 phases, however, primarily results in cells permanently terminating the cell cycle through senescence or death. The inability to re-enter the cell cycle is largely caused by P53-controlled mechanisms (Nurse and Bissett 1981; Lohka et al. 1988; Peters 1998). P53 activates the CDK inhibitor p21, which largely inhibits cyclin-CDK complexes in G1, to stop the onset of the S-phase. In the S and G2 phases, CHK2 degrades CDC2551, which promotes CDK1 phosphorylation under the control of WEE1 to prevent mitotic entry. P53 and ATM are less crucial for slowing or stopping cell cycle progression during the S and G2 phases due to some DNA replication checkpoint redundancy. DNA end resection at DSBs is regulated by the cell cycle, which has an impact on both the repair procedure and the DNA damage signalling cascade. The majority of DSB repair techniques employed during G1 focus on non-homologous end joining because DNA end resection is not occurring. However, repair through homologous recombination is made easier by the resection of DNA ends following DSBs during the S and G2 stages. The degree of the DNA damage determines how a cell will turn out because the concept of "severe" varies depending on the environment and the type of cell, judgments about a cell's fate are not uniform or always easy to predict. Processes for DSB repair are activated during G1 Phase of cell cycle. Apoptosis, permanent cell cycle stoppage, and senescence are the three events that cells can experience. The cell cycle arrest is either reversible (quiescence) or irreversible (senescence) if the cell does not undergo apoptosis during the pre-replicative G1 phase. In contrast, a cell is more likely to irreversibly exit the cell cycle through senescence or apoptosis when it is arrested for an extended period of time in the S or G2 phases. The inability to re-enter the cell cycle is largely caused by P53-controlled mechanisms (Crasta et al. 2006).

4.14 Therapeutic Agents

Targeting checkpoint controls to create novel therapeutic approaches for this disease offer a number of opportunities given that the breakdown of regular cell cycle regulation is a characteristic of cancer (Mir et al. 2022d). These techniques involve targeting medicines, arresting proliferating cells at specific stages of the cell cycle that may make them more susceptible to treatment with other therapeutic agents like radiation, and inducing checkpoint arrest that results in cytostasis and ultimately apoptosis towards particular cell cycle regulatory components. The process of causing DNA damage and thereafter apoptosis is one of the most well-known chemotherapy strategies. Cell cycle arrest can occur at both the G1/S and G2/M checkpoints in response to substances like cisplatin and nitrogen mustard, which cause DNA cross-links and chromosome breakage. Cyclin/cdk2 and cyclin/cdk4 complexes are inhibited in p21, and as a result Rb is hypo phosphorylated (Zhan et al. 1993; Guillot et al. 1997). Up-regulation of p21 also causes PCNA to be sequestered, which aids in G1/S arrest. DNA damage can trigger the G2/M checkpoint either through p53-dependent or independent mechanisms (Agarwal et al. 1995; Guillot et al. 1997). Phosphorylation of both cdk1 and p21 are necessary for entrance into M and can take part in the G2/M checkpoint for DNA damage because they are unable to stop and fix their damaged DNA, tumour cells with inactive p53 can circumvent the G1/S checkpoint and show increased sensitivity to DNA-damaging substances like cisplatin (Fan et al. 1997; Mir and Agrewala 2008). Taxol and vinca alkaloids, two microtubule inhibitors, interfere with normal tubulin polymerization/depolymerization and mitotic spindle formation (Schiff and Horwitz 1980; Gorbsky 1997). As a result, cells either start a p53-dependent arrest at the radiosensitive mitotic spindle assembly checkpoint (Schiff and Horwitz 1980) or proceed through M and become aneuploid and arrest in G1 (Andreassen et al. 1996; Cahill et al. 1998; Mir 2015). These medications cause G2/M arrest, which is accompanied by stability of the cyclin B/cdc2 complexes. Treatment of tumour cells with microtubule inhibitors may experience apoptosis after G1 or G2 arrest (Woods et al. 1995). Radiosensitizers made from microtubule inhibitors have also demonstrated efficacy in clinical settings. Combining chemotherapy and radiation therapy with taxol, a drug that prevents cells from completing the mitotic spindle assembly checkpoint can increase a tumour's sensitivity to radiation treatment (Liebmann et al. 1994; Chen et al. 1997). Radiosensitizers made from microtubule inhibitors have also demonstrated efficacy in clinical settings (Mir et al. 2022d). Combined Taxol chemotherapy/radiation therapy, prevents cells from dividing at the mitotic spindle assembly checkpoint, which can increase the sensitivity of cancers that are resistant to radiation treatment (Linke et al. 1996; Sofi et al. 2022).

4.15 Summary

The cell cycle represents a sequence of coordinated events that allow the cells to grow and divide. The cell cycle machinery is driven by the systemized action of cyclins and CDKs. The combined activity of these proteins drives the cell cycle progression. The fidelity of cell cycle is maintained by cell cycle checkpoints that operate as a surveillance mechanism and ensure the faithful replication and repair of genome. These checkpoints delay the cell cycle progression in response to irreparable DNA damage. The fidelity of this process is destroyed by mutations that prevent apoptosis and compromise cell cycle exit. These mutations disrupt the signalling pathways and their downstream counterparts CDKs and cyclins. CDK activity is the most targeted activity due to their major role in cell cycle progression, they are anti-proliferative and arrest cells in G1 or G2/M phase and also trigger apoptosis. Cell cycle checkpoints which play a pivotal role in driving cell cycle need to be defective for a cell to become cancerous. Cancer cells continue to divide, despite the accumulation of genetic errors as DNA damage checkpoints are compromised in the cell cycle.

4.16 Further Readings

The readers can further read about the role of CDKs in breast cancer by going through the following papers

- https://doi.org/10.1080/13543784.2022.2097067
- https://www.ncbi.nlm.nih.gov/pmc/articles/PMC8167670/

For more insights about the topic, we would suggest detailed findings from the books of (Mir MA, 2022) https://doi.org/10.1016/C2021-0-02565-7, https://doi.org/10.1016/C2014-0-02898-5 (Mir MA, 2021) https://doi.org/10.52305/WXJL6770, from cancer.net website, https://www.cancer.net/cancer-types/breast-cancer/types-treatment. Also, the readers can have a look upon the following visual presentations for the better conceptual understanding of CDKs and their role in breast cancer

https://youtu.be/0Sj3rbJPeXQ
https://youtu.be/RXsWAvdWG0s
https://youtu.be/YA67P2k2d6A

References

Agarwal ML et al (1995) p53 controls both the G2/M and the G1 cell cycle checkpoints and mediates reversible growth arrest in human fibroblasts. Proc Natl Acad Sci 92(18):8493–8497
Andreassen PR et al (1996) Chemical induction of mitotic checkpoint override in mammalian cells results in aneuploidy following a transient tetraploid state. Mutat Res 372(2):181–194
Atherton-Fessler S et al (1993) Mechanisms of p34cdc2 regulation. Mol Cell Biol 13(3):1675–1685
Bartek J, Lukas J (2003) Chk1 and Chk2 kinases in checkpoint control and cancer. Cancer Cell 3(5):421–429
Bodrug SE et al (1994) Cyclin D1 transgene impedes lymphocyte maturation and collaborates in lymphomagenesis with the myc gene. EMBO J 13(9):2124–2130
Bray F et al (2021) The ever-increasing importance of cancer as a leading cause of premature death worldwide. Cancer 127(16):3029–3030
Brugarolas J et al (1995) Radiation-induced cell cycle arrest compromised by p21 deficiency. Nature 377(6549):552–557
Busino L et al (2003) Degradation of Cdc25A by β-TrCP during S phase and in response to DNA damage. Nature 426(6962):87–91
Cahill DP et al (1998) Mutations of mitotic checkpoint genes in human cancers. Nature 392(6673):300–303
Chehab NH et al (2000) Chk2/hCds1 functions as a DNA damage checkpoint in G1 by stabilizing p53. Genes Dev 14(3):278–288
Chehab NH et al (1999) Phosphorylation of Ser-20 mediates stabilization of human p53 in response to DNA damage. Proc Natl Acad Sci 96(24):13777–13782
Chen MD et al (1997) Phase I trial of taxol as a radiation sensitizer with cisplatin in advanced cervical cancer. Gynecol Oncol 67(2):131–136
Crasta K et al (2006) Cdk1 regulates centrosome separation by restraining proteolysis of microtubule-associated proteins. EMBO J 25(11):2551–2563
Donehower LA et al (1992) Mice deficient for p53 are developmentally normal but susceptible to spontaneous tumours. Nature 356(6366):215–221
Dyson N (1998) The regulation of E2F by pRB-family proteins. Genes Dev 12(15):2245–2262
Easton J et al (1998) Disruption of the cyclin D/cyclin-dependent kinase/INK4/retinoblastoma protein regulatory pathway in human neuroblastoma. Cancer Res 58(12):2624–2632
El-Deiry WS et al (1993) WAF1, a potential mediator of p53 tumor suppression. Cell 75(4):817–825

Fan S et al (1997) Cells lacking CIP1/WAF1 genes exhibit preferential sensitivity to cisplatin and nitrogen mustard. Oncogene 14(18):2127–2136
Fisher R et al (2013) Cancer heterogeneity: implications for targeted therapeutics. Br J Cancer 108(3):479–485
Gorbsky GJ (1997) Cell cycle checkpoints: arresting progress in mitosis. BioEssays 19(3):193–197
Gu Y et al (1993) Inhibition of CDK2 activity in vivo by an associated 20K regulatory subunit. Nature 366(6456):707–710
Guillot C et al (1997) p21WAF1/CIP1 response to genotoxic agents in wild-type TP53 expressing breast primary tumours. Oncogene 14(1):45–52
Hanahan D, Weinberg RA (2000) The hallmarks of cancer. Cell 100(1):57–70
Harbour JW, Dean DC (2000) The Rb/E2F pathway: expanding roles and emerging paradigms. Genes Dev 14(19):2393–2409
Harper JW et al (1993) The p21 Cdk-interacting protein Cip1 is a potent inhibitor of G1 cyclin-dependent kinases. Cell 75(4):805–816
Hartwell LH, Weinert TA (1989) Checkpoints: controls that ensure the order of cell cycle events. Science 246(4930):629–634
Hayles J et al (1994) Temporal order of S phase and mitosis in fission yeast is determined by the state of the p34cdc2-mitotic B cyclin complex. Cell 78(5):813–822
Hoffmann I et al (1994) Activation of the phosphatase activity of human cdc25A by a cdk2-cyclin E dependent phosphorylation at the G1/S transition. EMBO J 13(18):4302–4310
Hunter T, Pines J (1994) Cyclins and cancer II: cyclin D and CDK inhibitors come of age. Cell 79(4):573–582
Iyer DR, Rhind N (2017) The intra-S checkpoint responses to DNA damage. Genes 8(2):74
Jinno S et al (1994) Cdc25A is a novel phosphatase functioning early in the cell cycle. EMBO J 13(7):1549–1556
Johnson DG (1995) Regulation of E2F-1 gene expression by p130 (Rb2) and D-type cyclin kinase activity. Oncogene 11(9):1685–1692
Kamb A (1998) Cyclin-dependent kinase inhibitors and human cancer. In: Cyclin dependent kinase (CDK) inhibitors. Springer, New York, pp 139–148
Kastan MB, Bartek J (2004) Cell-cycle checkpoints and cancer. Nature 432(7015):316–323
Kastan MB et al (1992) A mammalian cell cycle checkpoint pathway utilizing p53 and GADD45 is defective in ataxia-telangiectasia. Cell 71(4):587–597
Kim J-S et al (2005) Independent and sequential recruitment of NHEJ and HR factors to DNA damage sites in mammalian cells. J Cell Biol 170(3):341–347
King RW et al (1994) Mitosis in transition. Cell 79(4):563–571
Kinzler KW, Vogelstein B (1996) Lessons from hereditary colorectal cancer. Cell 87(2):159–170
Koh J et al (1995) Tumour-derived p16 alleles encoding proteins defective in cell-cycle inhibition. Nature 375(6531):506–510
Kops GJPL et al (2005) On the road to cancer: aneuploidy and the mitotic checkpoint. Nat Rev Cancer 5(10):773–785
Lapointe J et al (1996) A p18 mutant defective in CDK6 binding in human breast cancer cells. Cancer Res 56(20):4586–4589
Leach FS et al (1993) Amplification of cyclin genes in colorectal carcinomas. Cancer Res 53(9): 1986–1989
Liebmann J et al (1994) In vitro studies of Taxol as a radiation sensitizer in human tumor cells. J Natl Cancer Inst 86(6):441–446
Linke SP et al (1996) A reversible, p53-dependent G0/G1 cell cycle arrest induced by ribonucleotide depletion in the absence of detectable DNA damage. Genes Dev 10(8):934–947
Lohka MJ et al (1988) Purification of maturation-promoting factor, an intracellular regulator of early mitotic events. Proc Natl Acad Sci 85(9):3009–3013
Lovec H et al (1994) Cyclin D1/bcl-1 cooperates with myc genes in the generation of B-cell lymphoma in transgenic mice. EMBO J 13(15):3487–3495

Mir MA, Aisha S, Sofi S (2022a) Introduction to various types of cancers, chapter-1. In: Role of tumor microenvironment in breast cancer and targeted therapies. Elsevier, San Diego, pp 1–30; ISBN 978-0-443-18696-7. https://doi.org/10.1016/B978-0-443-18696-7.00010-5

Mir MA, Aisha S, Sofi S, Rasheid S (2022b) The tumor microenvironment, chapter-2. In: Role of tumor microenvironment in breast cancer and targeted therapies. Elsevier, San Diego, pp 31–58; ISBN 978-0-443-18696-7. https://doi.org/10.1016/B978-0-443-18696-7.00007-5

Mir MA, Sofi S, Aisha S (2022c) Role of cancer-associated fibroblasts in tumor microenvironment, chapter-3. In: Role of tumor microenvironment in breast cancer and targeted therapies. Elsevier, San Diego, pp 59–86; ISBN 978-0-443-18696-7. https://doi.org/10.1016/B978-0-443-18696-7.00002-6

Mir MA, Mir AY, Jan U, Dar MA, Zahoor ul Haq Shah M (2022d) Role of cancer-associated fibroblasts in tumor microenvironment, chapter-4. In: Role of tumor microenvironment in breast cancer and targeted therapies. Elsevier, San Diego, pp 87–112; ISBN 978-0-443-18696-7. https://doi.org/10.1016/B978-0-443-18696-7.00004-X

Mir MA, Mir AY (2022) Role of regulatory T cells in cancer, chapter-5. In: Role of tumor microenvironment in breast cancer and targeted therapies. Elsevier, San Diego, pp 113–136; ISBN 978-0-443-18696-7. https://doi.org/10.1016/B978-0-443-18696-7.00001-4

Mir MA, Mir AY, Mushtaq T (2022e) Role of tumor-associated macrophages in the breast tumor microenvironment, chapter-6. In: Role of tumor microenvironment in breast cancer and targeted therapies. Elsevier, San Diego, pp 137–170; ISBN 978-0-443-18696-7. https://doi.org/10.1016/B978-0-443-18696-7.00003-8

Mir MA, Gul A (2022) The extracellular matrix in breast cancer, chapter-8. In: Role of tumor microenvironment in breast cancer and targeted therapies. Elsevier, San Diego, pp 194–220; ISBN 978-0-443-18696-7. https://doi.org/10.1016/B978-0-443-18696-7.00006-3

Mira MA, Haq BUL (2022) Targeting tumor microenvironment for breast cancer treatment, chapter-10. In: Role of tumor microenvironment in breast cancer and targeted therapies. Elsevier, San Diego, pp 249–298; ISBN 978-0-443-18696-7. https://doi.org/10.1016/B978-0-443-18696-7.00008-7

Mailand N et al (2000) Rapid destruction of human Cdc25A in response to DNA damage. Science 288(5470):1425–1429

Mailand N et al (2002) Regulation of G2/M events by Cdc25A through phosphorylation-dependent modulation of its stability. EMBO J 21(21):5911–5920

Malumbres M, Barbacid M (2001) To cycle or not to cycle: a critical decision in cancer. Nat Rev Cancer 1(3):222–231

Malumbres M, Barbacid M (2005) Mammalian cyclin-dependent kinases. Trends Biochem Sci 30(11):630–641

Malumbres M, Barbacid M (2009) Cell cycle, CDKs and cancer: a changing paradigm. Nat Rev Cancer 9(3):153–166

Malumbres M et al (2009) Cyclin-dependent kinases: a family portrait. Nat Cell Biol 11(11): 1275–1276

Malumbres M et al (2004) Mammalian cells cycle without the D-type cyclin-dependent kinases Cdk4 and Cdk6. Cell 118(4):493–504

Martín-Caballero J et al (2001) Tumor susceptibility of p21 Waf1/Cip1-deficient mice. Cancer Res 61(16):6234–6238

Massagué J (2004) G1 cell-cycle control and cancer. Nature 432(7015):298–306

Matsushime H et al (1992) Identification and properties of an atypical catalytic subunit (p34PSK-J3/cdk4) for mammalian D type G1 cyclins. Cell 71(2):323–334

Meacham CE, Morrison SJ (2013) Tumour heterogeneity and cancer cell plasticity. Nature 501(7467):328–337

Mehraj U et al (2021) Tumor microenvironment promotes breast cancer chemoresistance. Cancer Chemother Pharmacol 87(2):147–158

Mehraj U et al (2022) Expression pattern and prognostic significance of CDKs in breast cancer: an integrated bioinformatic study. Cancer Biomark 34:505–519

Mir MA (2015) Developing costimulatory molecules for immunotherapy of diseases. Academic Press, London. https://doi.org/10.1016/C2014-0-02898-5. ISBN: 9780128025857
Mir MA, Agrewala JN (2008) Signaling through CD80: an approach for treating lymphomas. Expert Opin Ther Targets 12(8):969–979
Mir MA, Mehraj U (2019) Double-crosser of the immune system: macrophages in tumor progression and metastasis. Curr Immunol Rev 15(2):172–184
Mitra J et al (1999) Induction of p21 WAF1/CIP1 and inhibition of Cdk2 mediated by the tumor suppressor p16 INK4a. Mol Cell Biol 19(5):3916–3928
Molinari M et al (2000) Human Cdc25 A inactivation in response to S phase inhibition and its role in preventing premature mitosis. EMBO Rep 1(1):71–79
Morse L et al (1997) Induction of cell cycle arrest and B cell terminal differentiation by CDK inhibitor p18 INK4c and IL-6. Immunity 6(1):47–56
Motokura T et al (1991) A novel cyclin encoded by a bcl1-linked candidate oncogene. Nature 350(6318):512–515
Nasmyth K (1996) Putting the cell cycle in order. Science 274(5293):1643–1645
Nurse P, Bissett Y (1981) Gene required in G1 for commitment to cell cycle and in G2 for control of mitosis in fission yeast. Nature 292(5823):558–560
Nyberg KA et al (2002) Toward maintaining the genome: DNA damage and replication checkpoints. Annu Rev Genet 36:617
Okamoto A et al (1994) Mutations and altered expression of p16INK4 in human cancer. Proc Natl Acad Sci 91(23):11045–11049
Otterson GA et al (1994) Absence of p16INK4 protein is restricted to the subset of lung cancer lines that retains wildtype RB. Oncogene 9(11):3375–3378
Ouelle DE et al (1995) Alternative reading frames of the INK4a tumor suppressor gene encode two unrelated proteins capable of inducing cell cycle arrest. Cell 83(6):993–1000
Pardee AB (1989) G1 events and regulation of cell proliferation. Science 246(4930):603–608
Peng J et al (1998) Identification of multiple cyclin subunits of human P-TEFb. Genes Dev 12(5): 755–762
Pennycook BR, Barr AR (2020) Restriction point regulation at the crossroads between quiescence and cell proliferation. FEBS Lett 594(13):2046–2060
Peters J-M (1998) SCF and APC: the Yin and Yang of cell cycle regulated proteolysis. Curr Opin Cell Biol 10(6):759–768
Polyak K et al (1994) p27Kip1, a cyclin-Cdk inhibitor, links transforming growth factor-beta and contact inhibition to cell cycle arrest. Genes Dev 8(1):9–22
Pomerantz J et al (1998) The Ink4a tumor suppressor gene product, p19Arf, interacts with MDM2 and neutralizes MDM2's inhibition of p53. Cell 92(6):713–723
Qayoom H, Bhat BA (2020) U Mehraj U, Mir MA (2020) rising trends of cancers in Kashmir valley: distribution pattern, incidence and causes. J Oncol Res Treat 5(150):2
Qayoom H et al (2021) An insight into the cancer stem cell survival pathways involved in chemoresistance in triple-negative breast cancer. Future Oncol 17(31):4185–4206
Reynisdóttir I, Massagué J (1997) The subcellular locations of p15 (Ink4b) and p27 (Kip1) coordinate their inhibitory interactions with cdk4 and cdk2. Genes Dev 11(4):492–503
Richardson C, Jasin M (2000) Coupled homologous and nonhomologous repair of a double-strand break preserves genomic integrity in mammalian cells. Mol Cell Biol 20(23):9068–9075
Rickert P et al (1996) Cyclin C/CDK8 is a novel CTD kinase associated with RNA polymerase II. Oncogene 12(12):2631–2640
Roy R et al (1994) The MO15 cell cycle kinase is associated with the TFIIH transcription-DNA repair factor. Cell 79(6):1093–1101
Rubin SM et al (2020) Integrating old and new paradigms of G1/S control. Mol Cell 80(2):183–192
Sangfelt O et al (1997) Induction of Cip/Kip and Ink4 cyclin dependent kinase inhibitors by interferon-α in hematopoietic cell lines. Oncogene 14(4):415–423
Schiff PB, Horwitz SB (1980) Taxol stabilizes microtubules in mouse fibroblast cells. Proc Natl Acad Sci 77(3):1561–1565

Schottenfeld D, Fraumeni JF Jr (2006) Cancer epidemiology and prevention. Oxford University Press, New York

Shao Z, Robbins PD (1995) Differential regulation of E2F and Sp1-mediated transcription by G1 cyclins. Oncogene 10(2):221–228

Shechter D et al (2004) Regulation of DNA replication by ATR: signaling in response to DNA intermediates. DNA Repair 3(8–9):901–908

Sherr CJ (1996) Cancer cell cycles. Science 274(5293):1672–1677

Sherr CJ (2000) The Pezcoller lecture: cancer cell cycles revisited. Cancer Res 60(14):3689–3695

Sherr CJ, Roberts JM (1999) CDK inhibitors: positive and negative regulators of G1-phase progression. Genes Dev 13(12):1501–1512

Sherr CJ, Roberts JM (2004) Living with or without cyclins and cyclin-dependent kinases. Genes Dev 18(22):2699–2711

Shieh S-Y et al (2000) The human homologs of checkpoint kinases Chk1 and Cds1 (Chk2) phosphorylate p53 at multiple DNA damage-inducible sites. Genes Dev 14(3):289–300

Siegel RL et al (2020) Cancer statistics, 2016. CA Cancer J Clin 66:7–30. https://doi.org/10.3322/caac.21332

Sofi S et al (2022) Cyclin-dependent kinases in breast cancer: expression pattern and therapeutic implications. Med Oncol 39(6):1–16

Sørensen CS, Syljuåsen RG (2012) Safeguarding genome integrity: the checkpoint kinases ATR, CHK1 and WEE1 restrain CDK activity during normal DNA replication. Nucleic Acids Res 40(2):477–486

Sørensen CS et al (2003) Chk1 regulates the S phase checkpoint by coupling the physiological turnover and ionizing radiation-induced accelerated proteolysis of Cdc25A. Cancer Cell 3(3): 247–258

Sung H et al (2021) Global cancer statistics 2020: GLOBOCAN estimates of incidence and mortality worldwide for 36 cancers in 185 countries. CA Cancer J Clin 71(3):209–249

Tanaka H et al (2002) Cytoplasmic p21Cip1/WAF1 regulates neurite remodeling by inhibiting Rho-kinase activity. J Cell Biol 158(2):321–329

Toyoshima H, Hunter T (1994) p27, a novel inhibitor of G1 cyclin-Cdk protein kinase activity, is related to p21. Cell 78(1):67–74

Wang TC et al (1994) Mammary hyperplasia and carcinoma in MMTV-cyclin D1 transgenic mice. Nature 369(6482):669–671

Weinert TA, Hartwell LH (1988) The RAD9 gene controls the cell cycle response to DNA damage in Saccharomyces cerevisiae. Science 241(4863):317–322

Wölfel T et al (1995) A p16INK4a-insensitive CDK4 mutant targeted by cytolytic T lymphocytes in a human melanoma. Science 269(5228):1281–1284

Won K-A et al (1992) Growth-regulated expression of D-type cyclin genes in human diploid fibroblasts. Proc Natl Acad Sci 89(20):9910–9914

Woods CM et al (1995) Taxol-induced mitotic block triggers rapid onset of a p53-independent apoptotic pathway. Mol Med 1(5):506–526

Zhan Q et al (1993) Induction of cellular p53 activity by DNA-damaging agents and growth arrest. Mol Cell Biol 13(7):4242–4250

Zhang Y et al (1998) ARF promotes MDM2 degradation and stabilizes p53: ARF-INK4a locus deletion impairs both the Rb and p53 tumor suppression pathways. Cell 92(6):725–734

Cell Cycle Dysregulation in Breast Cancer

5

Manzoor Ahmad Mir, Sameer Ullah Khan, and Shariqa Aisha

5.1 Introduction

Cell cycle dysregulation causes cellular hypertrophy and tumor cell proliferation in breast carcinoma. Elevated oncogenic action (c-Ras, c-ErbB-2, c-Myc) and lack of tumor suppressor activity are common examples of aberrant gene expression (p53, pRB), (Mehraj et al. 2022a; Mir et al. 2022a). Furthermore, abnormal cell cycle mediator expression leads to the transition of healthy mammary cells (Sofi et al. 2022b). The cell cycle of mammals (Fig. 5.1) is divided into four phases: the S phase, which involves DNA formation, M phase (mitosis), which involves exact division of cells, as well as two gap or growth stages (G1, G2), that involve the replication of requisite cellular constituents. G0 refers to a non-proliferating, resting phase. Mitogen-induced signal transduction coordinates the stepwise (G1, S, G2, M) advancement via the cell cycle by orchestrating expression of kinase holoenzymes. Improper cyclin-dependent kinase (CDk) activation can lead to a failure of checkpoint regulation that results in aberrant cellular growth (Mir et al. 2022b; Mehraj et al. 2021).

G1, S, G2, and M are the four stages of the cell cycle that cells go through (Sherr 2000). CDKs play an important role in cell cycle progression (Mehraj et al. 2022b). CDks are a type of serine/threonine kinase having active subunits which forms a complex with cyclin proteins. There are 15 cyclins; cyclin A to cyclin T, and nine CDks, cdk1 to cdk9 (Tassan et al. 1994; Mir & Mir 2022; Morgan 1997) Binding of Cdk1 (cdc2) with cyclin A and B influences the G2–M shift, binding of cdk2 with

M. A. Mir (✉) · S. Aisha
Department of Bioresources, School of Biological Sciences, University of Kashmir, Srinagar, J&K, India
e-mail: drmanzoor@kashmiruniversity.ac.in

S. U. Khan
Pharmacology Division, CSIR-Indian Institute of Integrative Medicine, Srinagar, J&K, India

M. A. Mir (ed.), *Therapeutic Potential of Cell Cycle Kinases in Breast Cancer*,
https://doi.org/10.1007/978-981-19-8911-7_5

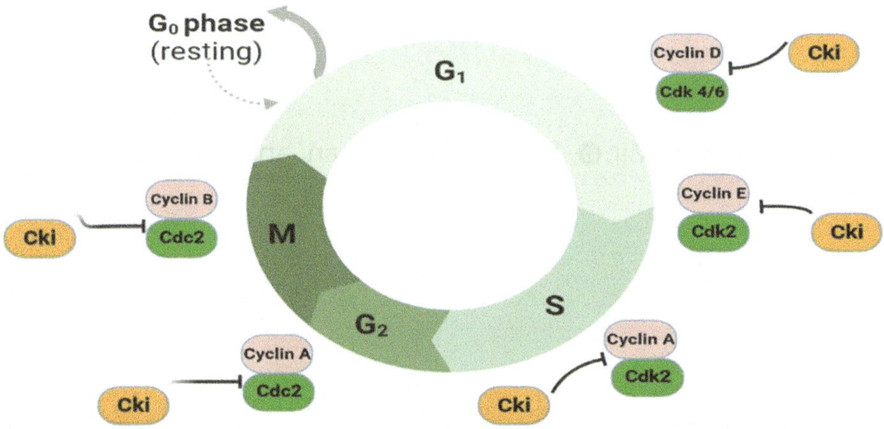

Fig. 5.1 The mammalian cell cycle regulation and cyclin-dependent kinase inhibitors

cyclins A, D, and E promotes the G1–S shift and S-phase shift, binding of cdk4 and cdk6 with cyclin D stimulates cell cycle development via G1, and binding of cdk7 with cyclin H triggers additional cyclin/CDk complexes (Buolamwini 2000; Mir et al. 2022a). Different molecular processes control the action of the holoenzymes formed by cyclins and their CDk catalytic subunits, including dephosphorylation and phosphorylation of the kinase subunits (Solomon 1993; Mir and Mehraj 2019). Myc and E2F stimulate the expression of Cdc25 in G1, which is required for S-phase entry (Foster et al. 2001; Mir et al. 2022g; Mir et al. 2020). The Rb family proteins, p107 and pRb, regulate the E2F family of transcriptional regulators, which regulate the expression of genes required for the progression of cell cycle (Mir et al. 2022c). (Weinberg 1995). Furthermore, interplay with particular CDk inhibitors and correlation with specific regulatory proteins play a role in CDk activity regulation. CDk inhibitors are divided into two categories (Sherr 1996). In the first place, inhibitors of cdk4 (INK4) include p15, p16, p19, and p18, which inhibit cyclin D-associated kinases specifically. The kinase inhibitor protein family (KIP) p27, p21, and p57, which attach to and downregulate the activity of the cyclin A/cdk2 and cyclin E/cdk2 complexes, are the second category (Pines 1995; Graña and Reddy 1995; Mehraj et al. 2022c) (Fig. 5.2).

During oncogenesis, the cell cycle becomes dysregulated due to a loss of regular restriction point command. Cyclins (D and E) overexpression, pRB down-regulation, as well as lowered CKI activity is all common observations. In human tumors, aberrations of the cyclin D1/Cdk4/p16/pRB axis are second only to p53 aberrations in terms of frequency (Sherr 1996; Mehraj et al. 2022a). Cyclin D1 is abundantly expressed in 30–45% of breast tumors (Pestell et al. 1999), and it adds to the tumorigenesis of regular mammary cells in collaboration with other misexpressed genes. In mouse models overexpressing cyclin D1, its involvement as a "driver oncogene" had also been proved (Wang et al. 1994; Mir et al. 2022f;

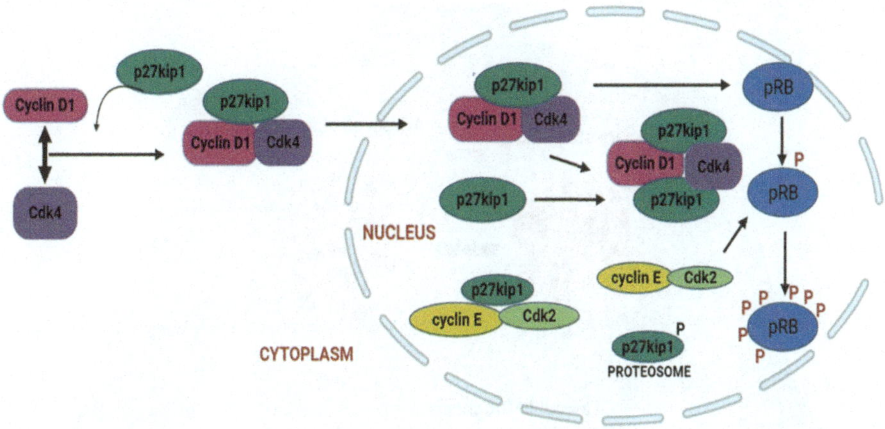

Fig. 5.2 In the presence of an assembly factor, the cyclin D1 gene product binds its catalytic subunit partner (Cdk4). Numerous subunits of a Cdk activating kinase (CAK) phosphorylate the cyc D1-Cdk4 holoenzyme. The tumor suppressor pRB can then be phosphorylated by activated cyclin D1-Cdk4. pRB is phosphorylated by cyclin ECdk2 as well. If p27Kip1 hinders cyclin E-Cdk2 but not cyclin D1-Cdk4, cyc D1 stimulation may enhance S-phase entry by titrating p27Kip1 from an inhibitory complex with cyclin E-Cdk2

Mehraj et al. 2022d). Numerous oncogenes, such as Ras activating mutants, Dbl, Rac, pp60src, and Neu, stimulate cyclin D1 abundance by increasing cyclin D1 promoter's activity (Lee et al. 1999; Albanese et al. 1995; Westwick et al. 1998).

The phosphorylation as well as inhibition of pRB (Weinberg 1995) and also cyclin D1-mediated sequester of CKIs are thought to be the major mechanisms of the transformation of cells. Unlike cyclin D1's increased expression in human BC, cyclins D2 and D3 are not linked to the development of breast tumors (Fig. 5.3).

5.2 Oncogenic Factors of Cell Cycle

5.2.1 Cyclin D

The cyclin D, proto-oncogene, is an important regulator of the cell cycle's shift from G1 to S. It generates functional cyclin D-CDK4/6 complexes by binding to CDK4 and CDK6, which subsequently phosphorylates the retinoblastoma protein (Rb) to enhance the progression of cell cycle (Kato et al. 1993). Cyclin D could potentially influence the activity of histone deacetylase enzymes and other transcriptional factor proteins (Coqueret 2002). Cyc D is destroyed within the cell mostly through the action of the 26S proteasomal degradation in a ubiquitin-dependent or Skp2 F-box protein-dependent way as it has a half-life of about 24 min (Diehl et al. 1997; Yu et al. 1998). The phosphorylation and inhibition of NRF1 as well as mitochondrial transcription factor A by the D1-CDK4/6 complex also can damage mitochondrial functioning (mtTFA). A previous study found a biological relationship between

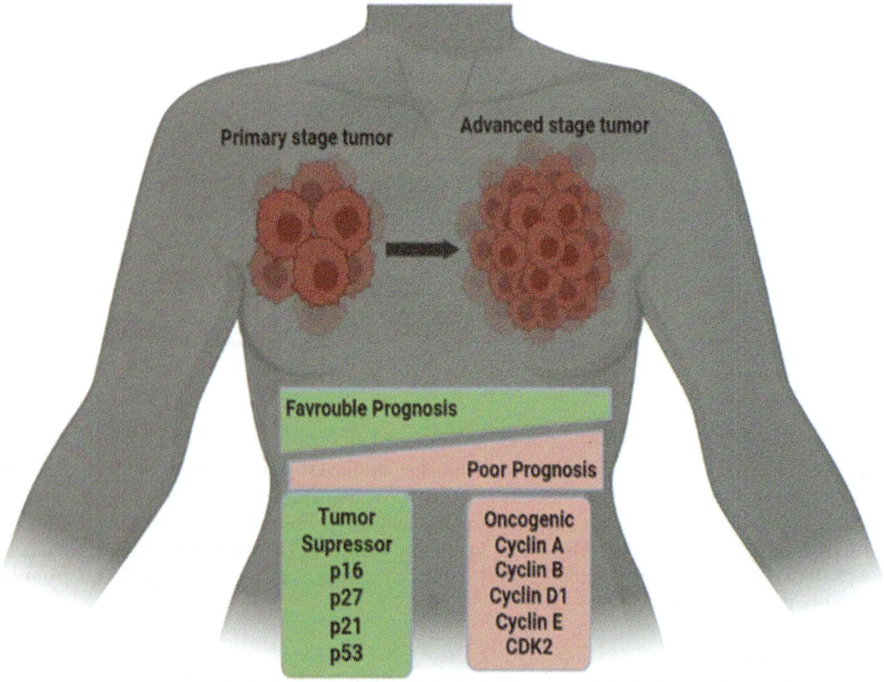

Fig. 5.3 Illustration demonstrating how the expression levels of genes acting as tumor-suppressive and oncogenic markers change during the development of tumors. The overall prognosis for breast malignancy and the progression of cancer into late stages are affected by these changes in expression

cyclin D1 and mitochondrial function control via nuclear respiratory factor 1 inhibition (Wang et al. 2006). Earlier accumulated findings support the significance of cyclin D1 in breast cancer carcinogenesis (Hall and Peters 1996; Yamamoto et al. 2006). In BC, cyclin D upregulation and gene multiplication have been also associated to a worse prognosis and developing resistance to hormonal treatment (Table 5.1) (Hodges et al. 2003; Kenny et al. 1999). Overexpression of the cyclin D1 gene as well as amplification in copy number were found in 20% and 50% of female BC patients, correspondingly (Mohammadizadeh et al. 2013; Barnes and Gillett 1998; Velasco-Velázquez et al. 2011). Moreover, cyc D1 expression was shown to be elevated in 67.5% of invasive ductal carcinoma patients (Ravikumar and Ananthamurthy 2014), and it has been found to be closely linked with the expression of estrogen receptor (ER) and progesterone receptor (PR). Furthermore, in invasive ductal & moderately differentiated cases of breast cancer, immunohistochemistry (IHC) positive for cyc D1 was linked with a considerably worse prognosis (Assem et al. 2017). Furthermore, in vitro and clinical study evidence revealed elevated cyclinD1 expression of genes and amplification in 45–50% of cases of BC (Buckley et al. 1993). Other in vitro investigations discovered genetic changes in the cyc D1 gene as well as expression of mRNA in the ER(−) MDA-MB-453 cell lines, which

Table 5.1 Oncogenic cell cycle elements clinical significance in breast carcinoma patients with various molecular subtypes

Marker	Expression	Outcomes	Receptor status	References
Cyclin D	Elevated expression	Increased risk of relapse, metastasis	ER−/ER+	Zhang et al. (1994)
	Elevated expression	A high-grade tumor	ER−/ER+/PR−/PR+/HER2+	Mohammadizadeh et al. (2013)
	Elevated expression	Proliferation is very high.	ER−/ER+/PR−/PR+/TNBC	Ravikumar and Ananthamurthy (2014)
	Elevated expression	Proliferation is very high.	ER−/ER+/PR−/PR+	Assem et al. (2017)
	Elevated expression	Proliferation is very high.	ER−/ER+	Buckley et al. (1993)
	Elevated expression	Proliferation is very high.	ER−/ER+/PR−/PR+	Zukerberg et al. (1995)
Cyclin A	Elevated expression	Recurrence, short disease-free survival	ER−/ER+	Baldini et al. (2006)
	Elevated expression	Poor prognosis	ER−/ER+	Michalides et al. (2002)
	Elevated expression	Short relapse duration	ER−/ER+	Poikonen et al. (2005)
	Elevated expression	Low survival rate, increased recurrence rate	PR−/ER−/PR+/ER+	Nielsen et al. (1996)
	Elevated expression	Shorter distant metastasis-free survival	PR−/ER−/PR+/ER+	Sieuwerts et al. (2006)
	Elevated expression	Worse prognosis	ER−/ER+	Donnellan et al. (2001)
	Elevated expression	Greater tumor grade, strong proliferation index	HER2−/HER2+	Potemski et al. (2009)
Cyclin B	Elevated expression	Lower survival	PR+/ER+/PR−/ER−/HER2−/HER2+	Niméus-Malmström et al. (2010)
	Elevated expression	Decreased overall survival, Disease-free survival, lymphatic infiltration	PR+/ER+/PR−/ER−/HER2−/HER2+	Sun et al. (2017)

could be linked to tumorigenesis (Lebwohl et al. 1994). Infiltrating BC with ER/PR positive was also studied for cyclin D1 proteins expression (Zukerberg et al. 1995). In 66% of breast infiltrating duct carcinomas, aberrant expression of cyclin D1 was found, implying a role for the protein in breast tumor migration (Khan et al. 2013). The cyclin D gene was discovered to be overexpressed in 82% of female breast cancers, and amplification of gene was detected in percent of instances, according to Zhang and coworkers (Rosendahl et al. 2015).

In research utilizing a mouse breast tumor virus model of breast carcinoma, amplification of CCND1 gene was found in 40% of breast cancer specimens with

Fig. 5.4 Gene amplification cyclin D1

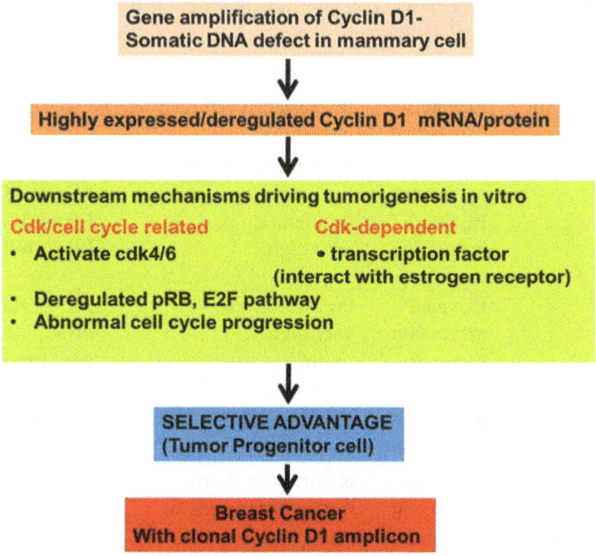

significant IHC staining. Furthermore, in ER-positive cases of breast cancer, the research discovered aberrant upregulation of cyc Dl and a reverse inhibitory growth effect following anti-hormonal treatment, suggesting a possible antitumor strategy (Wilcken et al. 1997; Mir et al. 2022b). Kenny et al. discovered that ER-positive BC women had increased expression of cyclin D1 while also having a higher risk of recurrence, metastases, and early mortality (Kenny et al. 1999; Mir et al. 2022d). Furthermore, the findings revealed that gene duplication of CCND1 is a powerful indicator of anti-hormonal treatment efficacy in breast carcinoma patients under the age of 50 (Jirström et al. 2005). Furthermore, overexpression of the cyc D1 gene was linked to ER(+) invasive lobular BC with lymph node metastases in another research (Courjal et al. 1996), indicating that it is an indication of poor prognosis. Another study found that amplification of the gene cyc D1 in the MCF-7 breast carcinoma cell lines was accountable for hyper-proliferation in the absence of growth factors (Zwijsen et al. 1996). Increased expression of cyc D1 was seen in both ER(+) and ER (−) breast tumor tissues in another investigation (Umekita et al. 2002), but only the ER-negative category had reduced OS and RFS Fig. 5.4.

In over 100 high-grade breast tumors, a link between elevated cyc D1-related increase and phosphorylation of Rb was discovered (Loden et al. 2002). In addition, a separate investigation found a clear link among cyc D gene multiplication and increased expression in ER-positive and basal-like breast carcinoma subgroups, implying that cyc D1 was an independent determinant of outcome in ER(+) breast malignancies (Elsheikh et al. 2008). Overexpression of cyc D1 was linked to a worse clinical prognosis and significantly shorter survival of female with BC in the ABCSG Trials 05 and 06 (Rudas et al. 2008) (Mir et al. 2022e). The presence of cyclin D1 in proliferating cancer without atypia, atypical ductal hypertrophy, reduced ductal DCIS, elevated DCIS, and metastatic cancer was confirmed in a separate study. Cyclin D1 was shown to be much greater in proliferative illness than healthy mammary epithelium, or even greater in DCIS compared to proliferative illness (Alle et al. 1998).

Furthermore, in the ER(+) breast carcinoma group, another study found a link among increased cyclinD1 gene expression as well as high-grade tumor progression, higher Ki-67 expression, and shorter survival (Heiss et al. 2010). Most of ILCs had cyc D1 protein upregulation, indicating that it plays a function in aggressive lobular carcinoma development (Oyama et al. 1998). Additional study found that ER-positive individuals with medium expression of cyclin D1 responded from anti-hormonal medication (tamoxifen), while those with significant expression of cyclin D1 did not, indicating that cyc D1 could be used as a predictor of tamoxifen susceptibility (Ahnström et al. 2005). Additional findings indicate that inhibiting the expression of cyc D1 may slow the progression and development of tamoxifen-resistant cancers (Kilker and Planas-Silva 2006). Cisplatin, a medication that targets cyclin D1, promoted cell damage or growth inhibition in ER-positive MCF-7 BC cells by lowering cyclin D levels (Yde and Issinger 2006). Researchers discovered that CCND1 exhibited greater amplification in high-grade infiltrating DC compared to low-grade IDC utilizing the techniques of FISH and immunohistochemistry (IHC) (Simpson et al. 1997). Increased expression of cyclin D1 has been linked to receptor status, indicating that expression of cyclin D1 may be used as a biomarker for favorable prognosis (Tobin and Bergh 2012; Boström et al. 2009; Mehraj et al. 2022d).

Furthermore, cyc D2 expression was shown to be extremely infrequent in BC patients compared to healthy human breast epithelial cells (Jirström et al. 2005; Dai et al. 2013), and its significance in cancer has yet to be determined (Zhang 1999). Although cyclin D3 had been found to be upregulated in breast tumor tissues, evidence on its link to illness outcomes is limited (Kilker et al. 2004; Zhang 1999; Sutherland and Musgrove 2004). Moreover, higher protein levels of cyc D1 and cyc D3 deposition have been seen in breast tumor tissues (Russell et al. 1999). Another investigation found gene amplification of cyclin D1 in 64 of 82 BC patients and cyclin D3 gene amplification in 36 of 86 patients (Husdal et al. 2006). The expression of cyclin D1 was examined in various molecular breast carcinomas, with favorable cyclin D1 staining being more intense in the ER (+)/PR (+) subtype compared to TNBC cases, and negative staining of cyc D1 being observed in HER2-positive subtypes. Furthermore, TNBC patients with reduced expression of cyclin D1 exhibited higher grade tumor, tumor stage, and higher positive lymph nodes having lymphovascular infiltration, suggesting that expression of cyclin D1 may be an important factor to evaluate in BC treatment (Lengare et al. 2020). Dicer expression was increased in luminal A and basal-like BC subtypes when cyclin D1 was overexpressed (Yu et al. 2013). In another research, reduced cyclin D levels were linked to a reduction in MDA-MB-231 cell motility, which was caused by a drop in filamin A protein phosphorylation (Zhong et al. 2010).

5.2.2 Cyclin A

Protein cyc A makes complexes both with CDK1 as well as CDK2, which function both in S to G2 phase shift and also in G2 to M phase shift of the cell cycle (Rudas et al. 2008). During the S phase, the cyc-A-CDK complex phosphorylates the elements of the DNA synthesis machinery, hence commencing replication (Rudas et al. 2008; Sofi et al. 2022a). During the mitotic phase, cyclinA/CDK2 regulates nuclear mitotic and centrosomal processes. Nevertheless, it is expected to promote the stability of different cyclin molecules (Rudas et al. 2008). The elevated expression of the cyc A gene had been observed in various forms of human cancers, notably BC, suggesting that cyclin A could possibly act as a prognostic marker for the illness (Table 5.1). Cyclin A microinjection into the Xenopus oocytes as well as mammalian cells promotes breast cancer epithelium cells and induces the cell cycle to enter the M phase (Boström et al. 2009; Joung et al. 2005; Ravazoula et al. 2003; Ates et al. 2011). There is a high statistical association among cyc A gene duplication and cyc A protein level in a large number of malignancies (Husdal et al. 2006). The data indicated that measuring cyc A and/or E2-promoter binding factor 1 (E2F1) expression patterns in association with Ki-67 could be a valuable technique for improving prognosis in individuals with lymph node (−) BC (Baldini et al. 2006). Other analyses revealed that cyclin A is an important predictive variable and indicator of both tumor relapse and tamoxifen response (Michalides et al. 2002). Finally, as compared to breast carcinoma individuals with favorable prognoses, abundantly expressed cyclin A was found to be substantially connected with early recurrence, greater risk, and lower overall chance of survival. As a result, cyclin A may be a potent biomarker for tumor growth and prognosis in breast malignancy (Poikonen et al. 2005).

5.2.3 Cyclin E

The rate-limiting component for the G1 to S phase cell cycle shift is assumed to be cyc E protein, a regulatory component for CDK-2 (Leng et al. 1997). In healthy cells, the cytokine E proteins as well as its accompanying kinase (CDK2) are activated in a well-regulated manner. The cyclin E and CDK complex, on the other hand, stays activated during the cell cycle in rapidly proliferating tumor epithelium cells (Hwang and Clurman 2005). Dysregulation of the cyclin E gene has been linked to the development of breast cancer tumors (Wilcken et al. 1997; Russell et al. 1999; Zwijsen et al. 1996). Previous research has shown that breast tumor tissues had greater amounts of cyclin E gene duplication (Table 5.1) (Husdal et al. 2006). In female breast tumor cells, another research discovered an 8-fold increase of the cyclin E genes as well as a 64-fold amplification of its mRNA, indicating that cyclin E expression is abnormal during carcinogenesis (Keyomarsi et al. 2002). In addition, a multivariable study linked high cyclin E levels to poor clinical outcomes, revealing that individuals with high cyclin E concentrations had a higher relative risk than those with lower cyclin E levels (Keyomarsi et al. 2002). Furthermore, a link among gene expression of cyclin E and ER-positive condition was discovered in women with BC. In other investigations (Davidoff et al. 1991; Porter et al. 2006), cyclin E expression was shown to be higher in the ER-negative group and was linked to an elevated risk of mortality and recurrence, showing that cyclin E is involved in ER-independent tumor growth. Increased cyc E expression in breast carcinoma cells has also been linked to HER2-positive tumors, ER-negative tumors, and high-grade tumors with high proliferative indexes (Donnellan et al. 2001; Potemski et al. 2006). When compared to unexpressed cyclin E patient populations, gene copy number or mRNA increased expression of cyclin E was linked to decreased treatment advantages and reduced rates of progression-free survival in a cohort research of 34 HER2-positive patient populations treated with trastuzumab (Herceptin)-based therapy. Furthermore, cyclin E expression was linked to a bad outcome and was found to be tightly linked to cyclin D1 as well as p27Kip1 expression (Han et al. 2003). In the node-positive breast carcinoma group, increased expression of cyclinE assessed by IHC was also a substantial indicator of poor outcome and was related with a greater probability of mortality, as shown in a subsequent multivariate study (Potemski1Abcdefg et al., 2009).

5.2.4 Cyclin B

Progression of cell cycle from G2 to Mitosis is regulated by two forms of mammalian cyclin B that form complexes with CDK1 kinases (Dorée and Galas 1994). According to the existing evidence, breast tumor individuals have cyc B gene overexpression and amplification at both the protein and RNA levels (Table 5.1) (Husdal et al. 2006). Big tumor size, higher tumor grade, ER (−) /PR (−) status, and HER2 (+) status have all been linked to higher expression (Aaltonen et al. 2009). Its higher expression had likewise been connected to a younger age at assessment and increased cyc E, cyc A, and Ki-67 expression levels (Patil et al. 2011). Both univariate and multivariable studies revealed that cyclin B1 upregulation is associated with a higher breast tumor mortality rate, implying that it is a significant predictive factor (Aaltonen et al. 2009).

The importance of cyclin B protein with clinicopathological features in breast carcinoma individuals was studied in a meta-analysis. Highly expressed cyclin B was linked to lower disease-free survival (DFS), disease-specific survival (DSS), and overall survival (OS) rates, as well as a positive relationship with lymphatic infiltration (Sun et al. 2017). While in absence of cyclin B, Androic et al. (Yang et al. 2013) reported apoptosis activation and growth suppression in multiple breast tumor cell lines, including MDA-MB-231, MCF-7, SK-BR-3, and BT-474. In breast tumor cell lines, siRNA-mediated reduction of cyclin B resulted in G2/M cell cycle stage arrest (Sun et al. 2017; Androic et al. 2008). Positive cyclin B1 staining was found to be associated with higher tumor stage, larger tumor size, lymph node-positive, young age, and stronger Ki-67 expression in HER2-positive metastatic breast tissues used for the study. As a result of its link to an invasive phenotype, cyclin B1 could be regarded a significant independent breast tumor predictive factor (Aaltonen et al. 2009).

5.3 Deregulation of CDKs in Breast Cancer

Out-of-control cell growth is one of the characteristics of cancer that is because of the deregulation of CDK/cyclins, which revokes many protections and damages cell cycle checkpoints. Until lately, an extensive study revealed the most common dysregulated activating modifications of the CDK/cyclins that resulted in the diverse BC phenotypes, which are listed in Table 5.2.

The CDK4/6–RB pathway, which is involved in the cell cycle's G1/S phase transformation, is significant in BCs. Generally, CDK4/6/cyclin D1 is the major regulator of RB phosphorylation in order to enhance cell growth. Dysregulation of the CDK4/6–cyclinD/INK4/pRB/E2F cascade or its regulators is predicted to promote to carcinogenesis and BC persistence (Santo et al.). Interestingly, CIP/KIP and INK4 family proteins decrease, and also CDK4/6 amplification, was found in BC (Asghar et al. 2015). According to a recent investigation, various BC subtypes exhibit distinct cell cycle checkpoint genetic changes. A dataset-based genome sequencing research of 482 metastatic BC cases revealed that 27.4% of CDK 4 and 6–RB pathway genetic dysregulation includes the expression of a specific

Table 5.2 Biological roles of CDKs in BCs

CDKs	CDK associate	Biological role in Breast cancer	References
CDK1	Cyclin A/B	Links with cell death of MYC-driven TNBC	Horiuchi et al. (2012) and Marais et al. (2010)
CDK2	Cyclin A/E	Associates with BC or TNBC phenotype	Marais et al. (2010) and Nie et al. (2019)
CDK3	Cyclin C	Correlated with BC invasion, cell proliferation, apoptosis, and cell migration	Zhang et al. (2017) and Cao et al. (2017)
CDK4/6	Cyclin D	Contributes to the onset and persistence of BC carcinogenesis	Malumbres and Barbacid (2001) and Santo et al. (2015)
CDK5	p39 and p35	Associated with ROS-mediated apoptosis in BC; necessary for TGF-β1-induced EMT	Pozo et al. (2013) and Dorand et al. (2016)
CDK7	Cyclin H	In TNBC, it regulates transcriptional addiction to a critical group of genes.	Wang et al. (2015) and Li et al. (2017)
CDK8	Cyclin C	In BC, it reacts to adjunctive treatment and is linked to tumor progression.	Firestein et al. (2008) and Crown (2017)
CDK9	Cyclin T	After neoadjuvant chemotherapy, a predictive biomarker in individuals with BC	Schlafstein et al. (2018), Del Re et al. (2019), and Mehraj et al. (2022b)
CD10	Cyclin M	Resistance to hormonal treatment; linked to lymph node metastases.	Guen et al. (2017) and You et al. (2015)

gene mutation or the combination of various gene mutations (Dukelow et al. 2015). In ER+ BC, in which the cyc D1–CDK 4 and6–RB complexes serve as the estrogen effector, estrogen can accelerate cell cycle transition from the G1 to the S phase. In a nutshell, binding of estrogen to ER-alpha causes cyc D1 transcription, while stimulation of CDK4/6 as well as phosphorylation of RB cause cell cycle progression via the checkpoint, resulting in the onset of the cell signal, which stimulates the activation of various receptor-driven genes, associated in proliferation of cells as well as survival. Augmentation of Cyc D1 is found in roughly 15% of BCs, especially ER+ BCs (Arnold and Papanikolaou 2005). Furthermore, estrogen receptor 1 (ESR1) proteins expression is frequently elevated in ER+ BCs, as is PIK3CA expression, both of which lead to progression of cell cycle via the protein kinase B AKT/mTOR signal transduction pathway. When compared to various BC subtypes, like TNBC and HER2+, ER+ BC is genetically more stable, with a major dependence on estrogen signaling, and it usually has adequate functioning of the RB and p53 tumor suppressor mechanisms. The CDK4/6–RB pathway is also involved in HER2-induced cell proliferation (Spring et al. 2017; Sofi et al. 2022b). The activation of the cyc D1–CDK 4 and 6 pathway contributes to a tumorigenic state and leads to the development and persistence of carcinogenesis in HER2+ BC in mouse models of clinical BC (Dukelow et al. 2015). CDK4, and also erb-b2 receptor tyrosine kinase 2 (ERBB2), tumor protein p53, PTEN, PIK3CA, and cyc D1, are elevated in HER2+ BC. On the other hand, TNBC genomic, proteomic and clinical RB mechanism data show RB1 deletion or mutation in 20% of instances as well as cyclin E1 overexpression in 9% of instances, elevated expression of CDKN2A, reduced expression of RB1, and elevated tumorigenic rate, and also regular changes in DNA damage responsive genes like tumor suppressor BC 1 (BRCA1) (Robinson et al. 2013; Fedele et al. 2019). Elevated cyclin E expression is linked with a worse prognosis in TNBC and is associated with a negative PR and ER status (Jabbour-Leung et al. 2016). TNBC also stimulates B-Raf proto-oncogene, (BRAF), PIK3CA, KRAS proto-oncogene, GTPase (KRAS), EGFR and PTEN deletion, leading in aberrant PI3K/Akt/mTOR or Raf/MAPK/ERK signaling pathways. The frequency of PIK3CA alteration in TNBC, on the other hand, is just 8.3% (Stemke-Hale et al. 2008; Khan et al. 2022b). The integrity of the cell cycle that is regulated by the Rb/E2F/CDK4/6 axis, is impaired in TNBC owing to recurrent deletion or mutations of RB1. Individuals with TNBC are frequently thought to be worse candidates for CDK suppression. According to a preclinical experiment, TNBC is particularly responsive to a CDK2/9 inhibitor, implying that there could be unknown variables linking the CDK complex in TNBC growth (Matutino et al. 2018; Singh et al. 2017). However, new research has found that the activation of many SAC genes, including BUB1, TKK, MAD2, AURKB, and DNA repair enzymes, is changed in TNBC, possibly because of genetic instability in TNBC (Thu et al. 2018).

Other CDKs, including CDK2, are also elevated, which often leads to the proliferation and/or upregulation of its counterparts cyc A or cyc E in BC (Santo et al.). CDK1 and its related cyclins, cyc A2 and B1, are frequently engaged in mitosis progression, while cyclin B1 expression is elevated in BC (Aaltonen et al. 2009). Nevertheless, there is no clear evidence associating genetic changes that disrupt CDK1 function to the onset of BC. According to one research, the deletion of CDK12 protein enhances the phenotype of TNBC because CDK12 deficiency causes DNA repair deficiencies (Naidoo et al. 2018).

5.4 Tumor Suppressive Proteins of the Cell Cycle

5.4.1 p16

Due to the inhibitory effect on CDK, the p16, commonly called as INK4A/MTS-1/CDKN2A, is mostly used in oncology (Li et al. 2011). The widespread occurrence of single nucleotide polymorphism (SNP) alterations and deletions of the p16 genes in breast tumor cells suggests that the gene plays a key function in carcinogenesis (Baker and Reddy 2012). The p16 protein component interacts to and deactivates CDK4/6/cyclin D complexes, causing Rb protein deactivation and cell cycle halt (Baker and Reddy 2012). Altered expression of p16 gene is most prevalent anomaly in human BC, according to archived mammary cancers of various histological types (Table 5.3) (Geradts and Wilson 1996). In addition, expression of p16 was revealed to be aberrant in ER-negative, pre-menopause women with breast carcinoma compared to ER-positive patient populations. The aberrant expression of p16 identified by these investigators was linked to a higher proliferation index (Muhammad et al. 2012). According to a previous study, aberrant expression of p16 could be a marker of poor endocrine therapy responsiveness (Cui et al. 2012). Additional study discovered positive p16 protein expression in women with breast carcinoma with a luminal A type, and increased expression was linked to breast tumor progression from DCIS to IDC (Shan et al. 2013). Abou-Bakr and colleagues (Abou-Bakr and Eldweny 2013) looked at expression of p16 in grade 3 basal-like breast cancer having histological characteristics that matched IDC. According to the findings, the protein p16 showed a highest IHC level in basal-like cancer that was later linked to lung and brain metastases (Abou-Bakr and Eldweny 2013). Reduced expression of p16 was reported in resistant TNBC cancer in a research by Arima and colleagues (Arima et al. 2012). Both positive-p16 and negative-p16 cells in invasive lobular carcinoma stroma cells indicated high nodal activation, rapid relapse, and metastatic tendency. Furthermore, restoring p16 expression in stromal fibroblasts stopped cancer cells from migrating and invading. As a result of these findings, high stromal expression of p16 has been suggested as a therapeutic option for preventing nodal or distant metastases (Harbhajanka et al. 2019; Khan et al. 2022a).

5.4.2 p21

The p21 which is an inhibitor of CDK also called as WAF1/CIP1/SDI1/MDA-6 stimulates CDK 4 as well as the proliferative cell nuclear antigen, allowing the cell to enter G1 stage. Upregulation of p21WAF1/CIP1 led in cell cycle's G1 stage halt and significantly inhibited tumor development in both in vitro and in vivo research

Table 5.3 Tumor suppressor cell cycle elements clinical significance in breast carcinoma patients with various molecular subtypes

Marker	Expression	Outcomes	Receptor status	References
p21	Elevated expression	Tumor grade high, big tumor size, lymph node status positive, Increased Ki-67 expression	PR+/ER+/PR−/ER−/HER2−/HER2+	Aaltonen et al. (2009)
	Elevated expression	Better prognosis	PR+/ER+/PR−/ER−	Göhring et al. (1999)
	Elevated expression	Good survival	PR+/ER+/PR−/ER−/HER2+	Domagala et al. (2001)
	Elevated expression	Good survival	PR+/ER+/PR−/ER−/HER2−/HER2+	Fayed et al. (2012)
	Elevated expression	Big size of tumor, tumor grade high, Lymph node metastases	PR+/ER+/PR−/PR+/HER2−/HER2+	Wei et al. (2015)
p27	Decreased expression	Tumor grade high, absence of tumor differentiation, Worse prognosis	PR+/ER+/PR−/ER−/HER2−/HER2+	Alkarain et al. (2004)
	Elevated expression	Good prognosis	ER−/ER+	Tsuchiya et al. (1999)
	Elevated expression	Better prognosis	PR+/ER+/HER2+	Traub et al. (2006)
	Decreased expression	Worse prognosis	PR+/ER+/PR−/ER−/HER2−/HER2+	Newman et al. (2001)
	Elevated expression	Prolonged DFS and OS	PR+/ER+/PR−/ER−	Mirchandani et al. (2011)
	Decreased expression	Worse prognosis	PR+/ER+/PR−/ER−	Leivonen et al. (2001)
p16	Elevated expression	Proliferation index high	PR+/ER+/PR−/ER−/HER2−/HER2+	Cui et al. (2012)
	Elevated expression	Good prognosis	PR+/ER+/PR−/ER−	Emig et al. (1998)
	Elevated expression	Good prognosis	PR−/ER−/HER2−	Pan et al. (2017)
	Elevated expression	Good prognosis	PR+/ER+/PR−/ER−	Al-Joudi et al. (2008)
	Elevated expression	Progression of disease	PR+/ER+/PR−/ER−	Shan et al. (2013)
	Decreased expression	Metastases	PR+/ER+/PR−/ER−/HER2−/HER2+	Harbhajanka et al. (2019)
Wild-type p53	Elevated expression	Good prognosis	PR+/ER+/PR−/ER−/HER2−/HER2+	Salmani et al. (2018)
	Elevated expression	Good disease-free survival	PR+/ER+/PR−/ER−	Yang et al. (2013)

(continued)

Table 5.3 (continued)

Marker	Expression	Outcomes	Receptor status	References
	Elevated expression	Better prognosis	PR+/ER+/PR−/ER−/HER2−/HER2+/Basal like	Bertheau et al. (2013)
	Elevated expression	Better prognosis	PR+/ER+/PR−/ER−	Li et al. (2019)
	Elevated expression	Poor prognosis	PR−/ER−/HER2−	Lee et al. (2011)
Mutant p53	Elevated expression	Early recurrence	PR+/ER+/PR−/ER−/HER2−/HER2+	Marchetti et al. (2003)
	Elevated expression	Worse prognosis	PR+/ER+/PR−/ER−	Loo et al. (2019)
	Elevated expression	Less 5-years relapse free survival	PR+/ER+/PR−/ER−	Kikuchi et al. (2013)

models (Table 5.3). According to data from lymph node-negative breast carcinoma individuals, the existence of p21 signals the existence of a tumor suppressor that could improve survival of patients (Göhring et al. 1999). In the other investigation, positive p21 cancer cell nuclei were found in greater than 30% of BC, which was related with a poor tumor grade and node-negative condition (Domagala et al. 2001). The results clearly showed that p21WAF1/CIP1 expression of genes might be employed as a crucial predictive marker for breast tumor, enabling therapeutic choices to be tailored to individual tumor patients more effectively (Fayed et al. 2012). E p21WAF1/CIP1 was shown to be upregulated in larger tumors in individuals with greater tumor dedifferentiation stages, greater lymph node metastasis, and poorer disease-free survival chances after resection (Wei et al. 2015). Furthermore, an in vitro investigation that immunostained ER(+) and ER (−) breast tumor cell lines for p21 expression discovered a strong link among p21WAF1/CIP1 with ER expression (Chen et al. 2000; Fritah et al. 2005).

Furthermore, p21WAF1/CIP1 has several functions in breast carcinoma. For example, regulating TGFβ/Smad signaling via expression of p21WAF1/CIP1 increased cell migration and demonstrated a link with OS and DFS in breast carcinoma individuals (Dai et al. 2012). In the cytoplasm elevated concentrations of p21WAF1/CIP1 in invasive breast carcinoma cells were linked to higher levels of p53 and worse prognoses in one study (Winters et al. 2001). p21WAF1/CIP1 phosphorylation by AKT1 disturbed its interaction with proliferative cell nuclear antigen and caused its buildup in cytoplasm, according to various studies. p21WAF1/CIP1 accumulates in breast tumor cells and promotes ERBB2-mediated growth as well as tumorigenesis (Li et al. 2002). Furthermore, in reaction to plasmacytoma variant translocation 1, an Ln non-coding RNA, decreased expression of p21WAF1/CIP1 promoted EMT, increased cell survival, and increased migratory potential in MDA-MB-468 and MDA-BA- 231 breast carcinoma cell lines (Wang et al. 2018). Other research employing breast tumor mice models found that infiltration is associated by an increase of p21WAF1/CIP1, suggesting an oncogenic role for this protein (Qian et al. 2013). In MCF-7 cells, the amplification of p21WAF1/CIP1 had been also linked to a poor responsiveness to tamoxifen therapy (Pérez-Tenorio et al. 2006). In SUM159 TNBC cells, Akt-dependent phosphorylation of p21WAF1/CIP1 increased doxorubicin sensitivity (Vincent et al. 2012; Wadhwa et al. 2020). In other research, p21WAF1/CIP1 was found to prevent apoptosis in breast carcinoma cells. Suppression of CDKs by upregulation of p21WAF1/CIP1 in BC cells lowered cell susceptibility to infrared-induced cell death (Sohn et al. 2006).

5.4.3 p27

The tumor suppressor p27 that is recognized to correlate the stimulation of the CDK2-cyc E complex with the buildup of CDK 4-cyc-D, which begins the cells escape from the cell cycle in reaction to antimitogenic signals (Neganova and Lako 2008). Reduced expression of the p27 gene is highly associated with increased histological grade and characteristics associated with weaker tumor differentiation (Table 5.3) (Alkarain et al. 2004).

In the majority of lymphatic node-negative breast carcinoma individuals, low p27 protein content are also associated with worse clinical outcomes (Alkarain et al. 2004). Several pieces of data indicate that TGF-β, rapamycin, and cAMP are involved in p27-induced G1 cell cycle phase halt (Courjal et al. 1996; Russell et al. 1999). Elevated levels of p27 expression in human breast tumor cells were also found to be inversely associated to the level of malignancy in the female breast in earlier research. Furthermore, elevated p27 expression in women with breast carcinoma was found to be substantially linked with ER-positive status and negatively linked to short survival. Reduced p27 expression was found to be significantly linked with a poorer clinical outcome in a univariate Kaplan–Meier analysis (Traub et al. 2006). In low p27-expressed samples, a flow cytometry research employing resistant breast carcinoma cells revealed a larger S-phase proportion and higher CDK2 activation, which was reverted following exogenous p27 injection (Nahta et al. 2004; Sofi et al. 2022c).

Immunohistochemistry of breast tumors revealed that p27 reduced expression was associated with HER2 gene amplification in initial breast tumors, which could be useful in prospective patient selection for HER2-positive/neu mab treatment (Newman et al. 2001). Tamoxifen therapy induced cell cycle halt in MCF-7 owing to an elevation of levels of p27, according to a different research (Cariou et al. 2000). Decreased expression of p27 was linked with an elevated histological grade, an advanced TNM phase (tumor size, lymph node status, and metastatic status), and negative hormonal receptor status in other studies of p27 expression (Mirchandani et al. 2011; Leivonen et al. 2001). Docetaxel-resistant breast carcinoma cells (MDA-MB-231 and MCF-7 cell lines) also had lower p27 expression (Brown et al. 2004). Reduced p27 expression was linked to higher tumor stage, mitosis, and nuclear pleomorphism, as well as reduced tubule development in ER-negative or ductal/no special subtype tumors in additional univariate research (Barnes et al. 2003).

Increased p27 expression was found to be an individual predictor of better relapse-free or overall survival, leading to the suggestion that it might be used as an individual predictor of hormonal treatment response (Pohl et al. 2003). A retrospective immunohistochemical study of 216 breast tumors indicated that p27-negative individuals had a worse outcome than individuals in other groups, indicating that p27 expression can help distinguish breast cancer individuals who might benefit from adjuvant treatment (Nohara et al. 2001). Furthermore, reduced p27 immunoreactivity was linked to greater tumor stage, greater HER2-positive amplification, more lymph node-positive populations, reduced thymidylate synthase expression, stronger expression of Ki-67, and worse DFS in lymphatic node-negative patients (Spataro et al. 2003). Reduced expression of p27Kip1 was linked to a reduced overall survival rate in hormonal receptor-positive cancer treated with adjuvant treatment (DOX and CP) (Porter et al. 2006). Expression of p27Kip1 was also shown to be inversely correlated to the level of aggressiveness in breast tumors (Fredersdorf et al. 1997). The expression pattern of p27Kip1 was measured in breast carcinoma individuals in Taiwan, and both univariate and multivariable studies revealed that decreased expression of p27Kip1 was associated with a shorter survival time in PR/ER-positive tumors. As a result, in Taiwan, p27Kip1 could be regarded an effective prognostic marker for breast carcinoma (Chu et al. 1999).

Additional meta-analysis investigation found a link among elevated expression of p27 and DFS, OS, and RFS in lymph node (−) or lymph node (+) BC individuals (Guan et al. 2010). Furthermore, the Austrian Breast and Colorectal Cancer Study Group Trial 06 registered initial-stage BC individuals with an ER/PR (+) status for the purpose of evaluating the expression of p27Kip1 and observing its effect on clinicopathological characteristics in female obtaining adjuvant tamoxifen for 5 years. In comparison to women having lower p27 expression, higher p27Kip1 expression was significantly related with prolonged DFS and OS (Filipits et al. 2009).

5.4.4 p53

The tumor suppressor protein p53 is implicated in organizing cell responses to a number of stressors, like oncogenic stimulation, hypoxia, as well as DNA damage (Zilfou and Lowe 2009). In healthy cells, p53 stimulates apoptosis in reaction to mitogenic stimulus. The same activation of cell death by p53 had been investigated in the reaction to antitumor treatment. Transformation of cell is facilitated by a mutant form of p53 which does not react accurately to oncogenic stimulation, leading in tumor initiation (Zilfou and Lowe 2009). According to immunohistochemistry analysis of p53 expression in early breast carcinoma tissues, p53 amplification was linked to advanced-stage tumors, metastatic dissemination, and reduced progesterone receptor levels (Table 5.3) (Davidoff et al. 1991). Breast tumor patients had an enhanced cytoplasmic buildup of p53 as well. When contrasted to the control sample, these patients' samples showed substantial proliferation, with mean Ki-67 fractions rising to 75% and a 74% higher median S-phase proportion (Emig et al. 1998).

Yang and colleagues (Dorée and Galas 1994) estimated DFS and its connection with p53 expression using invasive ductal carcinoma specimens. Expression of p 53 was found to be a predictor of DFS using Cox modeling and multivariate analyses (Yang et al. 2013). Furthermore, positive p53 expression has been associated to a worsening prognosis in various studies. A Kaplan–Meier evaluation of TNBC IDC specimens, revealed that p53 (+) expression was associated with poor survival rates, with patients having a 2.2 times increased death rate than p53 (−) individuals. Likewise, IHC testing of p53 amplification in altered radical mastectomy tissues from TNBC patients revealed worse overall survival percentages when contrasted to the patient population with minimal p53 expression. Furthermore, p53 upregulation was found to have the highest prognostic relevance in TNBC individuals (>50 years) in a multivariate study (Lee et al. 2011).

The findings from a retrospective analysis of a significant population of luminal/HER2 (−) BC patients revealed that 50% expression of p53 (found in 9% of individuals) was related with poorer DFS. As a result, p53 amplification has been labeled as a predictive factor for adverse outcomes (Kikuchi et al. 2013). Another study found a link between p53 upregulation and ER status in ER (+) and ER (−) invasive breast carcinoma. Increased p53 expression was related with OS and DFS in ER (−) breast cancers compared to ER (+) breast tumors (Coates et al. 2012).

The p53 gene had also been examined in all BC subtypes, including Lum A, Lum B, TNBC, HER2+ and basal-like, with the results indicating that expression of p53 was greater in the HER2-positive as well as TNBC subgroups than those in the luminal A or luminal B subgroups. Increased expression of p53 in HER2-positive and TNBC subgroups was linked to early-onset, high-grade cancers as well as a higher proliferation index (Abubakar et al. 2019). Elevated expression of p53 was associated with higher tumor stage (p 0.006), lymphovascular infiltration (p 0.003), and lymphocytic infiltration (p 0.004) in invasive breast cancer stage II and III specimens. These findings suggest that p53 amplification in more invasive breast carcinoma types is associated with a worse outcome and weakened immune responses (Muhammad et al. 2012).

Patients with highly expressed p53 in TNBC had a higher overall survival rate than those with p53-negative TNBC who received neoadjuvant treatment (Bae et al. 2020). In high-grade cancers with nodal metastases, Another analysis revealed that p53 amplification was negatively connected with ER/PR expression and strongly associated with HER2 (+) higher expression (Mehta et al. 2019). Epithelial p53 activity was assessed in a randomized II stage clinical study on lymph node (+) individuals who underwent cyclophosphamide 4 cycles plus one dosage of DOXO adjuvant treatment (utilizing mAbs DO7 and 1801). This research found that high p53 IHC was linked with poor OS and RFS in lymph node (+) individuals following univariate analyses (Lara et al. 2011).

5.4.5 Mutant p53

Marchetti and coworkers discovered a "Arg72Pro"p53 variant in 23 percent of early BC patients in a research. When compared to individuals having wild-type p53 status, those who tested positive for the Arg72Pro variation relapse within 10 months of a median DFS (Marchetti et al. 2003). Utilizing the PAb 1801 monoclonal antibodies, Lenora W.M. and colleagues discovered a greater nuclear expression of mutated p53 in young breast carcinoma patients. The researchers used Kaplan–Meier plotter as well as a log-rank test to find a link between mutant p53 expression and worse prognosis in different ethnic groups. In a separate investigation, TNBC individuals with aberrant mRNA expression of mutated p53 were found to have a worse 5-year recurrence-free rate of survival. As a result, mutated p53 could be utilized as a predictive biomarker in TNBC patients (Kim et al. 2016).

5.5 Summary

Numerous oncogenic and tumor-suppressive proteins implicated in cell cycle control and progressions have previously been identified in diverse subtypes of female breast carcinoma. Although it is well known that many genetic changes are necessary for carcinogenesis, greater research into the exact and sequential processes associated how these influence therapeutic outcomes continue to inspire novel therapeutic techniques for more successful cancer therapies. The study of differentially expressed gene products, growth regulators, and other biological agents that promote cell multiplication and tumor development has provided some insights into cancer development processes. These discoveries have paved the way for the development of specific medicines aimed at disrupting specific biological pathways. Transgenic mice models have supplied crucial data that has aided our comprehension of breast tumor processes and treatments. In the future, improvements will most likely be made in part by using genomic microarray analysis technologies to determine genetic propensity to tumor formation. This technique will almost certainly be crucial in the identification of critical novel treatments.

5.6 Further Readings

The readers can further read about the role of CDKs in breast cancer by going through the following papers

https://discovery.ucl.ac.uk/id/eprint/1472740/
https://www.jmedsciences.com/doi/JMEDS/pdf/10.5005/jp-journals-10045-00138

For more incites about the topic we would suggest detailed findings from the books of (Mir 2022) https://doi.org/10.1016/C2021-0-02565-7, https://doi.org/10.1016/C2022-0-00074-X (Mir 2021) https://doi.org/10.52305/WXJL6770

The following visual presentations are also available for the readers to view for a better conceptual grasp of CDKs and their function in breast cancer

https://www.youtube.com/watch?v=1mo80kTZgW4
https://www.youtube.com/watch?v=BIBpEP0jDm4
https://www.youtube.com/watch?v=GoMiAbEmHEI

References

Aaltonen K, Amini R-M, Heikkilä P, Aittomäki K, Tamminen A, Nevanlinna H, Blomqvist C (2009) High cyclin b1 expression is associated with poor survival in breast cancer. Br J Cancer 100:1055–1060

Abou-Bakr AA, Eldweny HI (2013) P16 expression correlates with basal-like triple-negative breast carcinoma. Ecancermedicalscience 7:317

Abubakar M, Guo C, Koka H, Sung H, Shao N, Guida J, Deng J, Li M, Hu N, Zhou B (2019) Clinicopathological and epidemiological significance of breast cancer subtype reclassification based on p53 immunohistochemical expression. NPJ Breast Cancer 5:1–9

Ahnström M, Nordenskjöld B, Rutqvist LE, Skoog L, Stål O (2005) Role of cyclin d1 in erbb2-positive breast cancer and tamoxifen resistance. Breast Cancer Res Treat 91:145–151

Al-Joudi FS, Iskandar ZA, Rusli J (2008) The expression of p53 in invasive ductal carcinoma of the breast: a study in the north-east states of Malaysia. Med J Malays 63:96–99

Albanese C, Johnson J, Watanabe G, Eklund N, Vu D, Arnold A, Pestell RG (1995) Transforming p21ras mutants and c-ets-2 activate the cyclin d1 promoter through distinguishable regions (∗). J Biol Chem 270:23589–23597

Alkarain A, Jordan R, Slingerland J (2004) P27 deregulation in breast cancer: prognostic significance and implications for therapy. J Mammary Gland Biol Neoplasia 9:67–80

Alle KM, Henshall SM, Field AS, Sutherland RL (1998) Cyclin d1 protein is overexpressed in hyperplasia and intraductal carcinoma of the breast. Clin Cancer Res 4:847–854

Androic I, Krämer A, Yan R, Rödel F, Gätje R, Kaufmann M, Strebhardt K, Yuan J (2008) Targeting cyclin b1 inhibits proliferation and sensitizes breast cancer cells to taxol. BMC Cancer 8:1–11

Arima Y, Hayashi N, Hayashi H, Sasaki M, Kai K, Sugihara E, Abe E, Yoshida A, Mikami S, Nakamura S (2012) Loss of p16 expression is associated with the stem cell characteristics of surface markers and therapeutic resistance in estrogen receptor-negative breast cancer. Int J Cancer 130:2568–2579

Arnold A, Papanikolaou A (2005) Cyclin d1 in breast cancer pathogenesis. J Clin Oncol 23:4215–4224

Asghar U, Witkiewicz AK, Turner NC, Knudsen ES (2015) The history and future of targeting cyclin-dependent kinases in cancer therapy. Nat Rev Drug Discov 14:130–146

Assem M, Youssef EA, Rashad RM, Yahia MA-H (2017) Immunohistochemical expression of cyclin d1 in invasive ductal carcinoma of human breast. Onco Targets Ther 2:80–87

Ates C, Sevinçli S, Pohl T (2011) Electromagnetically induced transparency in strongly interacting Rydberg gases. Phys Rev A 83:041802

Bae SY, Lee JH, Bae JW, Jung SP (2020) Differences in prognosis by p53 expression after neoadjuvant chemotherapy in triple-negative breast cancer. Ann Surg Treat Res 98:291

Baker SJ, Reddy EP (2012) Cdk4: a key player in the cell cycle, development, and cancer. Genes Cancer 3:658–669

Baldini E, Camerini A, Sgambato A, Prochilo T, Capodanno A, Pasqualetti F, Orlandini C, Resta L, Bevilacqua G, Collecchi P (2006) Cyclin A and E2F1 overexpression correlate with reduced disease-free survival in node-negative breast cancer patients. Anticancer Res 26:4415–4421

Barnes A, Pinder SE, Bell JA, Paish EC, Wencyk PM, Robertson JFR, Elston CW, Ellis IO (2003) Expression of p27kip1 in breast cancer and its prognostic significance. J Pathol 201:451–459
Barnes DM, Gillett CE (1998) Cyclin d1 in breast cancer. Breast Cancer Res Treat 52:1–15
Bertheau P, Lehmann-Che J, Varna M, Dumay A, Poirot B, Porcher R, Turpin E, Plassa L-F, De Roquancourt A, Bourstyn E (2013) P53 in breast cancer subtypes and new insights into response to chemotherapy. Breast 22:s27–s29
Boström P, Söderström M, Palokangas T, Vahlberg T, Collan Y, Carpen O, Hirsimäki P (2009) Analysis of cyclins a, b1, d1 and e in breast cancer in relation to tumour grade and other prognostic factors. BMC Res Notes 2:1–8
Brown I, Shalli K, McDonald SL, Moir SE, Hutcheon AW, Heys SD, Schofield AC (2004) Reduced expression of p27 is a novel mechanism of docetaxel resistance in breast cancer cells. Breast Cancer Res 6:1–7
Buckley MF, Sweeney KJ, Hamilton JA, Sini RL, Manning DL, Nicholson RI, Watts CK, Musgrove EA, Sutherland RL (1993) Expression and amplification of cyclin genes in human breast cancer. Oncogene 8:2127–2133
Buolamwini JK (2000) Cell cycle molecular targets in novel anticancer drug discovery. Curr Pharm Des 6:379–392
Cao T, Xiao T, Huang G, Xu Y, Zhu JJ, Wang K, Ye W, Guan H, He J, Zheng D (2017) Cdk3, target of mir-4469, suppresses breast cancer metastasis via inhibiting Wnt/β-catenin pathway. Oncotarget 8:84917
Cariou S, Donovan JCH, Flanagan WM, Milic A, Bhattacharya N, Slingerland JM (2000) Down-regulation of p21waf1/cip1 or p27kip1 abrogates antiestrogen-mediated cell cycle arrest in human breast cancer cells. Proc Natl Acad Sci 97:9042–9046
Chen X, Danes C, Lowe M, Herliczek TW, Keyomarsi K (2000) Activation of the estrogen-signaling pathway by p21waf1/cip1 in estrogen receptor-negative breast cancer cells. J Natl Cancer Inst 92:1403–1413
Chu J-S, Huang C-S, Chang K-J (1999) P27 expression as a prognostic factor of breast cancer in Taiwan. Cancer Lett 141:123–130
Coates AS, Millar EKA, O'toole SA, Molloy TJ, Viale G, Goldhirsch A, Regan MM, Gelber RD, Sun Z, Castiglione-Gertsch M (2012) Prognostic interaction between expression of p53 and estrogen receptor in patients with node-negative breast cancer: results from IBCSG Trials VIII and IX. Breast Cancer Res 14:1–12
Coqueret O (2002) Linking cyclins to transcriptional control. Gene 299:35–55
Courjal F, Louason G, Speiser P, Katsaros D, Zeillinger R, Theillet C (1996) Cyclin gene amplification and overexpression in breast and ovarian cancers: evidence for the selection of cyclin d1 in breast and cyclin e in ovarian tumors. Int J Cancer 69:247–253
Crown J (2017) Cdk8: a new breast cancer target. Oncotarget 8:14269
Cui SP, Wang HL, Peng W, Liu HJ, Hou L, Zhang B (2012) Aberrant expression and correlative analysis of p16 in breast cancers. Beijing Da Xue Xue Bao Yi Xue Ban 44:755–759
Dai M, Al-Odaini AA, Arakelian A, Rabbani SA, Ali S, Lebrun J-J (2012) A novel function for p21cip1 and acetyltransferase p/caf as critical transcriptional regulators of tgfβ-mediated breast cancer cell migration and invasion. Breast Cancer Res 14:1–22
Dai M, Al-Odaini AA, Fils-Aimé N, Villatoro MA, Guo J, Arakelian A, Rabbani SA, Ali S, Lebrun JJ (2013) Cyclin d1 cooperates with p21 to regulate tgfβ-mediated breast cancer cell migration and tumor local invasion. Breast Cancer Res 15:1–14
Davidoff AM, Herndon JE 2nd, Glover NS, Kerns BJ, Pence JC, Iglehart JD, Marks JR (1991) Relation between p53 overexpression and established prognostic factors in breast cancer. Surgery 110:259–264
Del Re M, Bertolini I, Crucitta S, Fontanelli L, Rofi E, De Angelis C, Diodati L, Cavallero D, Gianfilippo G, Salvadori B (2019) Overexpression of tk1 and cdk9 in plasma-derived exosomes is associated with clinical resistance to cdk4/6 inhibitors in metastatic breast cancer patients. Breast Cancer Res Treat 178:57–62

Diehl JA, Zindy F, Sherr CJ (1997) Inhibition of cyclin d1 phosphorylation on threonine-286 prevents its rapid degradation via the ubiquitin-proteasome pathway. Genes Dev 11:957–972

Domagala W, Welcker M, Chosia M, Karbowniczek M, Harezga B, Bartkova J, Bartek J, Osborn M (2001) P21/waf1/cip1 expression in invasive ductal breast carcinoma: relationship to p53, proliferation rate, and survival at 5 years. Virchows Arch 439:132–140

Donnellan R, Kleinschmidt I, Chetty R (2001) Cyclin E immunoexpression in breast ductal carcinoma: pathologic correlations and prognostic implications. Hum Pathol 32:89–94

Dorand RD, Nthale J, Myers JT, Barkauskas DS, Avril S, Chirieleison SM, Pareek TK, Abbott DW, Stearns DS, Letterio JJ (2016) Cdk5 disruption attenuates tumor pd-l1 expression and promotes antitumor immunity. Science 353:399–403

Dorée M, Galas S (1994) The cyclin-dependent protein kinases and the control of cell division. FASEB J 8:1114–1121

Dukelow T, Kishan D, Khasraw M, Murphy CG (2015) Cdk4/6 inhibitors in breast cancer. Anti-Cancer Drugs 26:797–806

Elsheikh S, Green AR, Aleskandarany MA, Grainge M, Paish CE, Lambros MBK, Reis-Filho JS, Ellis IO (2008) CCND1 amplification and cyclin D1 expression in breast cancer and their relation with proteomic subgroups and patient outcome. Breast Cancer Res Treat 109:325–335

Emig R, Magener A, Ehemann V, Meyer A, Stilgenbauer F, Volkmann M, Wallwiener D, Sinn HP (1998) Aberrant cytoplasmic expression of the p16 protein in breast cancer is associated with accelerated tumour proliferation. Br J Cancer 78:1661–1668

Fayed YM, El-Sheikh SA-M, Abulkheir ILH, Khorshed INA-H (2012) Immunohistochemical expression of p27 in ductal carcinoma of breast and its correlation with her2/neu expression and hormonal status. Egypt J Pathol 32:33–41

Fedele P, Sanna V, Fancellu A, Cinieri S (2019) A clinical evaluation of treatments that target cell cycle machinery in breast cancer. Expert Opin Pharmacother 20:2305–2315

Filipits M, Rudas M, Heinzl H, Jakesz R, Kubista E, Lax S, Schippinger W, Dietze O, Greil R, Stiglbauer W (2009) Low p27 expression predicts early relapse and death in postmenopausal hormone receptor–positive breast cancer patients receiving adjuvant tamoxifen therapy. Clin Cancer Res 15:5888–5894

Firestein R, Bass AJ, Kim SY, Dunn IF, Silver SJ, Guney I, Freed E, Ligon AH, Vena N, Ogino S (2008) Cdk8 is a colorectal cancer oncogene that regulates β-catenin activity. Nature 455:547–551

Foster JS, Henley DC, Bukovsky A, Seth P, Wimalasena J (2001) Multifaceted regulation of cell cycle progression by estrogen: regulation of Cdk inhibitors and Cdc25A independent of cyclin d1-Cdk4 function. Mol Cell Biol 21:794–810

Fredersdorf S, Burns J, Milne AM, Packham G, Fallis L, Gillett CE, Royds JA, Peston D, Hall PA, Hanby AM (1997) High level expression of p27kip1 and cyclin d1 in some human breast cancer cells: inverse correlation between the expression of p27kip1 and degree of malignancy in human breast and colorectal cancers. Proc Natl Acad Sci 94:6380–6385

Fritah A, Saucier C, Mester J, Redeuilh G, Sabbah M (2005) P21 waf1/cip1 selectively controls the transcriptional activity of estrogen receptor α. Mol Cell Biol 25:2419–2430

Geradts J, Wilson PA (1996) High frequency of aberrant p16(ink4a) expression in human breast cancer. Am J Pathol 149:15–20

Göhring U-J, Schöndorf T, Kiecker VR, Becker M, Kurbacher C, Scharl A (1999) Immunohistochemical detection of H-ras protooncoprotein p21 indicates favorable prognosis in node-negative breast cancer patients. Tumor Biol 20:173–183

Graña X, Reddy EP (1995) Cell cycle control in mammalian cells: role of cyclins, cyclin dependent kinases (CDKs), growth suppressor genes and cyclin-dependent kinase inhibitors (CKIs). Oncogene 11:211–220

Guan X, Wang Y, Xie R, Chen L, Bai J, Lu J, Kuo MT (2010) P27kip1 as a prognostic factor in breast cancer: a systematic review and meta-analysis. J Cell Mol Med 14:944–953

Guen VJ, Gamble C, Lees JA, Colas P (2017) The awakening of the cdk10/cyclin m protein kinase. Oncotarget 8:50174

Hall M, Peters G (1996) Genetic alterations of cyclins, cyclin-dependent kinases, and cdk inhibitors in human cancer. Adv Cancer Res 68:67–108

Han S, Park K, Bae BN, Kim KH, Kim HJ, Kim YD, Kim HY (2003) Prognostic implication of cyclin e expression and its relationship with cyclin d1 and p27kip1 expression on tissue microarrays of node negative breast cancer. J Surg Oncol 83:241–247

Harbhajanka A, Lamzabi I, Bitterman P, Reddy VB, Ghai R, Gattuso P (2019) Correlation of p16 expression on cancer and stromal cells with clinicopathologic and immunohistochemical features of lobular breast carcinoma. Appl Immunohistochem Mol Morphol 27:658–662

Heiss C, Schanz A, Amabile N, Jahn S, Chen Q, Wong ML, Rassaf T, Heinen Y, Cortese-Krott M, Grossman W (2010) Nitric oxide synthase expression and functional response to nitric oxide are both important modulators of circulating angiogenic cell response to angiogenic stimuli. Arterioscler Thromb Vasc Biol 30:2212–2218

Hodges LC, Cook JD, Lobenhofer EK, Li L, Bennett L, Bushel PR, Aldaz CM, Afshari CA, Walker CL (2003) Tamoxifen functions as a molecular agonist inducing cell cycle-associated genes in breast cancer cells. Mol Cancer Res 1:300–311

Horiuchi D, Kusdra L, Huskey NE, Chandriani S, Lenburg ME, Gonzalez-Angulo AM, Creasman KJ, Bazarov AV, Smyth JW, Davis SE (2012) MYC pathway activation in triple-negative breast cancer is synthetic lethal with CDK inhibition. J Exp Med 209:679–696

Husdal A, Bukholm G, Bukholm IRK (2006) The prognostic value and overexpression of cyclin A is correlated with gene amplification of both cyclin A and cyclin E in breast cancer patient. Anal Cell Pathol 28:107–116

Hwang HC, Clurman BE (2005) Cyclin E in normal and neoplastic cell cycles. Oncogene 24:2776–2786

Jabbour-Leung NA, Chen X, Bui T, Jiang Y, Yang D, Vijayaraghavan S, McArthur MJ, Hunt KK, Keyomarsi K (2016) Sequential combination therapy of cdk inhibition and doxorubicin is synthetically lethal in p53-mutant triple-negative breast cancer. Mol Cancer Ther 15:593–607

Jirström K, Stendahl M, Rydén L, Kronblad Å, Bendahl P-O, Stål O, Landberg G (2005) Adverse effect of adjuvant tamoxifen in premenopausal breast cancer with cyclin d1 gene amplification. Cancer Res 65:8009–8016

Joung YH, Lim EJ, Lee MY, Park J-H, Ye S-K, Park EU, Kim SY, Zhang Z, Lee KJ, Park DK (2005) Hypoxia activates the cyclin d1 promoter via the jak2/stat5b pathway in breast cancer cells. Exp Mol Med 37:353–364

Kato JY, Matsushime H, Hiebert SW, Ewen ME, Sherr CJ (1993) Direct binding of cyclin d to the retinoblastoma gene product (pRb) and pRb phosphorylation by the cyclin D-dependent kinase CDK4. Genes Dev 7:331–342

Kenny FS, Hui R, Musgrove EA, Gee JMW, Blamey RW, Nicholson RI, Sutherland RL, Robertson JFR (1999) Overexpression of cyclin d1 messenger rna predicts for poor prognosis in estrogen receptor-positive breast cancer. Clin Cancer Res 5:2069–2076

Keyomarsi K, Tucker SL, Buchholz TA, Callister M, Ding YE, Hortobagyi GN, Bedrosian I, Knickerbocker C, Toyofuku W, Lowe M (2002) Cyclin E and survival in patients with breast cancer. N Engl J Med 347:1566–1575

Khan N, Syed DN, Ahmad N, Mukhtar H (2013) Fisetin: a dietary antioxidant for health promotion. Antioxid Redox Signal 19:151–162

Khan SU, Fatima K, Malik F (2022a) Understanding the cell survival mechanism of anoikis-resistant cancer cells during different steps of metastasis. Clin Exp Metastasis 39:715

Khan SU, Pathania AS, Wani A, Fatima K, Mintoo MJ, Hamza B, Paddar MA, Bhumika W, Anand LK, Maqbool MS (2022b) Activation of lysosomal mediated cell death in the course of autophagy by mtorc1 inhibitor. Sci Rep 12:1–13

Kikuchi S, Nishimura R, Osako T, Okumura Y, Nishiyama Y, Toyozumi Y, Arima N (2013) Definition of p53 overexpression and its association with the clinicopathological features in luminal/her2-negative breast cancer. Anticancer Res 33:3891–3897

Kilker RL, Hartl MW, Rutherford TM, Planas-Silva MD (2004) Cyclin d1 expression is dependent on estrogen receptor function in tamoxifen-resistant breast cancer cells. J Steroid Biochem Mol Biol 92:63–71

Kilker RL, Planas-Silva MD (2006) Cyclin d1 is necessary for tamoxifen-induced cell cycle progression in human breast cancer cells. Cancer Res 66:11478–11484

Kim J-Y, Park K, Jung HH, Lee E, Cho EY, Lee KH, Bae SY, Lee SK, Kim SW, Lee JE (2016) Association between mutation and expression of tp53 as a potential prognostic marker of triple-negative breast cancer. Cancer Res Treat 48:1338

Lara JF, Thor AD, Dressler LG, Broadwater G, Bleiweiss IJ, Edgerton S, Cowan D, Goldstein LJ, Martino S, Ingle JN (2011) P53 expression in node-positive breast cancer patients: results from the cancer and leukemia group B 9344 trial (159905). Clin Cancer Res 17:5170–5178

Lebwohl DE, Muise-Helmericks R, Sepp-Lorenzino L, Serve S, Timaul M, Bol R, Borgen P, Rosen N (1994) A truncated cyclin D1 gene encodes a stable mRNA in a human breast cancer cell line. Oncogene 9:1925–1929

Lee D-S, Kim SH, Suh YJ, Kim S, Kim HK, Shim BY (2011) Clinical implication of p53 overexpression in breast cancer patients younger than 50 years with a triple-negative subtype who undergo a modified radical mastectomy. Jpn J Clin Oncol 41:854–866

Lee RJ, Albanese C, Stenger RJ, Watanabe G, Inghirami G, Haines GK, Webster M, Muller WJ, Brugge JS, Davis RJ (1999) pp60v-src induction of cyclin d1 requires collaborative interactions between the extracellular signal-regulated kinase, p38, and Jun kinase pathways: a role for camp response element-binding protein and activating transcription factor-2 in pp60v-src signaling in breast cancer cells. J Biol Chem 274:7341–7350

Leivonen M, Nordling S, Lundin J, Von Boguslawski K, Haglund C (2001) P27 expression correlates with short-term, but not with long-term prognosis in breast cancer. Breast Cancer Res Treat 67:15–22

Leng X, Connell-Crowley L, Goodrich D, Harper JW (1997) S-phase entry upon ectopic expression of g1 cyclin-dependent kinases in the absence of retinoblastoma protein phosphorylation. Curr Biol 7:709–712

Lengare PV, Sinai Khandeparkar SG, Joshi AR, Gogate BP, Solanke SG, Gore SH (2020) Immunohistochemical expression of cyclin d1 in invasive breast carcinoma and its correlation with clinicopathological parameters. Indian J Pathol Microbiol 63:376–381

Li B, Chonghaile TN, Fan Y, Madden SF, Klinger R, O'Connor AE, Walsh L, O'Hurley G, Udupi GM, Joseph J (2017) Therapeutic rationale to target highly expressed cdk7 conferring poor outcomes in triple-negative breast cancer. Cancer Res 77:3834–3845

Li J-P, Zhang X-M, Zhang Z, Zheng L-H, Jindal S, Liu Y-J (2019) Association of p53 expression with poor prognosis in patients with triple-negative breast invasive ductal carcinoma. Medicine 98:e15449

Li J, Poi MJ, Tsai M-D (2011) Regulatory mechanisms of tumor suppressor p16ink4a and their relevance to cancer. Biochemistry 50:5566–5582

Li Y, Dowbenko D, Lasky LA (2002) Akt/pkb phosphorylation of p21cip/waf1 enhances protein stability of p21cip/waf1 and promotes cell survival. J Biol Chem 277:11352–11361

Loden M, Stighall M, Nielsen NH, Roos G, Emdin SO, Östlund H, Landberg G (2002) The cyclin d1 high and cyclin E high subgroups of breast cancer: separate pathways in tumorogenesis based on pattern of genetic aberrations and inactivation of the prb node. Oncogene 21:4680–4690

Loo LWM, Gao C, Shvetsov YB, Okoro DR, Hernandez BY, Bargonetti J (2019) Mdm2, mdm2-c, and mutant p53 expression influence breast cancer survival in a multiethnic population. Breast Cancer Res Treat 174:257–269

Malumbres M, Barbacid M (2001) To cycle or not to cycle: a critical decision in cancer. Nat Rev Cancer 1:222–231

Marais A, Ji Z, Child ES, Krause E, Mann DJ, Sharrocks AD (2010) Cell cycle-dependent regulation of the forkhead transcription factor foxk2 by cdk· cyclin complexes. J Biol Chem 285:35728–35739

Marchetti P, Cannita K, Ricevuto E, De Galitiis F, Di Rocco ZC, Tessitore A, Bisegna R, Porzio G, De Rubeis GP, Ventura T (2003) Prognostic value of p53 molecular status in high-risk primary breast cancer. Ann Oncol 14:704–708

Matutino A, Amaro C, Verma S (2018) Cdk4/6 inhibitors in breast cancer: beyond hormone receptor-positive her2-negative disease. Ther Adv Med Oncol 10:1758835918818346

Mehraj U, Dar AH, Wani NA, Mir MA (2021) Tumor microenvironment promotes breast cancer chemoresistance. Cancer Chemother Pharmacol 87:147–158

Mehraj U, Aisha S, Sofi S, Mir MA (2022a) Expression pattern and prognostic significance of baculoviral inhibitor of apoptosis repeat-containing 5 (birc5) in breast cancer: a comprehensive analysis. Adv Cancer Biol-Metastasis 4:100037

Mehraj U, Alshehri B, Khan AA, Bhat AA, Bagga P, Wani NA, Mir MA (2022b) Expression pattern and prognostic significance of chemokines in breast cancer: an integrated bioinformatics analysis. Clin Breast Cancer 22:567

Mehraj U, Qayoom H, Sofi S, Farhana P, Asdaq SMB, Mir MA (2022c) Cryptolepine targets top2a and inhibits tumor cell proliferation in breast cancer cells-an in vitro and in silico study. Anti Cancer Agents Med Chem 22:3025

Mehraj U, Sofi S, Alshehri B, Mir MA (2022d) Expression pattern and prognostic significance of cdks in breast cancer: an integrated bioinformatic study. Cancer Biomark 34:505–519

Mehta P, Nema S, Narang S (2019) Role of p53 and ki-67 in prognostication of carcinoma breast. Indian J Pathol Oncol 6:261–265

Michalides R, Van Tinteren H, Balkenende A, Vermorken JB, Benraadt J, Huldij J, Van Diest P (2002) Cyclin a is a prognostic indicator in early stage breast cancer with and without tamoxifen treatment. Br J Cancer 86:402–408

Mir MA, Sofi S, Aisha S (2022a) Role of cancer-associated fibroblasts in tumor microenvironment, chapter-3. In: Role of tumor microenvironment in breast cancer and targeted therapies. Elsevier, San Diego, pp 59–86; ISBN 978-0-443-18696-7. https://doi.org/10.1016/b978-0-443-18696-7.00002-6

Mir MA, Mir AY, Jan U, Dar MA, Shah MZUH (2022b) Role of cancer-associated fibroblasts in tumor microenvironment, chapter-4. In: Role of tumor microenvironment in breast cancer and targeted therapies. Elsevier, San Diego, pp 87–112; ISBN 978-0-443-18696-7. https://doi.org/10.1016/b978-0-443-18696-7.00004-x

Mir MA, Mir AY (2022) Role of regulatory t cells in cancer, chapter-5. In: Role of tumor microenvironment in breast cancer and targeted therapies. Elsevier, San Diego, pp 113–136; ISBN 978-0-443-18696-7. https://doi.org/10.1016/b978-0-443-18696-7.00001-4

Mir MA, Mir AY, Mushtaq T (2022c) Role of tumor-associated macrophages in the breast tumor microenvironment, chapter-6. In: Role of tumor microenvironment in breast cancer and targeted therapies. Elsevier, San Diego, pp 137–170; ISBN 978-0-443-18696-7. https://doi.org/10.1016/b978-0-443-18696-7.00003-8

Mir MA, Aisha S, Sofi S (2022d) Introduction to various types of cancers, chapter-1. In: Role of tumor microenvironment in breast cancer and targeted therapies. Elsevier, San Diego, pp 1–30; ISBN 978-0-443-18696-7. https://doi.org/10.1016/b978-0-443-18696-7.00010-5

Mir MA, Mehraj U (2019) Double-crosser of the immune system: macrophages in tumor progression and metastasis. Curr Immunol Rev 15:172–184

Mir MA, Qayoom H, Mehraj U, Nisar S, Bhat B, Wani NA (2020) Targeting different pathways using novel combination therapy in triple negative breast cancer. Curr Cancer Drug Targets 20: 586–602

Mir MA, Aisha S, Sofi S, Rasheid S (2022e) The tumor microenvironment, chapter-2. In: Role of tumor microenvironment in breast cancer and targeted therapies. Elsevier, San Diego, pp 31–58; ISBN 978-0-443-18696-7. https://doi.org/10.1016/b978-0-443-18696-7.00007-5

Mir M, Jan S, Mehraj U, Qayoom H, Nisar S (2022f) Immuno-onco-metabolism and therapeutic resistance. In: Immuno-oncology crosstalk and metabolism. Springer, Singapore, pp 45–89; ISBN 9789811662263. https://link.springer.com/chapter/10.1007/978-981-16-6226-3_3

Mir MA, Aisha S, Mehraj U (2022g) Chapter 3 - Current therapeutics and treatment options in TNBC. In: Mir M (ed) Combinational therapy in triple negative breast cancer. Academic press, London, pp 61–94. ISBN: 978-0-323-96136-3. https://doi.org/10.1016/B978-0-323-96136-3.00007-8

Mirchandani D, Roses DF, Inghirami G, Zeleniuch-Jacquotte A, Cangiarella J, Guth A, Safyan RA, Formenti SC, Pagano M, Muggia F (2011) Loss of p27kip1 expression in fully-staged node-negative breast cancer: association with lack of hormone receptors in t1a/b, but not t1c infiltrative ductal carcinoma. Anticancer Res 31:4401–4405

Mohammadizadeh F, Hani M, Ranaee M, Bagheri M (2013) Role of cyclin d1 in breast carcinoma. J Res Med Sci 18:1021

Morgan DO (1997) Cyclin-dependent kinases: engines, clocks, and microprocessors. Annu Rev Cell Dev Biol 13:261

Muhammad EM, Ahmad AN, Guirguis MN, Ali A-EM (2012) Immunohistochemical p53 expression in breast carcinoma with correlation to clinico-pathological parameters. Med J Cairo Univ 80:179–189

Nahta R, Takahashi T, Ueno NT, Hung M-C, Esteva FJ (2004) P27kip1 down-regulation is associated with trastuzumab resistance in breast cancer cells. Cancer Res 64:3981–3986

Naidoo K, Wai PT, Maguire SL, Daley F, Haider S, Kriplani D, Campbell J, Mirza H, Grigoriadis A, Tutt A (2018) Evaluation of cdk12 protein expression as a potential novel biomarker for DNA damage response–targeted therapies in breast cancer. Mol Cancer Ther 17:306–315

Neganova I, Lako M (2008) G1 to s phase cell cycle transition in somatic and embryonic stem cells. J Anat 213:30–44

Newman L, Xia W, Yang HY, Sahin A, Bondy M, Lukmanji F, Hung MC, Lee MH (2001) Correlation of p27 protein expression with her-2/neu expression in breast cancer. Mol Carcinog 30:169–175

Nie L, Wei Y, Zhang F, Hsu Y-H, Chan L-C, Xia W, Ke B, Zhu C, Deng R, Tang J (2019) Cdk2-mediated site-specific phosphorylation of ezh2 drives and maintains triple-negative breast cancer. Nat Commun 10:1–15

Nielsen NH, Arnerlöv C, Emdin SO, Landberg G (1996) Cyclin E overexpression, a negative prognostic factor in breast cancer with strong correlation to oestrogen receptor status. Br J Cancer 74:874–880

Niméus-Malmström E, Koliadi A, Ahlin C, Holmqvist M, Holmberg L, Amini RM, Jirström K, Wärnberg F, Blomqvist C, Fernö M (2010) Cyclin b1 is a prognostic proliferation marker with a high reproducibility in a population-based lymph node negative breast cancer cohort. Int J Cancer 127:961–967

Nohara T, Ryo T, Iwamoto S, Gon G, Tanigawa N (2001) Expression of cell-cycle regulator p27 is correlated to the prognosis and er expression in breast carcinoma patients. Oncology 60:94–100

Oyama T, Kashiwabara K, Yoshimoto K, Arnold A, Koerner F (1998) Frequent overexpression of the cyclin d1 oncogene in invasive lobular carcinoma of the breast. Cancer Res 58:2876–2880

Pan Y, Yuan Y, Liu G, Wei Y (2017) P53 and ki-67 as prognostic markers in triple-negative breast cancer patients. PLoS One 12:e0172324

Patil VW, Singhai R, Patil AV, Gurav PD (2011) Triple-negative (ER, PgR, HER-2/neu) breast cancer in Indian women. Breast Cancer 3:9

Pérez-Tenorio G, Berglund F, Esguerra Merca A, Nordenskjöld B, Rutqvist LE, Skoog L, Stål O (2006) Cytoplasmic p21waf1/cip1 correlates with akt activation and poor response to tamoxifen in breast cancer. Int J Oncol 28:1031–1042

Pestell RG, Albanese C, Reutens AT, Segall JE, Lee RJ, Arnold A (1999) The cyclins and cyclin-dependent kinase inhibitors in hormonal regulation of proliferation and differentiation. Endocr Rev 20:501–534

Pines J (1995) Cyclins and cyclin-dependent kinases: theme and variations. Adv Cancer Res 66: 181–212

Pohl G, Rudas M, Dietze O, Lax S, Markis E, Pirker R, Zielinski CC, Hausmaninger H, Kubista E, Samonigg H (2003) High p27kip1 expression predicts superior relapse-free and overall survival for premenopausal women with early-stage breast cancer receiving adjuvant treatment with tamoxifen plus goserelin. J Clin Oncol 21:3594–3600

Poikonen P, Sjöström J, Amini RM, Villman K, Ahlgren J, Blomqvist C (2005) Cyclin a as a marker for prognosis and chemotherapy response in advanced breast cancer. Br J Cancer 93: 515–519

Porter PL, Barlow WE, Yeh IT, Lin MG, Yuan XP, Donato E, Sledge GW, Shapiro CL, Ingle JN, Haskell CM (2006) P27 kip1 and cyclin E expression and breast cancer survival after treatment with adjuvant chemotherapy. J Natl Cancer Inst 98:1723–1731

Potemski P, Kusińska R, Pasz-Walczak G, Piekarski JH, Watała C, Płuciennik E, Bednarek AK, Kordek R (2009) Prognostic relevance of cyclin E expression in operable breast cancer. Med Sci Monit 15:40

Potemski P, Kusinska R, Watala C, Pluciennik E, Bednarek AK, Kordek R (2006) Cyclin e expression in breast cancer correlates with negative steroid receptor status, her2 expression, tumor grade and proliferation. Carcinogenesis 6:7

Pozo K, Castro-Rivera E, Tan C, Plattner F, Schwach G, Siegl V, Meyer D, Guo A, Gundara J, Mettlach G (2013) The role of cdk5 in neuroendocrine thyroid cancer. Cancer Cell 24:499–511

Qian X, Hulit J, Suyama K, Eugenin EA, Belbin TJ, Loudig O, Smirnova T, Zhou ZN, Segall J, Locker J (2013) P21cip1 mediates reciprocal switching between proliferation and invasion during metastasis. Oncogene 32:2292–2303

Ravazoula P, Batistatou A, Aletra C, Ladopoulos J, Kourounis G, Tzigounis B (2003) Immunohistochemical expression of glucose transporter glut 1 and cyclin d1 in breast carcinomas with negatine lymph nodes. Eur J Gynaecol Oncol 24:544–546

Ravikumar G, Ananthamurthy A (2014) Cyclin d1 expression in ductal carcinoma of the breast and its correlation with other prognostic parameters. J Cancer Res Ther 10:671

Robinson TJW, Liu JC, Vizeacoumar F, Sun T, Maclean N, Egan SE, Schimmer AD, Datti A, Zacksenhaus E (2013) Rb1 status in triple negative breast cancer cells dictates response to radiation treatment and selective therapeutic drugs. PLoS One 8:e78641

Rosendahl AH, Perks CM, Zeng L, Markkula A, Simonsson M, Rose C, Ingvar C, Holly JMP, Jernström H (2015) Caffeine and caffeic acid inhibit growth and modify estrogen receptor and insulin-like growth factor I receptor levels in human breast cancer. Clin Cancer Res 21:1877–1887

Rudas M, Lehnert M, Huynh A, Jakesz R, Singer C, Lax S, Schippinger W, Dietze O, Greil R, Stiglbauer W (2008) Cyclin d1 expression in breast cancer patients receiving adjuvant tamoxifen-based therapy. Clin Cancer Res 14:1767–1774

Russell A, Thompson MA, Hendley J, Trute L, Armes J, Germain D (1999) Cyclin d1 and d3 associate with the SCF complex and are coordinately elevated in breast cancer. Oncogene 18: 1983–1991

Salmani H, Hosseini A, Azarnezhad A, Ahmad H (2018) Pten and p53 gene expressions in breast cancer specimens and their clinicopathological significance. Middle East J Cancer 9:105–111

Santo L, Siu KT, Raje N (2015) Targeting cyclin-dependent kinases and cell cycle progression in human cancers. Semin Oncol 42:788–800

Schlafstein AJ, Withers AE, Rudra S, Danelia D, Switchenko JM, Mister D, Harari S, Zhang H, Daddacha W, Ehdaivand S (2018) Cdk9 expression shows role as a potential prognostic biomarker in breast cancer patients who fail to achieve pathologic complete response after neoadjuvant chemotherapy. Int J Breast Cancer 2018:6945129

Shan M, Zhang X, Liu X, Qin Y, Liu T, Liu Y, Wang J, Zhong Z, Zhang Y, Geng J (2013) P16 and p53 play distinct roles in different subtypes of breast cancer. PLoS One 8:e76408

Sherr CJ (1996) Cancer cell cycles. Science 274:1672–1677

Sherr CJ (2000) The pezcoller lecture: cancer cell cycles revisited. Cancer Res 60:3689–3695

Sieuwerts AM, Look MP, Meijer-Van Gelder ME, Timmermans M, Trapman AMAC, Garcia RR, Arnold M, Goedheer AJW, De Weerd V, Portengen H (2006) Which cyclin e prevails as

prognostic marker for breast cancer? Results from a retrospective study involving 635 lymph node–negative breast cancer patients. Clin Cancer Res 12:3319–3328
Simpson JF, Quan DE, O'Malley F, Odom-Maryon T, Clarke PE (1997) Amplification of ccnd1 and expression of its protein product, cyclin d1, in ductal carcinoma in situ of the breast. Am J Pathol 151:161
Singh U, Chashoo G, Khan SU, Mahajan P, Nargotra A, Mahajan G, Singh A, Sharma A, Mintoo MJ, Guru SK (2017) Design of novel 3-pyrimidinylazaindole cdk2/9 inhibitors with potent in vitro and in vivo antitumor efficacy in a triple-negative breast cancer model. J Med Chem 60: 9470–9489
Sofi S, Mehraj U, Qayoom H, Aisha S, Almilaibary A, Alkhanani M, Mir MA (2022a) Targeting cyclin-dependent kinase 1 (cdk1) in cancer: molecular docking and dynamic simulations of potential cdk1 inhibitors. Med Oncol 39:1–15
Sofi S, Mehraj U, Qayoom H, Aisha S, Asdaq SMB, Almilaibary A, Mir MA (2022b) Cyclin-dependent kinases in breast cancer: expression pattern and therapeutic implications. Med Oncol 39:1–16
Sofi S, Mehraj U, Qayoom H, Aisha S, Asdaq SMB, Almilaibary A, Mir MA (2022c) Cyclin-dependent kinases in breast cancer: expression pattern and therapeutic implications. Med Oncol 39:106
Sohn D, Essmann F, Schulze-Osthoff K, Jänicke RU (2006) P21 blocks irradiation-induced apoptosis downstream of mitochondria by inhibition of cyclin-dependent kinase–mediated caspase-9 activation. Cancer Res 66:11254–11262
Solomon MJ (1993) Activation of the various cyclin/cdc2 protein kinases. Curr Opin Cell Biol 5: 180–186
Spataro VJ, Litman H, Viale G, Maffini F, Masullo M, Golouh R, Martinez-Tello FJ, Grigolato P, Shilkin KB, Gusterson BA (2003) Decreased immunoreactivity for p27 protein in patients with early-stage breast carcinoma is correlated with her-2/neu overexpression and with benefit from one course of perioperative chemotherapy in patients with negative lymph node status: results from international breast cancer study group trial v. Cancer 97:1591–1600
Spring LM, Zangardi ML, Moy B, Bardia A (2017) Clinical management of potential toxicities and drug interactions related to cyclin-dependent kinase 4/6 inhibitors in breast cancer: practical considerations and recommendations. Oncologist 22:1039–1048
Stemke-Hale K, Gonzalez-Angulo AM, Lluch A, Neve RM, Kuo W-L, Davies M, Carey M, Hu Z, Guan Y, Sahin A (2008) An integrative genomic and proteomic analysis of pik3ca, pten, and akt mutations in breast cancer. Cancer Res 68:6084–6091
Sun X, Zhangyuan G, Shi L, Wang Y, Sun B, Ding Q (2017) Prognostic and clinicopathological significance of cyclin b expression in patients with breast cancer: a meta-analysis. Medicine 96: e6860
Sutherland RL, Musgrove EA (2004) Cyclins and breast cancer. J Mammary Gland Biol Neoplasia 9:95–104
Tassan J-P, Schultz SJ, Bartek J, Nigg EA (1994) Cell cycle analysis of the activity, subcellular localization, and subunit composition of human cak (cdk-activating kinase). J Cell Biol 127: 467–478
Thu KL, Soria-Bretones I, Mak TW, Cescon DW (2018) Targeting the cell cycle in breast cancer: towards the next phase. Cell Cycle 17:1871–1885
Tobin NP, Bergh J (2012) Analysis of cyclin d1 in breast cancer: a call to arms. Curr Breast Cancer Rep 4:171–173
Traub F, Mengel M, Lück HJ, Kreipe HH, Von Wasielewski R (2006) Prognostic impact of skp2 and p27 in human breast cancer. Breast Cancer Res Treat 99:185–191
Tsuchiya A, Zhang GJ, Kanno M (1999) Prognostic impact of cyclin-dependent kinase inhibitor p27kip1 in node-positive breast cancer. J Surg Oncol 70:230–234
Umekita Y, Ohi Y, Sagara Y, Yoshida H (2002) Overexpression of cyclind1 predicts for poor prognosis in estrogen receptor-negative breast cancer patients. Int J Cancer 98:415–418

Velasco-Velázquez MA, Li Z, Casimiro M, Loro E, Homsi N, Pestell RG (2011) Examining the role of cyclin d1 in breast cancer. Future Oncol 7:753–765

Vincent AJ, Ren S, Harris LG, Devine DJ, Samant RS, Fodstad O, Shevde LA (2012) Cytoplasmic translocation of p21 mediates nupr1-induced chemoresistance: nupr1 and p21 in chemoresistance. FEBS Lett 586:3429–3434

Wadhwa B, Paddar M, Khan S, Mir S, Clarke PA, Grabowska AM, Vijay DG, Malik F (2020) Akt isoforms have discrete expression in triple negative breast cancers and roles in cisplatin sensitivity. Oncotarget 11:4178

Wang C, Li Z, Lu Y, Du R, Katiyar S, Yang J, Fu M, Leader JE, Quong A, Novikoff PM (2006) Cyclin d1 repression of nuclear respiratory factor 1 integrates nuclear DNA synthesis and mitochondrial function. Proc Natl Acad Sci 103:11567–11572

Wang L, Wang R, Ye Z, Wang Y, Li X, Chen W, Zhang M, Cai C (2018) Pvt1 affects EMT and cell proliferation and migration via regulating p21 in triple-negative breast cancer cells cultured with mature adipogenic medium. Acta Biochim Biophys Sin 50:1211–1218

Wang TC, Cardiff RD, Zukerberg L, Lees E, Arnold A, Schmidt EV (1994) Mammary hyperplasia and carcinoma in mmtv-cyclin d1 transgenic mice. Nature 369:669–671

Wang Y, Zhang T, Kwiatkowski N, Abraham BJ, Lee TI, Xie S, Yuzugullu H, Von T, Li H, Lin Z (2015) Cdk7-dependent transcriptional addiction in triple-negative breast cancer. Cell 163:174–186

Wei C-Y, Tan Q-X, Zhu X, Qin Q-H, Zhu F-B, Mo Q-G, Yang W-P (2015) Expression of cdkn1a/p21 and tgfbr2 in breast cancer and their prognostic significance. Int J Clin Exp Pathol 8:14619

Weinberg RA (1995) The retinoblastoma protein and cell cycle control. Cell 81:323–330

Westwick JK, Lee RJ, Lambert QT, Symons M, Pestell RG, Der CJ, Whitehead IP (1998) Transforming potential of Dbl family proteins correlates with transcription from the cyclin d1 promoter but not with activation of Jun NH2-terminal kinase, p38/mpk2, serum response factor, or c-jun. J Biol Chem 273:16739–16747

Wilcken NR, Prall OW, Musgrove EA, Sutherland RL (1997) Inducible overexpression of cyclin d1 in breast cancer cells reverses the growth-inhibitory effects of antiestrogens. Clin Cancer Res 3: 849–854

Winters ZE, Hunt NC, Bradburn MJ, Royds JA, Turley H, Harris AL, Norbury CJ (2001) Subcellular localisation of cyclin b, cdc2 and p21waf1/cip1 in breast cancer: association with prognosis. Eur J Cancer 37:2405–2412

Yamamoto M, Tamakawa S, Yoshie M, Yaginuma Y, Ogawa K (2006) Neoplastic hepatocyte growth associated with cyclin d1 redistribution from the cytoplasm to the nucleus in mouse hepatocarcinogenesis. Mol Carcinog 45:901–913

Yang P, Du CW, Kwan M, Liang SX, Zhang GJ (2013) The impact of p53 in predicting clinical outcome of breast cancer patients with visceral metastasis. Sci Rep 3:1–6

Yde CW, Issinger O-G (2006) Enhancing cisplatin sensitivity in mcf-7 human breast cancer cells by down-regulation of bcl-2 and cyclin d1. Int J Oncol 29:1397–1404

You Y, Li H, Qin X, Zhang Y, Song W, Ran Y, Gao F (2015) Decreased cdk10 expression correlates with lymph node metastasis and predicts poor outcome in breast cancer patients-a short report. Cell Oncol 38:485–491

Yu Z-K, Gervais JLM, Zhang H (1998) Human cul-1 associates with the skp1/skp2 complex and regulates p21cip1/waf1 and cyclin d proteins. Proc Natl Acad Sci 95:11324–11329

Yu Z, Wang L, Wang C, Ju X, Wang M, Chen K, Loro E, Li Z, Zhang Y, Wu K (2013) Cyclin d1 induction of dicer governs microrna processing and expression in breast cancer. Nat Commun 4: 1–10

Zhang P (1999) The cell cycle and development: redundant roles of cell cycle regulators. Curr Opin Cell Biol 11:655–662

Zhang S-Y, Caamano J, Cooper F, Guo X, Klein-Szanto AJP (1994) Immunohistochemistry of cyclin d1 in human breast cancer. Am J Clin Pathol 102:695–698

Zhang Z, Huang A, Zhang A, Zhou C (2017) Hur promotes breast cancer cell proliferation and survival via binding to cdk3 mrna. Biomed Pharmacother 91:788–795

Zhong Z, Yeow W-S, Zou C, Wassell R, Wang C, Pestell RG, Quong JN, Quong AA (2010) Cyclin d1/cyclin-dependent kinase 4 interacts with filamin A and affects the migration and invasion potential of breast cancer cells. Cancer Res 70:2105–2114

Zilfou JT, Lowe SW (2009) Tumor suppressive functions of p53. Cold Spring Harb Perspect Biol 1: a001883

Zukerberg LR, Yang W-I, Gadd M, Thor AD, Koerner FC, Schmidt EV, Arnold A (1995) Cyclin d1 (prad1) protein expression in breast cancer: approximately one-third of infiltrating mammary carcinomas show overexpression of the cyclin d1 oncogene. Mod Pathol 8:560–567

Zwijsen RM, Klompmaker R, Wientjens EB, Kristel PM, Van Der Burg B, Michalides RJ (1996) Cyclin d1 triggers autonomous growth of breast cancer cells by governing cell cycle exit. Mol Cell Biol 16:2554–2560

Molecular Subtypes of Breast Cancer and CDk Dysregulation

6

Manzoor Ahmad Mir and Ifshana Mohi Ud Din

6.1 Introduction

BC is a varying disease with a wide range of cell composition, molecular changes and clinical manifestations. The prognosis and response of cancer treatment is influenced by many characteristics which are: histological grade, tumour type and size, LN metastasis, ER, PR, and HER-2. BC is the most commonly seen cancer around the globe (Siegel et al. 2020). Approximately, 250,000 new BC cases were reported in the United States (US) in 2014, accounting for approximately 14% of all new cancer diagnoses (Economopoulou et al. 2015). In Great Britain, more than 50,000 cases of BC were confirmed, which is about 15% of all new cancer cases. BC is a broad term that includes a variety of disorders with distinct histological features, dissemination patterns, treatment responses, sufferers' outcomes and imaging features. BC affects females of all age groups, killing around 42,000 people in the USA in 2019 and dying from cancer in females, despite decades of research and advances in drug and diagnostic imaging (Khan et al. 2022). This is the second most usual cause. In spite of the fact that mortality due to BC has reasonably lowered as a result of currently accessible treatments, it is estimated that over 450,000 people die each year from the disease (Al-Rikabi and Husain 2012). The molecular subset of BC is a strong prognostic and predictor, depending on histological grade and LN metastasis. Therefore, categorising BC into molecular subtypes is a crucial part of treatment decisions. Long-established immunohistochemistry (IHC) markers like ER, HER-2, PR are important for molecular subtyping (Fig. 6.1) (Johnson et al. 2021). GE profiling using complementary DNA microarrays has emerged as a new tool for therapeutically essential molecular classification. The BC can be grouped

M. A. Mir (✉) · I. M. U. Din
Department of Bioresources, School of Biological Sciences, University of Kashmir, Srinagar, J&K, India
e-mail: drmanzoor@kashmiruniversity.ac.in

M. A. Mir (ed.), *Therapeutic Potential of Cell Cycle Kinases in Breast Cancer*,
https://doi.org/10.1007/978-981-19-8911-7_6

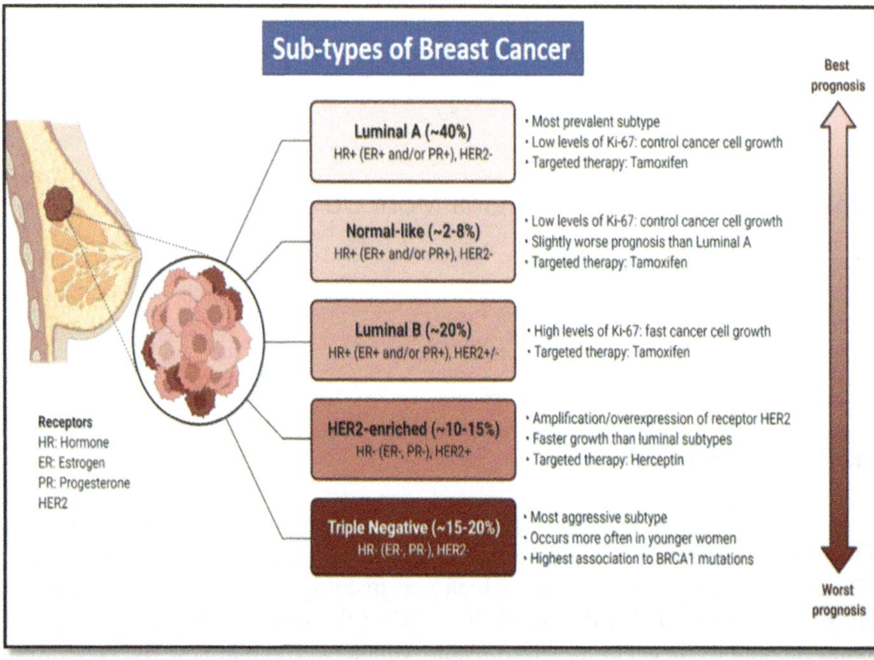

Fig. 6.1 Molecular subtypes of breast carcinoma

into molecular subtypes on behalf of GE patterns: luminal like, HER-2 enriched and basal like. The most often and common kind of BC is luminal-like tumours, which account for 60–70% of all cancers. The ER and various genes dominated by the epithelium cells that bound the lumen of TDLU of the duct where more BC develop are highly expressed. The oncogene Erb-B2 (HER-2) is present in low quantity more often (Goyal and Trivedi 2020).

BC patients with HER-2-enriched tumours constitute around 15% of total cases (Fig. 6.2). These are characterised by Erb-B2 oncogene overexpression and also modest levels of ER expression. Basal cell-like tumours make up for about 12–16% of all BCs and dominate numerous of the matching genes as TDLU basal myoepithelial cells (Johnson et al. 2021). They frequently do not express ER, as do many of the genes involved in estrogen receptor expression. According to a study by the American Cancer Society, around 73% of BC diagnoses in the USA are Luminal A, around 11% are Luminal B, around 12% are Triple Negative, and around 4% are HER-2 enriched. Further studies have shown that subgroups are associated with different clinical features. The E-R+ luminal subtype, for example, is currently divided into a couple of subgroups, A and B, and having each a distinct prognosis (Van't Veer et al. 2002).

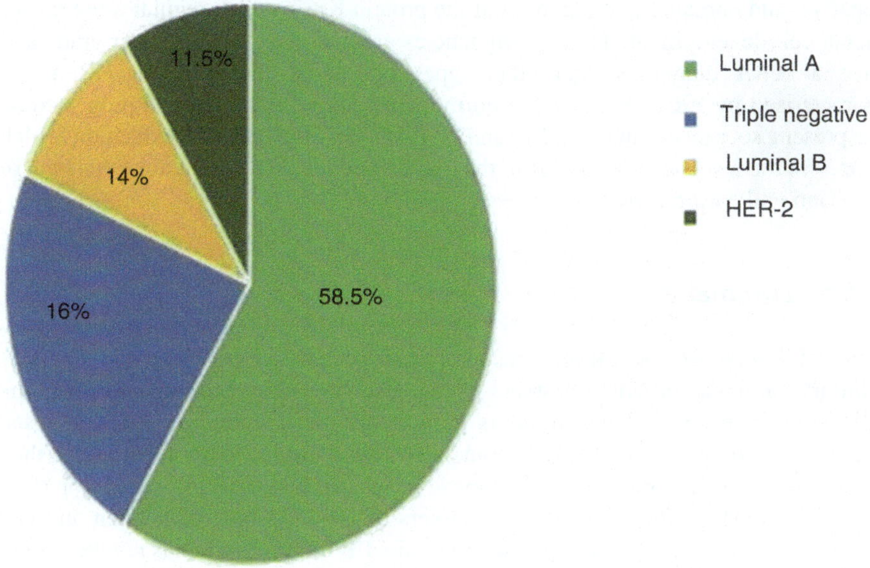

Fig. 6.2 Statistical data of molecular subtypes of Brest carcinoma in KAUH (2012–2018)

6.2 Genetic Expression

The estrogen receptor expression is a distinguishing feature of the luminal subtype. The term "luminal" derives from the resemblance linking the genes dominated by the tumours and the genes dominated by luminal breast epithelium cell (Perou et al. 2000). ER, PR, and more genes connected with Estrogen receptor stimulation are expressed in a majority of luminal cancers. BC luminal subtypes are the most common and can be classified into two groups A and B. GE patterns and clinical diagnosis of A and B luminal breast tumours differ significantly (Sørlie et al. 2003).

6.2.1 Luminal A

HER-2-negative tumours are also called luminal A tumours. It is a protein that the human body produces constantly on regular basis. BC with standard levels of HER-2 protein is called HER-2 negative. From a genetic point of view, HER-2 plays a crucial part in repairing of healthy mammary cells and also cell formation. Unlike Luminal-A tumours, luminal-B tumours are not that common and account for around 20% of all malignancies, compared to 40% for luminal A tumours (Anurag et al. 2020). These tumours are well-differentiated cancers and include morphologically undefined renal tubules, mucinous, classical lobular carcinomas and neuroendocrine carcinomas. Luminous BC has progesterone receptors as well as estrogen, is HER-2-

negative, and contains small amount of the protein Ki67, which regulates the rate of cancer cell development. These malignancies grow slower, have a lower grade and have a better diagnosis than other types of cancer. People with HR + are characterised by ER or PR-positive hormone receptors. Estrogen and progesterone are present so cancers that are ER and/or PR + develop faster. Medication which depresses the levels of ER and PR in the human body is effective against this type of BC (Gao and Swain 2018).

6.2.2 Luminal B

Luminal B-like BC is estrogen receptor-positive, HER-2-positive, and contains arbitrary amounts of Ki67 protein. It can also be positive or negative for the progesterone receptor. These tumours grow faster and are slightly less diagnosed than type A tumours. Luminal B tumours differ from luminal A tumours in that they express more proliferative and/or cell-cycle genes and have less PR expression (Prat et al. 2015). The Ki67 protein, IHC markers of growing nuclear antigen and cell proliferation are importantly expressed in luminal-B tumours which is not the case in luminal A cancer. In contrast to luminal A tumours, luminal B cancers have a high frequency of p53 mutations (Sørlie 2004). Luminal B tumours are poorly differentiated and usually highly malignant.

6.2.2.1 Clinical Implications and Management

Several researches have conveyed that ER+ cancers- luminal-A and B subtypes have two unique outcomes. Those suffering from luminal-B tumour had notably brief survival and ill health free survival than those with luminal A BC. Luminal-A tumours have the best diagnosis of all BC subtypes, but luminal-B, HER-2-rich, and basal subtypes do not show good clinical outcomes (Lahsaee 2018). Patients with luminal-B BC are poorly diagnosed with respect to luminal-A tumours due to overexpression of cell cycle and proliferative genes in these cancers (Sørlie 2004). A molecular subtype approach to BC treatment was accepted by the St Gallen expert consensus group in 2011 (Goldhirsch et al. 2011). IHC markers are used in these treatment algorithms to deliver comprehension into a tumour inherent molecular subtype and to guide therapy. We describe substitutes of four molecular types of BC, semi-quantitative IHC expression of Ki-67, HER-2, PR, and ER is currently employed (Qi et al. 2021). Luminal-B cancers are characterised from luminal-A tumours in clinical practise by a soaring Ki-67 expression (around 14%), soaring histologic grade and lower PR expression (under 20%). IHC surrogates, on the other hand, do not necessarily reflect the genuine intrinsic molecular subtype. Luminal-A and B tumours are recognised as HER-2- by IHC, ER+, PR+, and all HER-2+ tumours are classed as "HER-2 subtype" in any case of HR status. The "luminal B HER-2+" subtype includes tumours that are ER+, PR+, and HER-2+. Hormone therapy should be part of the treatment plan for all those suffering from luminal BC (Fig. 6.3).

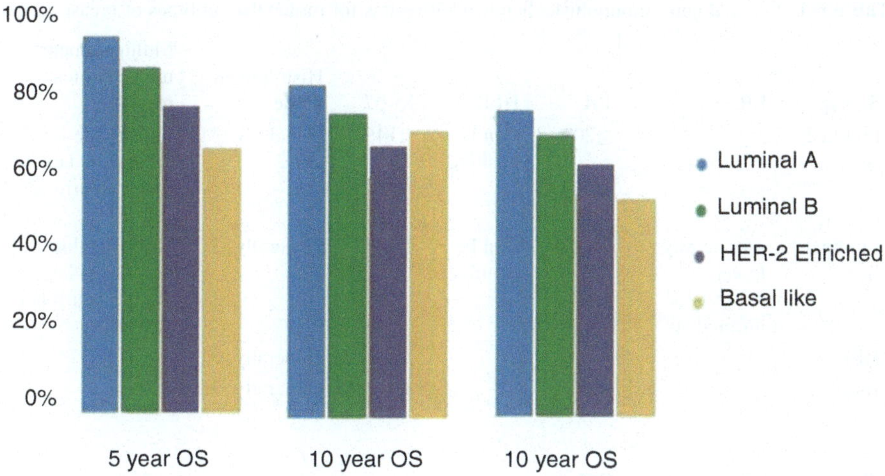

Fig. 6.3 Patients with invasive breast cancer are grouped by molecular subtype in the results

6.2.3 HER-2 Enriched Subtype

Genetic Expression E-R-negative, P-R-negative, and HER-2-positive. BCs are rich in HER-2. Tumours rich in HER-2-are usually successfully treated with targeted therapies that target the HER-2 protein, but develop earlier than luminal tumours and have a poor prognosis. HER-2 is transmembrane receptor (tyrosine kinase receptor) that holds together to extracellular signals and initiates a cascade that regulates cell proliferation, survival and differentiation. Between 12% and 20% of all BCs highly express the HER-2 protein and/or exhibit amplification of the HER-2 gene, leading to aggressive tumour development and poor clinical diagnosis (Tang and Tse 2016). The HER-2 oncogene is a known BC prognostic factor associated with shortening DFS and OS (Sørlie et al. 2003). Peru used HER-2 oncogene overexpression, specific gene selection overexpression, and low ER expression to classify HER-2-rich molecular subtypes. However, overexpression of proliferative genes such as Ki67 and proliferative nuclear antigens is not seen in the HER-2-enriched subtype. Nonetheless, HER-2-rich tumours have a worse prognosis than luminal malignancies, demonstrating the importance of the HER-2 oncogene. Surprisingly, proliferative genes such as Ki67 and proliferating nuclear antigens are not upregulated in HER-2-rich subtypes. However, HER-2-rich tumours have a worse prognosis than luminal cancer, emphasising the importance of the HER-2 oncogene. Tumours expressing HER-2 are E-R and P-R negative, but HER-2 positive. HER-2-positive BC cells have an abnormally high number of copies of the HER-2 gene and produce the HER-2 protein receptor found in BC cells. When the HER-2 receptors are functioning properly, they control how healthy breast cells regenerate, grow, and heal. As the cell proliferates, the receptors divide the cell and encourage it to proliferate rapidly and out of control. This is because cells overdose human epidermal growth factor 2, a substance that promotes cell development. Excessive

Table 6.1 Clinical and immunohistochemical surrogates for molecular subtypes of breast cancer

Subtype	ER	PR	HER-2	Ki-67	Histological grade	Multiparameter molecular test results
Luminal A	+	≥20%	Can be positive	<14%	Generally, 1or 2	Favourable prognosis, i.e., lower recurrence scores
Luminal B	+(generally lower relative to luminal A	<20%	Can be positive	>14%	Generally, 3	Unfavourable prognosis, i.e., higher recurrence scores
HER-2	–	–	+	Any	Generally, 3	
Basal	–	–	–	Any	Generally, 3	

HER-2-positive genes in BC tissue are often examined to determine if patients will benefit from targeted therapy options that prevent HER-2 from promoting the development of cancer cells (Kim and Koo 2020).

Symptoms of HER-2-positive BC are quite common to those of other types of BC. Lumps in the breast, changes in breast shape, pain, edema, and abnormal drainage are some of the common symptoms of BC. Many females with HER-2-positive BC first receive neoadjuvant chemotherapy with drugs that directly target HER-2. Some females, especially females with small tumours, undergo surgery first, followed by adjuvant therapy, including both chemotherapy and HER-2-targeted therapy. Endocrine therapy may also be given to females whose cancer is "hormone receptor positive", that is, females who need estrogen to grow. Doctors can perform tests to determine if BC falls into this category (Oh and Bang 2020) (Table 6.1).

6.2.3.1 Clinical Implications and Management

Detecting HER-2 positivity in BC gives critical diagnostic and forecastic information that helps guide HER-2-directed therapy and enhances clinical outcomes. Clinically, HER-2 positiveness is explained as full and robust membrane staining in fewer than 10% of cells by IHC. All IHC 3+ tumours are HER-2+, all IHC 1+ tumours are HER-2-, and all IHC 2+ tumours are ambiguous. Reflex HER-2 testing with fluorescent in situ hybridisation is triggered when the results are ambiguous (FISH). The ratio of HER-2 gene amplification to the chromosomes 17 probe defines the FISH results (CEP17). E-R, P-R, HER-2+ is the most often utilised IHC surrogate for the HER-2-enriched subtype (Table 6.1). However, the outcome of IHC and FISH, which clinically define tumours as HER-2+, do not necessarily match the outcomes of GEPs, which characterise tumours as HER-2-enriched, and HER-2+ illness is heterogeneous. "HER-2+" is used all over this review to refer to tumours that are useful for HER-2 expression by IHC, whereas "HER-2-enriched" refers to cancers identified by GEP(Godoy-Ortiz et al. 2019).

One of the most important outcomes in the treatment of BC is the success of anti-HER-2 therapies for females with early and advanced HER-2 + BC. Trastuzumab

(Herceptin, Genentech, San Francisco, CA) is the most commonly used anti-HER-2 therapy, but studies have shown that other anti-HER-2 drugs such as pertuzumab, neratinib, lapatinib and T-DM1 are also effective (Upton et al. 2021). In 2005, the first study where chemotherapy was compared to trastuzumab in patients with surgical HER-2 + disease showed improved DFS and about 33% reduction in mortality risk in patients receiving trastuzumab (Ahmed et al. 2015). Trastuzumab in combination with chemotherapy has a longer time to disease progression, a longer duration of response, and a lower one-year mortality rate in females with metastatic HER-2 + cancer than chemotherapy alone years later in 2012. It has been found to be associated with high median survival. In spite of continued advances in anti-HER-2 treatment, the prognosis for HER-2 + disease is poor. Because people with HER-2 + disease have a very broad clinical outcome, oncologists need diagnostic techniques to lead treatment. The traditional approach to females with HER-2 + malignancies is Neoadjuvant chemotherapy who can be operated on by IHC because the response to the therapy provides diagnostic and forecastive information (Pereira et al. 2019).

6.2.4 Triple-Negative or Basal-Like BC

Estrogen receptor-negative, progesterone receptor-negative, and HER-2-negative BCs are known as triple-negative or basal cell-like BCs. The chances of Triple-negative BC in young females with the BRCA1 mutation is more. Females with triple-negative BC are considered more aggressive than Luminal-A or Luminal-B BC (van Barele et al. 2021).

6.2.4.1 Genetic Expression

Many BC genes are dominated by either of the two types of epithelium cells, basal cells or luminal cells found in human breast tissue, according to Perou et al. (2000). High expression of keratin 5, keratin 17, integrin B4, laminin, and growth-related genes is a GE feature of basal epithelial cells, and thus basal-like BC (BLBC) subtypes. The p53 gene is mutated in most BLBC cancers. Most other genes co-expressed with ER are absent in these malignancies (Sørlie 2004). Basal-type BC accounts for about 15% of all invasive BCs and is usually high-grade, large at detection, and subject to local and distant recurrence (Tang and Tse 2016). Basal cell-like subtypes are the most usual of triple-negative BC (TNBC), accounting for around 70–80% of all TNBC cases. Triple-negative BC is a diverse group of tumours and the persisting 20–30% of TNBC can be divided into at least six distinct subgroups based on GEP (Teschendorff et al. 2007). TNBC is defined as the absence of IHC expression of ER, PR, or HER-2. TNBC is defined as a tumour in which less than 1% of the nucleus expresses ER and PR and HER-2 expression is 0 to 1 + or IHC2 + and FISH negative (Johnson et al. 2021). E-R, P-R, and HER-2 are the most commonly used IHC surrogate in BLBC. The majority of TNBC is ductal carcinoma that is not morphologically characterised. However, adenoid cystic carcinoma, secretory cancer, metaplastic cancer, and medullary thyroid cancer are also different forms of TNBC (Tang and Tse 2016). Several ongoing studies have shown many TNBCs as in all endogenous subtypes. It suggests that it can be subdivided into subtypes. In addition, the basal cell-like group is the most unique of the four endogenous subtypes of BC. BC tumours are thought to be derived from common luminal progenitor cell lines.

6.2.4.2 Clinical Implications and Management

Basal cell-like TNBC has absence of IHC expression, a major biomarker suitable for selected therapies, Estrogen receptor, progesterone receptor and HER-2. As a result, treatment of basal cell-like TNBC has not been selected, leaving chemotherapy as the only dependent and neoadjuvant treatment option. TNBC shows a better relationship between post-NACT pCR and better DFS and OS, as well as HER-2 + cancer. All TNBC tumours larger than 5 mm and/or with LN metastasis should receive neoadjuvant chemotherapy. Patients with residual lesions after completion of NACT may benefit from adjuvant chemotherapy to improve DFS and OS. BLBC is highly sensitive to chemotherapy because of its absence of ER expression, soaring grade, soaring proliferation index. Most neoadjuvant studies show that BLBC has a higher pCR among others. In spite of the 80% pCR rate reported by BLBC, the diagnosis remains abysmal. In spite of the relatively high pCR, the relatively abysmal diagnosis is called the "triple negative paradox".

6.3 Other Subtypes Under Investigation

Since the publication of Peruvian paper in 2000, further studies have revealed greater molecular heterogeneity in BC. Several new subtypes have been suggested, including (i) Claudin-low, (ii) Molecular Apocrine, and (iii) luminal HER-2 subtypes (Tang and Tse 2016). Expression of claudin 1, 3, 4, 7, 8 is low in claudin BC. These tumours are usually TNBC with a abysmal diagnosis. Molecular apocrine BC is defined by androgen receptor expression without ER expression and is often TNBC or HER-2. Ultimately, the "luminal HER-2" group is a new luminal-like subtype that exhibits E-R expression and is HER-2-positive by IHC in about 50% of patients. Compared to traditional intraluminal BC, these tumours are more malignant, have low P-R expression, relapse early, have more LN metastases, and are more responsive to endocrine therapy. In addition, it will be lower compared to ER, PR, and HER-2 + tumours.HER-2 intraluminal BC has a further aggressive clinical course identified by localised and early recurrence.

6.4 CDK Dysregulation in BC

The tumour microenvironment contains a number of elements that aid cancer growth and hinder anti-tumour responses (Mehraj et al. 2021). It has been hypothesised that targeting these cancer-promoting elements in the tumour microenvironment could be a powerful immunotherapeutic method for cancer treatment. Cyclin-Dependent

Table 6.2 Different cyclins and CDKs which take part in cell cycle

Phase	Cyclin	CDK
G0	C	CDK3
G1	D, E	CDK4, CDK2, CDK6
S	A, E	CDK2
G2	A	CDK2, CDK1
M	B	CDK1

Kinases (CDKs) have been proposed as a novel prospective target for cancer therapy among the different tumour supportive variables (Sofi et al. 2022). In conjunction with cyclins, these factors play an important role in cell cycle progression. CDK dysregulation has been linked to enhanced cell proliferation in diverse of malignancies, encompassing BC. As a result, the development and usage of CDK inhibitors in the treatment of BC has been linked to promising results. However, it is unknown which CDK inhibition method is the most successful for BC treatment (Deng et al. 2018). Because selective CDK1 blockade, either alone or in combination with other therapies, has been linked to effective anti-cancer effects, CDK1 may be the best CDK target for BC therapy.

Cyclins and CDKs play important roles in regulating cell cycle transition because they are required for cell cycle G1, S, G2, and M phase progression. To control the activity of cyclins and CDK inhibitors, CDK, a serine/threonine kinase, binds with both of these molecules. CDK activity is often dysregulated in cancer cells and is a promising target for cancer treatment. Human cells have 20 CDKs and 29 cyclins (Malumbres and Barbacid 2009). Cell cycle transition and cell division are directly regulated by CDK1, CDK2, CDK3, CDK4, CDK6, and CDK7, while cell cycle-related gene transcription is mediated by CDK 7–11 (Ding et al. 2020). At different times during the cell cycle, several cyclin-dependent kinases (CDKs) have diverse functions. In a certain cell environment during G1 phase, CDK4 and CDK6 assemble into a complex with one of the three D-type cyclins (D1, D2, or D3) (Table 6.2).

6.5 Role of CDK4/6 in Cell Cycle Control

The G1, G2, S, and M stages of mammalian cell cycle are traditionally classified into four different stages. The interaction of different cyclins with their cognate CDKs tightly controls the ordered movement between these phases at checkpoints. CDK is a well-conserved family of threonine protein kinases/serine with at least 12 loci known to encode them. Various regulatory CDKs containing three interphase CDKs (CDK_2, CDK_4, CDK_6), single-threaded split CDK (CDK_1, formerly called as CDC_2), and CDK7, component of the CDK activation complex, and transcription CDKs are all members of this family (CDK8, CDK9). Cyclins are a large family of proteins that are divided into four types (A, B, D, and E types of cyclins) Unlike CDKs and serve as regulatory subunits for the CDK cyclin holoenzyme. In spite of the vast quantity of CDKs and cyclins, only some of them are significantly included in the development of BC. Suppression of cell cycle progression is usually

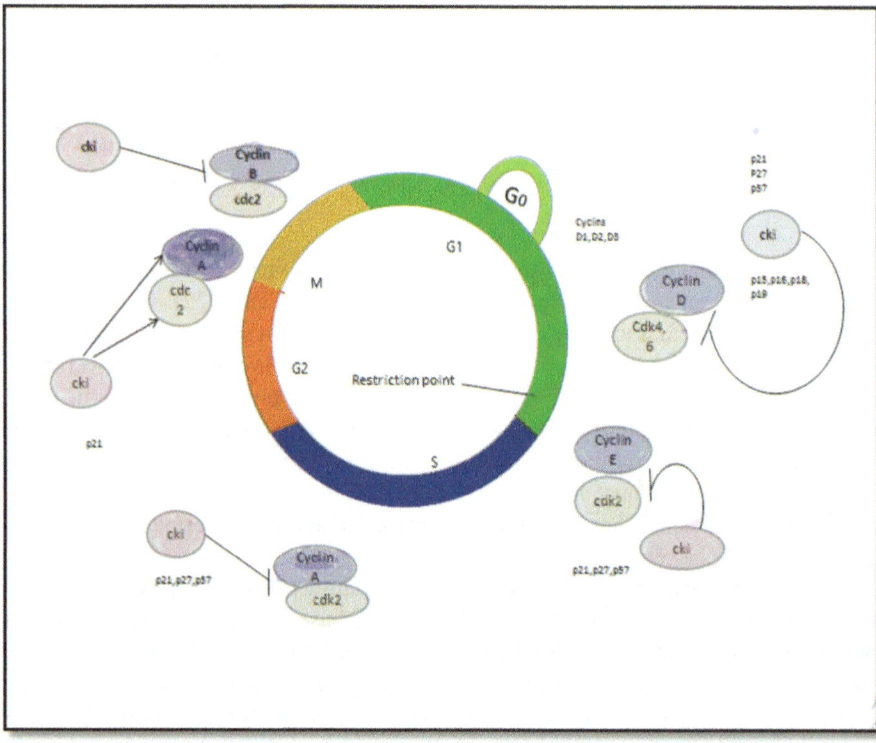

Fig. 6.4 The mammalian cell cycle includes passing through specific checkpoints in an organised manner for proper progression. By mitogen-induced CAK recruitment, the cyclin D1-CDk4 complex is brought into full holoenzyme activity. The PRB protein is phosphorylated by the cyclin D1-CDK4 complex, which causes cyclin E-Cdk2 to phosphorylate it later and release free E2F. Genes involved in the activation of S phase entrance are induced by the phosphorylation of pRB and the relaxation of downregulation by pRB

maintained by so-called pocket proteins and also the retinoblastoma gene product (pRb) that sequester the E2F family of transcription factors (Huun et al. 2017). Rest cells, on the other hand, when they pass the cell cycle, produce cyclin D1 in retaliation to adhesion cues and particular mitogen. After the formation of the activation complex with CDK4/CDK6, the newly generated cyclin D1 initiates phosphorylation of pRb. Transcription inhibition of E2F by pRb is abolished by phosphorylation of the cyclin D1: CDK4/6 complex. Cyclin E, a protein that binds to CDK2 and phosphorylates pRb and other G1/S checkpoint mediators. This process establishes a practical feedback loop in which cells irreversibly pass through the G1-S transition (also known as the "restriction point") and carry on the cell cycle without the use of mitogen (Deng et al. 2018). In the late S phase, cyclin A2 activates CDK2 and can switch from S phase to G2 phase. Finally, type A and type B cyclins stimulate CDK1 to accelerate the initiation and progression of mitotic processes (Fig. 6.4) (Goel et al. 2017).

Pan-CDK inhibitors were at first found to be ineffective in suppressing cancer cells and were related with serious side effects. Nonetheless, in later years, the success of tolerable, potent and selective CDK inhibitors has rekindled interest in this class of selected drugs. Long-term therapy with CDK4/6 inhibitors suppresses cell cycle genes and at the same time stimulates further genes responsible in various functions. In fact, inhibition of CDK4/6 is associated with stimulation of cell proliferation genes which are inhibited by endocrine therapy (Pernas et al. 2018), Strengthen the discussion for merging the two medicines. Palbociclib (PD0332991), Ribociclib (LEE011), and abemaciclib (LY2835219) are triplet highly selective inhibitors of CDK4 and CDK6 that bind to ATP cleft with low toxicity.

Dual function of $P27^{Kip1}$. In the context of a building factor, the cyclin D1 gene product contacts the regulatory element partner (Cdk4). An enzyme called a Cdk activating kinase (CAK), which itself is made up of multiple subunits, phosphorylates the cyclin D1-Cdk4 holoenzyme. The tumour suppressor Prb can then be upregulated by activated cyclin D1-Cdk4. Additionally phosphorylating pRB is Cyclin E-Cdk2. There is debate about the contribution of the p21 CKI family, represented by p27kip1, to the regulation of the cyclin D1-Cdk complex's activity. In some cases, p27kip1 is believed to reside in the complex's activity (Kim et al. 2008). In other situations, p27kip1 does not participate in the complex's assembly-factor activity. Cyclin D1 induction may facilitate S phase entrance via inoculating p27kip1 from an antagonistic complex with cyclin E-Cdk2 if p27kip1 inhibits cyclin E-Cdk2 but not cyclin D-Cdk4 (Wood et al. 2019). These results imply that p27kip1 activity may be influenced by stoichiometry or cell type (Fig. 6.5).

6.5.1 Palbociclib

Palbociclib (PD0332991, IBRANCE®, Pfizer) is a potent orally vigorous CDK4/6 inhibitor that binds to ATP fissures. Palbociclib was sensitive to ER-positive BC cells in an in vitro study operating a panel of BC cell lines, significantly reduced cell cycle progression due to G1 arrest, and prevented pRb hyperphosphorylation. Overall, susceptibility to the effects of palbociclib on cell cycle and growth inhibition was associated with increased Rb and cyclin D1 and decreased p16 (Kwapisz 2017). Non-luminal/basal BC cells showed the lowest activity, except for those with human epidermal growth factor receptor 2 (HER-2) amplification. In ER-positive and HER-2-amplified cell lines, palbociclib was synergistic with anti-estrogen tamoxifen and anti-HER-2 treatment trastuzumab, respectively. Palbociclib may also improve tamoxifen sensitivity in BC cell lines that have developed resistance to the drug (Malorni et al. 2018) and induce cellular senescence in hormone therapy-resistant cell lines. While most preclinical palbociclib studies focus on HR-positive BC, a few preclinical studies of HER-2-positive or triple-negative BC (TNBC) have shown promise. Palbociclib was more sensitive and additive in HER-2-positive BC models and primary human explants when used in combination with adtrastuzumab emtansine (TDM1). CDK4/6 and cyclin D complex formation is aided by external

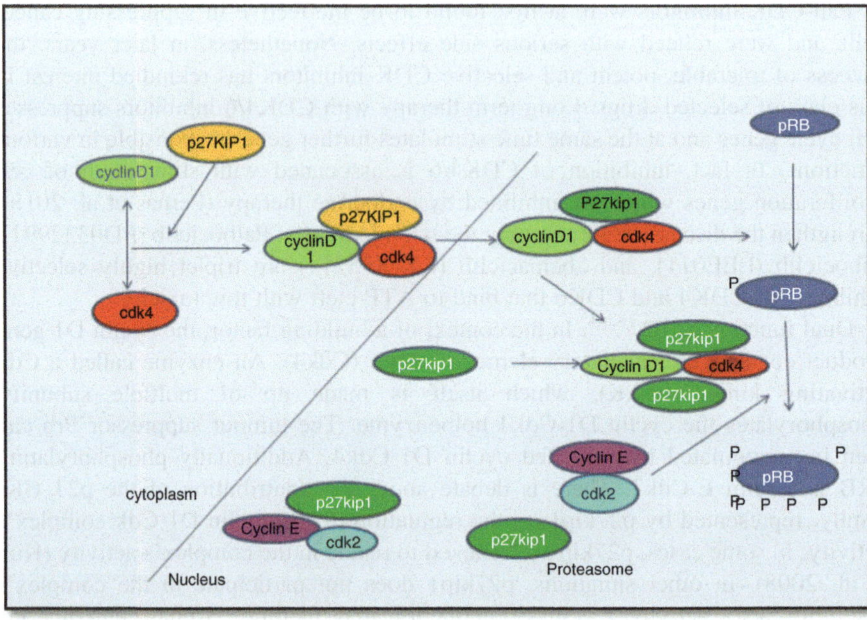

Fig. 6.5 Role of different cyclin/CDK complexes in the progression of cell cycle

mitogenic signals. The release of the E2F transcription factor, the transition from G1 to S phase, and the hyperphosphorylation of RB1 are all made possible by the CDK4/6-cyclin D complex, which promotes cell development (Zhang et al. 2020). Cell growth is inhibited by CDK4/6 inhibitors like palbociclib, Ribociclib, or abemaciclib because they prevent the phosphorylation of RB1 while it is still linked to the E2F transcription factor (Fig. 6.6).

6.5.2 Ribociclib

Ribociclib (LEE011; KISQALI; Novartis Pharmaceuticals Corp.) is an oral small molecule CDK4 and CDK6 inhibitor that completely dephosphorylates Rb, sequesters E2F transcription factors, and arrests Rb-positive cells' progression through the G1 cell cycle. In four ER-positive xenograft models, Ribociclib (Hortobagyi 2018) alone or in combination with letrozole or fulvestrant reduced tumour growth in vivo. Ribociclib and BYL719 (a PI3K inhibitor; alpericib) together improved tumour antigen presentation, cell cycle arrest, DNA damage, replication stress, and immunogenic cell death in the TNBC model. Ribociclib and BYL719 improved innate and adaptive immune system activation and cytotoxicity in immunocompetent mice while lowering the number of immunosuppressive monocytic bone marrow suppressor cells (MDSCs) in the tumour environment (Kwapisz 2017).

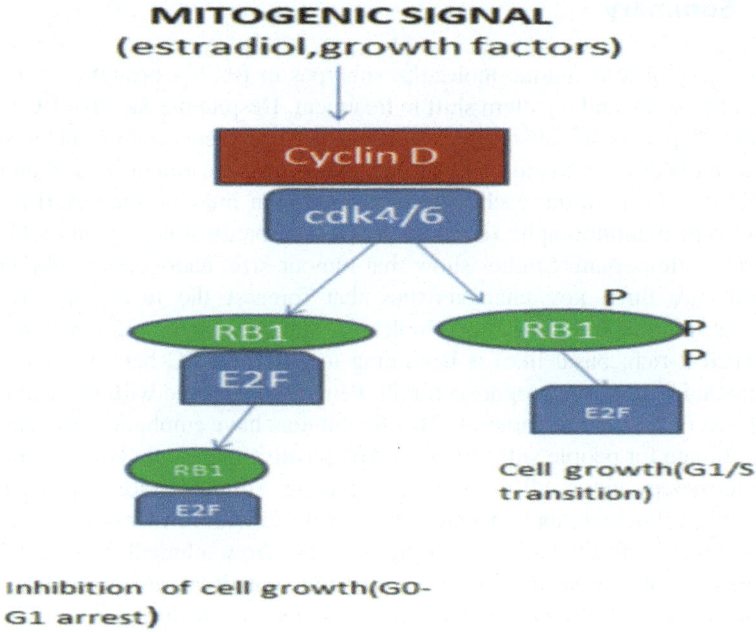

Fig. 6.6 Diagrammatic representation of the function of CDK4/6 inhibitors in cancer cells

6.5.3 Abemaciclib

Eli Lilly and Company's abemaciclib (LY2835219; VERZENIOTM) binds to ATP clefts and creates hydrogen bonds with the kinase's conserved catalytic residue (Lys43). It does not bind more selectively than Ribociclib or palbociclib, in this case. Abemaciclib prevents Rb from being phosphorylated along with CDK4 and CDK6, which causes Rb-competent cells to enter a G1 arrest and limit their ability to proliferate. In preclinical animals, Abemaciclib has a higher affinity for CDK4 than palbociclib or Ribociclib, which might account for the different toxicity profile. In vivo, ER-positive, HER-2-positive, and biomarker-selected TNBC xenografts demonstrated tumour growth suppression comparable to abemaciclib (Kwapisz 2017). In HER-2-positive BC models, the cyclin-D1-CDK4 pathway fosters resistance to anti-HER-2 therapy, and by boosting resistant tumour cells' dependence on EGFR-kinases, it resensitise resistance to anti-HER-2 therapy (Ahmed et al. 2015). Abemaciclib has been shown in preclinical research to pass the blood-brain barrier in a rodent model (Wander et al. 2022). In MDRMCFER-positive BC cells, it has been demonstrated that abemaciclib increases the intracellular accumulation of chemotherapeutic drugs as a result of decreased ABCB1/ABCG2 transport activity. It was discovered that abemaciclib capacity to reverse MDR is unrelated to CDK4/6 inhibition or Rb pathway phosphorylation blocking (Wander et al. 2022).

6.6 Summary

The discovery of four unique molecular subtypes in BC has brought about a new period in BC study and a pattern shift in treatment. Despite the fact that BC is still a horrifying diagnosis for all females, coordinated treatments help females survive longer with the disease, avoiding cytotoxicity and harsh treatments that often lead to comorbidities. In addition, each subtype has its own imaging characteristics, and importance of mammography (common method of breast imaging) in early observation stands important. Studies show that tumour size, endogenous subtypes and LN status are three key characteristics that forecast the result of early-stage BC. Recognition of the four unique molecular subtypes of BC (Luminal-A, Luminal-B, HER-2-rich, basal like) is beginning to elucidate BC heterogeneity and is more targeted to improve prognosis for all. Females diagnosed with BC will head to the progress of selected therapies. CDK4/6 inhibitors have emphatically changed the therapy climate for people suffering from HR-positive metastatic BC. Contrasted to endocrine therapy alone, all three Food and Drug Administration-approved drugs (palbociclib, Ribociclib, and abemaciclib) are related with further developed results and adequate harmfulness. If ongoing studies show clinical utility, CDK4/6 inhibitors may play a part in neoadjuvant or adjuvant situations in the near future. Unfortunately, resistance has evolved over time and new techniques are now being sought to hold up or control resistance. Studies are presently planned and of great importance to recognise victims/sufferers who can be successfully cured using endocrine therapy alone.

6.7 Further Readings

The authors can look for the following articles for further understanding of the given topic

(i) https://www.sciencedirect.com/science/article/pii/S1097276520307231
(ii) https://breast-cancer-research.biomedcentral.com/articles/10.1186/s13058-015-0661-5
(iii) https://www.mdpi.com/2227-9059/10/2/366

For more insights about the topic, we would suggest detailed findings from the books of (Mir MA, 2022) https://doi.org/10.1016/C2021-0-02565-7, https://doi.org/10.1016/C2014-0-02898-5 (Mir MA, 2021) https://doi.org/10.52305/WXJL6770, from cancer.net website, https://www.cancer.net/cancer-types/breast-cancer/types-treatment

For diagrammatic illustrations, descriptive tables, (Lazzeroni, 2012) http://www.eurekaselect.com/article/49928. The readers can also look for the following visual presentations for a better understanding of the topic:

(i) https://youtu.be/R_Y_du4Z3Zo
(ii) https://youtu.be/qKUPoovs92I
(iii) https://youtu.be/YA67P2k2d6A
(iv) https://youtu.be/poQEkyVhGMo

References

Ahmed S et al (2015) HER2-directed therapy: current treatment options for HER2-positive breast cancer. Breast Cancer 22(2):101–116

Al-Rikabi A, Husain S (2012) Increasing prevalence of breast cancer among Saudi patients attending a tertiary referral hospital: a retrospective epidemiologic study. Croat Med J 53(3):239

Anurag M et al (2020) Immune checkpoint profiles in luminal B breast cancer (Alliance). J Natl Cancer Inst 112(7):737–746

Deng J et al (2018) CDK4/6 inhibition augments antitumor immunity by enhancing T-cell activation. Cancer Discov 8(2):216–233

Ding L et al (2020) The roles of cyclin-dependent kinases in cell-cycle progression and therapeutic strategies in human breast cancer. Int J Mol Sci 21(6):1960

Economopoulou P et al (2015) Beyond BRCA: new hereditary breast cancer susceptibility genes. Cancer Treat Rev 41(1):1–8

Gao JJ, Swain SM (2018) Luminal a breast cancer and molecular assays: a review. Oncologist 23(5):556–565

Godoy-Ortiz A et al (2019) Deciphering HER2 breast cancer disease: biological and clinical implications. Front Oncol 9:1124

Goel S et al (2017) CDK4/6 inhibition triggers anti-tumour immunity. Nature 548(7668):471–475

Goldhirsch A et al (2011) Strategies for subtypes—dealing with the diversity of breast cancer: highlights of the St. Gallen International Expert Consensus on the Primary Therapy of Early Breast Cancer 2011. Ann Oncol 22(8):1736–1747

Goyal N, Trivedi MC (2020) Breast cancer classification and identification using machine learning approaches. Mater Today. https://doi.org/10.1016/j.matpr.2020.10.666

Hortobagyi GN (2018) Ribociclib for the first-line treatment of advanced hormone receptor-positive breast cancer: a review of subgroup analyses from the MONALEESA-2 trial. Breast Cancer Res 20(1):1–11

Huun J et al (2017) Effects of concomitant inactivation of p53 and pRb on response to doxorubicin treatment in breast cancer cell lines. Cell Death Discov 3(1):1–6

Johnson KS et al (2021) Molecular subtypes of breast cancer: a review for breast radiologists. J Breast Imaging 3(1):12–24

Khan MI et al (2022) Anticancer properties of medicinal plants and their bioactive compounds against breast cancer: a review on recent investigations. Environ Sci Pollut Res 29:1–34

Kim HM, Koo JS (2020) Clinicopathologic characteristics of breast cancer according to the infiltrating immune cell subtypes. Int J Mol Sci 21(12):4438

Kim SJ et al (2008) Determination of the specific activity of CDK1 and CDK2 as a novel prognostic indicator for early breast cancer. Ann Oncol 19(1):68–72

Kwapisz D (2017) Cyclin-dependent kinase 4/6 inhibitors in breast cancer: palbociclib, ribociclib, and abemaciclib. Breast Cancer Res Treat 166(1):41–54

Lahsaee S (2018) Estrogen receptor alpha signalling regulates the expression of the taxane-response biomarker PRP4K. Exp Cell Res 340:125

Malorni L et al (2018) Palbociclib as single agent or in combination with the endocrine therapy received before disease progression for estrogen receptor-positive, HER2-negative metastatic breast cancer: TREnd trial. Ann Oncol 29(8):1748–1754

Malumbres M, Barbacid M (2009) Cell cycle, CDKs and cancer: a changing paradigm. Nat Rev Cancer 9(3):153–166

Mehraj U et al (2021) The tumor microenvironment as driver of stemness and therapeutic resistance in breast cancer: new challenges and therapeutic opportunities. Cell Oncol 44:1–21
Oh D-Y, Bang Y-J (2020) HER2-targeted therapies—a role beyond breast cancer. Nat Rev Clin Oncol 17(1):33–48
Pereira BA et al (2019) Tissue engineered human prostate microtissues reveal key role of mast cell-derived tryptase in potentiating cancer-associated fibroblast (CAF)-induced morphometric transition in vitro. Biomaterials 197:72–85
Pernas S et al (2018) CDK4/6 inhibition in breast cancer: current practice and future directions. Ther Adv Med Oncol 10:1758835918786451
Perou CM et al (2000) Molecular portraits of human breast tumours. Nature 406(6797):747–752
Prat A et al (2015) Clinical implications of the intrinsic molecular subtypes of breast cancer. Breast 24:S26–S35
Qi P et al (2021) Concordance of the 21-gene assay between core needle biopsy and resection specimens in early breast cancer patients. Breast Cancer Res Treat 186(2):327–342
Siegel RL et al (2020) Cancer statistics, 2020. CA Cancer J Clin 70(1):7–30
Sofi S et al (2022) Cyclin-dependent kinases in breast cancer: expression pattern and therapeutic implications. Med Oncol 39(6):1–16
Sørlie T (2004) Molecular portraits of breast cancer: tumour subtypes as distinct disease entities. Eur J Cancer 40(18):2667–2675
Sørlie T et al (2003) Repeated observation of breast tumor subtypes in independent gene expression data sets. Proc Natl Acad Sci 100(14):8418–8423
Tang P, Tse GM (2016) Immunohistochemical surrogates for molecular classification of breast carcinoma: a 2015 update. Arch Pathol Lab Med 140(8):806–814
Teschendorff AE et al (2007) An immune response gene expression module identifies a good prognosis subtype in estrogen receptor negative breast cancer. Genome Biol 8(8):R157
Upton R et al (2021) Combining CD47 blockade with trastuzumab eliminates HER2-positive breast cancer cells and overcomes trastuzumab tolerance. Proc Natl Acad Sci 118(29):e2026849118
Van't Veer LJ et al (2002) Gene expression profiling predicts clinical outcome of breast cancer. Nature 415(6871):530–536
van Barele M et al (2021) Estrogens and progestogens in triple negative breast cancer: do they harm? Cancers 13(11):2506
Wander SA et al (2022) Targeting CDK4 and 6 in cancer therapy: emerging preclinical insights related to Abemaciclib. Oncologist 27:811
Wood DJ et al (2019) Differences in the conformational energy landscape of CDK1 and CDK2 suggest a mechanism for achieving selective CDK inhibition. Cell Chem Biol 26(1):121–130
Zhang Z et al (2020) CDK4/6 inhibition blocks cancer metastasis through a USP51-ZEB1-dependent deubiquitination mechanism. Signal Transduct Target Ther 5(1):1–13

Breast Tumor Microenvironment and CDKs 7

Manzoor Ahmad Mir and Abrar Yousuf Mir

7.1 Introduction

Around the globe, the most frequently investigated tumor in females is breast tumor, and is the main cause of death associated with females (Mir 2022a; Benson and Jatoi 2012). In 2012, 1.7 million fresh cases were estimated that led to 521,900 deaths. Additionally in 2018, greater than two million new BC cases were diagnosed resulting in 630,000 deaths (Bray et al. 2018). Predominantly, because of owning advancements in medication and detection processes, a remarkable decreased mortality rate has been observed, although it still accounts for 10–15% of all malignant deathliness in females notably because of advanced cancers and hindrance to comprehensive therapy (Benson and Jatoi 2012). Consequently, researchers are looking for potential targets for treatment, for instance, the tumor microenvironment (Cha and Koo 2020; Mir MA, 2022). We know well that a tumor is the result of communication between natural immune cells and cancerous cells, resulting in a fully formed tumor from a single normal cell. Immune cells mainly present in the tumor microenvironment include DCs, Natural Killer cells, T cells, mast cells, and TAMs, all these immune cells play an vital role in cancer development (Mehraj et al. 2021a).

Stephen Paget pioneer cancer researcher put forward the "seed and soil" hypothesis in 1889, claiming that cancer cells act as "Seeds" that could cause cancer growth only when there is the existence of a conducive microenvironment (Paget 1889). Despite the fact, that cancer detection and intrusion strategies have traditionally focused on cancer-cell-intrinsic factors. The latest findings focus more on immune cells like Macrophages, mast cells, neutrophils and endothelial cells, fibroblasts,

M. A. Mir (✉) · A. Y. Mir
Department of Bioresources, School of Biological Sciences, University of Kashmir, Srinagar, J & K, India
e-mail: drmanzoor@kashmiruniversity.ac.in

M. A. Mir (ed.), *Therapeutic Potential of Cell Cycle Kinases in Breast Cancer*,
https://doi.org/10.1007/978-981-19-8911-7_7

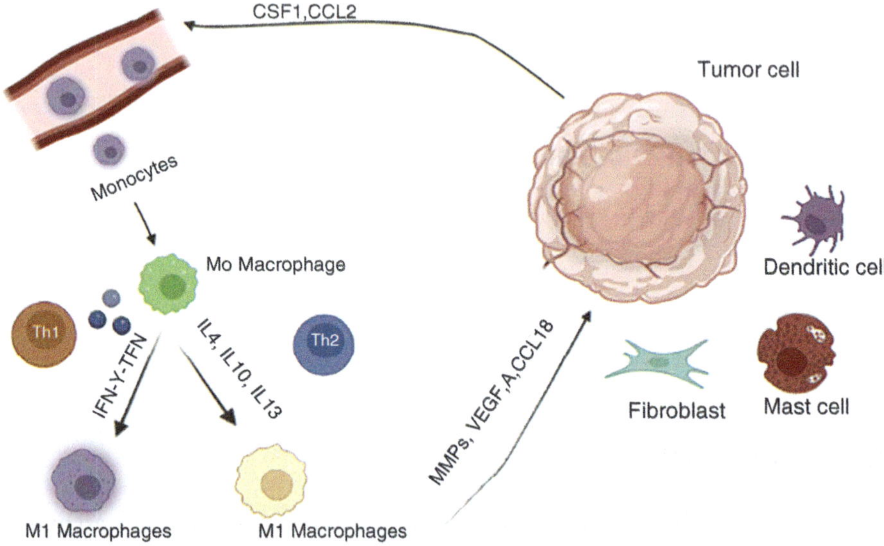

Fig. 7.1 Overview mechanism of recruited macrophages inside breast tumor microenvironment

adipocytes, and perivascular cells existing in TME (Williams et al. 2016). Macrophages, within the tumor, are termed tumor-associated macrophages, and represent the utmost prominent cells in the breast tumor microenvironment accounting for greater than 50% of the cell volume in maximum solid malignancy (Vitale et al. 2019; Mehraj et al. 2021b). Elie Metchnikoff in 1882, discovered macrophages as phagocytic immune cells while examining starfish embryos, and demonstrated that macrophages come up with the first line of defense, put a stop to diseases, upgrade wound healing and maintain tissue stability, distinguish between self and non-self antigens as well as solve inflammation processes. (Vitale et al. 2019).

A higher degree of cellular plasticity has been depicted by macrophages when triggered via an innumerable number of signals in TME, and respond hurriedly to take part in both inborn immune response as well as adaptive immune response. Signaling proteins like cytokines, chemokines, as well as other enzymes present in the TME, stimulate chemotaxis in responding to adjacent cells and TME greatly influences the functional properties of macrophages in the tumor microenvironment (Chanmee et al. 2014). TAM is built by resident macrophages and accumulation of disseminating leukocytes in breast tumors (Franklin et al. 2014; Mir 2022). With the help of "M-CSF" or "CSF1" (monocyte-colony-stimulating factor) transform recruited monocytes into non-polarized macrophages (MO) (Fig. 7.1) (Martinez et al. 2006). MO macrophages are extremely plastic, in response to external stimuli they can change their physical composition. Classification of macrophage density within a tumor can be categorized based on its functional scale (Mosser and Edwards 2008). According to the function of macrophages, they are broadly classified into two phenotypes such as –the M1/M2 phenotype, the M1 phenotype has pro-inflammatory and antitumor capability and the M2 phenotype led to anti-inflammatory as well as tumor-promoting function (Chanmee et al. 2014). These two macrophages are considered a functional spectrum in this classification (Mosser and Edwards 2008). Th-1 secrete chemokines like "interferon-γ (IFN-γ), either tumor necrosis factor (TNF)" for recruitment of macrophages like M1, and are also termed as classically activated macrophage, leading to the secretion of proinflammatory chemokines namely tumor necrosis factor (TNF) as well as IL-2 (interleukin), along with oxygen radical along with nitrogen intermediates (Biswas and Mantovani 2010). Contrary to macrophages like M2, which are intensified by cytokines like interleukin-4, interleukin-10, as well as interleukin-13, expressed by T helper cell type 2 (Th2), exhibit pro-tumor properties, and M2-are, termed as alternatively activated macrophage (Martinez and Gordon 2014). Current studies revealed that M2-like macrophages show comparable functionality with TAMs and show response to cytokines like "interleukin- 4 (IL-4), interleukin −10 (IL-10), transforming growth factor-beta (TGF-β), and interleukin -13 (IL-13)" stimulating re-grow of tissues (Rhee 2016). For recruiting macrophages cancer cells release cytokines in the TME and corresponding M2-like TAMs secrete pro-tumor chemokines to enhance tumor cell proliferation (Mantovani et al. 2002). Similarly, Breast tumor growth and development are predisposed by cancer microenvironments including effects of stroma, macrophages, and CDKs. Deregulation of cyclin-dependent kinases (CDKs), which play a central role in regulation of cell cycle, contributes to the expansion of the disease, including breast cancer. Cyclins and cyclin-dependent kinases have been recognized as an important part of cancer cell division and spread.

7.2 The Tumor Microenvironment

The knowledge of TME in breast carcinoma is speedily advancing. Breast carcinoma expansion is not only determined by the particular kind of breast cancer or genetic episodes within cancer cells, but also by the constitution of the TME (Bussard et al. 2016). Cancers are made of tumor parenchyma, matrix cells, and inflammatory mediators (Hanker et al. 2020). The matrix constitution of the initial stage cancer is measured to be the most significant agent showing breast tumor development. The microenvironment's complex and diverse interplay between carcinoma cells and stromal cells, such as tissue-resident and peripherally recruited immune cells, fibroblasts, and endothelial cells, among others, has a significant impact on BC development (Gao et al. 2019).

7.2.1 TME Components

Breast cancer TME is divided into three categories: soluble, physical components, and cellular (Table 7.1) (Soysal et al. 2015). Regional (breast), and metastatic compartments, local (intra-tumoral) can be distinguished among the cellular components (Soysal et al. 2015). The properties of cancerous cells and carcinoma-

Table 7.1 Tumor microenvironment Components with delineated functions

Component	Local	Regional	Metastatic
Cellular components	Lymphocytes, Cancer cells, Th-cells, Cytotoxic T-cells, Regulatory T-cells, Macrophages, Neutrophils, Dendrites .	Adipocytes Myoepithelial cells, Endothelial cells.	Immune cells Lymphatics Blood Peripheral immune cells, Distant organs
Soluble factors		Metalloproteinase Cytokines, Interferon-γ,Interleukins, Lysyl oxidase Matrix, TNF, MCSF, G- factors.	
Others		02 levels, PH	

infiltrating inflammatory cells such as lymphocytes, plasma cells, dendrites, big phage cells, and neutrophils are referred to as the local compartment (Salgado et al. 2015). The interaction between tumorous cells and adjacent cells in the stroma, particularly at the infiltrating edge, involves stromal fibroblasts, adipocytes, endothelial and vascular/lymphatic endothelial cells, and endothelial and vascular/lymphatic endothelial cells, is referred to as the regional compartment (Soysal et al. 2015). Different soluble and physical factors also play a role in tumor development in the breast and at far-off locations; these include enzymes, cytokines, and growth factors.

7.2.2 Composition

The breast tumor- microenvironment may be studied at three distinct degrees: local (intra-tumor), regional (in the breast), as well as a distant (metastatic) levels, each and everyone, is bounded by several kinds of cells like leukocytes, adipocytes, fibroblasts, endothelial and basket cells, extracellular matrix (ECM)]. They also comprise dissolved factors for instance growth factors, enzymes, chemokines, hormones, as well as other physical observables such as pH, and oxygen percentage as well (Coleman et al. 2013; Mir 2021).

7.2.3 Local Microenvironment

For the healthy development and proliferation of the breast glands, interferences between stromal and epithelial cells are necessary. Biological stroma keeps epithelial polarity and suppresses excessive cell proliferation and cancer (Folgueira et al.

2013). By the way of illustration basket cells (myoepithelial cells) turned out to be natural inhibitors in the mammary tumor and act as doorkeepers of tumor development, because these cells form the "Basement membrane" and denote an obstruction in the vicinity of luminal epithelial cells (Hu and Polyak 2008). A breast tumor xenograft study revealed a reduction of myo-epithelial cells and enhance the transformation of DCIS "Ductal carcinoma in situ" (DCIS) into "invasive carcinoma" (Hu and Polyak 2008). Escape and the release study of the Ductal tumor in situ carcinoma -to-foreign tumor transformation turned out to be represented by these two hypotheses (Hu and Polyak 2008). "escape" hypothesis revealed, that alteration at the gene level enables cancer epithelial cells to escape the tissue adjoining to the vessels, whereas the "release" hypothesis revealed out, aberrant micro-environment causes BM destruction and metastasis of cancer epithelial- cells inside stroma. The essential circumstances of the in situ-to-invasive transformation in breast tumors, is most likely the combination of these two models, underlining the importance of alterations in epithelial and stromal compartments for influencing cancer development and metastasis. In breast tumor cells, healthy myoepithelial cells also show suppression in growth, escape, and angiogenesis as well (Barsky and Karlin 2005). the associated tumor stroma, via paracrine signaling, stimulates myofibroblast as well as fibroblast to influence cancer progression and tumor spread by generating a receptive microenvironment. Even though tumor-associated stroma simulates a healing wound in many ways with fibroblast proliferation and remodeling of ECM but with no physiological controls (Dvorak 1986). Correspondingly, several studies demonstrated that IL-6 (interleukin-6) an inflammatory cytokine improves tumor growth and spread, urge on breast tumor hematopoietic cells (Jiang and Shapiro 2014). For instance, by recruiting "type 4 C-X-C chemokine receptor (CXCR4)"-stem and precursor cells the chemotactic cytokine "C-X-C motif" ligand cxcl-twelve is physically grave to start tissue renewal and replacement (Cojoc et al. 2013). Tumor-associated stromal cell migration and multiplication are stimulated by "CXCR4/CXCL12" signaling in the mammary TME and the release of "matrix metalloproteinases" (MMP) and subsequent tissue reshaping, furthermore multiple investigations have shown that -CXCL12 - has a straight consequence on cancer cell migration and cancer penetrate and also epithelial-mesenchymal transformation through CXCR4 production in tumor cells (Cojoc et al. 2013). However, the high expression of CXCR4 in the breast carcinoma compartment is associated with lymphoma and reduced analytical diagnosis.

7.2.4 Metastatic Microenvironment

Tumor spread is a complicated mechanism, whereas breast cancer cells exudates via narrow –blood-vessels to settle down into the new microenvironment (Coleman et al. 2013). However, cancer cells admit into a "sedentary" state for a long duration of time or initiate design micro- metastasis. Emphasize that, from the early tumor, chemokines and cytokines are produced and recruit hematopoietic-mediated cells that are secreted into the blood circulation and afterward supposed to generate pre-metastasis niche even earlier cancer cell mobilization (Kakonen and Mundy 2003). Intriguingly, it was illustrated that fibroblast and tumor cells have been seen to migrate to metastasis areas (Place et al. 2011). Bone metastasis is a potentially studied example and increases awareness about metastasis. Including osteoclasts, bone-developing cells, and blood stem cells "RANKL" production is increased when breast cancer cells release several kinds of growth factors and cytokines which also lead to the formation of osteoclast and enhance bone- reabsorption (Coleman et al. 2013), destruction of bones that secrete tumor-associating factors, leading to further bone damage, illustrate the self-sufficient process (Kakonen and Mundy 2003).

Currently, it has been demonstrated that RANKL is correlated with the development of pulmonary metastasis through "CD4+ regulatory T -cells" (T regs), promoting the contribution of well-defined lymphocytes, perhaps essential for boosting metastasis. Furthermore, mice models displayed a better opportunity in breast tumor for multi-directional metastatic cancer, spreading of tumor cells apart from early cancer to the bone besides from bone to distinctive parts and come again to the original site, indicating that the osseous microenvironment is considered a vital administrator in the cancer development and metastasis (Coleman et al. 2013).

7.3 Breast Cancer Cell-Stromal Interactions

7.3.1 Primary Site

Cells in stroma are in continual interaction with both breast cancer cells and their released components in the primary breast TME (Canavese et al. 2012). Alteration in expression of genes (such as reprogramming of basic functions and epithelial-mesenchymal transition, EMT) occurs not just in breast cancer cells, but also in the adjacent cells of stroma as a result of these interactions. Alterations in the main breast TME enhance both BC cell survival and metastasis in the long run (Watanabe et al. 1995). Even though it is beyond the scope of this study to discuss the advances gained in our knowledge of how cancer cells and surrounding "normal" cells interact, tunneling nanotubes deserve special attention.

7.3.2 Tunneling Nanotubes

Tunneling nanotubes are cytoplasmic extensions that reach hundreds of micrometers in length and link animal cells (Ma et al. 2000). The "horizontal movement" of cellular material, such as miRNAs, proteins, vesicles, autophagosomes, and even mitochondria, is enabled by these structures, which have been identified in a variety of cell systems both in vivo and in vitro (Okuda et al. 2000). One could extrapolate that these open connections between cancer cells and surrounding "normal" cells can muddle the identity of specific cells involved in the tumor network, making it difficult to mount a targeted immune response and, at the same time, increasing tumors' resistance to radiation and chemotherapy. Cancer cells were shown to actively communicate with several cell types inside the TME, including fibroblasts, endothelial cells, adipocytes, and immune cells, although the specific processes are unclear.

7.3.2.1 Immune Cells

For preserving homeostasis, tissue macrophage tissue performs the main function of destroying micro-organisms via a phagocytic process. Either from bone marrow macrophages are obtained and are termed recruited macrophages or macrophages are obtained from yolk-sac and are termed resident macrophages (Cha and Koo 2020). Frequently, sub-population of macrophages are characterized as classically activated macrophages (M1) associated with pro-inflammatory or we can say anti-tumor properties, that help them to find and kill tumor cells via phagocytosis and cytotoxicity behavior (Prenen and Mazzone 2019). On the other side, tissue healing and development can be done by alternatively activated macrophages (M2) because they can maintain anti-inflammatory characteristics (Atri et al. 2018). In in-vitro conditions macrophage is polarized into M1-like macrophages with the help of "Tumor-necrosis -factor alpha," "interferon-gamma," and "lipopolysaccharide" maintains cancer growth, cancer spread as well as mediates Th1 response (Weagel et al. 2015; Mehraj et al. 2021b).Interleukin-4, interleukin-10, and interleukin-13 on the other hand produce M2-type macrophages (Atri et al. 2018) and are important for inhibition of immune response and tissue remodeling, and proper angiogenesis development (Weagel et al. 2015) as illustrated in Fig. 7.1. In TME higher levels of M1 macrophages are correlated with diminished tumor violent behavior, as supported by several studies, instead of a higher number of M2 macrophages are associated with cancer development and weak outcome of tumor (Komohara et al. 2014). Additionally, macrophage polarization is induced by hormones, and cytokines in TME, (Aschenbrenner and Schultze 2017). Despite the fact several studies have displayed the cryptic result on the polarization of macrophages in TME, that display TAMs maybe act as both anti-tumoral and pro-tumoral, depending upon the macrophage phenotype (Cha and Koo 2020). But once TAMs gain M2 phenotype in TME after being linked with tumor cells, effector T cells as well as other cells, via the inhibition of adaptive immunity, tumor progression can be suppressed and induce angiogenesis and tumor repair (Mantovani et al. 2009). TAM phenotype particularity relies on tumor development, TAMs mostly contain the M2 phenotype which is pro-angiogenic (Qian and Pollard 2010). In the initial phase of the tumor, TAMs gap the "M1-phenotype" to induce anti-tumor activity and suppress the tumourigenesis., but in an advanced phase of malignancy, TAMs are redirected to the M2 type macrophages and promote tumor progression (Chen et al. 2021). In the course of tumor progression, the higher concentration of interleukin-12 and lower concentration of interleukin-10, polarized M1 macrophages infiltrate the tumor and boost immune response, promoting tumor cell destruction. On the other hand, at the advanced period of cancer proliferation, TAMs are polarized into the M2 phenotype, stimulated by reduced interleukin-12 expression levels and higher interleukin-10 expression levels leading to a reduced tumoricidal effect (Cha and Koo 2020).

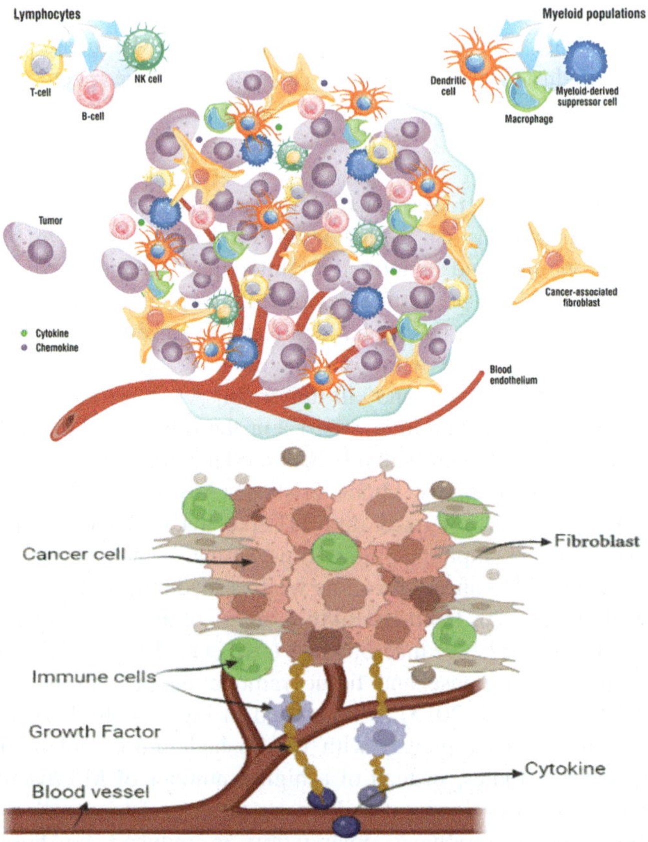

Fig. 7.2 Tumor microenvironment progression

M2-type macrophages produce a tumor micro-environment that regulates the tumor progression, their life span, and angiogenesis (Williams et al. 2016; Chanmee et al. 2014). Further study manifested that tumor-associated macrophages in mammary tumors have more M2-type macrophages, it has been revealed from different studies that chemicals released from breast tumor cells can polarize macrophages into M2-like. Apart from the above-discussed cytokines, other signals like "hypoxia-inducible-factor-1 (HIF-1), HIF-2," a "Nuclear-factor–kappa beta" (NF-kβ) also perform a significant part inside TME to re-polarize TAMs, by promoting HIF-1 and HIF-2 signals, TAMs favor to colonize themselves to less vascularized tumor cells and modify to the hypoxia microenvironment (Lewis and Pollard 2006). Once the hypoxia environment gets activity then it up-regulates the CXC chemokine receptor 4 (CXCR4) production and also increases the expression level of chemokine ligand 12 (CXCL12), which plays an essential part in metastasis (Müller et al. 2001), along with activation of another factor such as Nuclear factor–kappa beta (NF-kβ) also involved in modulation of TAMs transcriptional activities. TAMs reduce IL-12 expression levels, and responsible for defective NF-kβ stimulation, and also increased the IL-10 expression level (Sica et al. 2000). Disability in the switch-on of NF-kβ, by the over-exposure of nuclear p50 NF-kβ homodimers, suppresses the transcription of the pro-inflammatory gene, generating factors such as "interleukin-1, interleukin-12, and Tumor necrosis factor-α as well as nitric oxide (NO)" Fig.7.2 (Sica et al. 2000).

Tumor-associated macrophages make up a large inflammatory part in the invasion of breast cancer (O'Sullivan and Lewis 1994). Chemokines and cytokines are tumor-associated growth factors, aid in the recruitment of monocytes as well as macrophages into tumor sites (Leek and Harris 2002) "Chemokine (C-C motif) ligand-2 (CCL2)," called "monocyte-chemoattractant-protein 1 (MCP-1)," is secreted by the tumor as well as stromal cells (Ueno et al. 2000), it is considered one of the best cytokine involved in migration of TAMs into cancer site thus it is related with poor diagnosis in breast-cancer victims (Tsuyada et al. 2012). It has been demonstrated that CCL2 upgrade pulmonary malignancy in animal breast cancer study, by recruiting CCR2 expressing –monocytes (Qian et al. 2011). After CCL2-CCR2 switch on, they enhance the CCL3 secretion from macrophages, promoting breast cancer metastasis (Kitamura et al. 2015). Other prominent factors that attract TAMs to the breast tumor microenvironment are "CCL5" also termed "Regulated-upon-Activation" and RANTES secreted by normal T cells. Malignant epithelial cells also secrete CCL5 in breast cancer which is related to improved disease development.

CCL5 receptors are highly expressed by macrophages and infiltrate the TME in response to CCL5 secreted by tumor cells (An et al. 2019). CCL5 can also change the properties of tumor-associated macrophages by boosting the phenotype of colon cancer (Halama et al. 2016).Tumor cells secrete an additional number of substances for recruitment of macrophages like "Colony-stimulating-factor 1" termed as "macrophage-colony-stimulating factor" and "Granulocyte-macrophage colony-stimulating factor" and rest factors released by cells in breast cancer (Leek and Harris 2002; Fu et al. 1992). In breast cancer mouse models, macrophages express poor production of CSF-1 and enhance permeate of the Colony-stimulating factor – 1 receptor (CSF-1R) (Lin et al. 2001). Heparin-binding glycoprotein vascular endothelial growth factors are another cytokine linked with macrophage recruitment. This growth factor interplays an essential function in physiological and pathologic improvement of new blood vessels, and is considered a powerful mitogen, beyond that, it is also examined as the main component in angiogenic processes and plays a significant role in this, and higher expression of this cytokine is found to be in a variety of human malignancies (Table 7.1) (Ferrara and Davis-Smyth 1997). After activating one kind of VEGF receptor (VEGF-R1) makes VEGF chemotactic for macrophages and monocytes invitro (flflt-1) (Mir 2021; Sawano et al. 2001). Moreover, in the case of mice embryonic angiogenesis, macrophages lacking the VEGF-receptor expressed dramatically decreased migration concerning VEGF (Hiratsuka et al. 1998). In the current study, in breast cancer, we found a link between high VEGF expression and innumerable infiltrating macrophages (Leek et al. 2000) and evaluated increased VEGF production secreted by tumor cells recruit monocytes within TME and more eminently direct the migration of TAMs inside tumors. For tumor angiogenesis, another tissue factor, that is vascular permeability factor (VPF), is responsible for attracting monocytes inside TME (Clauss et al. 1990). In agreement herewith, higher levels of VEGF factor inside cancerous cells are linked with scavenger cells permeate in breast cancer in humans (Leek et al. 2000). The establishment of the Hypoxia state in TME leads to one more well-defined part, switching on macrophage activation and assemblage as well as putting a stop to the movement of macrophages out of this zone (Leek et al. 2000). In a breast cancer mouse model, TAMs are attracted into the hypoxia region mediated by "hypoxia-induced *Semaphorin* 3A" through the phosphorylation of VEGF-receptor (VEGF-R1). In tumor angiogenesis both VEGF and hypoxia signaling play a significant job and are shown to be an important trademark of cancer, indicating linkages between tumor angiogenesis and TAMs (Hanahan and Weinberg 2000). By adding further complexity to TAMs, Three lineages of macrophages are present at least that emerge at distinct periods of growth and remain upto maturity (Guerriero 2018). Every tissue of the body consists of 5–20% of resident macrophages originating from the yolk sac, formed during embryogenesis. In the course of homeostatic change, like tumor angiogenesis, several phenotypes of macrophage are recruited from bone marrow, spleen, and blood reservoir (Schulz et al. 2012) as well as from resident progenitors or by local recruitment. In different tissues, there is a distinct expression profile of macrophages at the transcriptional level and is demonstrated that malignant sites vary from the primary tumor region and thus may need to be diagnosed distinctly (Guerriero 2018). "Endothelial- monocyte -activating polypeptide II (EMAP II)" is another well-determined pro-inflammatory cytokine that recruits monocytes and macrophages (Kao et al. 1994). Knies et al. studies demonstrated, that "EMAP II-mRNA" and its messenger protein, "Pro-EMAP II," are produced via various kinds of malignancy (Table 7.1), Although mature-cytokine is predominantly produced by dead tumor cells lysate invitro and also persist at the apoptosis site in the mouse embryo development (Knies et al. 2000). This is due to the process of converting pro-form into a mature form of protein depending on cleavage with the help of protease enzyme and its release is upregulated during necrosis (Knies et al. 2000) after the secretion of mature EMAP II protein, aiding the movement of macrophages to apoptosis and necrosis sites where they are supposed to be capable of removing dead and necrotic debris.This is because tumor is associated with many necrosis sites, and there is an increased number of TAMs with increased necrosis areas (Leek et al. 1999), it is feasible because EMAP II protein plays a significant role in the recruitment of monocytes to some tumors. A wide variety of cell types secrete "endothelin's 1–3 (ET-1, -2, and -3)" which are small neuro-peptide and bioactive proteins and function like that of chemokines and show their effect by binding with two "7-trans-membrane-G protein-coupled receptors," "ET-RA and ET-RB," respectively. In vitro endothelin along with their receptors are greatly produced in various cell lines as well as in human tumors (Kusuhara et al. 1990) (Table 7.1). ET-1 act as a chemoattractant for "Human-monocytes-binding to ET-RA receptor," although ET-2 act as a chemoattractant for macrophages binds to "ET-RB receptor," so these data imply that ET-1 may aid monocyte recruitment within a tumor, while

ET-2 may aid monocyte location within tumor bulk. Notably, descriptive studies revealed their expression level, with an assemblage of macrophages in a wide array of human carcinoma, and provided most of the data related to the function of chemotaxis to recruit monocytes as well as macrophages inside TME. Studies using knock-out mice, and neutralizing antibodies to disrupt the function of these molecules are uncommon, and they are now needed to evaluate whether every molecule is required for recruitment of monocytes and in vivo TAMs localization. However monocytes are recruited by chemoattractant to the tumor site and are conceivable, that phagocytic cells are recruited also by cell detritus discharged by cell death of malignant cells, so it is shown in cell cultures that cell detritus cannot recruit phagocytes (Bessis and de Boisfleury-Chevance 1984), although up to now it has been investigated, soluble factors secreted from apoptosis of malignant cells or either through destruction of extracellular matrix go (ECM) through necrosis, and may recruit phagocytic cells for instance. Partially degraded collagen proteins aid to be a chemo-attractant for monocytes and macrophages (Mehraj et al. 2021b).

7.4 T Regulatory Cell Infiltration in Tumor Microenvironment

Heavy T regulatory cell movement and presence in the tumor microenvironment have been observed and studied widely (Fig. 7.3) and it is largely associated with an unfavorable prognosis in several malignancy types(Fu et al. 2007).In colon cancer, T

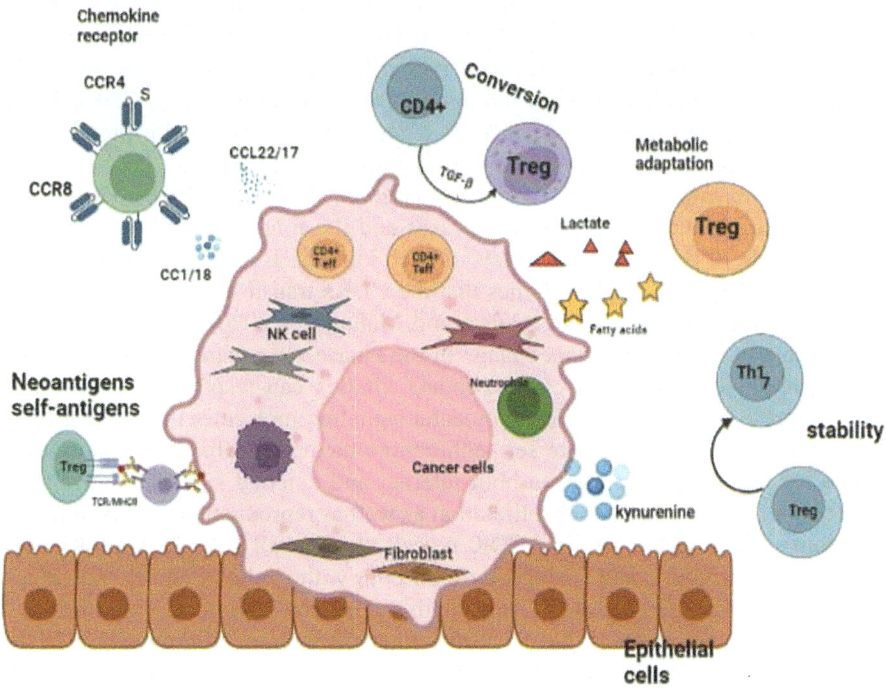

Fig. 7.3 T regulatory cell movement and presence in the tumor microenvironment

regulatory cells may suppress tumor-favoring inflammatory signals that are evoked by the gut microbial pool. The occurrence of T regulatory cells may also have difficulties in estimating advanced responses to immunotherapy. Advanced pre-treatment levels of circulating T regulatory cells were linked with better survival in skin carcinoma patients treated with "*ipilimumab*"(Martens et al. 2016). correspondingly, in non-small-cell lung carcinoma patients, an optimistic relationship has been seen between response to pd-one pdl- one blockade and occurrence of pdl-one+ T regulatory cells in the Tme (Wu et al. 2018). Several mechanisms could be used through the T regs to suppress anti-tumor immune responses. Molecules like the IL-2/Rα chain could divest local surroundings through an expression of high levels of Il-two and effector T cell role and contraction (Pandiyan et al. 2007). The change in IL-2 concentration by T regulatory cells affects natural killer cell (NK) function and homeostatic state (Gasteiger et al. 2013). Antigen-presenting cells are also controlled by T regulatory cells, by specific extension of dendritic cells, and activation of T regulatory cell depletion (Kim et al. 2007). This subduing is due to CT LA-4-dependent downward regulation of factor cd-eighty and cd-86 appearance by a course known as trans Endocytosis (Qureshi et al. 2011).

7.5 Cyclin-Dependent Kinases and Cell Cycle

Cell division, as well as cell enlargement, is controlled by a regular process of different phases called the cell cycle. There are main four steps/phases of the cell cycle including growth phase 1 (G1), DNA duplication phase (S phase), growth phase 2 (G2), and Mitotic phase (M phase). These four stages of the cell cycle are controlled by a series of protein complexes called CDK/cyclin complexes, these complexes aid the cell to proceed through all these phases normally. CDKs do not change all over the process, but there is a regulation in the Cyclin proteins that usually depends on the stage in the cell cycle present (Thu et al. 2018). Protein kinases are enzymes, possessing a highly conserved (Ding et al. 2020). The kinases are activated via the CDK/cyclin complexes and the four phases of the cell cycle control whole of the expression levels of such kinases. It has been proved experimentally via Crystallographic studies that the CDKs transit from inactive to active states and vice versa (Martínez-Alonso and Malumbres 2020). The interaction of cyclins causes a conformational change in the kinase's regulatory pocket, which is conformationally flexible. Cyclin-dependent kinases can respond in a variety of ways in presence of multiple developmental signaling molecules due to their flexibility. CDKs These kinases are serine/threonine mediators which phosphorylate a variety of proteins in reaction to both internal and external signals, controlling a variety of elements of cellular proliferation as well as reproduction. The coupling of a certain cyclin with a specific CDK reveals the overall selectivity of Kinases. CDK2, for example, could specifically bind both with cyclin A and E and behave differently depending on which Cyc it binds to. It thus shows that a specific CDK with a particular framework has a specific function within cell growth (Lu 2020). Furthermore, the cyclin/CDK loop controls the overall stability, phosphorylation,

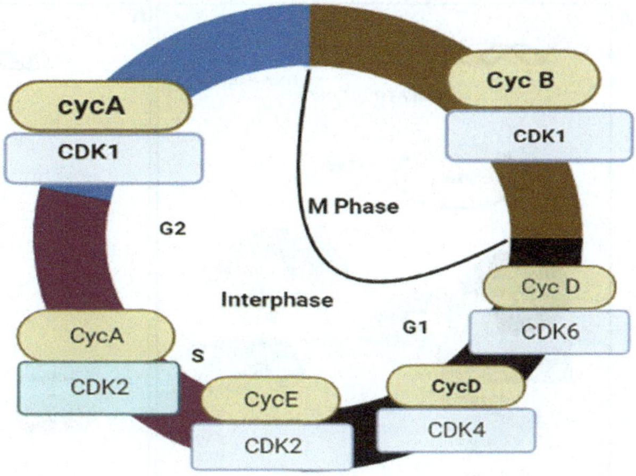

Fig. 7.4 CDKs are vital to cell cycle regulation: At each phase of the cell cycle, there occurs a specific association between CDKs and their cyclins to proceed further

and stimulation of CDKs during various stages of the cell division cycle. The relevance of such a combination stems from the fact that it is required for the advancement of cell growth concerning phosphorylation of numerous target loci, such as the tumor repressor protein retinoblastoma protein. In reaction to mitogenic cues and checks, the cyclins/CDKs switch from suppression to stimulation (Lu 2020). There are certain negative regulators of such complexes including the CKIs (Cyclin-dependent kinase inhibitors) such as INK4 proteins and CIP/KIP (CDK-interacting protein/kinase inhibitory proteins) (Fig. 7.4). APC/C (anaphase-promoting complex/ cyclosome) and SCF (Skp1–Cul1–F-box-protein) are two spindle protein molecules that regulate all the cell cycle phases and their transitions (Sivakumar and Gorbsky 2015). Dysfunction of any one of those pathways might cause cells to proliferate more quickly by disrupting the cell cycle, which is a characteristic of many malignancies, especially breast cancer.

7.6 Cell Cycle, CDKs, and Cancer

Tumors, especially breast cancer, are known for their cell growth dysregulation. Dysfunction of the cyclin/CDK complex disturbs cell cycle rhythm in a variety of cancers, resulting in tumor cells that continue to proliferate (Lee et al. 2019). The improper stimulation of CDKs, which is linked to cyclin gene amplification and upregulation, cell delocalization, or early cyclin production, plus suppression of INK4 or the CIP/KIP subfamily, results in cell cycle disruption, which causes tumors. Multiple investigations have found that cancerous cells mostly lack such cell cycle-regulating suppressive pathways. De Inactivation of tumor suppressor genes including Rb and TP53 (p53) or oncogene overexpression affects cyclin/CDK overexpression, culminating in unmanageable cell division cycle differentiation and expansion (Lee et al. 2019). Many tumor suppressors, including the

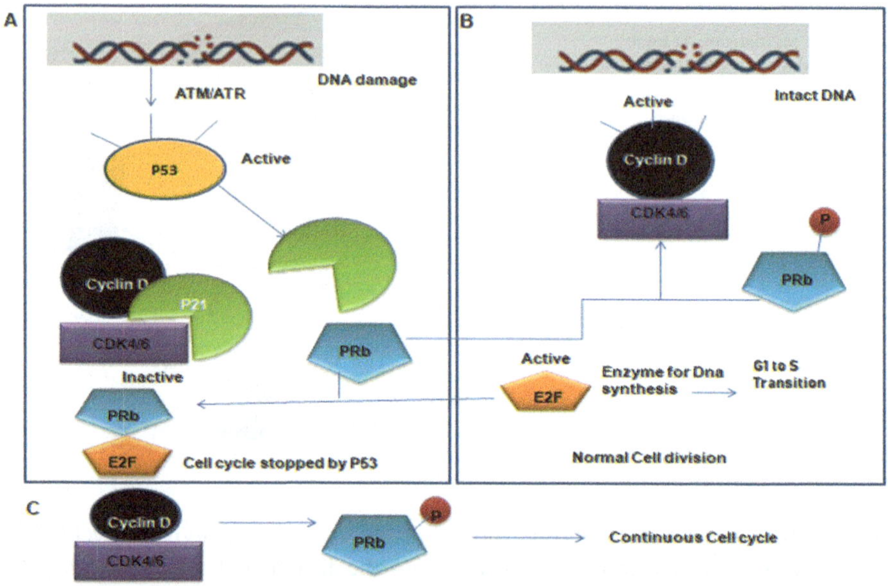

Fig. 7.5 Role of Rb protein in the cell cycle: (**a**) In its functional state, p53 leads to cell cycle arrest in those cells which are having DNA break, (**b**) The phosphorylation of pRb protein by CDK4/6/ cyclin- D complex leads to the release of pRb protein from E2F and thereby leads to the transcription of central genes responsible for G1/S transition by E2F, (**c**) In tumor cells augment in CDK4/6/cyclin-D leads to Rb protein hyper-phosphorylation, thus leading to unregulated cell division

Retinoblastoma (Rb) gene, which functions as a transcriptional inhibitor of E2F when this is dephosphorylated, adversely control cellular proliferation. Cyclin D/CDK4/6 complex phosphorylates pRb protein, causing it to be released off E2F as well as the expression of numerous genes, notably Polymerases, cyclin E, and A, that are essential for E2F's G1/S transitions (Fig. 7.5) Via attaching to the E2F transcription factor and inhibiting the production of genes necessary for the G1 to S transition, the pRb protein remains active in its hypo-phosphorylated form or pauses cell in the quiescent G0 stage. The phosphorylated status of Rb protein is modulated when the CDK/cyclin pair is dysregulated, changing the Rb protein's function and causing unregulated cellular proliferation. Likewise, uncontrolled stimulation of the tumor suppressor gene p53 promotes tumorigenesis (Yue et al. 2017). In the cells experiencing DNA damage, active p53 causes growth inhibition. p53 deactivation is common in a variety of tumors (Yue et al. 2017). Knockdown of the tumor suppressors such as Retinoblastoma or overexpression of oncogenes affects cyclin/ CDK elevation, leading to unregulated cell cycle continuation and multiplication (Wenzel and Singh 2018).

7.7 CDKS in BC Progression

Among women, the most widespread malignancies in breast cancer cause deaths at a higher rate due to tumors all over the world (Harbeck 2020). Regulation and development of breast tumorigenicity have been imputed by some of the CDKs. Metastasis and burden of tumors have been reduced by inhibiting the action of overexpressed CDKs (Bashour et al. 2017). Among cdk/Cyclin complexes, cdk4/6/Cyclin-D is the most significant and plays a vital role in the development and initiation of different cancers, BC is also included. It is revealed in various studies that during early and metastatic BC, cyclin D is overexpressed. Cyclin D1 and CDK4 both show different expressions like cyclin d-1 and cdk-4 show significantly elevated expression in lum-b and HER-2 BC, moderate expression in lum -A, and lowest expression in triple-negative breast cancer. Furthermore, cyclin D1/CDK4/6 contributes crucially to the phosphorylation of protein RB results in cell proliferation and also aids the progression (Lu 2020) of breast cancer tumors (Harbeck 2020). Moreover, numerous different CDKs were observed deregulated in most cancers, which includes BC.

For instance, upregulation of CDK2 in BC results in high expression of its cognates viz. cyclin- E and A (Santo et al. 2015). In the cell cycle performing as the core regulator from late G-1 up to the end of the S phase during the cell cycle is the essential role of CDK2. Cyclin partners, namely cyclins E-1or E-2 and A-2 activate it. Various regulatory functions like phosphorylating of smad-3, Rb protein, SMAD3, and some other proteins that order the synthesis of DNA have been displayed by CDK2 [45]. The studies have urged that a lot of cancers are amid over-expression of cdk-2 and are related to tumor cell production (Tadesse et al. 2020). The study conducted by Xiangming He and colleagues disclosed that cdk-2 plays a significant role before Christ's beginning and progression. It had been analyzed that inhibiting cdk-2 effectively de-accelerates breast carcinoma cell proliferation (Table 7.2).

7.8 CDKs in Breast Cancer Metastasis

Breast Carcinoma spreads effectively to BC patients with high mortality and survival rate. Breast tumor metastasis is the chief cause of death among breast cancer patients (Mehraj et al. 2021a). Almost 20/30% of early-level breast tumor sufferers develop distant metastasis, and about 90% of breast cancer deaths occur because of the troubles of spreading breast tumor. For this reason, it turns essential to throw mild on the position of CDKs in breast tumor spread. Breast tumor often reaches to the organs like the brain, liver, lungs, and bones. The aberrant changes in cyclins and their CDKs result in non-stop mobile division and the migration of tumor cells to different organs (Qureshi et al. 2021). The BC cells may be slothful. But, a few cancer agents, consisting of cyclin D1 and CDK4/6. could cause a systemic response, resulting in the metastasis of BC cells. This depicts the twin feature of cyclins that could engage with the CDKs and create a more competitive nature for

Table 7.2 Role of some CDK/Cyclin complexes in normal cells and BC cells

CDKs	Cyclin/ CDK complex	Role in normal cells	Role in BC Cells
CDK-1	Cyclin -A/ B	Aid in the M phase of the cell division	Plays role in apoptosis of MYC-driven TNBC
	Cyclin -A/ E	Associated to the G1 & S phase of the cell cycle	Plays role in developing BC or TNBC phenotype
	Cyclin -D	Aid in G1 & S phase shift of the cell cycle	Helps in BC initiation and regulation of tumorigenesis
	Cyclin -D	Aid in the G1 & S phase development of the cell cycle	Linked with the initiation of BC and control of tumorigenesis
	Cyclin- H	Linked with transcription of CAK and RNAP-II	Controls transcriptional addiction to a main cluster of genes in TNBC
	Cyclin- C	Assist to kick off transcription	Responds to adjuvant therapy in BC; associated with tumor progression
	Cyclin T	RNAP-II transcription; controls the elongation of transcription	A prognostic biomarker in BC patients following NACT

Dufy et al. (2015)

this ailment. The studies have discovered the role of CDK4/6 in Epithelial-Mesenchymal Transition (EMT), which is a function characteristic of cancer metastasis (Krajewski et al. 2020). The studies became executed using Zhen and co-employees and found out the function of CDK4/6 in breast tumor metastasis. CDK5 has been studied for its function in tgf-β1 precipitated emt in BC progression. Further, Adrian and co-people additionally revealed the position of cyclin-F in EMT (Krajewski et al. 2020). As a consequence, CDKs show their significant function in BC metastasis, in that way speeding the hostility of this illness amongst breast tumor sufferers.

7.9 Relative Numbers of Macrophages in Breast Cancer Progression

Apart from different sub-types of breast carcinoma, the abundance of macrophages varies, but It also varies with the developmental stages of cancer. In preclinical studies, it has been demonstrated that macrophages display early breast tumor distribution and progression of HER2+ breast cancer in mouse models, at that place cancer cells and myeloid cells secrete CCL2 and recruit "CD206+/Tie2+" macrophages to proliferate the illness. On the other hand, identical consequences were observed in "MMTV-PyMT luminal B breast cancer" of murine models, by hampering CCL2 production to prevent the TAMs recruitment inside the breast tumor site. Cause a reduction in metastasis, and persistent mice survival (Qian et al. 2011). Further preclinical models revealed that there is a remarkable link between "CSF-1" and metastatic breast tumors by utilizing the "mmtv-Pymt" ideal (Lin et al. 2001). In "DCIS (ductal -carcinoma in situ)" and in IDS (invasive-ductal-carcinoma) number and progression of macrophages are considerably elevated than the usual breast tissue. In breast cancer, TAMs displayed distinct transcriptomic trademarks from the normal breast tissue (Cassetta et al. 2019). In addition, M2-like macrophages are notably defined by authors as CD68+ macrophages (Esserman et al. 2006) have been demonstrated higher grades in comparison to DCI. The current study proposed by Gil- Del Alcazar et al. revealed that in-filtration of Immune cells in the case of "HER2+" and "TNBC" helps in disease proliferation when correlated with DCIS and IDCs (Gil Del Alcazar et al. 2017). It has been shown that IDS contains an abundance of macrophages compared to DCIS. Moreover, a higher frequency of macrophages is present in DCIS when associated with a higher amount of CD8+ T cells. Th1 and Th2 are supplemented with HER2+ IDCS while Th17 and T regulatory cells were enriched in TNBC IDCs, as revealed from the gene expression profile. The transition from DCIS to IDCs in TNBC tumors was associated with a substantial quantity of TILs, then in HER2+ tumors, although fewer numbers were seen in the functional state, well-defined exhaustion of T cells was seen in advanced stages of TNBC, still, more inquiry is needed to understand how macrophages take part in tumor proliferation at various stages.

7.10 Location of Macrophages in Breast Tumors

The placement of macro-phages inside the TME, besides disease state and breast tumor subtype, may be a forecast of their properties and are linked with scientific consequences. Nevertheless, scientists have yet to agree on the degree to which region-specific-TAM behavior may be used predictive indicator. Normally, macrophages in tumor stroma are correlated to inhibition of immune response, angiogenesis, and migration of tumor cells. Macrophages in cancer nests, on the other hand, are more heterogeneous among cell kinds, as a consequence, they are linked with worse "overall survival" (OS) as well as "Recurrence-free survival" (RFS) in breast tumor patients (Yang et al. 2018). Merdeck et al., described that stromal-tumor-associated macrophages are significantly related with tumor development but not tumor nest TAMs (Medrek et al. 2012). In 60 patients with invasive BC. It has been currently revealed the presence of CD60 and CD163 in tumor nests and tumor stroma. In tumor stroma, the highest concentration of $CD68^+$ TAMs was extremely correlated with giant tumor size and positive extra-nodal spread, moreover higher amount of "CD163+" TAMs in cancer stroma has been linked with good lympho-vascular invasion (LBV1) also termed vascular invasion(VI), extra-nodal extensive, and other molecular subtypes (Mwafy and El-Guindy 2020). The process is exchangeable TME modulates the behavior of macrophages and macrophages, in turn, modulate the tumor location by secreting several signals in breast cancer subtypes for instance macrophages focus on invasive tumor front (ITF) in HER2+ and basal-like subtypes, and utilize "TGF-β signaling" in condensing the extracellular matrix (ECM), and participating in mammary cancer progression (Acerbi et al. 2015). Furthermore, macrophages and TME both of them work altogether by stimulating each other's composition. Both macrophages and the degree of TME infiltration lead to additional complexity by targeting anti-cancer therapy in breast cancer.

7.11 Breast Cancer Cell Metastasis

In patients with breast tumors, metastasis is the mainly frequent cause of death amongst them. TAMs play a major part in breast tumor assault and spread. For the therapeutic method, TAMs are considered viable for targeting strategy (Chen et al. 2019). With the help of "chemokine (C-C motif) ligand -2. Chemokine(C-C) ligand-5 (CCL5), chemokine (C-C) ligand -18 (CCL18), along with CCL2," TAMs promotes breast cancer cell spread to bone and lung tissues, as part of its functional mechanism. Breast cancer cells secrete CCL2 and recruit chemokine "C-C -Receptor 2+ (CCR2+)" macro-phages to get together in lung and production of bone cells from osteoblast, thus take part in cancer cell colonization and building of metastatic niche, as a result inhibiting "CCL2-CCR2" could effectively prevent tumor metastasis. CCL5 released from breast tumor cells acts upon mononuclear macrophages' transition to TAMs whichever stimulates tumor spread as well as infiltration (An et al. 2019). Another factor significantly secreted by TAMs is CCL18, and its function is also linked with metastasis and reduced survival of patients. CCL18 is having a functional receptor that is "PYK2 N-terminal-domain-interacting receptor 1 (PITPNM3)," inhibiting cancer spread and invasive effect of CCL18. Between "malignant -phyllodes-tumors" (PT) in mammary and TAMs, Nie and colleagues discovered a feedback-loop of "CCL5-CCR5 and 'CCL18-PIPTNM3," which help preserve TAM phenotype and PT aggressiveness. For the further inhibition of tumor metastasis, they used a CCR5 inhibitor and CCL-18-monoclonal antibody and blocked the CCL-5-CCR-5 and CCL-18-PIPTNM-3 pathways. TAMs release both cellular cytokines and their surface receptor, both of which play a central part in exciting breast tumor metastasis. TAMs lead to the construction of epidermal growth factor (EGF) which activates its receptor that is epidermal growth factor receptors (EGFRs) in breast tumors, promoting metastasis and CSF-1 production. TAMs are recruited and activated by CSF-1 causing them to secrete more EGF, implying the presence of an "EGF/CSF-1" regulating a good link between TAMs and tumor cells. EGF causes breast cancer cells to infiltrate blood vessels, resulting in metastasis in blood vessels (Condeelis and Pollard 2006). TAMs secrete a cluster of "matrix-metalloproteinases" (MMPs)," like MMP-2, MMP-7, and MMP-9, which are linked with the destruction of matrix materials in the TME, stimulating the establishment of metastatic tumor microenvironment by facilitating tumor cell growth. Higher production of macrophages receptors also termed scavenger receptor increases tumor development and metastasis with collagenous structure (MARCO). MARCO is linked to epithelial-mesenchymal-transition (EMT), gene profiles that drive metastasis, and inhibiting MARCO production could effectively stop EMT (Georgoudaki et al. 2016).

7.12 Breast Cancer Angiogenesis

Angiogenesis is the development of new connective tissue vessels to help cancer grow and develop. Tumor-associated macrophages play a significant role in angiogenesis by being tightly linked to the high-density vascular system that emerged in breast tumors. "Vascular-endothelial- growth factor (VEGF)" is a vital source for TAMs in TME of breast tumors. Angiogenesis in breast tumors is triggered by an interaction between "VEGF" with "vascular endothelial growth factor receptors" (VEGFRs). As a result blocking interaction between VEGF/ VEGFR can dramatically reduce angiogenesis and tumor spread (Song et al. 2018). Macrophage-colony-stimulating factor also termed colony-stimulating factor-1, stimulates the recruitment of macrophages and distinguishes them into the M2 phenotype. "Colony-stimulating factor -1 receptor (CSF1-R)" suppressors during neoplasm formation in the breast glands can decrease TAMs, therefore reducing angiogenesis, metastasis, and reducing the risk of recurrence. In TME hypoxia is a trademark that stimulates angiogenesis and helps in macrophage recruitment (LaGory and Giaccia 2016). Upregulation of hypoxia-inducible factors (HIFs) in hypoxia conditions stimulates the transfer of macrophages into TAMs, which in turn act as transcriptional promotors of VEGF. VEGF promotes the hypoxia environment by stimulating angiogenesis which helps the tumor to receive oxygen and nutrients, for its development (Mir et al. 2022; Mir and Mehraj 2019). The earliest evidence supported that HIF signaling is linked with angiogenesis, suppression in its signaling slows down angiogenesis and tumor development. Suppression of HIF-2α signaling causes highly disordered blood vessels to develop and the hypoxia situation in the TME to worsen (LaGory and Giaccia 2016). In addition, hypoxia TME up-grade the production of "Activating-transcription-factor 4 (ATF4)," one family member of "ATF/cAMP-response-element-binding protein (CREB)," linked with the macrophage recruitment and development of angiogenesis which promotes cancer development indirectly (Liu et al. 2015); therefore, angiogenesis is up-regulated by TAMs in tumor malignancy.

7.13 Conclusion

Breast malignancy (BC) is the most frequently detected cancer in women globally and is also the common reason for tumor-associated fatality in females. It has been broadly accepted that the Tumor surrounding (TME) has a major influence on the aggressive deportment of malignant solid tumors. In TME, macrophages and other cells are the most common inflammatory mediator, other mediators are also present such as "chemokines, growth factors, pro-and anti-inflammatory cytokines." Furthermore, it has been demonstrated that cancer is linked with dysregulation of the cyclins and CDKs. Uncontrolled regulation of CDKs leads to the development of tumor. Different types of inhibitors have been identified which can control unregulated CDKs and thus have the potential to retard the growth of cancer. Many different types of CDKs have been identified and approved by FDA. In addition

Cells like TAMs is the result of the tumor microenvironment's reprogramming macrophages like tumor-mediated "exosomes, cytokines, and other immune cells," resulting in distinct TAM cytokine production. So, the association between other cells with macrophages in TME can be changed and generate a new place to promote the tumor cell's life span and its progression. So, TAMs perform an essential part in breast tumor progression by enhancing TME angiogenesis and metastasis, initiating tumor cells' steaminess, and energy metabolism, and supporting immune response inhibition. In TME several factors regulate the macrophage activity and its polarization. So as a result, TAMs are a crucial player in cancer development and the aim of generating a viable treatment strategy.

7.14 Glossary and Abbreviations

7.14.1 Glossary

The tumor microenvironment: is the area that surrounds the tumor, several factors are present in this area like endothelial cells, immune cells, growth factors, cytokines, chemokines, TAM, fibroblast, etc.

Cytokines: are small glycoproteins that regulate the immune cells to do their function by secreting various chemicals, that help immune cells to move at the site of inflammation, trauma, infection, etc., and act as signaling molecules between cell-to-cell transmission of immune response.

Vascular endothelial growth factor: secreted by macrophages and suppresses the activity of T cells by enhancing the recruitment of Treg, MDSCs and inhabits the lymphocyte activation and their differentiation, induces angiogenesis in cancer.

Colony-stimulating factor: in breast cancer, transfer of recruited monocytes into non-polarized macrophages (MO). MO-macro-phages are high-plastic, and alter their phenotype about outer-signals either 'M1-M2'. M1-phenotype has 'pro-inflammatory' and 'anti-tumor' capability. While M2 led to "anti-inflammatory as well as tumor-promoting" function.

Tumor angiogenesis: is the process of forming fresh blood vessels around cancer cells, for delivery of oxygen and other substance that nourish the tumor cells, crucial for tumor growth and development.

Programmed-Death Ligand-1: is an immune checkpoint, which generated a significant pathway for TAMs to promote the effector-T cell's tumor-killing function.

Interleukins are glycoproteins that are one of the groups of cytokines produced by leucocytes and promote immune response, also termed IL. It is categorized into four groups depending upon its structural features.

Tumor hypoxia: it is the condition, where the oxygen level is low, in the case of tumors oxygen quantity runs lower than control, hypoxia state turns malignant cell "metabolism" to undertake "aerobic glycolysis" from oxidative phosphorylation., ultimately accelerating growth.

Dendritic cells: are special kinds of cells that act as a connecting link between inborn and acquired immunity. The central function is to process antigenic molecules and present them on the cell surface of the T cells via MHC-II for additional immune action.

T-lymphocytes: T-lymphocytes are white blood cells that grow from stem cells in the bone marrow. They shield our body from pathogenic invasion and may help to fight cancer.

7.15 Further Readings

F For more incites about the topic we would suggest detailed findings from the books of (Mir MA, 2022) https://doi.org/10.1016/C2021-0-02565-7, https://doi.org/10.1016/C2022-0-00074-X (Mir MA, 2021) https://doi.org/10.52305/WXJL6770 or diagrammatic illustrations, descriptive tables, (lazzeroni, 2012), http://www.eurekaselect.com/article/49928https://www.researchgate.net/publication/342829505_Molecular_Complexity_of_Lymphovascular_Invasion_The_Role_of_Cell_Migration_in_Breast_Cancer_as_a_Prototype/figures?lo=1

See video links on Cancer, its various types, current new treatment possible options available, https://www.sciencedirect.com/science/article/pii/S205970292032278Xhttps://youtu.be/wIsdjfwPUxY, https://youtu.be/SVjJt984PlU

References

Acerbi I, Cassereau L, Dean I, Shi Q, Au A, Park C et al (2015) Human breast cancer invasion and aggression correlates with ECM stiffening and immune cell infiltration. Integr Biol 7:1120–1134

An G, Fand W, Huang S, Feng L, Bai J, Gu S et al (2019) Effects of CCL5 on the biological behavior of breast cancer and the mechanisms of its interaction with tumor-associated macrophages. Oncol Rep 42:2499–2511

Aschenbrenner AC, Schultze JL (2017) New "programmers" in tissue macrophage activation. Pflugers Arch - Eur J Physiol 469:375–383

Atri C, Guerfali FZ, Laouini D (2018) Role of human macrophage polarization in inflammation during infectious diseases. Int J Mol Sci 19:1801

Barsky SH, Karlin NJ (2005) Myoepithelial cells: autocrine and paracrine suppressors of breast cancer progression. J Mammary Gland Biol Neoplasia 10:249–260

Bashour SI, Doostan I, Keyomarsi K, Valero V, Ueno NT, Brown PH, Litton JK, Koenig KB, Karuturi M, Abouharb S (2017) Rapid breast cancer disease progression following cyclin-dependent kinase 4 and 6 inhibitor discontinuation. J Cancer 8:2004–2009

Benson JR, Jatoi I (2012) The global breast cancer burden. Future Oncol 8(6):697–702. https://doi.org/10.2217/fon.12.61

Bessis M, de Boisfleury-Chevance A (1984) Facts and speculation about necrotaxis (chemotaxis toward a dying cell). Blood Cells 10:5–22

Biswas SK, Mantovani A (2010) Macrophage plasticity and interaction with lymphocyte subsets: cancer as a paradigm. Nat Immunol 11:889–896

Bray FFJ, Soerjomataram I, Siegel RL, Torre LA, Jemal A (2018) GLOBOCAN estimates incidence and mortality worldwide for 36 cancers in 185 countries. CA Cancer J Clin 68: 394–424

Bussard KM, Mutkus L, Stumpf K, Gomez-manzano C, Marini FC (2016) Tumor-associated stromal cells as key contributors to the tumor microenvironment. Breast Cancer Res 18:1–11

Canavese M, Santo L, Raje (2012) Cyclin-dependent kinases in cancer: potential for therapeutic intervention. Cancer Biol Ther 13:451–457

Cassetta L, Fragkogianni S, Sims AH, Swierczak A, Forrester LM, Zhang H et al (2019) Human tumor-associated macrophage and monocyte transcriptional landscapes reveal cancer-specific reprogramming, biomarkers, and therapeutic targets. Cancer Cell 35:588–602

Cha YJ, Koo JS (2020) Role of tumor-associated myeloid cells in breast cancer. Cell 9:1785

Chanmee T, Ontong P, Konno K, Itano N (2014) Tumor-associated macrophages as major players in the tumor microenvironment. Cancers 6:1670–1690

Chen D, Zhang X, Li Z, Zhu B (2021) Metabolic regulatory crosstalk between tumor microenvironment and tumor-associated macrophages. Theranostics 11:1016

Chen Y, Song Y, Du W et al (2019) Tumor-associated macrophages: an accomplice in solid tumor progression. J Biomed Sci 26:78

Clauss M, Gerlach M, Gerlach H, Brett J, Wang F, Familletti PC et al (1990) Vascular permeability factor: a tumor-derived polypeptide that induces endothelial cell and monocyte procoagulant activity, and promotes monocyte migration. J Exp Med 172:1535–1545

Cojoc M, Peitzsch C, Trautmann F, Polishchuk L, Telegeev GD, Dubrovska A (2013) Emerging targets in cancer management: role of the CXCL12/CXCR4 axis. Onco Targets Ther 6:1347–1361

Coleman RE, Gregory W, Marshall H, Wilson C, Holen I (2013) The metastatic microenvironment of breast cancer: clinical implications. Breast 22:s50–s56

Condeelis J, Pollard JW (2006) Macrophages: obligate partners for tumor cell migration, invasion, and metastasis. Cell 124:263–266

Ding L, Cao J, Lin W, Chen H, Xiong X, Ao H, Yu M, Lin J, Cui Q (2020) The roles of cyclin-dependent kinases in cell-cycle progression and therapeutic strategies in human breast cancer. Int J Mol Sci 21:1960

Dufy MJ, Walsh S, Mcdermott EW, Crown J (2015) Biomarkers in breast cancer: where are we and where are we going. Adv Clin Chem 71:1–23

Dvorak HF (1986) Tumors: wounds that do not heal. Similarities between tumor stroma generation and wound healing. N Engl J Med 315:1650–1659

Esserman LJ, Kumar AS, Herrera AF, Leung J, Au A, Chen YY et al (2006) Magnetic resonance imaging captures the biology of ductal carcinoma in situ. J Clin Oncol 24:4603–4610

Ferrara N, Davis-Smyth T (1997) The biology of vascular endothelial growth factor. Endocrinol Rev 18:4–25

Folgueira MA, Maistro S, Katayama ML, Roela RA, Mundim FG, Nanogaki S, De Bock GH, Brentani MM (2013) Markers of breast cancer stromal fibroblasts in the primary tumor site associated with lymph node metastasis: a systematic review including our case series. Biosci Rep 33:e00085

Franklin RA et al (2014) The cellular and molecular origin of tumor-associated macrophages. Science 344(6186):921–925. https://doi.org/10.1126/science

Fu J, Xu D, Liu Z, Shi M, Zhao P, Fu B, Zhang Z, Yang H, Zhang H, Zhou C, Yao J, Jin L et al (2007) Increased regulatory T cells correlate with CD8 T-cell impairment and poor survival in hepatocellular carcinoma patients. Gastroenterology 132:2328–2339

Fu YX, Cai JP, Chin YH, Watson GA, Lopez DM (1992) Regulation of leukocyte binding to endothelial tissues by tumor-derived GM-CSF. Int J Cancer 50:585–588

Gao Y, Bado I, Wang H, Zhang W, Rosen JM, Zhang XH-F (2019) Metastasis organotropism: redefining the congenial soil. Dev Cell 49:375–391

Gasteiger G, Hemmers S, Firth MA, Le Floc'h A, Huse M et al (2013) IL-2-dependent tuning of NK cell sensitivity for target cells is controlled by regulatory T cells. J Exp Med 210:1167–1178

Georgoudaki AM, Prokopec KE, Boura VF et al (2016) Reprogramming tumor-associated macrophages by antibody targeting inhibits cancer progression and metastasis. Cell Rep 15: 2000–2011

Gil Del Alcazar CR, Huh SJ, Ekram MB, Trinh A, Liu LL, Beca F et al (2017) Immune escape in breast cancer during in situ to invasive carcinoma transition. Cancer Discov 7:1098–1115
Guerriero JL (2018) Macrophages: the road less traveled, changing anticancer therapy. Trends Mol Med 24:472–489
Halama N, Zoernig I, Berthel A, Kahlert C, Klupp F, Suarez-Carmona M et al (2016) Tumoral immune cell exploitation in colorectal cancer metastases can be targeted effectively by anti-CCR5 therapy in cancer patients. Cancer Cell 29:587–601
Hanahan D, Weinberg RA (2000) The hallmarks of cancer. Cell 100:57–70
Hanker AB, Sudhan DR, Arteaga CL (2020) Overcoming endocrine resistance in breast cancer. Cancer Cell 37:496–513
Harbeck N (2020) Breast cancer is a systemic disease optimally treated by a multidisciplinary team. Nat Rev Dis Primers 1:1–2
Hiratsuka S, Minowa O, Kuno J, Noda T, Shibuya M (1998) Flt-1 lacking the tyrosine kinase domain is sufficient for normal development and angiogenesis in mice. Proc Natl Acad Sci U S A 95:9349–9354
Hu M, Polyak K (2008) Microenvironmental regulation of cancer development. Curr Opin Genet Dev 18:27–34
Jiang X, Shapiro DJ (2014) The immune system and inflammation in breast cancer. Mol Cell Endocrinol 382:6673–6682
Kakonen SM, Mundy GR (2003) Mechanisms of osteolytic bone metastases in breast carcinoma. Cancer 97:834–839
Kao J, Houck K, Fan Y et al (1994) Characterization of a novel tumor-derived cytokine. Endothelialmonocyte activating polypeptide II. J Biol Chem 269:25106–25119
Kim JM, Rasmussen JP, Rudensky AY (2007) Regulatory T cells prevent catastrophic autoimmunity throughout the lifespan of mice. Nat Immunol 8:191–197
Kitamura T, Qian BZ, Soong D, Cassetta L, Noy R, Sugano G et al (2015) CCL2-induced chemokine cascade promotes breast cancer metastasis by enhancing retention of metastasis-associated macrophages. J Exp Med 212:1043–1059
Knies UE, Kröger S, Clauss M (2000) Expression of EMAP II in the developing and adult mouse. Apoptosis 5:141–151
Komohara Y, Jinushi M, Takeya M (2014) Clinical significance of macrophage heterogeneity in human malignant tumors. Cancer Sci 105:1–8
Krajewski A, Gagat M, Mikołajczyk K, Izdebska M, Żuryń A, Grzanka A (2020) Cyclin F downregulation afects epithelial-mesenchymal transition increasing proliferation and migration of the A-375 melanoma cell line. Cancer Manage Res 12:13085
Kusuhara M, Yamaguchi K, Nagasaki K et al (1990) Production of endothelin in human cancer cell lines. Cancer Res 50:3257–3261
LaGory EL, Giaccia AJ (2016) The ever-expanding role of HIF in tumor and stromal biology. Nat Cell Biol 18:356–365
Lee Y, Lahens N, Zhang S, Bedont J, Field JM, Sehgal A (2019) G1/S cell cycle regulators mediate effects of circadian dysregulation on tumor growth and provide targets for timed anticancer treatment. PLoS Biol 17
Leek RD, Harris AL (2002) Tumor-associated macrophages in breast cancer. J Mammary Gland Biol Neoplasia 7:177–189
Leek RD, Hunt NC, Landers RJ, Lewis CE, Royds JA, Harris AL (2000) Macrophage infiltration is associated with VEGF and EGFR expression in breast cancer. J Pathol 190:430–436
Leek RD, Landers RJ, Harris AL, Lewis CE (1999) Necrosis correlates with high vascular density and focal macrophage infiltration in invasive carcinoma of the breast. Br J Cancer 79:991–995
Lewis CE, Pollard JW (2006) Distinct role of macrophages in different tumor microenvironments. Cancer Res 66:605–612
Lin EY, Nguyen AV, Russell RG, Pollard JW (2001) Colony-stimulating factor 1 promotes progression of mammary tumors to malignancy. J Exp Med 193

Liu C, Li Z, Wang L et al (2015) 2015 activating transcription factor 4 promotes angiogenesis of breast cancer through enhanced macrophage recruitment. Biomed Res Int

Lu Y (2020) The role of cyclin-dependent kinases on the metastasis of breast cancer. *Novel Approach Cancer Study*

Ma T, Van Tyne BA, Wei Y, Garrett MD, Nelson D, Adams PD, Wang J, Qin J, Chow LT, Harper JW (2000) Cell cycle-regulated phosphorylation of p220(NPAT) by cyclin E/Cdk2 in Cajal bodies promotes histone gene transcription. Genes Dev 14:2298–2313

Mantovani A, Sica A, Allavena P, Garlanda C, Locati M (2009) Tumor-associated macrophages and the related myeloid-derived suppressor cells as a paradigm of the diversity of macrophage activation. Hum Immunol 70:325–330

Mantovani A, Sozzani S, Locati M, Allavena P, Sica A (2002) Macrophage polarization: tumor-associated macrophages as a paradigm for polarized M2 mononuclear phagocytes. Trends Immunol 23:549–555

Martens A, Wistuba-Hamprecht K, Geukesfoppen M, Yuan J, Postow MA et al (2016) Baseline peripheral blood biomarkers associated with clinical outcome of advanced melanoma patients treated with ipilimumab. Clin Cancer Res 22:2908–2918

Martinez FO, Gordon S (2014) The M1 and M2 paradigm of macrophage activation: time for reassessment. F1000Prime Rep 6:13

Martinez FO, Gordon S, Locati M, Mantovani A (2006) Transcriptional profiling of the human monocyte-to-macrophage differentiation and polarization: new molecules and patterns of gene expression. J Immunol 177:7303–7311

Martínez-Alonso D, Malumbres M (2020) Mammalian cell cycle cyclins. Semin Cell Dev Biol 107

Medrek C, Pontén F, Jirstrom K, Leandersson K (2012) The presence of tumor-associated macrophages in tumor stroma as a prognostic marker for breast cancer patients. BMC Cancer 12:306

Mehraj U, Ganai RA, Macha MA et al (2021a) The tumor microenvironment as driver of stemness and therapeutic resistance in breast cancer: new challenges and therapeutic opportunities. Cell Oncol 44:1209–1229

Mehraj U, Qayoom H, Mir MA (2021b) Prognostic significance and targeting tumor-associated macrophages in cancer: new insights and future perspectives. Breast Cancer 28:539–555

Mir MA, Mehraj U (2019) Double-crosser of the immune system: macrophages in tumor progression and metastasis. Curr Immunol Rev 15(2):172–184

Mir MA (2021) Combination therapies and their effectiveness in breast cancer treatment. Nova Biomedical Science Publishers USA, pp 1–411. https://doi.org/10.52305/WXJL6770. Book ISBN: 978-1-68507-195-0

Mir MA (2022a) Combinational therapy in triple negative breast cancer, 1st edn. Elsevier Inc. Publishers USA, pp 1–242. https://doi.org/10.1016/C2021-0-02565-7. ISBN: 9780323961363

Mir MA et al (2022) Role of tumor-associated macrophages in the breast tumor microenvironment, Chapter-6. In: Role of tumor microenvironment in breast cancer and targeted therapies. Academic Press, Elsevier, pp 137–169. https://doi.org/10.1016/B978-0-443-18696-7.00003-8, ISBN 9780443186967

Mir MA (2022b) Role of tumor microenvironment in breast cancer and targeted therapies, 1st edn. Elsevier Inc. Publishers USA, pp 1–350. https://doi.org/10.1016/C2022-0-00074-X, ISBN: 9780443186974

Mosser DM, Edwards JP (2008) Exploring the full spectrum of macrophage activation. Nat Rev Immunol 8:958–969

Müller A, Homey B, Soto H, Ge N, Catron D, Buchanan ME, Mcclanahan T, Murphy E, Yuan W, Wagner SN et al (2001) Involvement of chemokine receptors in breast cancer metastasis. Nature 410:50–56

Mwafy SE, El-Guindy DM (2020) Pathologic assessment of tumor-associated macrophages and their histologic localization in invasive breast carcinoma. J Egypt Natl Canc Inst 32:6

O'Sullivan C, Lewis CE (1994) Tumor-associated leucocytes: friends or foes in breast carcinoma. J Pathol 172:229–235

Okuda M, Horn HF, Tarapore P, Tokuyama Y, Smulian AG, Chan PK, Knudsen ES, Hofmann IA, Snyder JD, Bove KE et al (2000) Nucleophosmin/B23 is a target of CDK2/cyclin E in centrosome duplication. Cell 103:127–140

Paget S (1889) The distribution of secondary growths in cancer of the breast. Lancet 133:571–573

Pandiyan P, Zheng L, Ishihara S, Reed J, Lenardo MJ (2007) CD4+CD25+Foxp3+ regulatory T cells induce cytokine deprivation-mediated apoptosis of effector CD4+ T cells. Nat Immunol 8: 1353–1362

Place AE, Huh SJ, Polyak K (2011) The microenvironment in breast cancer progression: biology and implications for treatment. Breast Cancer Res 13:227

Prenen H, Mazzone M (2019) Tumor-associated macrophages: a short compendium. Cell Mol Life Sci 76:1447–1458

Qian B-Z, Li J, Zhang H, Kitamura T, Zhang J, Campion LR et al (2011) CCL2 recruits inflammatory monocytes to facilitate breast-tumor metastasis. Nature 475:222–225

Qian BZ, Pollard JW (2010) Macrophage diversity enhances tumor progression and metastasis. Cell 141:39–51

Qureshi MFH, Shah M, Lakhani M, Abubaker ZJ, Mohammad D, Farhan H, Zia I, Tafveez R, Khan ST, Rubina G (2021) Gene signatures of cyclin-dependent kinases: a comparative study in naïve early and advanced stages of lung metastasis breast cancer among pre and post-menopausal women. Genes Cancer 12

Qureshi OS, Zheng Y, Nakamura K, Attridge K, Manzotti C et al (2011) Trans-endocytosis of CD80 and CD86: a molecular basis for the cell-extrinsic function of CTLA-4. Science 3(332): 600

Rhee L (2016) Diverse macrophages polarization in tumor microenvironment. Arch Pharm Res 39: 1588–1596

Salgado R, Denkert C, Demaria S, Sirtaine N, Klauschen F, Pruneri G, Wienert S, van den Eynden G, Baehner FL, Penault-Llorca F et al (2015) The evaluation of tumor-infiltrating lymphocytes (TILs) in breast cancer: recommendations by an International TILs Working Group 2014. Ann Oncol 26:259–271

Santo L, Siu K, Raje N (2015) Targeting cyclin-dependent kinases and cell cycle progression in human cancers. Semin Oncol 42:788–800

Sawano A, Iwai S, Sakurai Y et al (2001) Vascular endothelial growth factor receptor 1, is a novel cell surface marker for the lineage of monocyte macrophages in humans. Blood 97:785–791

Schulz C, Perdiguero GE, Chorro L, Szabo-Rogers H, Cagnard N, Kierdorf K et al (2012) A lineage of myeloid cells independent of Myb and hematopoietic stem cells. Science 336:86

Sica A, Saccani A, Bottazzi B, Polentarutti N, Vecchi A, van Damme J, Mantovani A (2000) Autocrine production of IL-10 mediates defective IL-12 production and NF-kappa B activation in tumor-associated macrophages. J Immunol 164:762–767

Sivakumar S, Gorbsky G (2015) Spatiotemporal regulation of the anaphase-promoting complex in mitosis. Nat Rev Mol Cell Biol 16:82–94

Song Y, Tang C, Yin C (2018) Combination antitumor immunotherapy with VEGF and PIGF siRNA via systemic delivery of multi-functionalized nanoparticles to tumor-associated macrophages and breast cancer cells. Biomaterials 185:117–132

Soysal SD, Tzankov A, Muenst SE (2015) Role of the tumor microenvironment in breast cancer. Pathobiology 82:142–152

Tadesse S, Anshabo A, Portman N, Lim E, Tilley W, Caldon CE, Wang S (2020) Targeting CDK2 in cancer: challenges and opportunities for therapy. Drug Discov Today 25:406–413

Thu KL, Soria-Bretones I, Mak TW, Cescon DW (2018) Targeting the cell cycle in breast cancer: towards the next phase. Cell Cycle 17:1871–1885

Tsuyada A, Chow A, Wu J, Somlo G, Chu P, Loera S et al (2012) CCL2 mediates cross-talk between cancer cells and stromal fibroblasts that regulates breast cancer stem cells. Cancer Res 72:2768–2779

Ueno T, Toi M, Saji H, Muta M, Bando H, Kuroi K (2000) Significance of macrophage chemoattractant protein-1 in macrophage recruitment, angiogenesis, and survival in human breast cancer. Clin Cancer Res 6:3282–3289

Vitale I, Manic G, Coussens LM, Kroemer G, Galluzzi L (2019) Macrophages and metabolism in the tumor microenvironment. Cell Metab 30:36–50

Watanabe N, Broome M, Hunter T (1995) Regulation of the human WEE1Hu CDK tyrosine 15-kinase during the cell cycle. EMBO J 14

Weagel ESC, Liu PG, Robison R, O'neill K (2015) Macrophage polarization and its role in cancer. J Clin Cell Immunol 6:338

Wenzel ES, Singh A (2018) Cell-cycle checkpoints and aneuploidy on the path to cancer. In Vivo 32:1–5

Williams CB, Yeh ES, Soloff AC (2016) Tumor-associated macrophages: unwitting accomplices in breast cancer malignancy. NPJ Breast Cancer 2:15025

Wu SP, Liao RQ, Tu HY, Wang WJ, Dong ZY et al (2018) Stromal PD-L1-positive regulatory T cells and PD-1-positive CD8-positive T cells define the response of different subsets of non-small cell lung cancer to PD-1/PD-L1 blockade immunotherapy. J Thorac Oncol 13:521–532

Yang M, McKay D, Pollard JW, Lewis CE (2018) Diverse functions of macrophages in different tumor microenvironments. Cancer Res 78:5492–5503

Yue X, Zhao Y, Xu Y, Zheng M, Feng Z, Hu W (2017) Mutant p53 in cancer: accumulation, gain-of-function, and therapy. J Mol Biol 429:1595–1606

CDK Dysregulation in Breast Cancer: A Bioinformatics Analysis

8

Manzoor Ahmad Mir, Shazia Sofi, and Pir M. Ishfaq

8.1 Introduction

Breast cancer is one of the most aggressive and lethal types of transformation among women (Mehraj et al. 2021a, b; Mir 2021). However, the enormous improvements in screening techniques, early diagnosis, and treatment discoveries are to blame for the increasing survival rate (Mehraj et al. 2022a, b, c, d, e; Mir 2022). In this crucial area of pharmaceutical business research over the past ten years, there have been numerous acquisitions (Qayoom et al. 2021). The increased understanding of breast cancer made it possible to develop more intelligent therapies that could effectively target the disease and react to its milieu. This was made possible by advancements in molecular biology and pharmacology. Depending upon the intrinsic gene expression profiling, there are five main subtypes of breast cancer (BC): Luminal A, the most common subtype, maybe PR positive, ER positive or negative, HER2 negative, and EGFR and CK5/6 negative; luminal B (ER- or PR-positive and HER2-positive); basal-like (ER-, PR-, and HER2-negative, cytokeratin 5/6-positive, and/or epidermal growth factor receptor (EGFR) (Johnson et al. 2021; Mehraj et al. 2022a, b, c, d, e). These BC subtypes were classified according to distinct structural design, biological characteristics, prognosis, and clinical stages. The targeted medication has shown success in treating approximately 77% of BC patients who are receptor-positive (Hurvitz et al. 2013; Mir et al. 2020; Ghafouri et al. 2022). Unfortunately, because of the lack of appropriate tailored treatment, 15–25% of TNBC patients have poor outcomes. For patients with TNBC and the majority of BC, surgery combined with chemotherapy and radiotherapy is frequently advised (Al-Mahmood et al. 2018; Mehraj et al. 2022a, b, c, d, e). To enhance prognosis, stop cancer from progressing,

M. A. Mir (✉) · S. Sofi · P. M. Ishfaq
Department of Bioresources, School of Biological Sciences, University of Kashmir, Srinagar, J&K, India
e-mail: drmanzoor@kashmiruniversity.ac.in

M. A. Mir (ed.), *Therapeutic Potential of Cell Cycle Kinases in Breast Cancer*,
https://doi.org/10.1007/978-981-19-8911-7_8

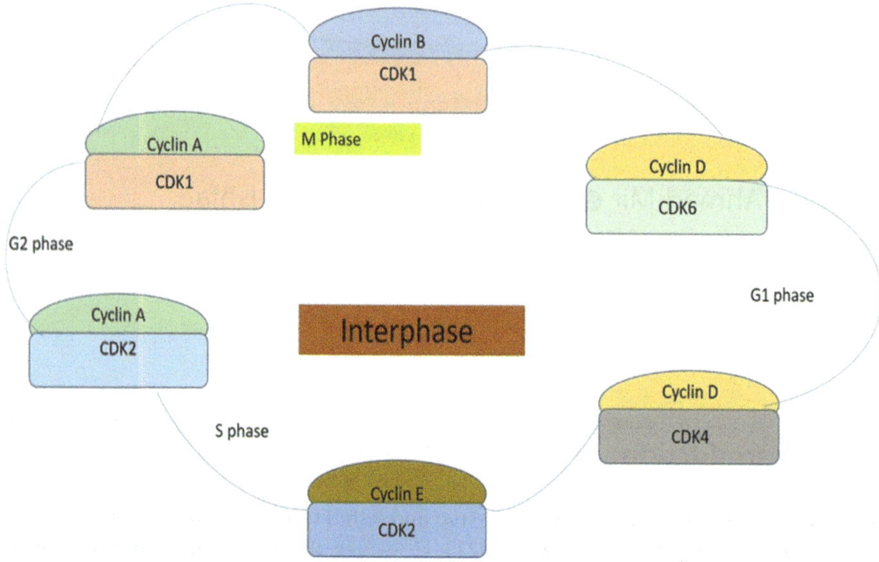

Fig. 8.1 Role of cyclin and cyclin dependent kinases in cell cycle

and create effective medicines, early diagnosis, accurate therapy, and prognosis are critically required. Currently, one exciting field of such research is the function of cell-cycle regulation (Sofi et al. 2022a, b).

The two most important and predominant physiological processes in tissue homeostasis are cell division and apoptosis. The cell-cycle process has four distinct ordered phases and is referred to as G1 phase, S phase, G2 phase, and M phase. Various cell cycle checkpoints are present that regulate transition from one phase to another to prevent chromosomal aberrations (Malumbres and Barbacid 2009; Mehraj et al. 2021a, b). Availability of growth factors and various intrinsic and extrinsic signals determine whether a cell will go for division or not. Otherwise, the cell will enter into a non-dividing phase in G1 called G0 phase. There is one checkpoint called restriction checkpoint in G1, passing of which guarantees cell division. Numerous cyclins and cyclin-dependent kinases (CDKs), a class of serine/threonine kinases, control the cell cycle (Mehraj et al. 2022a, b, c, d, e). To stabilize, activate, and phosphorylate CDKs in the designated phases, they assemble complexes with cyclins (Brown et al. 2015) (Fig. 8.1). By phosphorylating the target genes, such as the tumor suppressor protein retinoblastoma, the synthesis of cyclin/CDKs regulates the course of the cell cycle (Rb) (Thu et al. 2018). DNA damage usually leads to the inhibition of CDKs. CDK inhibitors (CKIs) are divided into two families namely INK4/6 family (including INK4a, INK4b, INK4c, INK4d) and CIP/KIP family (including p21, p27, and p57). These inhibitors control activation or inhibition of CDKs. The SCF, E3ubiquitin ligase and the anaphase-promoting complex/cyclosome (APC/C) control synthesis and degradation of multiple cyclins. Any type of genetic or epigenetic change in cell cycle regulatory proteins will lead to malignant transformations like breast cancer (Zhou et al. 2016). This represents a frontier in medical sciences for developing artificial CDK inhibitors as antitumor treatments (Gupta et al. 2019).

Using TCGA BRCA datasets accessible on the UCSC XENA and Gepia2 Web servers, we investigated the expression of CDKs in clinical samples of BC patients in this chapter. Here, we report that BC has a considerable deregulation in CDK expression. Additionally, it was discovered that unregulated CDK expression in BC patients affected both OS and RFS. Studies on enrichment demonstrated the importance of CDKs in neoplastic processes and suggested that regulating CDKs in conjunction with traditional medications might be a promising option to treat BC patients.

8.2 Expression Profiles of CDKs in Molecular Subtypes of Breast Cancer

Using UALCAN to evaluate the highly deregulated CDK expression patterns across various BC subclasses, age groups, and ethnicities, it was discovered that TNBC patients express high levels of CDK1, followed by Her2 enriched and luminal subtypes. Additionally, CDK1 was discovered to be substantially upregulated in women between the ages of 20 and 40 and of African-American origin (Fig. 8.2). Asian women and breast cancers that were enriched in HER2 had significantly lower levels of CDK18. However, there were no appreciable differences in CDK11A expression patterns among different BC subclasses, age, groups, or nationalities.

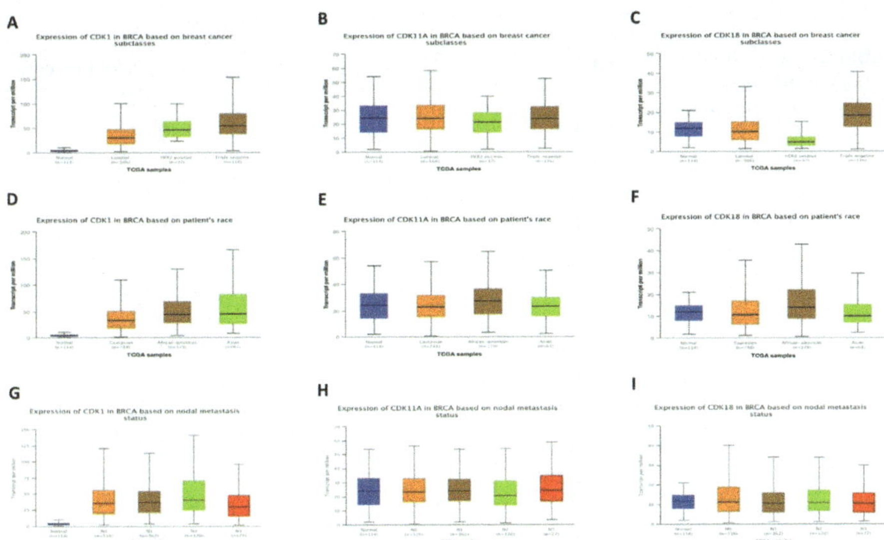

Fig. 8.2 Expression profiles of CDKs on the basis of molecular subtypes of breast cancer, age, ethnicities, and races of patients

8.3 Expression Analysis of CDKs in Breast Cancer

CDKs are important regulatory enzymes that play their role in cell cycle along with their cyclin partners (Asghar et al. 2015). The dysregulation of different CDKs is a hallmark of any cancer, including breast cancer. Most of the CDKs are highly overexpressed, while as some of them are downregulated in breast cancer (Ramachandiran et al. 2002). For instance, CDK1 is highly upregulated in breast cancer, while as CDK11A and CDK18 are highly downregulated (Table 8.1). Using different online portals, the expression profiles of various CDKs reveal that CDKs are highly upregulated in breast cancer. Among the various CDKs, CDK1 is of utmost importance, as it is the universal master kinase, that is conserved from yeast to humans (Sofi et al. 2022a, b). The expression profiles of various CDKs using UCSC XENA revealed the dysregulation of CDKs in breast cancer (Fig. 8.3) (Goldman et al. 2020).

Further, the dysregulation is highest in case of CDK1, as is evident by the fold change of 2.84 and a p-value of 2.82E-184 (Table 8.1).

8.4 CDK Expression and Various Clinicopathological Parameters

Age, ethnicity, tumor subtypes, and many other pathological characteristics are linked to CDK dysregulation. The relationship between common clinical-pathological traits in breast cancer patients and significantly disrupted CDKs—especially CDK1, CDK x11A, and CDK18—was investigated using the

Table 8.1 Log 2fold change of different deregulated CDKs in breast cancer

CDKs	Log 2(fold change)
CDK1	2.842
CDK2	0.352
CDK3	−1.323
CDK4	0.635
CDK5	1.475
CDK6	−0.423
CDK7	0.877
CDK8	0.181
CDK9	−0.341
CDK10	−0.997
CDK11A	−1.552
CDK15	−0.361
CDK16	0.475
CDK17	−0.414
CDK18	−1.777
CDK19	0.329
CDK20	0.223

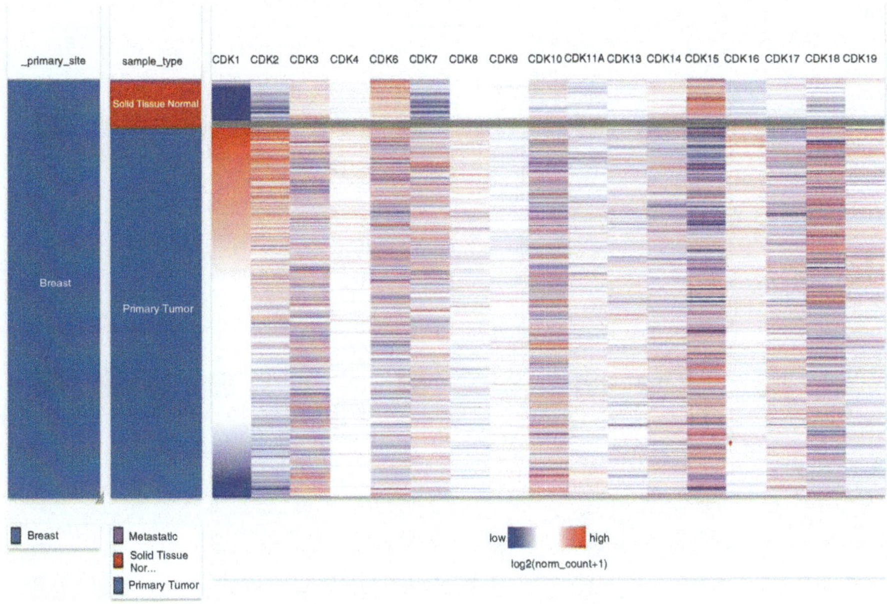

Fig. 8.3 Heat map of different deregulated CDKs in breast cancer

bc-GenEXMiner database (Jézéquel et al. 2012). Breast tumors (ER 235 and PR-negative) were shown to have considerably higher CDK1 expression than those with hormone receptors (*p* value-0.0001). On the other hand, breast tumors that were HER2-enriched had higher levels of CDK1 expression (*p* value-0.0001). It was found that greater CDK1 expression was associated with SBR3 in terms of SBR grade. Furthermore, patients with mutant p53 exhibited considerably higher amounts of CDK1 mRNA than those with wild-type p53. Patients with ER-negative breast cancer had significantly lower CDK11A expression than those with HER2-enriched tumors, who had much higher CDK11A expression.

However, there was little correlation with PR, SBR 246 grade, or p53 status and CDK11A expression. Contrarily, CDK18 was found to be markedly downregulated in breast cancer patients who expressed hormone receptors (ER, & PR), indicating a close association with hormone receptor negativity. In BC patients with HER2 amplification, CDK18 mRNA levels were found to be low, and they were found to be correlated with SBR3 grade. It was revealed that CDK18 expression and p53 status are closely related in BC patients with wild-type 53 (Fig. 8.4).

8.5 Protein-Protein Interaction of CDKs in Breast Cancer

The STRING v11 was used to construct related proteins using CDK-1, CDK11a, and CDK18 as input genes (Szklarczyk et al. 2015). The results found that CDKs interact with numerous other genes crucial to the development of breast cancer (Gupta et al.

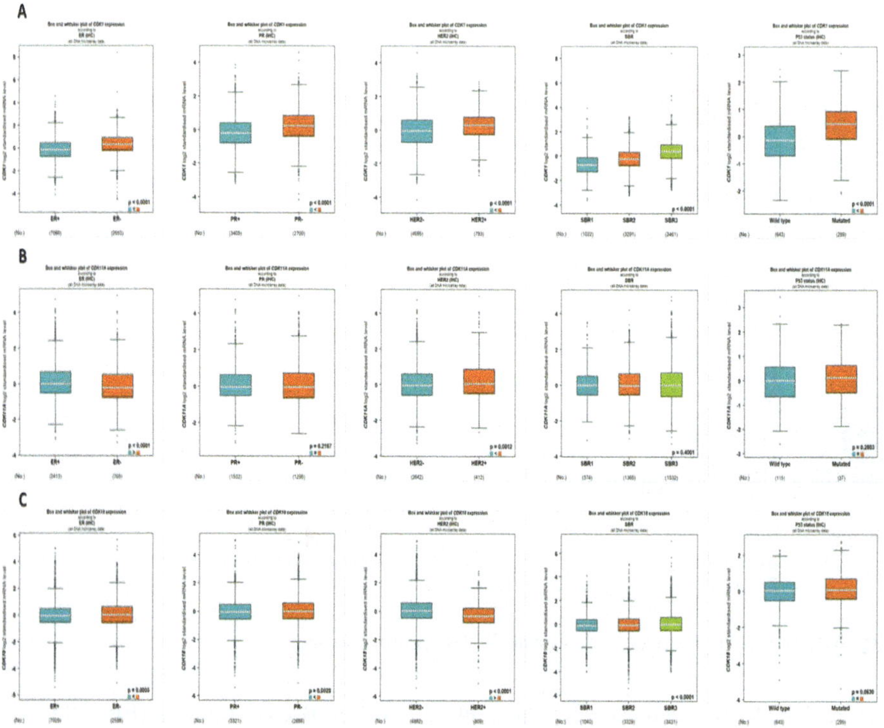

Fig. 8.4 bc-GenEXMiner investigation of unregulated CDKs demonstrated a strong correlation with clinicopathological factors of BC

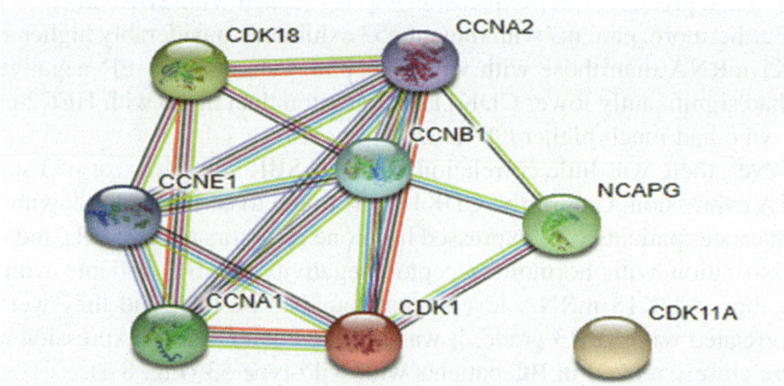

Fig. 8.5 Protein–Protein interaction of highly deregulated CDK1, CDK11A, and CDK18 with other proteins that are responsible for the progression of breast cancer

2019; Ding et al. 2020; Sun et al. 2020). The most significant proteins that play a substantial role in the carcinogenesis of breast cancer among these interacting proteins are CCNA2, CCNE1, and CCNA1 (Fig. 8.5).

8.6 Gene Ontology of CDKs

Using the FunRich program, a Gene Oncology (GO) enrichment study was carried out to determine the functional categories and distinctive biological characteristics of CDKs (Kuleshov et al. 2016). Cellular component (CC), biological process (BP), molecular function (MF), and biological pathway (BP) analysis of GO terms were filtered (Fig. 8.6). At a p-value of 0.05 or below, the top 10 GO words of the related genes in BP, CC, MF, and BP were deemed significant (Figure 8.6a–d). The GO keywords cell communication, ontology; BP, kinase binding, ontology; MF, cyclin-dependent protein kinase holoenzyme complex, ontology; CC, and cell cycle, ontology; BP were all implicated in the functions of these genes. The biological pathway analysis revealed that the genes linked with CDKs, which are connected to the FOXM1 transcription factor network, the regulation of retinoblastoma protein, and the E2F transcription factor network, were highly enriched for the cell cycle (Figure 8.6d). As a result, it is evident from the gene ontology that CDKs control various vital cellular functions that are crucial to the development of breast cancer.

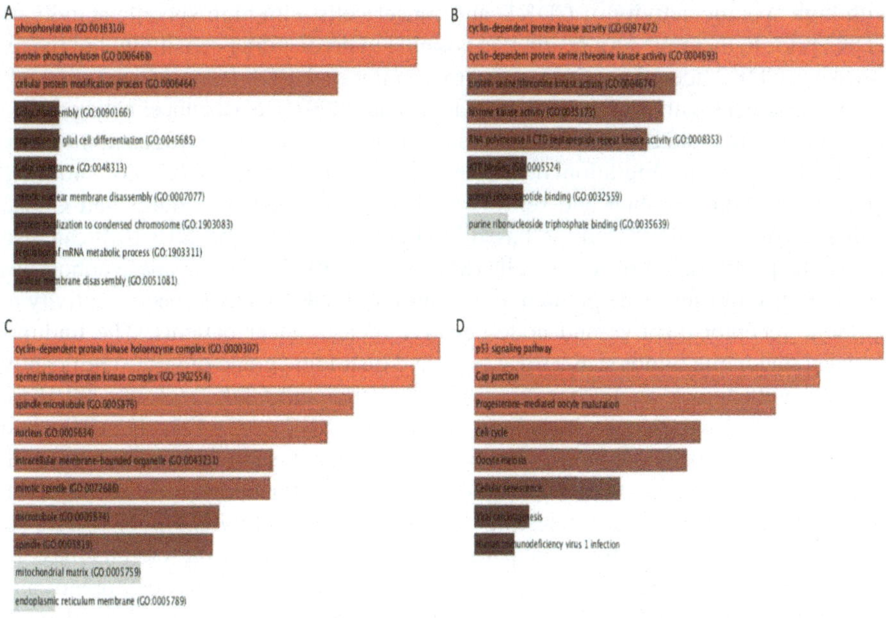

Fig. 8.6 Gene ontology and pathway analysis. (**a**–**c**) GO analysis of the BP, CC, and MF and, (**d**) KEGG pathway analysis of CDK1, CDK11, and CDK18 genes in BC

8.7 Prognostic Significance of CDKs in Breast Cancer

Breast cancer patients who have dysregulated CDKs had lower relapse-free survival and overall survival rates. It is interesting to note that there is a theory that breast cancer patients' prognosis may be influenced by the precise activity of CDK1 and CDK2 (Sofi et al. 2022a, b). Treatment of breast cancer patients with various chemotherapeutics has demonstrated that particular CDK1 and CDK2 activity can be utilized to forecast how well these patients would respond to chemotherapy (Li et al. 2021). Similar to this, Kim and colleagues have demonstrated that patients who have high CDK1 and CDK2 specific activity had poor 5-year relapse-free survival. They also demonstrated how patients with breast cancer can be divided into low-risk and high-risk categories based on this score. As a result, it is hypothesized that assessing CDK1 and CDK2 specific activity may be useful for predicting the course of disease (Kim et al. 2012) . Furthermore, it has been proposed that the distinct CDK1 and CDK2 activities can predict breast cancer patients' paclitaxel sensitivity (Kim et al. 2012). An in vitro sensitivity to paclitaxel cannot perfectly predict the same response in an in vivo setting, according to research by Nakayama and colleagues. Paclitaxel was effective in treating cancer cells in vitro with high specific activity of CDK1, and cancer cells with high specific activity of CDK2 prior to therapy. As a result, it is possible to predict drug sensitivity in vivo by assessing CDK2-specific activity prior to therapy and CDK1 specific activity following treatment with paclitaxel (Nakayama et al. 2009). Breast cancer cells are also resistant to other anti-cancer medications, including the antiestrogen tamoxifen, which slows the proliferation of cancer cells by blocking CDK2,4,6. Inducing apoptosis in both resistant and sensitive cell lines by silencing CDK1 and CDK2 with siRNA molecules or purine-based inhibitors (NU2058 and NU6102) can also reduce the proliferation of tumor cells (Martin et al. 2005). The Japanese cohort has also been used to show the predictive relevance of CDK1-CDK2-specific activity in hormone receptor-positive and node-negative breast cancer patients. The findings revealed that while tumor size, HER2, age, and histologic grade did not significantly correlate with tumor recurrence, specifically CDK1 and CDK2 activity did (Kim et al. 2008). Five genes, including CDK1, CCNB2, ACACB, PPARG, and MAD2L1, have been demonstrated by Ding and colleagues to be involved in regulation and serve as potential diagnostic indicators for ductal carcinoma in situ (Ding et al. 2017). It is interesting to note that the elimination of CDK1 is recommended as a valuable anticipatory biomarker to assess the effectiveness of anti-cancer therapy. Chemotherapy has been shown by Galindo-Moreno and colleagues to accelerate CDK1 degradation in MCF7 cancer cells through p62/HDAC6-mediated autophagy (Galindo-Moreno et al. 2017).

8.8 Role of CDKs in Breast Cancer

Uncontrolled cell growth that results in tumor clonality, in which a single cell replicates improperly, is a major contributor to the development of both benign and malignant tumors (Thu et al. 2018). In the context of CDKs and breast cancer, both varieties of tumors can be considered. Breast cancer metastasis would indicate a malignant tumor, which is considerably harder to remove locally than a benign tumor (Redig and McAllister 2013; Mehraj et al. 2022a, b, c, d, e). In order to restore control of the cell cycle, CDK inhibition has become a potential strategy. Cancer cells have been stopped in their tracks in the G1 phase of the cell cycle by inhibition of specific CDK pathways (Wenzel and Singh 2018). The specificity of CDKs is one characteristic that makes this a workable option. Cyclins are particular to their linked CDKs, a complexity illustrated by the fact that various cell types can go through mitosis at various periods (Sofi et al. 2022a, b). Globally, breast cancer affects people, primarily but not exclusively women. Unfortunately, patients' resistance to targeted and non-targeted medicines frequently leads to the failure of existing treatments. In light of this, CDK research has continued, with a focus on CDK4/6 (Deng et al. 2018). Cyclin D1 overexpression is linked to this specific CDK. Research on CDKs, such as CDK4 and CDK6, is being done to find strategies to block their activity and prevent the formation of malignant tissue without harming healthy tissue (Zhang et al. 2020a, b). Additionally, this method of treating breast cancer offers a potential means of avoiding cytotoxic anticancer drugs or chemotherapy-induced alopecia (CIA), a side effect of chemotherapy. Inhibiting CDKs, specifically CDK2, lowers the likelihood that such drugs may divide the epithelium of hair follicles, which results in hair loss (Ding et al. 2020). Treatments for breast cancer that involve CDK inhibition and an understanding of CDK function are safer and more efficient.

Due to their connections to breast cancer development, CDKs can also be employed to treat breast cancer by controlling proliferation with CDK inhibitors. For instance, based on the presence of hormone receptors in breast cancer cells that have undergone biopsy, breast cancer might be either hormone receptor positive or negative. The proliferation of cancer cells is fueled by hormone receptors like estrogen receptors; therefore, understanding this is crucial when treating breast cancer (Sofi et al. 2022a, b).

Because they prevent the growth of these receptors, selective CDK4/6 inhibitors have been demonstrated to be effective in treating breast tumors that are estrogen receptor-positive (Kang et al. 2014). In order to actively contribute to DNA replication and, by extension, the advancement of a cell through the cell cycle, CDK4/6 interacts with cyclin D. Breast cancer could arise if cells divide uncontrolled in the absence of such regulation. In other situations, cyclin D1, an oncogene that encodes cyclin D1, exhibits aberrant behavior, which promotes its overexpression and controls the transition from G1 to S (Mohammadizadeh et al. 2013). This would result in a lack of regulation in CDKs and an unregulated cell cycle because the inactivation of CDKs includes declining amounts of cyclin. Because it is overexpressed in a significant portion of cases, Cyclin D1 is intimately linked to breast cancer. The overexpression of the CCND1 gene is associated with the overexpression of the estrogen receptor (ER), both of which contribute significantly to the proliferation of cancer cells (Eeckhoute et al. 2006). The tumor suppressor protein pRb may become phosphorylated as a result of cyclin D1 activating CDKs, which then frees transcription factors like E2F that promote DNA synthesis (He et al. 2014). As a result of increased proliferation, breast cancer's pathophysiology is influenced. Cyclin D1 levels are influenced by CCND1 expression and can be utilized to provide patients with a more precise prognosis. Because they prevent the growth of these receptors, selective CDK4/6 inhibitors have been demonstrated to be effective in treating breast tumors that are estrogen receptor positive. In order to actively contribute to DNA replication and, by extension, the advancement of a cell through the cell cycle, CDK4/6 interact with cyclin D (Murphy and Dickler 2015). Breast cancer could arise if cells divide uncontrolled in the absence of such regulation. In other situations, cyclin D1, an oncogene that encodes cyclin D1, exhibits aberrant behavior, which promotes its overexpression and controls the transition from G1 to S (Eeckhoute et al. 2006). This would result in a lack of regulation in CDKs and an unregulated cell cycle because the inactivation of CDKs includes declining amounts of cyclin. Because it is overexpressed in a significant portion of cases, Cyclin D1 is intimately linked to breast cancer. The overexpression of the CCND1 gene is associated with the overexpression of the estrogen receptor (ER), both of which contribute significantly to the proliferation of cancer cells (Eeckhoute et al. 2006). The tumor suppressor protein pRb may become phosphorylated as a result of cyclin D1 activating CDKs, which then frees transcription factors like E2F that promote DNA synthesis. As a result of increased proliferation, breast cancer's pathophysiology is influenced. According to a mouse study, cyclin D1-deficient animals did not grow mammary carcinomas even when the ErbB-2 oncogene was turned on, which would typically cause the growth of cancer cells. These results provide strong evidence that cyclin D1 plays a key role in breast cancer when combined with the frequent overexpression of cyclin D1 in human breast cancer. Furthermore, a study that was conducted from January 2016 to June 2017 and involved numerous breast tumor patients discovered that enhanced cyclin D1 levels were present in 60% of cases and that given its relationship to other common breast cancer markers, cyclin D1 is a good prognostic indicator for the disease. Cyclin D1 can be overexpressed as a result of CCND1, which can improve prognosis. Cyclin E is another cyclin that may contribute to the emergence of breast cancer. Despite the fact that it is less frequently overexpressed than cyclin D1, the breakdown of its pathways results in the buildup of other products that disrupt the cell cycle. While cyclin E, like cyclin D1, can phosphorylate rPb to promote proliferation, it can also cause a cell to enter the S phase without the aid of rPb or E2F (Ding et al. 2020). In contrast to cyclin E overexpression, which causes breast cancer with higher rates of proliferation and worse outcomes, cyclin D1 overexpression is more frequently found in breast cancer(Ding et al. 2020). In both instances, overexpression of the cyclin does not coincide with overexpression of the gene, suggesting that changes in the breakdown pathway are most likely the cause of the issue (Ding et al. 2020).

Treatment Advancements Given that CDKs are known to have a role in controlling checkpoints and cell division, CDK pathways have long been of interest in the treatment of breast cancer. The goal is to identify the source of the cell cycle anomaly and create a remedy that will enable cells to reestablish control over proliferation. In the past, using CDK inhibitors to create medicines showed great promise, but it was plainly not a simple fix. Novel research has indicated that there is more possibility to understand the association between CDKs and breast cancer as a result of the introduction of new medications and combination therapies. It has been discovered that CDK 4/6 and CDK 4/6 inhibitors in particular regulate pathways important for the growth of tumor cells in breast cancer. Cyclin E and the E2F proteins participate in a positive feedback loop that phosphorylates and then hyper phosphorylates RB in the cyclin D1-CDK4 pathways. The phosphorylation of RB can be stopped by CDK 4/6 inhibitors, which causes the cell to be arrested in the G1 phase (Pernas et al. 2018). In some situations, controlling the cyclin D-CDK 4/6-retinoblastoma pathway may be preferable. For instance, endogenous proteins in cells, including the INK4 proteins, can reduce CDK 4/6 activity by attaching to the enzyme's catalytic subunits. However, the gene that codes for p16, one of these proteins, is deleted in some cells. As a result, CDK 4/6 activity is elevated at a baseline level, which presumably makes these cells more vulnerable to treatment with CDK 4/6 inhibitor (Knudsen and Witkiewicz 2016). Since ER+/HER2-positive breast cancer is the most often diagnosed kind, numerous therapies have been created that particularly target pathways in this subtype (Knudsen and Witkiewicz 2016). Treatments for this type of breast cancer are frequently less effective over time since it has a higher probability of recurrence. The purpose of CDK 4/6 inhibitors is to stimulate the RB tumor suppressor response; however, for unexplained reasons, they have also been proven to enhance breast cancer prognosis. Despite the fact that this might be connected to less well-studied breast cancer subgroups, these responses to CDK 4/6 inhibition aid in preventing recurrences in the future. Endocrine therapy and CDK 4/6 inhibition are complementary therapies that have been shown to reduce the development of tumor cells. Unexpectedly, the use of CDK 4/6 inhibitors led to the activation of genes that may have aided in continued cell development. Fortunately, it has been demonstrated that using endocrine medication blocks this reaction, reducing cycle D1 activity (Knudsen and Witkiewicz 2016). This role of endocrine therapy coupled with CDK 4/6 inhibitors is very important since one round of tumor cell development would result in another round of growth. Palbociclib, Ribociclib, and abemaciclib are three CDK 4/6 inhibitors that have received FDA approval and are each intended to complement a different type of therapy (Asghar et al. 2015). For the treatment of ER+/HER2-metastatic breast cancer, fulvestrant was specifically approved for use in conjunction with palbociclib. Because their mechanical pathways are redundant, metastatic tumors frequently respond rather effectively to combination therapy, as shown in Table 8.2.

A 2017 study that looked at all three inhibitors discovered that they were significant patient options because of their low toxicity but high efficacy and oral dosing. The effectiveness, safety, and pharmacology of palbociclib used in

Table 8.2 Hallmarks of metastasis and their implications

Hallmarks	Implications
Redundancy of mechanistic pathways	Need for combination therapy
Variable dormancy	Clinical trials must address delayed relapses
Contribution of cancer initiating cell	Incorporation of therapies that target stem cells.

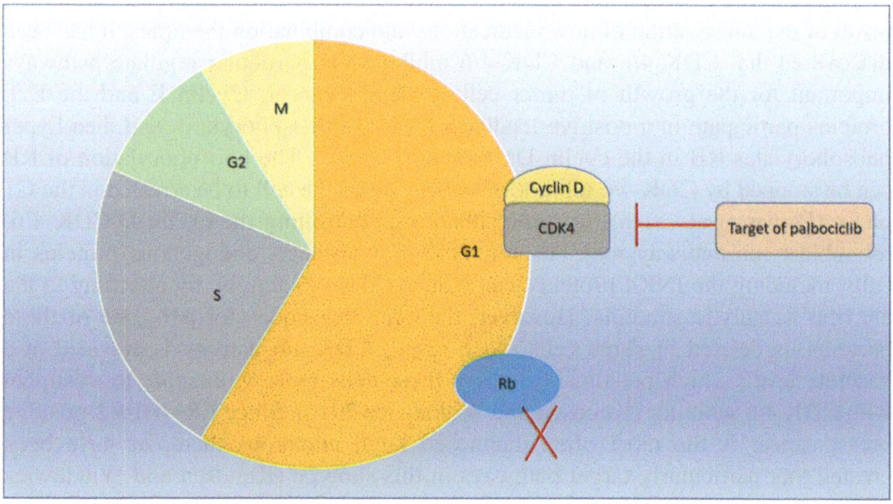

Fig. 8.7 Palbociclib an FDA-approved CDK inhibitor inhibits specifically CDK4 and CDK6, thus leading to cell cycle arrest at G1 phase

conjunction with endocrine therapy were closely examined in 2018; the findings revealed that this combinatorial therapeutic approach increased standard of life and progression-free survival in patients with metastatic breast cancer (Yu et al. 2006). Palbociclib was the only CDK4/6 inhibitor still in use in 2019 whose results from clinical studies enhanced overall survival (Fig. 8.7) (Ding et al. 2020). Promising findings from phase II clinical trials provided researchers cause to trust in the efficacy of palbociclib when it entered phase III trials to be examined with adjuvant endocrine therapy. In Phase II, postmenopausal, treatment-nave women with ER+/HER2-metastatic breast cancer either received just letrozole treatment or a letrozole and palbociclib combination treatment (Serra et al. 2019). Palbociclib has been shown to have clinical activity when used alone, and it also showed therapeutic activity when combined with the same endocrine medication administered to individuals sparingly before the condition progressed (Qayoom et al. 2020; Orbaugh et al. 2016).

Phase III testing of palbociclib included testing it not just with letrozole but also with fulvestrant. The data from these trials revealed that CDK 4/6 inhibitors especially palbociclib in combination with either letrozole or fulvestrant improved median progression-free survival (mPFS) by more than 10 months and showed signs of reversing endocrine resistance in patients (Malorni et al. 2018). According

Table 8.3 Similar adverse effects shown by Abemaciclib, palbociclib, and Ribociclib

Abemaciclib	Palbociclib	Ribociclib
Neutropenia	Neutropenia	Neutropenia
Leukopenia	Leukopenia	Leukopenia
Diarrhea	Infections	Diarrhea
Anemia	Anemia	Vomiting

to a 2019 study from the Siteman Cancer Center, resistance to CDK 4/6 inhibitors can happen via a variety of routes since ER+/HER2-breast cancer might be incalculable due to its molecular heterogeneity. As a result, various patients may require tailored treatment plans (Xi et al. 2019). For instance, patients who had previously been identified as being sensitive to endocrine therapy reacted to a palbociclib and fulvestrant combination treatment with a longer overall survival than the placebo and fulvestrant treatment (Turner et al. 2018). Fulvestrant has been demonstrated to nullify some of the negative side effects of selective estrogen receptor modulators like tamoxifen when used as adjuvant therapy (Johnston and Cheung 2010). Ribociclib is a primary combination medication licenced for use in the treatment of ER+/HER2-breast cancer especially patients of bone-only diseases, visceral metastasis, de novo diseases, and prior therapy (Hortobagyi 2018). Ribociclib was examined in phase III trials and proved to be an effective treatment in these subgroups. Neutropenia, leukopenia, abnormal liver function tests, infections, and vomiting were the most frequently reported adverse events (AEs) in these clinical trials; however, when ribociclib and letrozole were tested, the safety profile of the combination treatment was consistent across all subgroups, and mPFS and media clinical benefit response (mCBR) were higher in the ribociclib group compared to the placebo group (Hortobagyi 2018). In comparison to endocrine therapy alone, ribociclib was found to notably improve OS in a study of 672 patients when given in conjunction with tamoxifen or goserelin (Im et al. 2019). The estimated overall survival rate in the ribociclib group at 42 months was 70.2%, with a 95% confidence interval (CI) of 63.5 and 76.0. Estimated overall survival in the placebo group was 46.0%, with a 95% confidence interval (CI) of 32.0 to 58.9. Regarding the adverse effects, studies on the toxicity of both palbociclib and Ribociclib—even when used together to better understand their drug-drug interactions—have been conducted (Bellet et al. 2019). This is in part due to CDK4/6 inhibitors' capacity to be used over extended periods of time. This study examines how to control the adverse effects of each medication. Abemaciclib, a CDK4/6 inhibitor and the third and most recently created therapy, stops the advancement of the cell cycle by obstructing the phosphorylation of the retinoblastoma tumor suppressor protein (Palumbo et al. 2019). Because it may be administered orally to patients constantly and as monotherapy, abemaciclib is seen to be a good alternative (Martin and Goldstein 2018). Abemaciclib response rates in clinical trials ranged from 19.7% to 59.0% with significantly higher patient mPFS. Additionally, it showed palbociclib and ribociclib-like AEs (Horie et al. 1992). As seen in Table 8.3, all three therapies have demonstrated neutropenia and leukopenia as AEs, although they differ in other ways. For example, fatigue restricted the dosages of abemaciclib rather than neutropenia necessitating lower dosages. Abemaciclib has a bigger effect on the body than the other two therapies, maybe as a result of its higher CDK 4 specificity (Martin and Goldstein 2018). Abemaciclib treatment outcomes were observed in, a phase III clinical stage combining fulvestrant and abemaciclib. Out of the 669 recruited patients, 25.3% were found to be resistant to main endocrine therapy; the group receiving abemaciclib and fulvestrant had a clinical benefit rate of 72.2% and mPFS of 16.4 months compared to the placebo group's 56.1% and 9.3 months. Abemaciclib has also been investigated in relation to particular ER+/HER2-breast cancer characteristics, such as liver metastasis, CNS metastasis, and quicker tumor shrinkage. Additionally, this phase III setting viz.: MONARCH 2 demonstrated that those who received abemaciclib treatment postponed chemotherapy (Sledge et al. 2020).

8.9 Combination Therapy of CDK Inhibitors and PD1-PDL1 Antibodies

Cancer immunotherapy has become a potent and successful method of cancer treatment thanks to decades of research. Dr. Honjo discovered PD1 (programmed death receptor 1) and showed that T cells express PD1 in 1992. Dr. Chen discovered PDL1 (B7-H1) in 1999 and showed that immune and tumor cells express PDL1 at high levels. T cell death is induced by the interaction between PDL1 and PD1, and lymphocyte activation is adversely regulated. Therefore, inhibiting PD1-PDL1 immunological checkpoints encourages T cell activation, which helps T cells have a lethal effect on tumor cells. Even while blocking PD1-PDL1 immune checkpoints has been clinically effective in treating a number of malignancies, the majority of cancer patients still did not benefit from immunotherapy. Additionally, drug resistance could develop while treating PD1-PDL1 with targeted therapy. As a result, numerous studies have been carried out to determine how combination therapy tactics can increase the responsiveness of cancer patients to immunotherapy. Some CDK inhibitors can improve the anti-tumor immune response, according to recent studies. Some CDK inhibitors have shown strong anti-tumor effectiveness in preclinical and clinical trials when combined with PD1-PDL1 immunotherapy.

8.9.1 Dinaciclib Enhances Anti-PD1 Mediated Tumor Suppression

As previously mentioned, dinaciclib, a powerful CDK inhibitor of CDK1, 2, 5, 9, and 12, can cause apoptosis in a variety of tumor cells. According to Hossain et al., combined therapy using Dinaciclib and an anti-PD1 antibody had significant antitumor activity. Combination therapy has the potential to increase anti-tumor immune response and promote tumor regression since it can activate DC and trigger T cell infiltration. Additionally, Dinaciclib can cause immunogenic cell death (ICD) in conjunction with anti-PD1 antibodies to transform tumor cells into endogenous vaccines (Hossain et al. 2018). Together, these studies have opened up new possibilities for addressing pan-CDK inhibitor toxicity and side effects, which expands the range of potential applications for these drugs.

8.9.2 CDK4/6 Inhibitors Augment the Anti-Tumor Efficacy of PD1-PDL1 Immune Checkpoint Blockade

Fundamental cell cycle regulators CDK4 and CDK6, which are necessary for the onset and development of different cancers, are CDK4 and CDK6 (Qayoom et al. 2022). A number of solid cancers have shown notable activity against CDK4/6 pharmacological inhibitors. Goel et al. discovered that CDK4/6 inhibitors not only cause tumor cell cycle arrest but also foster anti-tumor immunity in rat tumor model research (Goel et al. 2017). On the one hand, CDK4/6 inhibitors promote the expression of endogenous retroviral elements in tumor cells, stimulating the synthesis of type III interferons while also improving the presentation of tumor antigens. Conversely, CDK4/6 inhibitors significantly reduce the growth of regulatory T cells. Treatment with CDK4/6 inhibitors dramatically increases the clearance of tumor cells mediated by cytotoxic T cells based on these two functions. Theoretically, this study supported the use of CDK4/6 inhibitors and PD1-PDL1 antibodies in combination therapy.

Another CDK4/6 inhibitor that has received clinical approval for the treatment of HR+ breast cancer is abemaciclib. According to a recent study by Schaer et al., Abemaciclib therapy can boost the expression of antigen presentation genes in breast cancer cells and can stimulate human T cell activation (Schaer et al. 2018). According to another research, abemaciclib monotherapy can boost T cell inflammation and slow the growth of tumors. Abemaciclib and anti-PDL1 antibody combination therapy can promote immune memory and tumor eradication. These findings indicated that Abemaciclib and an anti-PDL1 antibody could be used in combination therapy to successfully activate both innate and adaptive immune response. When used in combination, abemaciclib and an anti-PDL1 antibody have shown tremendous promise for use in clinical settings.

Zhang et al. looked into the regulating mechanisms of PDL1 expression and stability because the effectiveness of PDL1 antibody therapy depends on the protein abundance of PDL1 (Zhang et al. 2020a, b). They discovered that CDK4 participates in the control of PDL1. Another study further demonstrated the extraordinary anti-tumor effectiveness of the combination therapy using CDK4/6 inhibitors and anti-PDL1 antibodies (Deng et al. 2018). Together, our results support the therapeutic utility of combination therapies using CDK4/6 inhibitors and anti-PD1-PDL1 antibodies. Another combination therapy is now being tested (Dai et al. 2003; Gao et al. 2004; Lin et al. 2010; Kruse et al. 2011), in addition to the combination of CDK4/6 inhibitors and PDL1 antibodies. Future cancer treatments are predicted to heavily rely on combination therapies.

8.9.3 Other Combination Therapies

Neoantigen load and tumor T cell infiltration and clonal growth are reported to be enhanced in CDK12 mutant patients (Wu et al. 2018). Immune checkpoint treatment may be beneficial for a subclass of metastatic castration-resistant prostate cancer (mCRPC) defined by CDK12 inactivation (Antonarakis 2018). In fact, patients with metastatic prostate cancer who had CDK12 deficiency underwent a phase ll clinical trial (ClinicalTrials.gov Identifier: NCT03570619). Immune-checkpoint inhibitors, nivolumab, and ipilimumab were given to these individuals, followed by nivolumab monotherapy. SR-4835, a CDK12 and CDK13 selective inhibitor, was also found to trigger immunogenic cell death, which improved the anti-tumor effectiveness of PD1-PD-L1 immune checkpoint therapy in breast cancer, according to a recent study (Li et al. 2020). Additionally, the CDK7 and CDK9 inhibitors YKL-5-124 and MC18029 are being investigated in combination therapy with the previously mentioned PD1/PD-L1 (Zhang et al. 2018; Zhang et al. 2020a, b). These results suggested that PD1-PDL1 immunotherapy in combination with CDK12 inhibitors will be a successful cancer treatment method.

8.10 Summary

Breast cancer is a type of cancer that involves the dysregulation of many genes. CDKs are the key players that are dysregulated in breast cancer patients. Breast cancer patients who have dysregulated CDKs had lower relapse-free survival and overall survival rates. The dysregulation of different CDKs in breast cancer has become a target for the treatment of breast cancer patients. The expression profiles of various CDKs using UCSC XENA revealed the dysregulation of CDKs in breast cancer. The breast tumors that are HER2-enriched have higher levels of CDK1 expression. Further, greater CDK1 expression is associated with SBR3 in terms of SBR grade. Also, CDKs interact with numerous other genes crucial to the development of breast cancer. Gene ontology also revealed that CDKs control various vital cellular functions that are crucial to the development of breast cancer. Due to their connections to breast cancer development, CDKs can also be employed to treat breast cancer by controlling proliferation with CDK inhibitors. Therefore, regulating CDKs in conjunction with traditional medications might be a promising option to treat BC patients.

8.11 Further Reading

The readers can further read about the role of CDKs in breast cancer by going through the following papers

- https://doi.org/10.1080/13543784.2022.2097067
- https://www.ncbi.nlm.nih.gov/pmc/articles/PMC8167670/

For more incites about the topic we would suggest detailed findings from the books of (Mir 2022) https://doi.org/10.1016/C2021-0-02565-7, https://doi.org/10.1016/C2022-0-00074-X (Mir 2021) https://doi.org/10.52305/WXJL6770. Also, the readers can have a look upon the following visual presentations for the better conceptual understanding of CDKs and their role in breast cancer.

https://youtu.be/0Sj3rbJPeXQ
https://youtu.be/RXsWAvdWG0s
https://youtu.be/YA67P2k2d6A

References

Al-Mahmood S et al (2018) Metastatic and triple-negative breast cancer: challenges and treatment options. Drug Deliv Transl Res 8(5):1483–1507

Antonarakis ES (2018) Cyclin-dependent kinase 12, immunity, and prostate cancer. N Engl J Med 379(11):1087–1089

Asghar U et al (2015) The history and future of targeting cyclin-dependent kinases in cancer therapy. Nat Rev Drug Discov 14(2):130–146

Bellet M et al (2019) Palbociclib and ribociclib in breast cancer: consensus workshop on the management of concomitant medication. Ther Adv Med Oncol 11:1758835919833867

Brown NR et al (2015) CDK1 structures reveal conserved and unique features of the essential cell cycle CDK. Nat Commun 6(1):1–12

Dai Y et al (2003) Proteasome inhibitors potentiate leukemic cell apoptosis induced by the cyclin-dependent kinase inhibitor flavopiridol through a SAPK/JNK-and NF-κB-dependent process. Oncogene 22(46):7108–7122

Deng J et al (2018) CDK4/6 inhibition augments antitumor immunity by enhancing T-cell activation. Cancer Discov 8(2):216–233

Ding L et al (2020) The roles of cyclin-dependent kinases in cell-cycle progression and therapeutic strategies in human breast cancer. Int J Mol Sci 21(6):1960

Ding Z-H et al (2017) Docking of CDK1 with antibiotic drugs revealed novel therapeutic value in breast ductal cancer in situ. Oncotarget 8(37):61998

Eeckhoute J et al (2006) A cell-type-specific transcriptional network required for estrogen regulation of cyclin D1 and cell cycle progression in breast cancer. Genes Dev 20(18):2513–2526

Galindo-Moreno M et al (2017) Both p62/SQSTM1-HDAC6-dependent autophagy and the aggresome pathway mediate CDK1 degradation in human breast cancer. Sci Rep 7(1):1–10

Gao N et al (2004) Contribution of disruption of the nuclear factor-κB pathway to induction of apoptosis in human leukemia cells by histone deacetylase inhibitors and flavopiridol. Mol Pharmacol 66(4):956–963

Ghafouri SR et al (2022) Recently approved treatment options for patients with metastatic triple-negative and HER2-neu-positive breast cancer. J Investig Med 70(6):1329–1341

Goel S et al (2017) CDK4/6 inhibition triggers anti-tumour immunity. Nature 548(7668):471–475

Goldman MJ et al (2020) Visualizing and interpreting cancer genomics data via the Xena platform. Nat Biotechnol 38(6):675–678

Gupta P et al (2019) CDK inhibitors as sensitizing agents for cancer chemotherapy. Protein kinase inhibitors as sensitizing agents for chemotherapy. Elsevier, pp 125–149

He J et al (2014) Prognosis of lymph node-negative breast cancer: association with clinicopathological factors and tumor associated gene expression. Oncol Lett 8(4):1717–1724

Horie K et al (1992) Immunohistochemical localization of androgen receptor in the human ovary throughout the menstrual cycle in relation to oestrogen and progesterone receptor expression. Hum Reprod 7(2):184–190

Hortobagyi GN (2018) Ribociclib for the first-line treatment of advanced hormone receptor-positive breast cancer: a review of subgroup analyses from the MONALEESA-2 trial. Breast Cancer Res 20(1):1–11

Hossain DMS et al (2018) Dinaciclib induces immunogenic cell death and enhances anti-PD1–mediated tumor suppression. J Clin Invest 128(2):644–654

Hurvitz SA et al (2013) Current approaches and future directions in the treatment of HER2-positive breast cancer. Cancer Treat Rev 39(3):219–229

Im S-A et al (2019) Overall survival with ribociclib plus endocrine therapy in breast cancer. N Engl J Med 381(4):307–316

Jézéquel P et al (2012) Bc-GenExMiner: an easy-to-use online platform for gene prognostic analyses in breast cancer. Breast Cancer Res Treat 131(3):765–775

Johnson KS et al (2021) Molecular subtypes of breast cancer: a review for breast radiologists. J Breast Imaging 3(1):12–24

Johnston SJ, Cheung KL (2010) Fulvestrant-a novel endocrine therapy for breast cancer. Curr Med Chem 17(10):902–914

Kang J et al (2014) Targeting cyclin-dependent kinase 1 (CDK1) but not CDK4/6 or CDK2 is selectively lethal to MYC-dependent human breast cancer cells. BMC Cancer 14(1):1–13

Kim SJ et al (2008) Determination of the specific activity of CDK1 and CDK2 as a novel prognostic indicator for early breast cancer. Ann Oncol 19(1):68–72

Kim SJ et al (2012) Recurrence risk score based on the specific activity of CDK1 and CDK2 predicts response to neoadjuvant paclitaxel followed by 5-fluorouracil, epirubicin and cyclophosphamide in breast cancers. Ann Oncol 23(4):891–897

Knudsen ES, Witkiewicz AK (2016) Defining the transcriptional and biological response to CDK4/6 inhibition in relation to ER+/HER2-breast cancer. Oncotarget 7(43):69111

Kruse U et al (2011) Chemoproteomics-based kinome profiling and target deconvolution of clinical multi-kinase inhibitors in primary chronic lymphocytic leukemia cells. Leukemia 25(1):89–100

Kuleshov MV et al (2016) Enrichr: a comprehensive gene set enrichment analysis web server 2016 update. Nucleic Acids Res 44(W1):W90–W97

Li N et al (2021) Expression and prognostic value of transcription-associated cyclin-dependent kinases in human breast cancer. Aging (Albany NY) 13(6):8095

Li Y et al (2020) CDK12/13 inhibition induces immunogenic cell death and enhances anti-PD-1 anticancer activity in breast cancer. Cancer Lett 495:12–21

Lin TS et al (2010) Flavopiridol, fludarabine, and rituximab in mantle cell lymphoma and indolent B-cell lymphoproliferative disorders. J Clin Oncol 28(3):418

Malorni L et al (2018) Palbociclib as single agent or in combination with the endocrine therapy received before disease progression for estrogen receptor-positive, HER2-negative metastatic breast cancer: TREnd trial. Ann Oncol 29(8):1748–1754

Malumbres M, Barbacid M (2009) Cell cycle, CDKs and cancer: a changing paradigm. Nat Rev Cancer 9(3):153–166

Martin JM, Goldstein LJ (2018) Profile of abemaciclib and its potential in the treatment of breast cancer. Onco Targets Ther 11:5253

Martin LA et al (2005) The anti-estrogen ICI 182,780, but not tamoxifen, inhibits the growth of MCF7 breast cancer cells refractory to long-term estrogen deprivation through downregulation of ER and IGF signalling. Breast Cancer Res 7(2):1–2

Mehraj U et al (2021a) Tumor microenvironment promotes breast cancer chemoresistance. Cancer Chemother Pharmacol 87(2):147–158

Mehraj U et al (2021b) The tumor microenvironment as driver of stemness and therapeutic resistance in breast cancer: new challenges and therapeutic opportunities. Cell Oncol:1–21

Mehraj U et al (2022a) Expression pattern and prognostic significance of baculoviral inhibitor of apoptosis repeat-containing 5 (BIRC5) in breast cancer: a comprehensive analysis. Adv Cancer Biol-Metastasis 100037

Mehraj U et al (2022b) Expression pattern and prognostic significance of chemokines in breast cancer: an integrated bioinformatics analysis. Clin Breast Cancer

Mehraj U et al (2022c) Cryptolepine targets TOP2A and inhibits tumor cell proliferation in breast cancer cells-an in vitro and in silico study. Anti Cancer Agents Med Chem

Mehraj U et al (2022d) Expression pattern and prognostic significance of CDKs in breast cancer: an integrated bioinformatic study. Cancer Biomarkers(Preprint):1–15

Mehraj U et al (2022e) Adapalene inhibits the growth of triple-negative breast cancer cells by S-phase arrest and potentiates the antitumor efficacy of GDC-0941. Front Pharmacol 13:958443

Mir MA et al (2020) Targeting different pathways using novel combination therapy in triple negative breast cancer. Curr Cancer Drug Targets 20(8):586–602

Mir MA (2021) Combination therapies and their effectiveness in breast cancer treatment. Nova Biomedical Science Publishers USA, pp 1–411. https://doi.org/10.52305/WXJL6770. Book ISBN: 978–1–68507-195-0

Mir MA (2022) Role of tumor microenvironment in breast cancer and targeted therapies, 1st edn. Elsevier Inc. Publishers USA, pp 1–350. https://doi.org/10.1016/C2022-0-00074-X. ISBN: 9780443186974

Mohammadizadeh F et al (2013) Role of cyclin D1 in breast carcinoma. J Res Med Sci 18(12):1021

Murphy CG, Dickler MN (2015) The role of CDK4/6 inhibition in breast cancer. Oncologist 20(5): 483–490

Nakayama S et al (2009) Prediction of paclitaxel sensitivity by CDK1 and CDK2 activity in human breast cancer cells. Breast Cancer Res 11(1):1–10

Orbaugh K et al (2016) Palbociclib plus letrozole for the treatment of metastatic breast cancer: an illustrative case scenario. J Adv Pract Oncol 7(5):550

Palumbo A et al (2019) Abemaciclib: the newest CDK4/6 inhibitor for the treatment of breast cancer. Ann Pharmacother 53(2):178–185

Pernas S et al (2018) CDK4/6 inhibition in breast cancer: current practice and future directions. Ther Adv Med Oncol 10:1758835918786451

Qayoom H, Bhat BA, Mehraj U, Mir MA (2020) Rising trends of cancers in Kashmir valley: distribution pattern, incidence and causes. J Oncol Res Treat 5(150):2

Qayoom H et al (2021) An insight into the cancer stem cell survival pathways involved in chemoresistance in triple-negative breast cancer. Future Oncol 17(31):4185–4206

Qayoom H et al (2022) Expression patterns and therapeutic implications of CDK4 across multiple carcinomas: a molecular docking and MD simulation study. Med Oncol 39(10):1–13

Ramachandiran S et al (2002) Mitogen-activated protein kinases contribute to reactive oxygen species-induced cell death in renal proximal tubule epithelial cells. Chem Res Toxicol 15(12): 1635–1642

Redig AJ, McAllister SS (2013) Breast cancer as a systemic disease: a view of metastasis. J Intern Med 274(2):113–126

Schaer DA et al (2018) The CDK4/6 inhibitor abemaciclib induces a T cell inflamed tumor microenvironment and enhances the efficacy of PD-L1 checkpoint blockade. Cell Rep 22(11): 2978–2994

Serra F et al (2019) Palbociclib in metastatic breast cancer: current evidence and real-life data. Drugs Context 8

Sledge GW et al (2020) The effect of abemaciclib plus fulvestrant on overall survival in hormone receptor–positive, ERBB2-negative breast cancer that progressed on endocrine therapy—MONARCH 2: a randomized clinical trial. JAMA Oncol 6(1):116–124

Sofi S et al (2022a) Targeting cyclin-dependent kinase 1 (CDK1) in cancer: molecular docking and dynamic simulations of potential CDK1 inhibitors. Med Oncol 39(9):1–15

Sofi S et al (2022b) Cyclin-dependent kinases in breast cancer: expression pattern and therapeutic implications. Med Oncol 39(6):1–16

Sun B et al (2020) Inhibition of the transcriptional kinase CDK7 overcomes therapeutic resistance in HER2-positive breast cancers. Oncogene 39(1):50–63

Szklarczyk D et al (2015) STRING v10: protein–protein interaction networks, integrated over the tree of life. Nucleic Acids Res 43(D1):D447–D452

Thu KL et al (2018) Targeting the cell cycle in breast cancer: towards the next phase. Cell Cycle 17(15):1871–1885

Turner NC et al (2018) Overall survival with palbociclib and fulvestrant in advanced breast cancer. N Engl J Med 379(20):1926–1936

Wenzel ES, Singh ATK (2018) Cell-cycle checkpoints and aneuploidy on the path to cancer. In Vivo 32(1):1–5

Wu Y-M et al (2018) Inactivation of CDK12 delineates a distinct immunogenic class of advanced prostate cancer. Cell 173(7):1770–1782

Xi J et al (2019) Retrospective analysis of treatment patterns and effectiveness of palbociclib and subsequent regimens in metastatic breast cancer. J Natl Compr Cancer Netw 17(2):141–147

Yu Q et al (2006) Requirement for CDK4 kinase function in breast cancer. Cancer Cell 9(1):23–32

Zhang H et al (2018) Targeting CDK9 reactivates epigenetically silenced genes in cancer. Cell 175(5):1244–1258

Zhang H et al (2020a) CDK7 inhibition potentiates genome instability triggering anti-tumor immunity in small cell lung cancer. Cancer Cell 37(1):37–54

Zhang Z et al (2020b) CDK4/6 inhibition blocks cancer metastasis through a USP51-ZEB1-dependent deubiquitination mechanism. Signal Transduct Target Ther 5(1):1–13

Zhou Z et al (2016) Insights into APC/C: from cellular function to diseases and therapeutics. Cell Div 11(1):1–18

CDK1 Dysregulation in Breast Cancer 9

Manzoor Ahmad Mir and Burhan Ul Haq

9.1 Introduction

A family of protein kinases responsible for modulating the cell cycle are the CDKs (Poon 2016). CDKs require modulatory subunits to bind them, known as cyclins, to exert their effects. The latter are formed and destroyed at various cell cycle stages in a specific and timely manner, thus regulating the cell cycle properly. The relevance of complex Cdc2 (CDK1) has been discovered in the *Schizosaccharomyces pombe* (Sofi et al. 2022). The participation of CDK1 in the homologous recombination DNA double-stranded break repair mechanism is also known. In yeast and human cells, the cyclin-dependent kinase activity is required for eliminating DNA double-stranded breaks to form single strands during homologous recombination by recruiting endonucleases Sae2 or CtIP, respectively (Ira et al. 2004; Huertas and Jackson 2009). CDK1 is the most important CDK for maintaining cell cycle control in mammalian cells (Santamaría et al. 2007). During the G_1, NHEJ is primarily operational, and during S and G_2 phase, HR is in action in yeast and Cdk1 appears to play a crucial part in choosing between the aforementioned methods utilized primarily to repair the ds-breaks (Huertas Sánchez et al. 2008). The tumor suppressor BRCA2 is phosphorylated by CDK1 and CDK2 in humans to regulate its interaction with RAD51. During the S and G_2 phases, this connection promotes homologous recombination-dependent repair (Esashi et al. 2005). Deregulation of specific CDK-cyclin complexes is frequently observed as a result of tumor-associated mutations, resulting in either unscheduled cell cycle re-entry or persistent

M. A. Mir (✉)
Department of Bioresources, School of Biological Sciences, University of Kashmir, Srinagar, J&K, India
e-mail: drmanzoor@kashmiruniversity.ac.in

B. U. Haq
Department of Biotechnology, Central University of Kashmir, Ganderbal, J&K, India

M. A. Mir (ed.), *Therapeutic Potential of Cell Cycle Kinases in Breast Cancer*,
https://doi.org/10.1007/978-981-19-8911-7_9

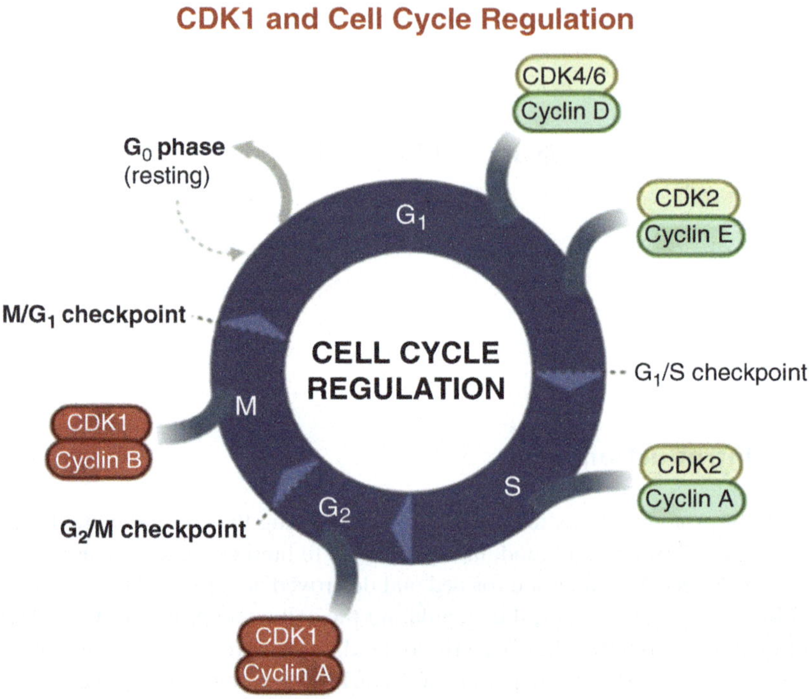

Fig. 9.1 The CDKs form specific pairs with their respective interacting partners during the various cell cycle phases, thus controlling its progression

proliferation. Furthermore, most human tumor cells have these two characteristics (Malumbres and Barbacid 2001). The checkpoints monitor the normal course of the cell cycle and detect any problems during DNA synthesis and chromosomal segregation. Furthermore, these active checkpoints cause the cell cycle to be arrested by regulating CDK function. The goal of stopping the cell cycle is to give cells enough time to fix their faults so that they do not proceed to the daughter cells that will be generated. Endogenous genotoxic factors, such as chemicals, free radicals, ionizing radiation, and exogenous products, can cause extensive alterations in the DNA molecule, and DNA damage checkpoints assist to protect cells from such attacks. Furthermore, when these changes occur, they are detected by a signaling pathway, resulting in CDK hindrance and, finally, cell cycle block (Bartek et al. 2004). If the repair process is inefficient due to massive DNA damage caused by checkpoint failure or poor repair machinery caused by genetic flaws in the same pathways, the cells may undergo programmed cell death (apoptosis) or enter senescence (Fig. 9.1).

On the other side, accumulating DNA mutations can lead to GIN (genomic instability), which can lead to cell transformation and thus cancer (Kastan and Bartek 2004). Unscheduled proliferation, chromosomal instability (CIN), and genomic instability (GIN) are the three most common disorders in the cell cycle. They are caused by either direct or indirect poor control of CDKs (Malumbres and Barbacid

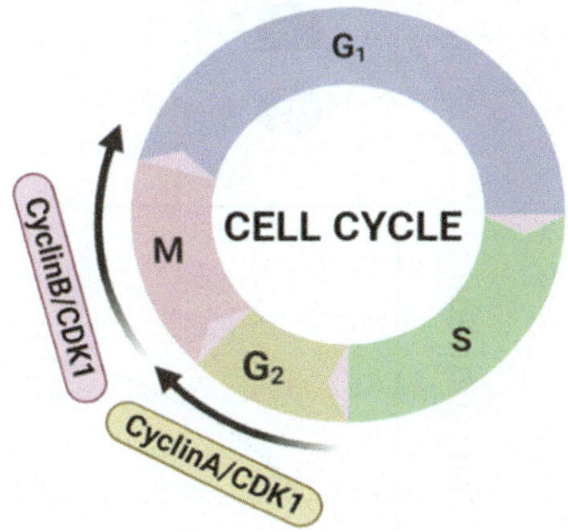

Fig. 9.2 The CDK1 with its interacting partners Cyclin B and Cyclin A is essential in cell cycle regulation and transition at the G2/M phase

2005). The SAC (spindle assembly checkpoint) is in charge of chromosomal separation after DNA duplication. This signaling system regulates CDK1 activity and protects against chromosomal segregation faults (Kops et al. 2005; Musacchio and Salmon 2007; Pérez de Castro et al. 2007). A defective SAC (spindle assembly checkpoint) can also lead to unequal DNA inheritance. If CIN (numerical chromosomal abbreviations) continue to accumulate and are not addressed, it may lead to tumor growth. A-type cyclins are known to activate CDK1 near the end of interphase to aid mitotic commencement. When the nuclear envelope degrades, the A-type cyclins are destroyed, allowing CDK1-cyclin B complexes to form (Fig. 9.2), which drive the cells into mitosis (Malumbres and Barbacid 2005).

Because of their critical function in the progression of the cell cycle, downregulation of CDKs may result in improper homeostasis in specific tissues. On the other hand, by initiating the untimely division in cells (progenitor or stem cells), the hyperactivated CDKs could also aid in the development of tumors (Malumbres and Barbacid 2009). Breast cancer can result from many causes, including mutations in genes that are concerned with the repair of DNA, TSG (p53), and the proto-oncogenes like HER-2, c-myc, as well as cyclin D (Gerger et al. 2007). When BRCA1 and BRCA2 genes are dysregulated, the chance of breast cancer increases. These are known to perform multiple functions, including obstructing the progression of the cell cycle at the S-phase by halting retinoblastoma and possibly CDK2 (Rahman and Stratton 1998; Hashemi et al. 2019). CDK1 is a member of the cell cycle-associated CDK family, which also includes CDK2, CDK4, and CDK6. These CDKs are known to have a direct role in the cell cycle progression and the various phases. Moreover, the cyclins show different concentrations during the different cell cycle phases (Fig. 9.3). The relative levels of CDKs are nearly untouched but their action can be modulated by changes in these cyclin concentrations.

The uncontrolled cellular growth represents one of the hallmarks of cancer. The cell cycle checkpoint disables and overrides multiple protections implicated in

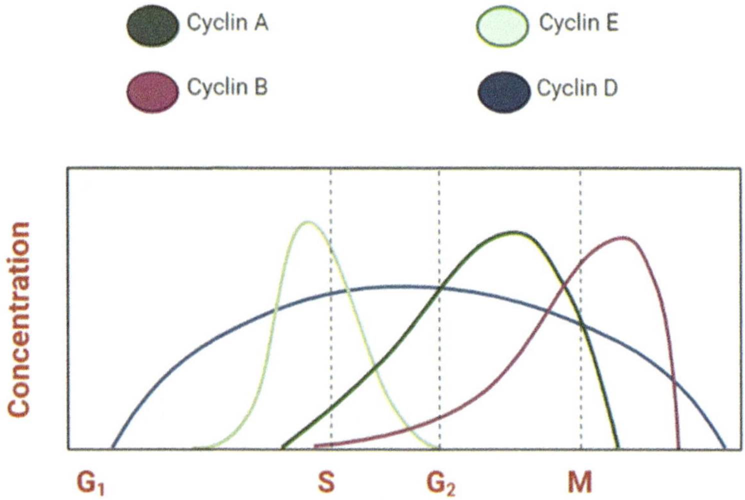

Fig. 9.3 The relative levels of different cyclins during various cell cycle phases. These specific cyclin concentrations are significant for the successful cell cycle progression along with their association with particular CDKs

Fig. 9.4 Activation of CDK1 is brought about by B-type cyclin (mainly B1 cyclin), which leads to mitotic entry after successful substrate phosphorylation. The CDK1 inactivation occurs post cyclin B1 destruction, ultimately leading to exit from mitosis

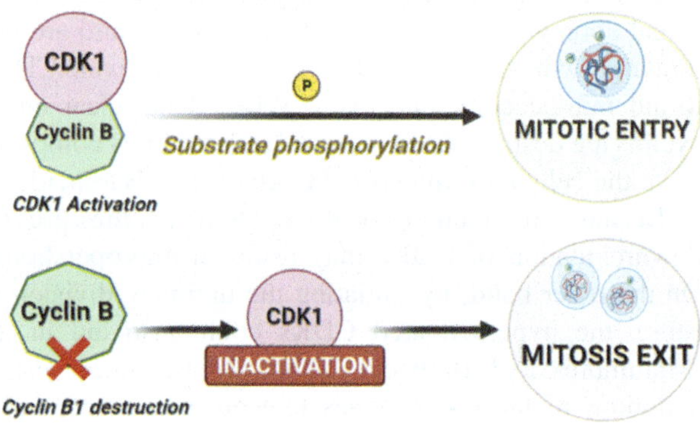

cyclin/CDKs dysregulation in the same way. In several solid tumors, including BC, uncontrolled cell proliferation is witnessed due to cell cycle dysregulation and genetic alterations in the proteins that are involved in the regulation of cell cycle (Hanahan and Weinberg 2011; Brigham et al. 2012). The CDK1, along with its interacting partners (A and B cyclins), controls progression at the S-G_2 and G_2-M phases of the cell cycle.as shown in Fig. 9.1. Through phosphorylation and dephosphorylation, the CDK1 aids in cell cycle regulation. For inducing apoptosis (programmed cell death), the activated CDK1 has an essential role (Malumbres and Barbacid 2009). Together with A and B-type cyclins, CDK1 kinase modulates the centrosome cycle and the mitotic onset (Fig. 9.4) and represents one of the central modulators of mitosis.

After the successful chromosome condensation and their alignment at the metaphase plate, the CDK1 function is turned down to permit the separation of sister chromatid via separase or separin activation (Musacchio and Salmon 2007). The decondensation of chromosomes, nuclear envelope reformation, and the process of cytokinesis all require this inactivation of CDK1 (Potapova et al. 2006). The CDK1-related cyclins have already been shown to be unstable, and their control is carried out via ubiquitination (Gavet and Pines 2010a, b). Thr14 and Tyr15 phosphorylation of CDK1 inhibits the CDK1/Cyclin B complex during the G_2 phase (Gould and Nurse 1989). Tyr15 phosphorylation prevents substrate phosphorylation by blocking CDK1's ATP-binding site (Li et al. 1995). Phosphorylation of Thr14 prevents ATP binding (Heald et al. 1993).

9.2 Role of CDK1

The CDK1 (also called CDC2 = cell division control protein 2) is a mitotic CDK. After duplicating the DNA, the chromosomal segregation is controlled by SAC (the spindle assembly checkpoint). This process regulates the activity of CDK1 and hinders any defects in the segregation of chromosomes (Kops et al. 2005; Musacchio and Salmon 2007; Pérez de Castro et al. 2007). On similar grounds, impaired SAC (spindle assembly checkpoint) could lead to an equal inheritance of DNA. If not repaired, it could aid in tumor progression due to the accumulation of CIN (numerical chromosomal abbreviations). The A-type cyclins are believed to activate CDK1 toward the interphase end to aid in the start of mitosis. After the degradation of nuclear envelope, the A-type cyclins are dissolved to aid in the CDK1-cyclin B complexes formation, which drives the cells through mitosis (Malumbres and Barbacid 2005). CDK activity has also been recruited for associating the BRCA1 to the MRN [Mre11-Rad50-Nbs1] complex during homologous combination (Chen et al. 2008). The CDK1, 2, 4, and 6 participate in the regulation of cell cycle progression, while CDK7, 8 9 participate in transcription (Izadi et al. 2020). Despite containing the complete complement of interphase CDKs, mice embryos with CDK1 absent do not show division, demonstrating that these CDKs are unable to compensate for the lack of CDK1 (Santamaría et al. 2007). Furthermore, using homologous recombination to replace Cdk1 with Cdk2 ends up causing early embryonic lethality (Satyanarayana et al. 2008), demonstrating that CDK2 cannot replace the role of CDK1, even when the Cdk1 locus is used for its expression. Cyclin A2 knockout causes early embryonic lethality (Murphy et al. 1997), implying that this cyclin's primary function is to trigger CDK1, the mitotic CDK. A transient delay in interphase was witnessed in human cell lines that lacked CDK1 although there was no hindrance to mitotic entry. Also the mitosis occurring afterward is characterized by several abnormalities (Lau et al. 2021).

Some of the documented functions of CDK1 include participation in Cell division, Checkpoint activation, DNA repair, Apoptotic process, DNA replication, and G2/M transition as shown in Fig. 9.5. In association with cyclin B, the resulting complex CDK1/Cyclin B aids in the progression of the cell cycle at the mitosis phase

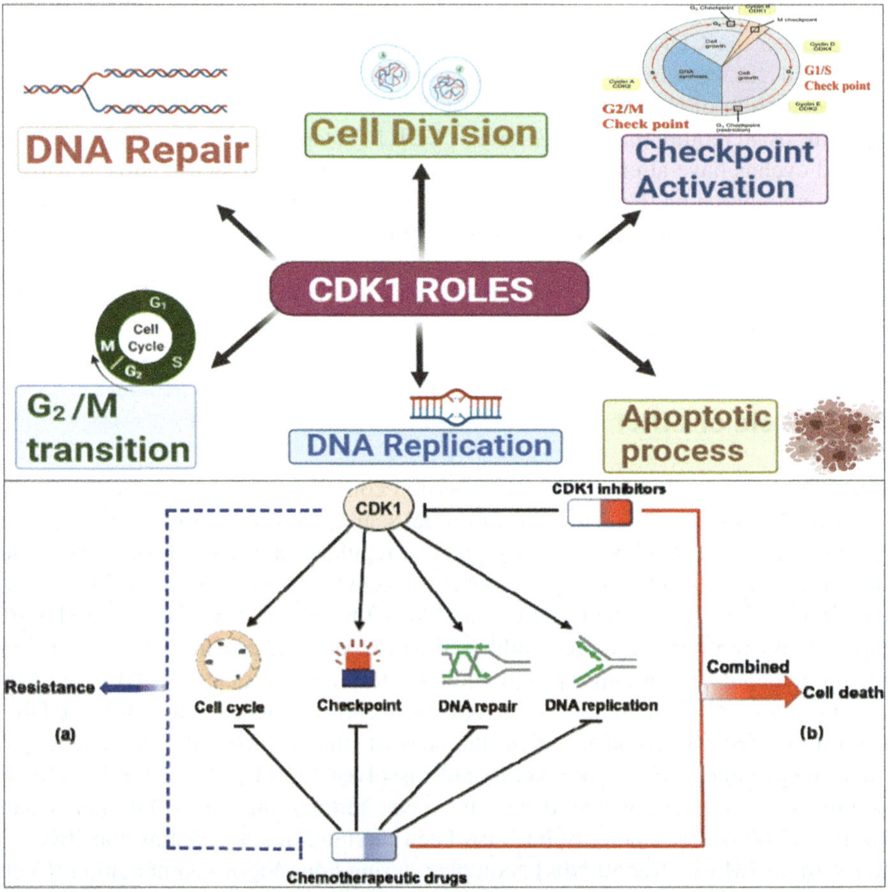

Fig. 9.5 Some of the functions of CDK1: Cell division, Checkpoint activation, DNA repair, Apoptotic process, DNA replication, and G2/M transition

(Draetta and Beach 1988) and the complex CDK2/Cyclin B modulates the G_1 to S transition (Endicott et al. 1999). As per a study's upshot, the CDK1 removal leads to inaccurate control at G_2/M. While as the CDK2 absence does not impact the progression of the cell cycle indicating that other CDKs can compensate for the CDK2 roles (Lau et al. 2021).

9.3 Dysregulation of CDK1 in BC

The uncontrolled cellular proliferation manifests as one of the cancer hallmarks. The same occurs through the cell cycle checkpoint disabling and overriding several safeguards involved with the cyclin/CDKs dysregulation or impairment. It has been seen that the CDK1 shows high expression in multiple cancers, like in the case of BC (Izadi et al. 2020). In the case of MYC-dependent BC patients, CDK1 inhibition is regarded as a potential therapeutic strategy (Izadi et al. 2020). Typically, the CDK1 and cyclin A/B aid the M phase of the cell cycle, but in BC cells, these participate in the programmed cell death of MYC-driven TNBC (Duffy et al. 2015). It has been analyzed through heat map studies that CDK1 and CDK2, 4, 5, and 8 display elevated expression relative to CDK6 and 9 in the case of primary tumors of BC (Sofi et al. 2022).

Roles of CDK1 in mitotic progression have been observed along with the overexpression of Cyclin A2 and B1 (Aaltonen et al. 2009). During the mitotic phase, CDK1/Cyclin B aids in the cell cycle progression (Draetta and Beach 1988). Cyclin B1 has been associated with higher promoter activity of Cyclin B1 as well as the $G_1/S/G_2$ cell cycle phases in numerous BC cell lines. Furthermore, enhanced CDK1/Cyclin B1 complex activity has been seen in T-47D and BT-549 cells during the G1 phase (Barrett et al. 2002). docosahexaenoic (DHA) and eicosapentaenoic (EPA) are omega-3 fatty acids that biochemically display anti-cancer effects, and these effects have been examined in several studies. Moreover, both DHA and EPA partially hindered the MDA-MB-231 BC cell proliferation via the CDK1/Cyclin B1 complex obstruction. The duration of G_2/M phases is increased after treating the MDA-MB-231 cells with EPA and DHA in the cell cycle. Downregulation of CDK1, Cyclin B1, and Cyclin A was also witnessed as a result of this, and Cyclin B1 phosphorylation was also suppressed and 25C phosphatase reduced, which is known to activate the CDK1 (Barascu et al. 2006). Moreover, as per the study on patients (Chinese Han Women), it was observed that the genetic polymorphisms of genes that code for CDK1 and cyclin B1 could significantly impart the susceptibility to the BC progression and survival in these patients (Li et al. 2013). This also indicated that for BC patients, CDK1 and CDK2 specific activity could be utilized as a prognosis factor. It was observed in a study that CDK1 and CDK2 specific activity could help predict the possible chemotherapy outcome in patients of this study (Kim et al. 2012). Poor five-year relapse-free survival has been observed in patients who exhibit heightened specific activity of CDK1 and CDK2, as shown by Kim and colleagues, and based on the same, the BC patients could be categorized as low and high-risk groups. Thus, a valuable way to predict the outcome of the disease is by monitoring the CDK1 and CDK2 specific activity (Kim et al. 2008). It has been seen that tumor cell growth can be suppressed by employing siRNA molecules or NU2058 and NU6102 (the purine-based inhibitors), both of which lead to CDK1 and CDK2 silencing. It has been observed that apoptosis is induced in both sensitive and resistant cell lines (Johnson et al. 2010).

9.4 The CDK1 and Breast Cancer

Many in vitro and in vivo studies concerned with the function of CDK1 in BC have been carried out; some of the studies have been included in Table 9.1 A heightened expression and activity was seen in $G_1/S/G_2$ phases of cell cycle for CDK1/Cyclin B1 in case of certain breast cancer cell lines through an in vitro investigation (Barrett

Table 9.1 Studies on CDK1 in BC

S. no.	Cell lines employed	Type of Study	Study upshots	References
1.	MDA-MB-231	In vitro	Downregulation of CDK1/cyclin B1 by docosahexaenoic and eicosapentaenoic acids triggered programmed cell death in BC cells	Barascu et al. (2006)
2.	T47D MCF-7 MDA-MB-468 MDA-MB-231	In vivo	The cancer cell susceptibility to paclitaxel therapy can be anticipated by determining the CDK1 & 2 specific activity	Nakayama et al. (2009)
3.	MCF-7 SK-BR-3 MDA-MB-231 HCC1937	In vitro	The susceptibility of cells to PARP inhibition was increased when CDK1 was downregulated	Xia et al. (2014)
4.	MDA-MB-231 HCC1937 MCF-10A HEK-293Tc	In vivo & in vitro	Cancer cell proliferation was greatly reduced when CDK1 was silenced by siRNA. By decreasing CDK1, miR-424 reduces cancer cell growth and stops the cell cycle.	Li et al. (2013)
5.	MDA-MB-231	In vivo & In vitro	Through inhibition of CDK1, the treatment of ER-positive MDA-MB-231 BC cells with an ERβ agonist inhibited growth of cells. In TNBC cells, silencing CDK1 and CDK7 decreased proliferation in an ERβ-independent way.	Reese et al. (2017)

et al. 2002). In another study, it was concluded that evaluating the CDK1 and 2 specific activity could be regarded a significant prognostic value for determining the outcomes in BC subjects, and multiple drugs were used in this in vivo study (Kim et al. 2008). In another in vitro study, Flavopiridol, siRNA were used and the disruption of CDK1 and 2, which in certain BC cell lines resulted in the arrest of cell cycle and apoptosis (Johnson et al. 2010). A study observed that the particular activity of CDK1 and 2 could possibly predict reaction toward the treatment being employed as well as the chances of recurrence and for this in vivo trial, the medicines Paclitaxel, 5-fluorouracil, Epirubicin, and Cyclophosphamide were employed (Kim et al. 2012). As per another study, the in vivo polymorphisms of CDK1 and CCNB1 genes increase the predisposition to BC, disease advancement as well as rate of survival (Li et al. 2013). The growth of TNBC displayed a suppression in SUM149, BT549 MCF-10A cell lines when CDK1 was downregulated by siRNA-laden nanoparticles (Liu et al. 2014). In an in vivo human investigation, increased CDK1 specific activity was linked to early and high recurrence rates (Kim et al. 2014).

In a study using the medications Aminophenazone, Pomalidomide, and Rosoxacin in an in vitro investigation, CDK1 was found to be a diagnostic marker in ductal carcinoma in situ (Ding et al. 2017). The observed phenotype in mouse model with gene-targeted CDK alleles (lacking CDK3) was that in the initial cell divisions, embryonic lethality was witnessed because of deficiency of CDK1 with Cdk1mut/mut type of genotype (Santamaría et al. 2007) and this $Cdk1^{mut}$ allele has been developed by using the insertion of gene trap vector and it represents loss-of-function strain. It has been seen in the study that when the roscovitine, a pan-CDK inhibitor, is administered sequentially preceding doxorubicin treatment is synthetically lethal in the triple-negative breast cancer cells. This inhibitor, when administered, halts the cell cycle in phase G_2/M, preparing them for DNA damage. It was observed that this combined treatment approach led to an enhancement in DNA double-stranded breaks and lowered the recruitment of proteins necessary for homologous recombination compared to the solo treatment by doxorubicin. It was also witnessed that by employing this combination therapy, there was a reduction in the tumor volume and an elevated survival was observed in comparison with the solo drug or related treatment in the case of xenograft studies (Jabbour-Leung et al. 2016). It was observed in a study that in the lack of CDK2, the CDK1 acts as G1-S CDK and binds to Cyclin E and when CDK1 is absent, CDK2 binds Cyclin B leading to mitotic entry (Lau et al. 2021) (Table 9.2). It was also seen that although performing all the mitotic roles of CDK1 by CDK2 was not enough in its normal concentrations, the CDK2 overexpression could overcome the mitotic abnormalities that occur due to lack of CDK1 (Lau et al. 2021). The results of CDK1 dysregulation include a robust growth of the tumor, heightened cancer cell proliferation rates, and chromosomal mutability (Barascu et al. 2006), as shown in Fig. 9.6.

Table 9.2 The biological role of CDK1 in BC

Cyclin-dependent kinases	CDK interacting Partners (Cyclins)	Attributed role(s) in Breast cancer	Reference
Cyclin-dependent kinase	*A/B*	Partners with programmed cell death of MYC-driven triple-negative breast cancer.	Chen et al. (2009), Marais et al. (2010), Horiuchi et al. (2012)

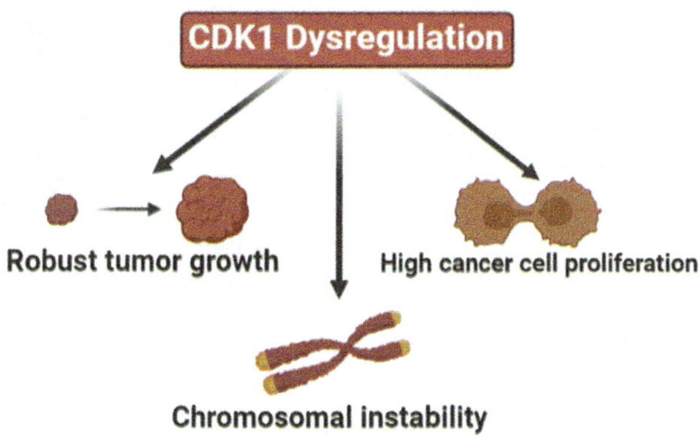

Fig. 9.6 The dysregulation of CDK1 could lead to many possible changes/events, including enhanced tumor growth, increased cancer cell proliferation, and chromosomal instability

It was revealed in a study that CDK2 generates less mitotic phosphorylation when compared to CDK1 leading to abnormal late mitotic events and a lack of both CDK1 and 2 results in total abolishment of mitotic entry. It was also elucidated in this study that in the absence of CDK1, the RPE1 human epithelial cell line is unable to undergo mitotic entry unlike cancer cell lines (Lau et al. 2021).

9.5 Therapeutic Implications

The inhibitors directed against CDKs are categorized either as Non-selective or Selective, i.e., either pan-inhibitors or against one single cyclin-dependent kinase, solely based on meticulousness against the CDKs (Ding et al. 2020). Various drugs that are CDK inhibitors in action have entered breast cancer clinical trials and are known to target cell modulators in the cancerous cells, thus furnishing a therapeutic window (Ding et al. 2020). Various pan-CDK inhibitors have been employed in trials to inhibit the activities of CDKS including CDK1. All pan-CDK inhibitors are non-specific in action and produce various undesirable toxicities too. Some of the inhibitors of CDK1 include the following:

1. Flavopiridol (a semi-synthetic flavone) (Kaur et al. 1992; Sedlacek et al. 1996).
2. Roscovitine (a synthetic purine) (Lin et al. 2010).
3. Dinaciclib (Paruch et al. 2010).
   These are all pan-CDK inhibitors (Fig. 9.7).

Roscovitine, Ro-3306, and Dinaciclib are the inhibitors that target CDK1/2 (Lin et al. 2018). Roscovitine inhibits the CDK1 and others (Table 9.3) by directly competing at the ATP-binding sites (Vassilev et al. 2006). Also, Ro-3306 blocks/prevents the G_2 to M transition, leading to programmed cell death of tumor cells after CDK1/2 inhibitor exposure for a long time (Xia et al. 2014; D'Andrea 2018). One of the examples is Flavopiridol, a semi-synthetic flavonoid obtained from rohitukine

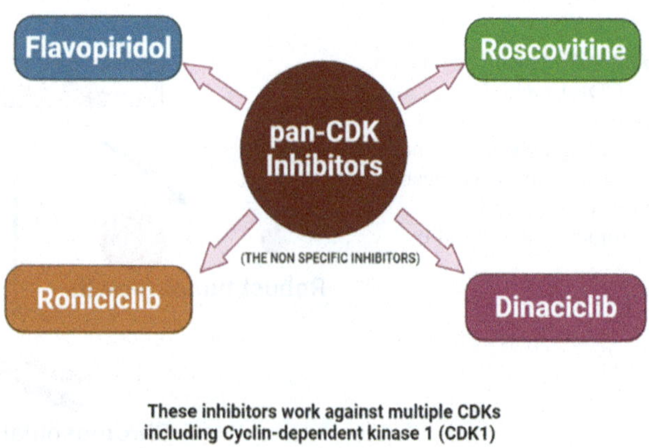

Fig. 9.7 Some of the pan-CDK inhibitors that are also effective against several CDKs including CDK1, and for BC Roscovitine and Dinaciclib trials have been conducted

Table 9.3 Examples of pan-CDK inhibitors along with their targets

S. No.	Drug	Inhibitor of	Developed by
1.	Flavopiridol (alvocidib)	CDK1,2,4,6,7,9.	Sanofi-Aventis
2.	R-roscovitine (Seliciclib/Roscovitine)	CDK1,2,5,7,9.	Cyclacel
3.	Dinaciclib	CDK1,2,5,9.	Merck
4.	Roniciclib	CDK1,2,3,4,7,9.	Bayer

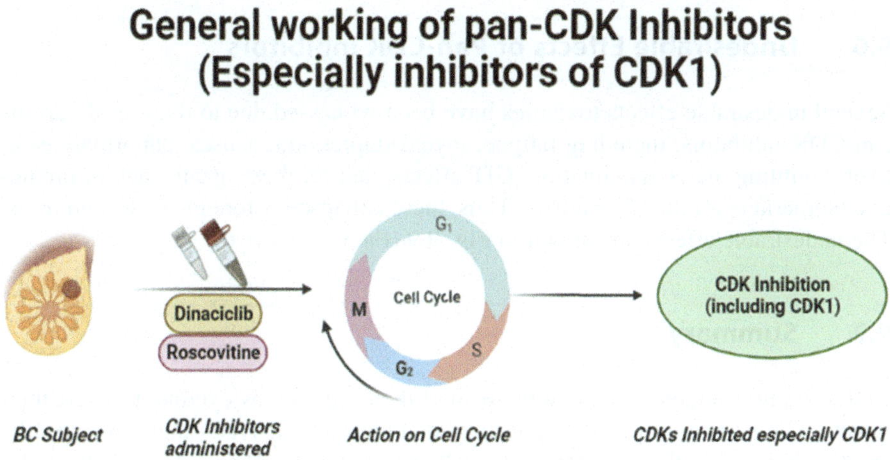

Fig. 9.8 The pan-CDK inhibitors' working model in general: The drugs like dinaciclib and roscovitine lead to the inhibition of multiple CDKs including CDK1

(a chromosome alkaloid). It exerts its anti-cancer effects by inhibiting CDK1, 2, 4, 6, 7, and 9 (Sedlacek et al. 1996; Shapiro 2006) Flavopiridol also known as Alvocidib, a first-generation pan-CDK inhibitor, the primary pan-CDK inhibitor employed in clinical trials. The activities of CDK1, 2, 4, 6, and 7,9 are primarily halted by Flavopiridol (Asghar et al. 2015). In the G_1 and G_2 phases, Flavopiridol leads to the arrest of the cell cycle and also induces cytotoxicity by blocking CDK7 and CDK9 and c-MYC transcription (Canavese et al. 2012).

The targets of Seliciclib are CDK1, CDK2, CDK5, CDK7, and CDK9 (Whittaker et al. 2004). This inhibitor surfaced as the initial orally available drug (from this class) to become part of the clinical trials due to its relative success in the pre-clinical stage, where its success leads to the onset of apoptosis in tumor cells (Shapiro 2006; Galons et al. 2010; Nanos-Webb et al. 2012). Another example from the pan-CDK inhibitors is provided by Dinaciclib, which is known to inhibit CDK1, 2, 5, and 9 with excellent Rb phosphorylation inhibitory potency (Fig. 9.8), thus showing a better therapeutic index in comparison with the Flavopiridol.

It must be mentioned that the palbociclib and abemaciclib display very low potency against CDK1, 2, 7, and 9, and with fulvestrant these have been marked for the second-line therapy (Chen et al. 2019). The drug that is orally administered and displays high potency with bioavailability and inhibits CDK4/6 selectivity is Ribociclib. This drug does not display significant activity against CDK2 and CDK1 (Sobhani et al. 2019). The ER+ breast cancer cell proliferation and migration is known to be inhibited by PL (**piperlongumine**), a novel CDK inhibitor discovered by Jeong et al. The PL is a natural product, and it is obtained from pepper. It hinders the CDK1 and CDK4/6 expression levels and leads to obstruction of the cell cycle at the G_2/M phase in order to stop tumorigenesis (Table 9.3) (Asghar et al. 2017).

9.6 Undesirable Effects of Pan-CDK Inhibitors

Several undesirable effects/toxicities have been witnessed due to the use of various pan-CDK inhibitors, including fatigue, myelosuppression, nausea, abnormalities in liver, vomiting, nerve dysfunction, GIT effects, and for these agents lack of predictive biomarkers for the BC patients. Thus, these collapsed before phase second trials. The undesirable effects are shown in Fig. 9.9 (Finn et al. 2016).

9.7 Summary

CDKs require modulatory subunits to bind them, known as cyclins, to exert their effects. The latter are formed and destroyed at various cell cycle stages in a specific and timely manner, thus regulating the cell cycle properly. The uncontrolled cellular proliferation manifests as one of the cancer hallmarks. The same occurs through the cell cycle checkpoint disabling and overriding several safeguards involved with the cyclin/CDKs dysregulation or impairment. In many solid cancers like BC,

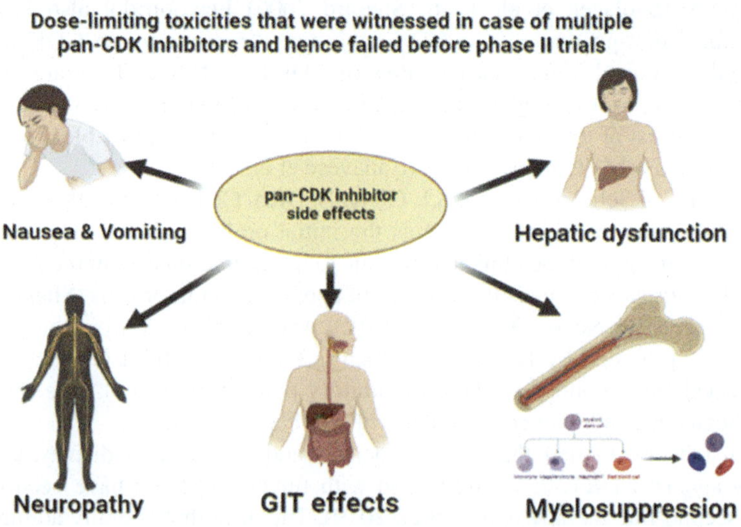

Fig. 9.9 Some side effects observed due to consumption of pan-CDK inhibitors in BC patients

uncontrolled cell proliferation is witnessed as a result of cell cycle dysregulation and the genetic changes in the proteins involved in cell cycle regulation. CDKs require modulatory subunits to bind them, known as cyclins, to exert their effects. The latter are formed and destroyed at various cell cycle stages in a specific and timely manner, thus regulating the cell cycle properly. Unscheduled proliferation, chromosomal instability (CIN), and genomic instability (GIN) are the three most common disorders in the cell cycle caused by either direct or indirect poor control of CDKs. Because of their critical participation in cell cycle progression, downregulation of CDKs may result in improper homeostasis in specific tissues. On the other hand, by initiating the untimely division in cells (progenitor or stem cells), the hyperactivated CDKs could also aid in the tumor development. CDK1 is a member of the cell cycle-associated CDK family, which also includes CDK2, CDK4, and CDK6. These CDKs play a direct role in the progression of cell cycle. Multiple studies have demonstrated that the CDK1 expression levels and function are dysregulated indicating its potential role in BC progression. It has also been witnessed that by either blocking or silencing the CDK1 could suppress the BC growth, particularly when combined with other anti-cancer agents. The results of CDK1 dysregulation include a robust growth of the tumor, heightened cancer cell proliferation rates, and chromosomal mutability. Furthermore, many studies have been previously carried out to examine the possible roles of CDK1 in the case of BC like the heightened expression of CDK1/Cyclin B1 was witnessed in G_1/S/G_2phases in breast cancer cell lines. The pan-CDK inhibitors employed for this treatment, however, come with multiple undesirable effects on BC patients. As such combination with other anti-cancer therapeutics for relatively superior outcomes could be a better option for BC patients.

9.8 Further Readings

The readers can have a look upon the following articles for the better understanding of the given topic:

(i) https://www.ncbi.nlm.nih.gov/pmc/articles/PMC3636749/
(ii) https://ascopubs.org/doi/10.1200/JCO.2005.05.064

The readers can also take a look upon the following visual presentations:

(i) https://youtu.be/foR2tZHj5Eo
(ii) https://youtu.be/tBoG9d0tGCE

For more insights about the topic, we would suggest detailed findings from the books of Mir MA (2022) https://doi.org/10.1016/C2021-0-02565-7, https://doi.org/10.1016/C2022-0-00074-X Mir MA (2021) https://doi.org/10.52305/WXJL6770, from cancer.net website, https://www.cancer.net/cancer-types/breast-cancer/types-treatment.

References

Aaltonen K et al (2009) High cyclin B1 expression is associated with poor survival in breast cancer. Br J Cancer 100(7):1055–1060

Asghar U et al (2015) The history and future of targeting cyclin-dependent kinases in cancer therapy. Nat Rev Drug Discov 14(2):130–146

Asghar US et al (2017) Single-cell dynamics determines response to CDK4/6 inhibition in triple-negative breast CancerCDK4/6 inhibition in triple-negative breast cancer. Clin Cancer Res 23(18): 5561–5572

Barascu A et al (2006) CDK1-cyclin B1 mediates the inhibition of proliferation induced by omega-3 fatty acids in MDA-MB-231 breast cancer cells. Int J Biochem Cell Biol 38(2):196–208

Barrett KL et al (2002) Cyclin b1 promoter activity and functional cdk1 complex formation in G1 phase of human breast cancer cells. Cell Biol Int 26(1):19–28

Bartek J et al (2004) Checking on DNA damage in S phase. Nat Rev Mol Cell Biol 5(10):792–804

Brigham et al (2012) Comprehensive molecular portraits of human breast tumours. Nature 490(7418): 61–70

Canavese M et al (2012) Cyclin dependent kinases in cancer: potential for therapeutic intervention. Cancer Biol Ther 13(7):451–457

Chen L et al (2008) Cell cycle-dependent complex formation of BRCA1· CtIP· MRN is important for DNA double-strand break repair. J Biol Chem 283(12):7713–7720

Chen Y-J et al (2009) A conserved phosphorylation site within the forkhead domain of FoxM1B is required for its activation by cyclin-CDK1. J Biol Chem 284(44):30695–30707

Chen X et al (2019) Latest overview of the cyclin-dependent kinases 4/6 inhibitors in breast cancer: the past, the present and the future. J Cancer 10(26):6608

D'Andrea ADJDR (2018) Mechanisms of PARP inhibitor sensitivity and resistance. DNA Repair (Amst) 71:172–176

Ding Z-H et al (2017) Docking of CDK1 with antibiotic drugs revealed novel therapeutic value in breast ductal cancer in situ. Oncotarget 8(37):61998

Ding L et al (2020) The roles of cyclin-dependent kinases in cell-cycle progression and therapeutic strategies in human breast cancer. Int J Mol Sci 21(6):1960

Draetta G, Beach DJC (1988) Activation of cdc2 protein kinase during mitosis in human cells: cell cycle-dependent phosphorylation and subunit rearrangement. Cell 54(1):17–26

Duffy MJ et al (2015) Biomarkers in breast cancer: where are we and where are we going? Adv Clin Chem 71:1–23

Endicott JA et al (1999) Cyclin-dependent kinases: inhibition and substrate recognition. Curr Opin Struct Biol 9(6):738–744

Esashi F et al (2005) CDK-dependent phosphorylation of BRCA2 as a regulatory mechanism for recombinational repair. Nature 434(7033):598–604

Finn RS et al (2016) Targeting the cyclin-dependent kinases (CDK) 4/6 in estrogen receptor-positive breast cancers. Breast Cancer Res 18(1):1–11

Galons H et al (2010) Cyclin-dependent kinase inhibitors: a survey of recent patent literature. Expert Opin Ther Pat 20(3):377–404

Gavet O, Pines JJDC (2010a) Progressive activation of CyclinB1-Cdk1 coordinates entry to mitosis. Dev Cell 18(4):533–543

Gavet O, Pines JJJOCB (2010b) Activation of cyclin B1–Cdk1 synchronizes events in the nucleus and the cytoplasm at mitosis. J Cell Biol 189(2):247–259

Gerger A et al (2007) A multigenic approach to predict breast cancer risk. Breast Cancer Res Treat 104(2):159–164

Gould KL, Nurse PJN (1989) Tyrosine phosphorylation of the fission yeast cdc2+ protein kinase regulates entry into mitosis. Nature 342(6245):39–45

Hanahan D, Weinberg RAJC (2011) Hallmarks of cancer: the next generation. Cell 144(5):646–674

Hashemi V et al (2019) The role of DEAD-box RNA helicase p68 (DDX5) in the development and treatment of breast cancer. J Cell Physiol 234(5):5478–5487

Heald R et al (1993) Human wee1 maintains mitotic timing by protecting the nucleus from cytoplasmically activated Cdc2 kinase. Cell 74(3):463–474

Horiuchi D et al (2012) MYC pathway activation in triple-negative breast cancer is synthetic lethal with CDK inhibition. J Exp Med 209(4):679–696

Huertas P, Jackson SPJJOBC (2009) Human CtIP mediates cell cycle control of DNA end resection and double strand break repair. J Biol Chem 284(14):9558–9565

Huertas Sánchez P et al (2008) CDK targets Sae2 to control DNA-end resection and homologous recombination

Ira G et al (2004) DNA end resection, homologous recombination and DNA damage checkpoint activation require CDK1. Nature 431(7011):1011–1017

Izadi S et al (2020) CDK1 in breast cancer: implications for theranostic potential. Anti Cancer Agents Med Chem 20(7):758–767

Jabbour-Leung NA et al (2016) Sequential combination therapy of CDK inhibition and doxorubicin is synthetically lethal in p53-mutant triple-negative breast CancerCDK inhibition sensitizes p53-mutated cells to doxorubicin. Mol Cancer Ther 15(4):593–607

Johnson N et al (2010) Pre-clinical evaluation of cyclin-dependent kinase 2 and 1 inhibition in anti-estrogen-sensitive and resistant breast cancer cells. Br J Cancer 102(2):342–350

Kastan MB, Bartek JJN (2004) Cell-cycle checkpoints and cancer. Nature 432(7015):316–323

Kaur G et al (1992) Growth inhibition with reversible cell cycle arrest of carcinoma cells by flavone L86-8275. J Natl Cancer Inst 84(22):1736–1740

Kim S et al (2008) Determination of the specific activity of CDK1 and CDK2 as a novel prognostic indicator for early breast cancer. Ann Oncol 19(1):68–72

Kim S et al (2012) Recurrence risk score based on the specific activity of CDK1 and CDK2 predicts response to neoadjuvant paclitaxel followed by 5-fluorouracil, epirubicin and cyclophosphamide in breast cancers. Ann Oncol 23(4):891–897

Kim SJ et al (2014) The cell cycle profiling-risk score based on CDK1 and 2 predicts early recurrence in node-negative, hormone receptor-positive breast cancer treated with endocrine therapy. Cancer Lett 355(2):217–223

Kops GJ et al (2005) On the road to cancer: aneuploidy and the mitotic checkpoint. Nat Rev Cancer 5(10):773–785

Lau HW et al (2021) Quantitative differences between cyclin-dependent kinases underlie the unique functions of CDK1 in human cells. Cell Rep 37(2):109808

Li S et al (1995) The cdc2-related kinase, PISSLRE, is essential for cell growth and acts in G2 phase of the cell cycle. Cancer Res 55(18):3992–3995

Li Y et al (2013) Association study of germline variants in CCNB1 and CDK1 with breast cancer susceptibility, progression, and survival among Chinese Han women. PLoS One 8(12):e84489

Lin TS et al (2010) Flavopiridol, fludarabine, and rituximab in mantle cell lymphoma and indolent B-cell lymphoproliferative disorders. J Clin Oncol 28(3):418

Lin ZP et al (2018) Targeting cyclin-dependent kinases for treatment of gynecologic cancers. Front Oncol 8:303

Liu Y et al (2014) Triple negative breast cancer therapy with CDK1 siRNA delivered by cationic lipid assisted PEG-PLA nanoparticles. J Control Release 192:114–121

Malumbres M, Barbacid MJNRC (2001) To cycle or not to cycle: a critical decision in cancer. Nat Rev Cancer 1(3):222–231

Malumbres M, Barbacid MJNRC (2009) Cell cycle, CDKs and cancer: a changing paradigm. Nat Rev Cancer 9(3):153–166

Malumbres M, Barbacid MJTIBS (2005) Mammalian cyclin-dependent kinases. Trends Biochem Sci 30(11):630–641

Marais A et al (2010) Cell cycle-dependent regulation of the forkhead transcription factor FOXK2 by CDK· cyclin complexes. J Biol Chem 285(46):35728–35739

Murphy M et al (1997) Delayed early embryonic lethality following disruption of the murine cyclin A2 gene. Nat Genet 15(1):83–86

Musacchio A, Salmon EDJNRMCB (2007) The spindle-assembly checkpoint in space and time. Nat Rev Mol Cell Biol 8(5):379–393
Nakayama S et al (2009) Prediction of paclitaxel sensitivity by CDK1 and CDK2 activity in human breast cancer cells. Breast Cancer Res 11(1):1–10
Nanos-Webb A et al (2012) Targeting low molecular weight cyclin E (LMW-E) in breast cancer. Breast Cancer Res Treat 132(2):575–588
Paruch K et al (2010) Discovery of dinaciclib (SCH 727965): a potent and selective inhibitor of cyclin-dependent kinases. ACS Med Chem Lett 1(5):204–208
Pérez de Castro I et al (2007) A census of mitotic cancer genes: new insights into tumor cell biology and cancer therapy. Carcinogenesis 28(5):899–912
Poon RYJCCO (2016) Cell cycle control: a system of interlinking oscillators. Methods Mol Biol:3–19
Potapova TA et al (2006) The reversibility of mitotic exit in vertebrate cells. Nature 440(7086):954–958
Rahman N, Stratton MRJAROG (1998) The genetics of breast cancer susceptibility. Annu Rev Genet 32:95
Reese JM et al (2017) ERβ inhibits cyclin dependent kinases 1 and 7 in triple negative breast cancer. Oncotarget 8(57):96506
Santamaría D et al (2007) Cdk1 is sufficient to drive the mammalian cell cycle. Nature 448(7155): 811–815
Satyanarayana A et al (2008) Genetic substitution of Cdk1 by Cdk2 leads to embryonic lethality and loss of meiotic function of Cdk2. Development 135(20):3389–3400
Sedlacek H et al (1996) Flavopiridol (L86 8275; NSC 649890), a new kinase inhibitor for tumor therapy. Int J Oncol 9(6):1143–1168
Shapiro GIJJOCO (2006) Cyclin-dependent kinase pathways as targets for cancer treatment. J Clin Oncol 24(11):1770–1783
Sobhani N et al (2019) Updates on the CDK4/6 inhibitory strategy and combinations in breast cancer. Cell 8(4):321
Sofi S et al (2022) Cyclin-dependent kinases in breast cancer: expression pattern and therapeutic implications. Med Oncol 39(6):1–16
Vassilev LT et al (2006) Selective small-molecule inhibitor reveals critical mitotic functions of human CDK1. Proc Natl Acad Sci U S A 103(28):10660–10665
Whittaker SR et al (2004) The Cyclin-dependent kinase inhibitor CYC202 (R-roscovitine) inhibits retinoblastoma protein phosphorylation, causes loss of cyclin D1, and activates the mitogen-activated protein kinase pathway. Cancer Res 64(1):262–272
Xia Q et al (2014) The CDK1 inhibitor RO3306 improves the response of BRCA-proficient breast cancer cells to PARP inhibition. Int J Oncol 44(3):735–744

Cdk4/Cdk6 Dysregulation in Estrogen-Positive Receptor Breast Cancers

10

Manzoor Ahmad Mir and Ulfat Jan

10.1 Introduction

The important physiological process for growth and development as well as maintenance of our body is cell division (Sofi et al. 2022a, b). There are various pathways that normally regulate this cell division via various mechanisms. During the developmental processes and throughout an individual's life, cell growth proceeds possibly at the proper location and time, and the cellular contents, along with each chromosome, must be perfectly copied (Susanti and Tjahjono 2021; Mehraj et al. 2022a, b, c). The whole cell cycle procedure is extremely preserved and strictly managed to regulate the normal cellular division as well as genome duplication. This cell division consists of 4 unique organized stages including (G0/G1), S (synthetic phase, i.e., DNA duplication), G2 (pre-mitotic phase), and M (mitotic phase) as well as various checkpoints that duly secure the normal replication process in the synthetic period of the cell division as well as precise chromosome aggregation into the new cells (Thu et al. 2018; Mehraj et al. 2022a, b, c). In mammals, this cycle is governed by various types of protein regulators called as protein kinases that usually control the whole process of cell cycle. CDKs (cyclin-dependent kinases) control whole of the cell's commitment, DNA replication as well as the arrival of the mitosis (Malumbres and Barbacid 2007; Mehraj et al. 2021). The interaction of regulatory elements termed cyclins is required for CDK functioning. Cyclins are generated and eliminated at precise moments during the cell cycle, allowing for appropriate regulation of kinase activity. CDKs and cyclins are encoded by numerous loci in human cells 13th & 25th loci, respectively (Malumbres and Barbacid 2005). Although there are few subtypes of the CDK–cyclin complexes that are

M. A. Mir (✉) · U. Jan
Department of Bioresources, School of Biological Sciences, University of Kashmir, Srinagar, J&K, India
e-mail: drmanzoor@kashmiruniversity.ac.in

M. A. Mir (ed.), *Therapeutic Potential of Cell Cycle Kinases in Breast Cancer*,
https://doi.org/10.1007/978-981-19-8911-7_10

Table 10.1 Representing some of the basic Cdk complexes and their role in the cell cycle

S. no.	Cell cycle stage	Cdk/Cyc complex	Function
01.	G1 phase	Cdk4/6/ Cyc D	Regulates cellular proliferation The main target of this complex is Rb. Rb usually inhibits the cell cycle progression. This Rb is phosphorylated by this complex and inactivates it, thereby helps in cell cycle continuation.
02.	G1/S phase	Cdk2/Cyc E	It helps in G1/S phase transition It also phosphorylates RB and thus promotes G1 progression of the cell cycles and also determines the initiation of the DNA replication It assists in the expression of Cyc A, thus permitting the progression toward the S phase of the cell cycles
03.	S phase	Cdk2/Cyc A	It initiates the DNA replication It also prevents the formation of new replication complexes, ensuring that DNA is copied only once per cell cycle. This occurs via phosphorylation of the CDC6, one of the replication machinery components
04.	G2/M phase	Cdk1/ CycA	Regulates the M-phase of the cycle and also activates and stabilizes the Cdk1/Cyc B complex
05.	M-phase	Cdk1/Cyc B	Also called as maturation promoting factor (MPF) and acts as a key controller of M-phase transition and also phosphorylates and activates various downstream protein kinases.

engaged in the cell cycle control as well as progression, only a fraction of CDK–cyclin complexes, however, is directly engaged in cell cycle control. Three inter-phase cyclin-dependent kinases (CDK 2, 4, and 6), one mitotic kinase (cyclin-dependent kinase-1, also known as cell division control protein 2 (CDC2)), plus 10 cyclins from 4 categories are included (the A-, B-, D-, and E-type cyclins). Cancer abnormalities typically de-regulate particular CDK–cyclin interactions, culminating either in unscheduled re-entrance within the cell cycle or prolonged growth, both of which are characteristics of most human cancer cells (Malumbres and Barbacid 2001; Sofi et al. 2022a, b) (Table 10.1).

Checkpoints that detect probable abnormalities during DNA synthesis and chromosomal segregation monitor proper cell cycle progression. Despite regulation of CDK activity, activation of these checkpoints leads toward the cell cycle arrest. Cell cycle arrest permits cell to correct such errors effectively, prohibiting them from being passed onto the new cells. The DNA damage checkpoint preserves the cells from extrinsic and intrinsic genotoxic agents (such as chemical toxins, free radicals, ionizing radiation, subcellular metabolic by-products, or medicinal therapies) that cause a variety of DNA mutations. Such changes are detected by a signal transduction pathway, which results in CDK blockage and proper arrest of the cell cycle (Bartek et al. 2004). The transition from G1 to S phase depends wholly and solely on the restriction point located between G1 and S phase of the cell cycle. This point decides whether a cell should enter the synthetic stage of the cell cycle or the cell should quit the cell division to stop at the G0 stage, making the G1 and G2 stages crucial control checkpoints. Several cyclins and CDKs, which are serine/threonine kinases, govern the cell cycle. In particular phases, CDKs form complexes with cyclins to stabilize, turn on, and phosphorylate CDKs (Malumbres 2014; Niu et al. 2019). Further, CDKs and the main role played by these key players in the cell cycle will be discussed in the next sections of this chapter.

10.2 Regulation of Cell Cycle by CDKs

CDKs regulate cell division in relation to extracellular and intracellular signals by acting as catalytic components in a heterodimeric complex with cyclins, which act as regulating elements. There are usually 20 CDKs and 29 cyclins in human cells (Canavese et al. 2012; Cao et al. 2014) (Fig. 10.1). The revelation of the cdc28 and cdc2 (human homologs of CDK1) in budding and fission yeast, respectively, revealed the role of CDKs in the cell growth (Hartwell 1974; Nurse 2002). There are various CDKs that promptly control all the four phases of the cellular division and play a crucial function in the cell cycle transition from one phase to another including CDK 1, 2, 3, 4, 5, 6 as well as 7. But the overall CDK activity varies in a circular pattern during the cell cycle (Cicenas and Valius 2011).

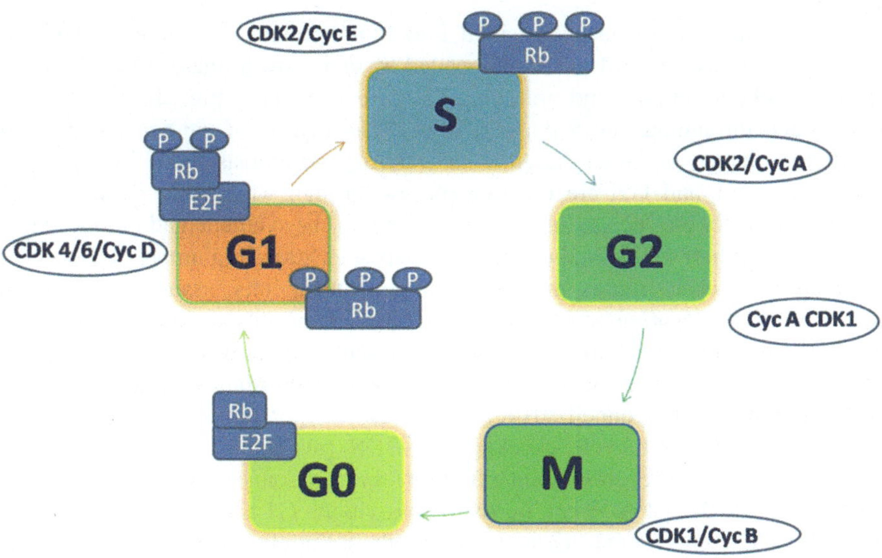

Fig. 10.1 A schematic illustration of the important CDK/Cyc complexes that are involved during the whole process of the cell cycle and they are involved in the cell cycle control as well as progression. Retinoblastoma (Rb) protein families represent the main targets. Their phosphorylation inhibits the binding as well as deactivation of the E2F TFs, their activation allows the transcription and translation of various genes that are necessary for synthetic phase of the cell cycle

Human cells must stimulate CDKs, which necessitates the attachment of the CDKs to a cyclin subunit to become catalytically operative, in order to begin DNA synthesis (S phase) (Massagué 2004). There are 3 main interphase CDKs (CDK6, CDK4, and CDK2), a mitotic CDK (CDK1, commonly called as cell division control protein 2 (CDC2)), and eleven cyclins from 4 groups make up this group (the B-, D-, E-, and A-type cyclins) (Table 10.2). Cancer mutation typically deregulates particular CDK–cyclin complexes, leading to the uncontrolled reentry into the cell growth or prolonged growth, both of which are characteristics of most human breast cancer cells (Malumbres and Barbacid 2001). Cdks are kinase proteins that are made up of an enzymatic component commonly called as Cdk and a regulatory subunit known as Cyclin. Even though human DNA consists of 12 Cdk loci, just 5 of these, Cdk4, Cdk6, Cdk2, Cdk3, and Cdk1, were directly linked to cell cycle control. While Cdk1 is thought being a mitotic kinase, rest of all Cdks are thought to be involved in cell division's early stages (interphase). But CDK4 has been discovered mutated in human cancers thus far and only in a handful of patients with inherited melanoma. CDK6 amplification has indeed been recorded as a result of genomic translocation in lymphomas, leukemias, and melanomas (Malumbres and Barbacid 2005; Chen et al. 2006; Qayoom et al. 2022). Furthermore, Cdk activity is misregulated in a considerable percentage of human malignancies due to mutations. These involve cyclin amplification and cyclin-dependent kinase inhibitor inactivation, which includes the representatives of the INK4 (inhibitor of kinase 4) and Cip/Kip subgroups (Malumbres and Barbacid 2001; Malumbres and Barbacid 2005).

The cell division cycle is classified into four fundamental phases: G1-phase, G2-phase, S-phase, and M-phase also called as first growth phase, second growth phase, synthetic phase, and mitotic phase, respectively (Fig. 10.2). DNA is synthesized throughout the S phase of this cycle, rest all of the cell components are divided into two similar daughter cells during the mitosis stage. The first two phases, i.e., G1 and G2, are the two phases through which that particular cell prepares itself for the successful accomplishment of the synthetic and mitotic stages sequentially (Sherr 1996; Malumbres and Barbacid 2009; Malumbres 2014). The completion of each step is monitored in the checkpoints, assuring the proper chromosome segregation and duplication in daughter cells therefore minimizing oncogenesis-inducing genetic instabilities. In circumstances where cellular damage is discovered, cells then undergo apoptosis (Sherr 1996; Malumbres and Barbacid 2009; Barnum and O'Connell 2014; Malumbres 2014). Additionally, it is followed by three primary cell cycle checkpoints: G1-S (restriction point), G2-M, and metaphase-to-anaphase checkpoint (Barnum and O'Connell 2014; Kolch et al. 2015; Roskoski Jr 2019). A cell needs to pass via the G1-S checkpoint to proceed the cell cycle, a checkpoint that is guided by the retinoblastoma-associated protein (RB1) and is normally controlled by changes in the fine line across the pro-mitotic and antimitotic cues. Mitogenic signaling is actually mandatory for entrance into the regular cell division cycle, but its significance diminishes once the cell enters the synthetic stage of the cell cycle (Blagosklonny and Pardee 2002; Mehraj et al. 2022a, b, c).

Table 10.2 representing the various types of CDKs and their cyclin partners along with their normal and abnormal functions after complex formation

S. no.	CDK family member	Associated cyclin partner	Normal function of complex	Abnormal function of complex
01.	CDK1	A1, A2, B1, B2	Acts as a central regulatory system and drives cells through G2 phase as well as mitosis	Development and progression of various cancer types Takes part in MYC-driven triple-negative BC
02.	CDK2	B3, E1, E2, A1, A2	Inhibition of proliferation rate and cell cycle arrest at GO/G1 & G2/M transitions	Associated with growth as well as progression of BC
03.	CDK3	E1, E2, C	Involved in G0/G1 and G1/S transitions CDK3 overexpression suppresses cell movements, invasion, and metastasis of BC cells	By promoting cell proliferation and development, it acts as a tumor promoter
04.	CDK4	D1, D2, D3	Aids in the G1/S phase transition in cell cycle CDK4/6 complex aids in growth and continuance of the division cycle	CDK4/D1 complex interacts with filamin A influencing metastasis and invasion of BC cells Aids in growth and development of the cancer especially breast cancer
05.	CDK5	D, E, G1		Growth as well as progression of multiple cancer types including breast cancer, etc.
06.	CDK6	D1, D2, D3	CDK4/6 complex aids in growth and progression of the cell cycle	Aids in growth and progression of the cancer especially breast cancer
07.	CDK7	H	Activates various CDKs via T-loop phosphorylation and thus drives cell cycle progression In cancer, CDK7 causes growth arrest, death, and transcriptional suppression	Overexpressed in ER+ breast cancer and increases cancer progression and metastasis
08.	CDK8	C	Regulates transcription via associating with mediator complex Controller of several cell cycle steps	Contributes toward tumorigenesis in various cancer types especially colorectal, breast, and hematological malignancies
09.	CDK9	K, T1, T2a, T2b	Key regulator of transcription It mainly controls the elongation and termination of the process of translation	Downregulates miR-874, a tumor suppressive microRNA that suppresses proliferation process by downregulating CDK9

(continued)

Table 10.2 (continued)

S. no.	CDK family member	Associated cyclin partner	Normal function of complex	Abnormal function of complex
10.	CDK10	M	Tumor suppressor Represses ciliogenesis G2-M transition	Major determinant of the resistance to endocrine therapy STAR syndrome
11.	CDK11	L	Regulates the transcription of RNA, splicing, and mitosis	Breast cancer growth and proliferation

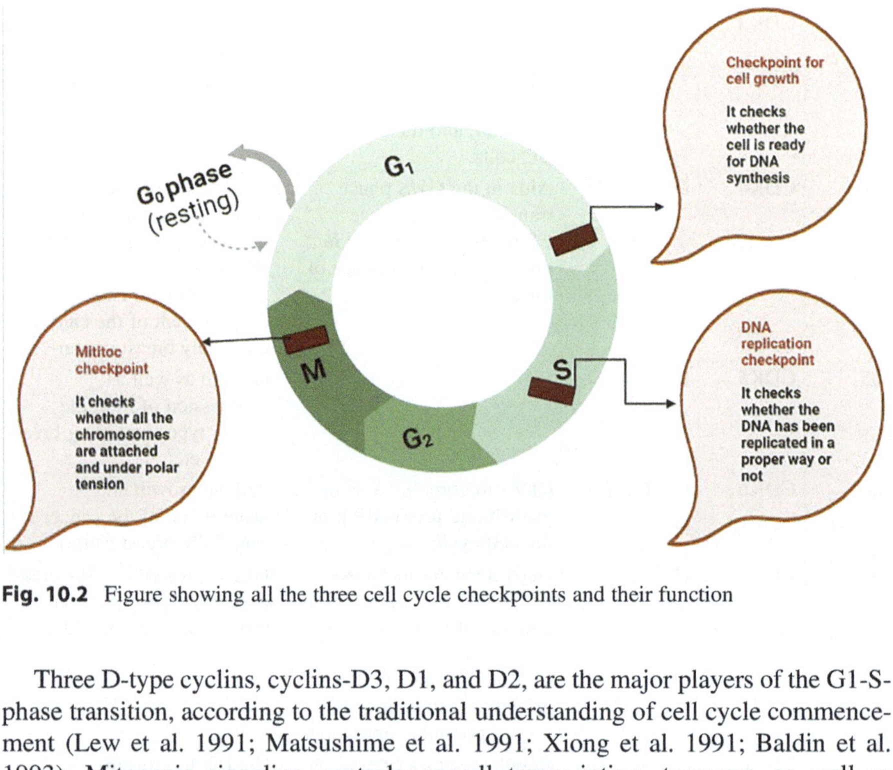

Fig. 10.2 Figure showing all the three cell cycle checkpoints and their function

Three D-type cyclins, cyclins-D3, D1, and D2, are the major players of the G1-S-phase transition, according to the traditional understanding of cell cycle commencement (Lew et al. 1991; Matsushime et al. 1991; Xiong et al. 1991; Baldin et al. 1993). Mitogenic signaling controls overall transcription, turnover, as well as nuclear transport of D-type cyclins and growth factor signaling controls overall expression levels of these cyclins (O'leary et al. 2016). Increased efficiency of D-type cyclins, which interact with and turn on CDK4/6, happens early in the cell cycle's G1 phase as a result of a pro-mitotic signaling balance. Such combination then phosphorylates RB1 as well as two RB1-like "pocket" proteins p130 and p107 (also known as retinoblastoma-like proteins 1 and 2, respectively) at a variety of sites (Matsushime et al. 1991; Kato et al. 1993; Meyerson et al. 1994). RB1 suppresses the transcriptional activity of the genes needed for division cycle continuation when it binds to the transactivation domain of the E2F transcription factor family of proteins in its hypophosphorylated form (O'leary et al. 2016).

10.3 CDK4/6's Dysregulation and Cell Cycle in BC

In around half of all invasive breast tumors, the tumor suppressor and CDK inhibitor p^{16INKA} are inactivated, which can happen through a variety of methods (Lee et al. 2012). Inactivation of the p16 gene is linked to abnormal cell growth, dedifferentiation, and pervasiveness (Lee et al. 2012). In spite of that p^{16INKA} is a tumor suppressor, its overexpression has been related to tumor growth (Davalos et al. 2010). Several malignancies increase cyclin D-dependent functionality, therefore prevent cellular senescence by a variety of methods, including p16 suppression, CDK4 expansion, CDK4 mutation with lack of INK4 binding, Cyclin D1 upregulation, or CCND1 translocation or overexpression (Satyanarayana and Kaldis 2009). Rb expression is lost in approximately 20–30% of breast tumors, hence the bulk of BCs are Rb proficient (Bosco and Knudsen 2007). Rb deficiency promotes tumor growth through loss of proliferative control as well as conversion to severe disease, this fact is very much common in TNBC, where it indicates a better diagnosis (Arima et al. 2008; Trere et al. 2009; Musgrove and Sutherland 2010). Conversely, Rb dysregulation in ER+ BC is a poor prognostic sign because this pattern is linked to a higher risk of metastatic spread (Ertel et al. 2010). Even if most BCs have normal Rb functioning, a number of additional events can cause the CDK4/6-cyclin D axis to become faulty, allowing cells to proliferate and tumors to form (Roberts et al. 2012). Cyclin D1 levels limit cell division in Rb-proficient malignancies through affecting Rb phosphorylation and activation (Millar et al. 2009). Antibodies against cyclin D1 suppress the estrogen-dependent G1-S progression. Estrogen receptor has a putative transcriptional target, i.e., Cyclin D1 (Yu et al. 2006). Elevated expressions of cyclin D1 are observed in around half of all mammary tumors, although its prognostic significance is unknown (Arnold and Papanikolaou 2005; Roy and Thompson 2006). Elevation in the expression levels of the Cyclin D1 gene (CCND1) is found in 15–20 percent of breast tumors that continues after the establishment of metastasis (Bartkova et al. 1994; Gillett et al. 1994; McIntosh et al. 1995). In preclinical studies, Rb deactivation has been associated with tamoxifen and fulvestrant tolerance (Bosco et al. 2007). Furthermore, experimental and clinical findings reveal that the cyclin D1 amplification stimulates the creation of cyclin D1-CDK4/6 complexes that in turn activates cyclin E1-CDK2 complexes, resulting in tamoxifen tolerance (Hui et al. 2002; Stendahl et al. 2004; Jirström et al. 2005; Rudas et al. 2008). Less knowledge is available regarding the mechanisms of AI treatment resistance, but cellular stress response stimulation as well as cell death is involved in tamoxifen and fulvestrant resistance (Riggins et al. 2005).

10.4 CDK6/4and its Relation with the Breast Cancer

BC, a widespread disease affecting over 1.3 million people each year and accounts for nearly about 23% of all types of cancers (Jemal et al. 2011). Changes within cell cycle's processes are seen as a "hallmark of cancer," that results in abnormal cell growth and proliferation (Hanahan and Weinberg 2011). Dysfunctional CDK4/6/D1 complex is involved in the onset and advancement of numerous cancerous diseases, including BC, according to numerous lines of evidence. Dysfunctional CDK4/6: cyclinD1 complex seems to be an initial step in mammary cancer progression and development, given that "upregulation" of cyclin D1 is commonly observed as early as invasive ductal carcinoma and is sustained in malignant tumors, and yet is invisible in the oldest tumors such as atypical ductal hyperplasias (Bartkova et al. 1994; Dickson et al. 1995). Uncontrollable cell growth is another one of cancer's hallmarks, and it's driven by CDK/cyclin dysfunction that bypasses cell cycle checkpoints and overcomes multiple safeguards. Previously, much study says that dysregulated CDK/cyclin activation was the most common cause of numerous breast cancer manifestations (Santo et al. 2015). CDK4 plays a pivotal role in ErbB-2 carcinogenesis but not for Wnt-induced oncogenes (Reddy et al. 2005). Findings reveal that the cyclin D1:CDK4/6 loop is crucial for BC preservation and BC advancement, add to the evidence for their participation in carcinogenic etiology. These studies are based on in vivo findings revealing ErbB2-driven tumor arrest and senescence regarding the cyclin D1 elimination or selective CDK4/6 inactivation (Choi et al. 2012). CDK4/6–RB pathway, which is crucial in cell cycle's G1/S phase transition, is significant in BCs. RB phosphorylation is controlled by cyclin D1/CDK4/6, which aids in cell growth. Downregulation of the CDK4/6–cyclin D/INK4/pRB/E2F axis, or its promoters, had been seen to play a pivotal role in tumorigenesis as well as BC retention (Santo et al. 2015). Although D-type cyclins are not required for breast growth, they are needed for effective tumor formation, as manifested due to the fact that mutant mice deficient in operative D1 type cyclin are immune to cancers caused by ErbB-2/HER2/neu and RAS oncogenes, while cyclin D3 deficient mice appear to be immune to Notch1-driven T-cell acute lymphoblastic leukemia (Yu et al. 2001; Landis et al. 2006; Choi et al. 2012). Moreover, it is observed that cyclin D1 and D3 are able to counterbalance each other in promoting tumor formation as well as tumor growth and metastasis (Zhang et al. 2011). Dysfunctional CDK4/6:cyclin D1 complex has been involved in genesis and also development of a number of different types of cancers, such as breast cancer, according to numerous lines of evidence. "Increased expression" of cyc D1 is typically observed as soon as invasive ductal carcinoma & retained in malignant tumors, although it is not found in the earliest lesions including anomalous ductal hyperplasia, suggesting that dysfunction of the cyclin D1:CDK4/6 axis is an initial step in BC development and metastasis (Bartkova et al. 1994, Dickson et al. 1995). Clinically, it has been seen that loss of INK4 and CIP/KIP family proteins, along with CDK4/6 overexpression, has been seen in BC (Koboldt et al. 2012; Asghar et al. 2015). According to a recent survey, distinct breast cancer subtypes have different molecular abnormalities in the cell cycle checkpoints (Koboldt et al. 2012).

A set of data based on cancer genome research of 482 aggressive BC patients found that 27.4 percent of CDK4/6–RB axis genetic dysregulation includes the expression of a specific gene change or various genetic changes in combinations (Dukelow et al. 2015). Numerous E2F targeting genes are responsible with cell cycle regulation, DNA duplication, and mitotic growth. Retinoblastoma proteins are transcriptional co-repressors that restrict the activation of several E2F target genes (Ren et al. 2002; Burkhart and Sage 2008). Hyperphosphorylation of Rb inhibits transcriptional suppression by lowering its affinity for E2F and causing the transfer of E2F transcriptional regulators, allowing transcription of CDK2, E-type cyclins, and many more proteins, which further form a complex capable of phosphorylating Rb as well as facilitating the S phase entry (Weinberg 1995; Sherr 1996; Sherr and McCormick 2002). Two classes of CDK inhibitors (CDKi) limit CDK4/6 kinase activities by adhering to an ATP-binding site of CDK and inhibiting subsequent CDK4/6-mediated phosphorylation of Rb; such intrinsic CDK4/6 restriction potently arrests cell growth but needs operative Rb protein (Asghar et al. 2015; Niu et al. 2019). CDK-interacting protein/kinase inhibitory protein (CIP/KIP) family, which includes proteins p21CIP1 (CDKN1A), p27KIP1 (CDKN1B), and p57KIP2 (CDKN2D), and the inhibitor of CDK4 (INK4) family, which includes proteins p16INK4A (CDKN2A), p14ARF (CDKN2A), INK4 family of proteins interact specifically with CDK4/6's isoforms, restricting their alliance to D-type cyclins as well as repressing kinase activity, whereas the CIP/KIP proteins both possess suppressive and activating impacts, meddling with the functioning of all CDK/Cyclin complexes (Hirai et al. 1995; Lim and Kaldis 2013; O'leary et al. 2016). Furthermore, RAS-RAF-MEK-ERK axis and HER2-PI3K-AKT pathway both play a pivotal role in cyclin D1 regulation as well as expression (Winston et al. 1996; Klein and Assoian 2008). Cyclin D/CDK4/6 axis was an appealing target for the development of treatment methods due to the coordinating function of CDK4/6/Cyc D in controlling the whole process of cell cycle. Creation of CDK4/6 inhibitors has become the most intriguing result in this context, because cancerous cells' susceptibility has been combated while keeping an acceptable toxicity profile (Piezzo et al. 2020).

10.5 CDK4/6 in Relation to ER+ Breast Cancer

Anticancer medications, including the anti-estrogen tamoxifen, considerably enhanced disease-free life in women with BC; nonetheless, a large minority of individuals seem to be either resistant to treatment or experience disease reappearance (Jensen and Jordan 2003; Lancet 2005). In breast cancer patients, recurrences at metastatic sites, particularly the lungs and bones, are the leading cause of death (Blanco et al. 1990; Chambers et al. 2002). Whereas the majority of clinical studies in the last decade have concentrated on the application of CDKIs (especially CDK4/6 inhibitors) in Estrogen positive and Human Epidermal Receptor2-negative BC, some previous research has proved that CDK4/6 suppression could be utilized to address HER2+ tumors. Finn and his colleagues discovered that luminal-type BC cells also express the estrogen receptor, together with luminal-type cells with HER2 modulation, seem to have been extremely susceptible to palbociclib than basal-type ER-negative cells. It was found by them in one of the first research on the functioning of the CDK4/6 anticancer drug palbociclib (Finn et al. 2009). The steroid and peptide growth factors promote proliferation in ER+ BC by stimulating the cyclin D-CDK4/6, which causes pRb hyperphosphorylation. The transcription factor E2 (E2F) is released in the presence of hyperphosphorylated pRb, and the cell enters the S phase of the cycle. CDK4/6 inhibitors are small molecules that stop pRb from hyperphosphorylating and causing G1 cell cycle arrest. Understanding the interactions among the steroid hormone system, peptide growth factor signaling, and CDK biology would likely improve our thinking about the pathways of the hormone resistance (Dean et al. 2010; Finn et al. 2016). Estrogens work by attaching to a particular estrogen receptor (ER), that is the part of the steroid and thyroid hormone receptor superfamily. This receptor is indeed a nuclear regulatory protein having a molecular weight of 66 kDa that acts as a hormone-activated transcriptional activator. The activation of receptors is assumed to be the result of ligand-induced conformational changes in the ER organization (Kumar and Chambon 1988; Tsai et al. 1988). ER-positive BC accounts for the majority of instances (Li et al. 2003), accounting for about 80% of all BCs (Lumachi et al. 2015). Estrogen receptors (ERs) comprise nuclear proteins that regulate gene expression, and about 80 percent of the total breast cancers are estrogen receptor positive (ER+), with 65 percent also being progesterone receptor (PR) positive.

Estrogen's effect in ER+ BC is regulated predominantly by ER. The estrogen receptor is a nuclear receptor that plays a role in a variety of developmental and physiological processes (Shao and Brown 2003). Stimulation of the estrogen receptor (ER) promotes carcinogenesis in various types of cancers, including BC, and the role of the ER is still unknown (Haque and Desai 2019). As a result, inhibiting the ER-alpha has become one of the most dominant techniques for mammary cancer prevention and treatment (Sommer and Fuqua 2001). The major technique for treating ER-positive BC is to use hormone therapies to prevent estrogen's activity, although this is restricted by the resistance development (Lewis et al. 2005). ER score can, however, indicate hormone responses to treatment. Such tumors are frequently linked to a higher overall life expectancy (Clahsen et al. 1999), relapses happen at a consistent rate for approximately 20 years (Pan et al. 2017). Estrogen receptor-negative BCs return in the first 3 to 5 years, but ER-positive breast cancers do not. Congenital or evolved hormonal tolerance is prevalent in ER-positive cancers, occurring in 40–50% of cases (Osborne and Schiff 2011). Estrogen, in particular, can speed up cell cycle transition from the G1 to the S phase in ER+ breast cancer, where the estrogen effector is the cyclin D1–CDK4/6–RB complex. In a nutshell, estrogen binding to ER-alpha causes cyclin D1 transcription, while CDK4/6 stimulation and phosphorylation of retinoblastoma cause cell cycle continuation via the checkpoint, resulting in the initiation of the cell cycle signal, which stimulates the expression of several receptor-driven genes that are engaged in cell growth, division, and survival (Fig. 10.3).

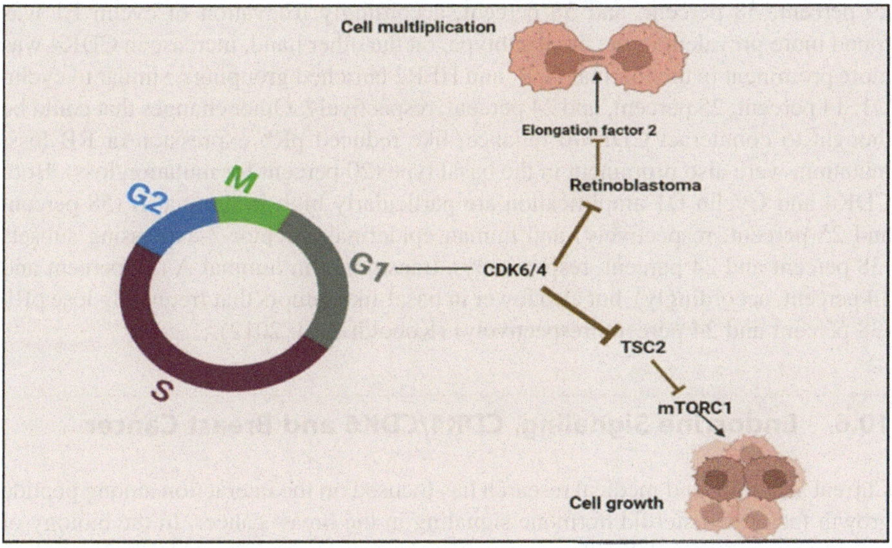

Fig. 10.3 Showing the relationship between CDK4/6 and cancer progression as well as growth

Elevated expressions of the Cyclin D1 is found in roughly fifteen percent of all BCs, mostly in estrogen-positive breast cancers (Arnold and Papanikolaou 2005). ER+ BCs also have higher levels of estrogen receptor 1 (ESR1) protein & phosphatidylinositol-4,5-bisphosphate 3-kinase catalytic subunit alpha (PIK3CA), both of these lead to progression in the cell cycle via the mitogenic protein kinase B (AKT)/mTOR signaling pathway (Koboldt et al. 2012).

When comparing with all other mammary cancer subgroups, such as HER2+ and TNBC, ER+ BC is almost genetically reliable, with a major dependence on estrogen signaling, and RB and p53 tumor suppressor genes are usually normal in this condition. The CDK4/6–RB axis is also engaged in HER2-induced cell growth (Spring et al. 2017). It has been indicated in the laboratory mouse models of human BC that the stimulation of the cyclin D1-CDK4/6 axis results in the tumorigenesis as well as promotes the development and persistence of carcinogenesis in HER2+ BC (Dukelow et al. 2015). In HER2+ BC, CDK4, erb-b2 receptor tyrosine kinase 2 (ERBB2, the gene encoding the HER2 receptor), tumor protein p53 (TP53), PIK3CA, and phosphatase and tensin homolog (PTEN), as well as cyclin D1, are all amplified. TNBC transcriptomic, analytical, and metagenomic RB signal transduction data, on the other hand, show RB1 mutation or deletion in twenty percent of cases, cyclin E1 overexpression in nine percent of cases, elevated expression of cyclin-dependent kinase inhibitor 2A (CDKN2A), minimal expression of RB1, and a high degree of proliferative frequency, as well as significant changes in DNA damage repair genes like tumor suppressor breast cancer 1 (BRCA1) (Robinson et al. 2013; Fedele et al. 2019). When cell cycle genes were examined by inherent subtype of breast cancer, cyclin D1 overexpression was discovered, commonly in the luminal A, B, and HER2 enriched categories, having rates of

29 percent, 58 percent, and 38 percent, accordingly. Elevation of cyclin E1 was found more prevalent in the basal subtype, on the other hand, increase in CDK4 was more prominent in the luminal A, B, and HER2 enriched groupings, similar to cyclin D1: 14 percent, 25 percent, and 24 percent, respectively. Other changes that could be thought to counteract CDK4/6 reliance, like reduced pRb expression or RB loss/ mutation, were also prominent in the basal type (20 percent for mutation/loss). Both CDK4 and Cyclin D1 amplification are particularly high in luminal B (58 percent and 25 percent, respectively) and human epidermal receptor-2-expressing subsets (38 percent and 24 percent, respectively), transitional in luminal A (29 percent and 14 percent, accordingly), but also lower in basal-like tumors that frequently lose pRb (38 percent and 24 percent, respectively) (Koboldt et al. 2012).

10.6 Endocrine Signaling, CDK4/CDK6 and Breast Cancer

Current scientific and medical research has focused on the interaction among peptide growth factor and steroid hormone signaling in the breast cancer. In the biology of roughly 60 percent and 20–25 percent of BCs, respectively, ER and HER-2 signaling seems to be recognized "makers" (Tinoco et al. 2013). Although treatment strategies to both subgroups are centered on their separate receptors, the two routes may merge and impose downstream consequences on the CDK4/6:cyc D axis. Regarding growth and survival, ER+ breast tumors rely heavily on estrogen signaling (Varma et al. 2007). By ER suppression, tumor cell survival is lowered and marks a cell cycle arrest in the G1 stage (Sutherland et al. 1983; Carroll et al. 2000). ER signaling is believed to significantly enhance cyclin D1 concentrations as well as amplify numerous signaling pathways, with the increase of CDK4/6 activity being the most common outcome (Watts et al. 1994; Foster et al. 2001). Hormone-based treatment approaches, unsurprisingly, are at the heart of the treatment options for the ER+ breast malignancies. Unfortunately, not all ER+ tumors react to such a treatment option, but resistance is common among those that do. Findings show that, at least in some of these malignancies, this is mediated by dysregulation of many additional mitogenic cascades (for instance, HER2, PI3K/AKT, and others), which can amplify cyclin D1:CDK4/ 6 signaling in an estrogen receptor-independent manner. Furthermore, as previously stated, cyclin D1 may stimulate estrogen receptor on its own, and the most of cyclin D1 upregulated BCs are ER+ (Buckley et al. 1993; Zwijsen et al. 1997). Such discoveries imply that CDK4/6-mediated:Cyc D signaling may play a crucial role in the estrogen independence of estrogen-positive BC cells (Dean et al. 2010).

10.7 Role of CDKs in Transcription

Cell division as well as RNA pol II (Polymerase-II)-dependent transcription is regulated by the CDKs. The activation requirements for cell cycle and transcriptional CDKs are comparable, but only the former appears to be sensitive to inhibitory phosphorylation. Furthermore, a rarely employed mechanism of transcriptional CDK regulation is controlled cyclin production and disintegration, which is a central characteristic of cell cycle control (Malumbres 2014). RNA pol II (RNAPII) catalyzes the initiation, pausing elongation and termination of messenger RNAs (mRNAs) in humans. It is made of a major component (Rpb1), having a C-terminal domain (CTD) loop of such an historically preserved heptapeptide (Tyr-Ser-Pro-Thr-Ser-Pro-Ser-Ser-Pro-Ser-Ser (Whittaker et al. 2017). Despite changes in its phosphorylation level, the CTD plays an important function in RNA replication and chromosome assembly in the management of transcriptional and co-transcriptional activities (Jeronimo et al. 2013; Suh et al. 2013). Several CDK/cyclin components, such as CDK1 or CDK2, and most transcriptional CDKs, including CDK7, CDK8, and CDK9 subtypes, catalyze Tyr1, Ser2, Thr4, Ser5, and Ser7 in the heptapeptide (Jeronimo et al. 2016). For the transcriptional initiation of the promoters, the CTD-RNAPII must be phosphorylated at Ser5 and Ser7. To enhance transcriptional elongation, Ser5 phosphorylation declines as starting transcription occurs, while Ser2 and Tyr1 phosphorylation elevate. Tyr1 is dephosphorylated first during transcriptional termination, then Ser5, Ser7, and Ser2, allowing the transcriptional process to be restarted (Galbraith et al. 2019). The pre-initiation complex is formed when the promoter is recognized and the DNA is unwound, resulting in effective transcription. As previously stated, RNAPII interacts with a huge multi-subunit mediator complex as well as other general transcription factors, and the process is launched by TATA binding protein of transcription factor II D (TFIID) attaching to the core promoter to create the pre-initiation complex (PIC). Type-C cyclins, a component of the mediator complex kinase module (MED), interact with CDK8 or CDK19, this complex module serves as a bimolecular link between DNA-bound TFs and the basic RNAPII pre-initiation complex transcriptional process at the promoter (Yin and Wang 2014; Allen and Taatjes 2015). The four-subunit kinase module of MED is composed of CDK8 (or CDK19), cyclin C, Med12, and Med13, which is usually related with transcription repression. MED phosphorylates cyclin H to prevent the formation of PICs, which inhibits TFIIH's action on CTD, and it phosphorylates CTD-RNAPII to prevent it from adhering to promoter DNA and preventing the formation of the PIC (Lim and Kaldis 2013). TFIIH, a complex structure which is part of a 10-subunit general transcription factor, is made up of the regulatory subunit cyclin H, the catalytic subunit CDK7, and the ring finger protein ménage a trois 1 (Mat1), which serves as a helicase, ATPase, and kinases, as well as being the last to be employed. The helicase activities at the gene promoter site unwinds the DNA, forming single-strand DNA in the RNAPII catalytic site. The CDK7 subunit's kinase activity in the TFIIH complex phosphorylates Ser5 and Ser7 of CTD-RNAPII, which aids transcription initiation and promoter clearance. Phosphorylated CTD also facilitates the binding of a capping enzyme, which catalyzes the insertion of a methylguanosine cap to the nascent mRNA's 5 end (Fisher and cancer 2012). CDK7/cyclin H acts as a CDK-activating kinase (CAK), phosphorylating as well as stimulating CDK9, the subunit of the positive transcription elongation factor b (P-TEFb), binding with the T-type cyclins (T1 and T2) to liberate the regulator from downstream blockage and encourage elongation. By phosphorylating NELF (negative elongation factor) as well as 5,6-dichloro-1-beta-D-ribofuranosylbenzimidazole sensitivity-inducing factor (DSIF) to unleash the elongation complex from obstruction and it also phosphorylates CTD at serine 2 to assist its RNA polymerization process, the stimulated CDK9/cyclin T tends to promote the extension of the pre-mRNA transcript (Peterlin and Price 2006; Larochelle et al. 2012). CDK7 phosphorylation of CTD is essential for P-TEFb activation, and CDK7 inhibition causes CDK9 phosphorylation of Ser2 to decrease (Viladevall et al. 2009). CDK12 itself and its homolog CDK13, together with related component cyclin K, have been linked to Ser2 phosphorylation at the CTD in latest research. CDK9 phosphorylates Ser2 sooner in transcription and then assigns the rest of the elongation phase to CDK12; although CDK12's function in C-terminal domain phosphorylation is gene-specific (Jeronimo et al. 2016). Furthermore, CDK12 is engaged in alternate exon splicing, which is essential for the physiological DNA damage response, forming a new connection among transcriptional machinery and cell cycle regulation (Blazek et al. 2011). CDK1 phosphorylates the CTD to stop transcription, however the etiology is still unknown (Malumbres 2014). CDK11/cyclin L (cycL) interacts with a different type of ELFs, including RNA polymerase elongation factor 2 (ELL2), general transcription factor II F (TFIIF), and general transcription factor IIS (TFIIS), to aid transcription elongation and chromatin transcription (FACT). Furthermore, CDK11/cyclin L regulates splicing of the RNA by phosphorylating factors involved in pre-mRNA splicing, including SC35 (Srfs2) and 9G8 (Srfs7) (Lim and Kaldis 2013). The dephosphorylation of Ser5 of CTD-RNAPII by sarcoplasmic calcium-binding protein 1 (SCP1) facilitates transcriptional cessation (Whittaker et al. 2017). Detailed research into the mechanism of dephosphorylation is needed; although certain CDK-counteracting phosphatases, such as Cdc14, are thought to be engaged in such process (Clemente-Blanco et al. 2011; Guillamot et al. 2011).

10.8 Targeting the CDK4/6:Cyc D Pathway Therapeutically

In women, one of the most commonly diagnosed types of cancer is BC, preceded by lung and colorectal cancer conversely in case of death rates. New BC cases and deaths are predicted to reach 2.08 million new cases and 627,000 deaths globally (Bray et al. 2018). For anti-HER2-targeted and hormone therapy, hormone receptor and HER2 (progesterone and estrogen receptor, respectively) are essential prognostic determinants and indicators. Both PR and ER are indications of hormone therapy sensitivity, and they are found in about 75% among all BC cases (Hammond et al. 2010; Tsang and Tse 2020). Breast cancer patients with targeted therapy had a longer life expectancy. However, preclinical investigations have indicated that hyperactivity of the CDK4/CDK6–cyclin D complex leads to excessive cell multiplication, resulting in pharmacological suppression of the complex, a promising treatment option (Pernas et al. 2018; Sobhani et al. 2019). Regarding the ER-positive breast cancer, CDK4/CDK6 inhibitors are the most reasonable treatment option. Rb function is always present in this form of malignancy, showing that the inhibitor's principal mechanism of action is still intact (Finn et al. 2009). These inhibitors have evolved as an appealing treatment option due to the role of this complex in controlling the cellular expansion and the ways through which this mechanism is thought to be stimulated in cancer (Roberts et al. 2012). A key theoretical problem seems to be that CDKs are involved in both normal and malignant cell proliferation, potentially producing a confined remedial window in which toxic behavior would limit the capacity to achieve therapeutically beneficial levels of exposure. Flavopiridol, a well-studied pan-CDK blocker to date, has shown modest clinical efficacy, owing to its complicated pharmacokinetics and severe off-target adverse effects (Shapiro 2006; Dickson 2014). It's probable that tumors with known cyclin D-CDK4/6-INK4-Rb pathway mutations are additionally susceptible to CDK6/4 suppression versus healthy tissues (Roberts et al. 2012). However, emerging scientific studies indicate that certain Cdks at least in malignancies overexpressing D-type Cyclins may be appropriate targets for therapeutic intervention. Cyclin D1, one of the several regulating components of Cdk4 and Cdk6, was knocked out, conferring resistance to breast tumorigenesis caused by oncogenes including Ras and ErbB2 (Yu et al. 2001). Because Cyclin D1 is required for appropriate breast development, the mice's susceptibility to mammary cancers may be attributable to a developmental problem instead of a lack of Cdk4 and Cdk6 activities. Furthermore, animals articulating a Cyclin D1 mutant that binds but does not activate Cdk4 or Cdk6 have been shown to be resistant to BC induced by ErbB2, although possessing regular breast gland growth and maturation (Landis et al. 2006; Yu et al. 2006). Palbociclib (PD-0332991; Pfizer), Ribociclib (LEE011; Novartis), and abemaciclib are three CDK4/6 inhibitors. Theses 3 inhibitors are either approved or in stage III of development (LY2835219; Lilly). These drugs have IC50 values of less than 40 nM for CDK4 and CDK6, but they have different IC50 values for other CDKs; these variances in sensitivity might alter both ideal dosage regimens and adverse reaction profiles. In addition, all three drugs have shown preclinical activity in a variety of Rb + cancer types (Fry et al. 2004; Baker et al. 2012; Dean et al. 2012; Kim et al. 2013; Gelbert et al. 2014; Witkiewicz et al. 2014) (Fig. 10.4).

Along with the above three approved drugs, other numerous CDKIs especially CDK4/6 inhibitors are under clinical trials in different stages for evaluation of their results either singly or in combination with other drugs especially in case of breast cancer, for example: PEARL (testing Palbociclib with hormonal therapy against chemotherapy in case of HR+ and HER2 metastatic breast cancers having resistance against aromatase inhibitors) under stage III trial (Sobhani et al. 2019). Several new CDKIs have previously been explored. Piperlongumine (PL) suppresses the advancement and metastasis of estrogen-positive BC cells, according to Jeong et al. Piperlongumine, a phytochemical isolated from pepper that inhibits Cdk4/6 & Cdk1 activity and promotes G2 to M stage cell cycle arrest to prevent cancer (Jeong et al. 2019). Quereda et al. discovered that SR-4835 inhibits triple-negative BC cell growth by operating as a selectively double blocker of Ckd13 and Cdk12 (Quereda et al. 2019). Along with, several other chemical and natural drugs are being addressed for clinical trials so as to discover a promising drug for cancer patients especially breast cancer.

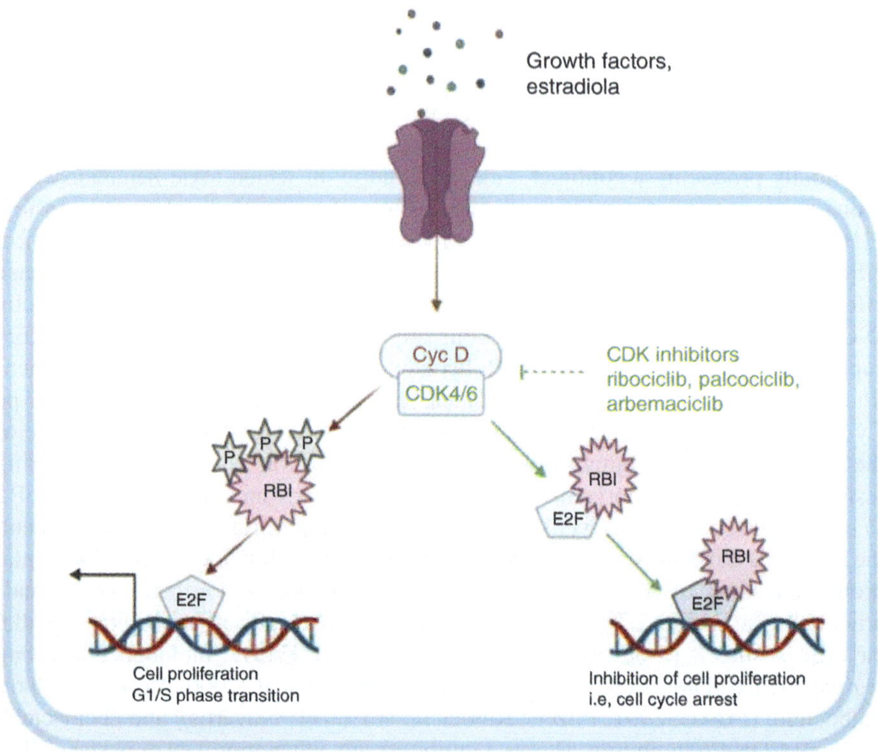

Fig. 10.4 Figure illustrating the mechanism of cell cycle arrest by the use of CDK inhibitors as well as how CDK4/6/CycD accounts for G1/S phase transition

10.9 Summary

Cell growth is defined as a complex system including kinases as well as other phosphoprotein phosphatases catalyzing phosphorylation and dephosphorylation. Cell cycle requires various CDKs and Cyclins that normally regulate all the four steps of the cell cycle. There are different kinds of CDKs and Cyclins, out of which CDK4/6/Cyc D, Cyclin A/E CDK2, Cyclin A/B CDK1 are essential to precede in each stage of the particular cycle. Cyclin CDK4/6 pathway acts as an important axis in normal cell cycle and plays a crucial role in the BC continuance as well as development. Palbociclib, the inhibitor CDK4/6, was sanctioned by the USFDA in 2015 for use in combination with letrozole as an early, hormone-based therapy for postmenopausal ER+/HER2+ aggressive BC. There are various available drugs either as natural CDKIs or chemical inhibitors that can be used against various types of breast cancer, but no drug is promising that can decrease the cancer progression and deaths. Due to various circumstances, need for other drugs rises due to increase in the BC patients worldwide as well as due to resistance for some commonly used drugs. In future, we are expecting some more and valuable drugs and drug targets that can decrease the death rate of cancer patients and can increase their survival rates.

10.10 Further Readings

Modern textbooks on cell cycle and its regulation and related topics include those by Kaldis and Pagano (2006). Also go through section II, chapter 4 of Hurvitz and McCann (2018), it informs us about the HER2-positive breast cancer, and chap. 95 of part III of Niederhuber et al. (2019). Video lectures by Sanofi are very helpful to understand the metastasis of the breast cancer.

https://youtu.be/LVeHGJbX3nQ

El-Ahmad Y, Tabart M, Halley F, et al. (2019) J Med Chem 63:512–528. https://doi.org/10.1021/acs.jmedchem.9b01293

Shomali M, Cheng J, Sun F, et al. (2021) Mol Cancer Ther 20:250–262. https://doi.org/10.1158/1535-7163.mct-20-0390

Video lecture by Dana-Farber Cancer Institute · 28-Nov-2019. https://www.youtube.com/watch?v=XnWKzq92i40

For more insights about the topic, we would suggest detailed findings from the books of Mir MA (2022) https://doi.org/10.1016/C2021-0-02565-7, (Mir MA, 2021) https://doi.org/10.52305/WXJL6770

References

Allen BL, Taatjes DJJNRMCB (2015) The Mediator complex: a central integrator of transcription. Nat Rev Mol Cell Biol 16(3):155–166

Arima Y et al (2008) Rb depletion results in deregulation of E-cadherin and induction of cellular phenotypic changes that are characteristic of the epithelial-to-mesenchymal transition. Cancer Res 68(13):5104–5112

Arnold A, Papanikolaou AJJOCO (2005) Cyclin D1 in breast cancer pathogenesis. J Clin Oncol 23(18):4215–4224

Asghar U et al (2015) The history and future of targeting cyclin-dependent kinases in cancer therapy. Nat Rev Drug Discov 14(2):130–146

Baker SJ et al (2012) CDK4: a key player in the cell cycle, development, and cancer. Genes Cancer 3(11–12):658–669

Baldin V et al (1993) Cyclin D1 is a nuclear protein required for cell cycle progression in G1. Genes Dev 7(5):812–821

Barnum KJ, O'Connell MJ (2014) Cell cycle regulation by checkpoints. Methods Mol Biol 1170:29–40. Cell cycle control, Springer

Bartek J et al (2004) Checking on DNA damage in S phase. Nat Rev Mol Cell Biol 5(10):792–804

Bartkova J et al (1994) Cyclin D1 protein expression and function in human breast cancer. Int J Cancer 57(3):353–361

Blagosklonny MV, Pardee ABJCC (2002) The restriction point of the cell cycle. Cell Cycle 1(2): 102–109

Blanco G et al (1990) Prognostic factors in recurrent breast cancer: relationships to site of recurrence, disease-free interval, female sex steroid receptors, ploidy and histological malignancy grading. Br J Cancer 62(1):142–146
Blazek D et al (2011) The cyclin K/Cdk12 complex maintains genomic stability via regulation of expression of DNA damage response genes. Genes Dev 25(20):2158–2172
Bosco EE, Knudsen ESJCC (2007) RB in breast cancer: the crossroads of tumorigenesis and treatment. Cell Cycle 6(6):667–671
Bosco EE et al (2007) The retinoblastoma tumor suppressor modifies the therapeutic response of breast cancer. J Clin Invest 117(1):218–228
Bray F et al (2018) Global cancer statistics 2018: GLOBOCAN estimates of incidence and mortality worldwide for 36 cancers in 185 countries. CA Cancer J Clin 68(6):394–424
Buckley MF et al (1993) Expression and amplification of cyclin genes in human breast cancer. Oncogene 8(8):2127–2133
Burkhart DL, Sage JJNRC (2008) Cellular mechanisms of tumour suppression by the retinoblastoma gene. Nat Rev Cancer 8(9):671–682
Canavese M et al (2012) Cyclin dependent kinases in cancer: potential for therapeutic intervention. Cancer Biol Ther 13(7):451–457
Cao L et al (2014) Phylogenetic analysis of CDK and cyclin proteins in premetazoan lineages. BMC Evol Biol 14(1):1–16
Carroll JS et al (2000) A pure estrogen antagonist inhibits cyclin E-Cdk2 activity in MCF-7 breast cancer cells and induces accumulation of p130-E2F4 complexes characteristic of quiescence. J Biol Chem 275(49):38221–38229
Chambers AF et al (2002) Dissemination and growth of cancer cells in metastatic sites. Nat Rev Cancer 2(8):563–572
Chen H-H et al (2006) Identification and characterization of the CDK12/cyclin L1 complex involved in alternative splicing regulation. Mol Cell Biol 26(7):2736–2745
Choi YJ et al (2012) The requirement for cyclin D function in tumor maintenance. Cancer Cell 22(4):438–451
Cicenas J, Valius M (2011) The CDK inhibitors in cancer research and therapy. J Cancer Res Clin Oncol 137(10):1409–1418
Clahsen P et al (1999) The utility of mitotic index, oestrogen receptor and Ki-67 measurements in the creation of novel prognostic indices fornode-negative breast cancer. Eur J Surg Oncol 25(4): 356–363
Clemente-Blanco A et al (2011) Cdc14 phosphatase promotes segregation of telomeres through repression of RNA polymerase II transcription. Nat Cell Biol 13(12):1450–1456
Davalos AR et al (2010) Senescent cells as a source of inflammatory factors for tumor progression. Cancer Metastasis Rev 29(2):273–283
Dean J et al (2010) Therapeutic CDK4/6 inhibition in breast cancer: key mechanisms of response and failure. Oncogene 29(28):4018–4032
Dean JL et al (2012) Therapeutic response to CDK4/6 inhibition in breast cancer defined by ex vivo analyses of human tumors. Cell Cycle 11(14):2756–2761
Dickson MAJCCR (2014) Molecular pathways: CDK4 inhibitors for cancer therapy. Clin Cancer Res 20(13):3379–3383
Dickson C et al (1995) Amplification of chromosome band 11q13 and a role for cyclin D1 in human breast cancer. Cancer Lett 90(1):43–50
Dukelow T et al (2015) CDK4/6 inhibitors in breast cancer. Anti-Cancer Drugs 26(8):797–806
Ertel A et al (2010) RB-pathway disruption in breast cancer: differential association with disease subtypes, disease-specific prognosis and therapeutic response. Cell Cycle 9(20):4153–4163
Fedele P et al (2019) A clinical evaluation of treatments that target cell cycle machinery in breast cancer. Expert Opin Pharmacother 20(18):2305–2315
Finn RS et al (2009) PD 0332991, a selective cyclin D kinase 4/6 inhibitor, preferentially inhibits proliferation of luminal estrogen receptor-positive human breast cancer cell lines in vitro. Breast Cancer Res 11(5):1–13

Finn RS et al (2016) Targeting the cyclin-dependent kinases (CDK) 4/6 in estrogen receptor-positive breast cancers. Breast Cancer Res 18(1):1–11
Fisher RPJG (2012) The CDK network: linking cycles of cell division and gene expression. Genes Cancer 3(11–12):731–738
Foster JS et al (2001) Multifaceted regulation of cell cycle progression by estrogen: regulation of Cdk inhibitors and Cdc25A independent of cyclin D1-Cdk4 function. Mol Cell Biol 21(3):794–810
Fry DW et al (2004) Specific inhibition of cyclin-dependent kinase 4/6 by PD 0332991 and associated antitumor activity in human tumor xenografts. Mol Cancer Ther 3(11):1427–1438
Galbraith MD et al (2019) Therapeutic targeting of transcriptional cyclin-dependent kinases. Transcription 10(2):118–136
Gelbert LM et al (2014) Preclinical characterization of the CDK4/6 inhibitor LY2835219: in-vivo cell cycledependent/independent anti-tumor activities alone/in combination with gemcitabine. Invest New Drugs 32(5):825–837
Gillett C et al (1994) Amplification and overexpression of cyclin D1 in breast cancer detected by immunohistochemical staining. Cancer Res 54(7):1812–1817
Guillamot M et al (2011) Cdc14b regulates mammalian RNA polymerase II and represses cell cycle transcription. Sci Rep 1(1):1–7
Hammond MEH et al (2010) American Society of Clinical Oncology/College of American Pathologists guideline recommendations for immunohistochemical testing of estrogen and progesterone receptors in breast cancer (unabridged version). Arch Pathol Lab Med 134(7): e48–e72
Hanahan D, Weinberg RAJC (2011) Hallmarks of cancer: the next generation. Cell 144(5):646–674
Haque M, Desai KVJFIE (2019) Pathways to endocrine therapy resistance in breast cancer. Front Endocrinol (Lausanne) 10:573
Hartwell LHJBR (1974) Saccharomyces cerevisiae cell cycle. Bacteriol Rev 38(2):164–198
Hirai H et al (1995) Novel INK4 proteins, p19 and p18, are specific inhibitors of the cyclin D-dependent kinases CDK4 and CDK6. Mol Cell Biol 15(5):2672–2681
Hui R et al (2002) Constitutive overexpression of cyclin D1 but not cyclin E confers acute resistance to antiestrogens in T-47D breast cancer cells. Cancer Res 62(23):6916–6923
Hurvitz S, McCann K (2018) HER2-positive breast cancer, 1st edn. Elsevier, pp 1–247. ISBN: 978-0-323-58122-6. https://doi.org/10.1016/C2017-0-01382-4
Jemal A et al (2011) Global cancer statistics. CA Cancer J Clin 61(2):69–90
Jensen EV, Jordan VCJCCR (2003) The estrogen receptor: a model for molecular medicine. Clin Cancer Res 9(6):1980–1989
Jeong CH et al (2019) Piperlongumine induces cell cycle arrest via reactive oxygen species accumulation and IKKβ suppression in human breast cancer cells. Antioxidants (Basel) 8(11): 553
Jeronimo C et al (2013) The writers, readers, and functions of the RNA polymerase II C-terminal domain code. Chem Rev 113(11):8491–8522
Jeronimo C et al (2016) The RNA polymerase II CTD: the increasing complexity of a low-complexity protein domain. J Mol Biol 428(12):2607–2622
Jirström K et al (2005) Adverse effect of adjuvant tamoxifen in premenopausal breast cancer with cyclin D1 gene amplification. Cancer Res 65(17):8009–8016
Kaldis P, Pagano M (2006) Cell Division, a new open access online forum for and from the cell cycle community. Cell Div 1(1):1. https://doi.org/10.1186/1747-1028-1-1
Kato J-Y et al (1993) Direct binding of cyclin D to the retinoblastoma gene product (pRb) and pRb phosphorylation by the cyclin D-dependent kinase CDK4. Genes Dev 7(3):331–342
Kim S et al (2013) Abstract PR02: LEE011: an orally bioavailable, selective small molecule inhibitor of CDK4/6–reactivating Rb in cancer. AACR Mol Cancer Ther 12(11_Supplement): PR02
Klein EA, Assoian RKJJOCS (2008) Transcriptional regulation of the cyclin D1 gene at a glance. J Cell Sci 121(23):3853–3857

Koboldt D et al (2012) Comprehensive molecular portraits of human breast tumours. Nature 490(7418):61–70
Kolch W et al (2015) The dynamic control of signal transduction networks in cancer cells. Nat Rev Cancer 15(9):515–527
Kumar V, Chambon PJC (1988) The estrogen receptor binds tightly to its responsive element as a ligand-induced homodimer. Cell 55(1):145–156
Lancet EBCTCGJT (2005) Effects of chemotherapy and hormonal therapy for early breast cancer on recurrence and 15-year survival: an overview of the randomised trials. Lancet 365(9472): 1687–1717
Landis MW et al (2006) Cyclin D1-dependent kinase activity in murine development and mammary tumorigenesis. Cancer Cell 9(1):13–22
Larochelle S et al (2012) Cyclin-dependent kinase control of the initiation-to-elongation switch of RNA polymerase II. Nat Struct Mol Biol 19(11):1108–1115
Lee JJ et al (2012) Methylation and immunoexpression of p16INK4a tumor suppressor gene in primary breast cancer tissue and their quantitative p16INK4a hypermethylation in plasma by real-time PCR. Korean J Pathol 46(6):554
Lew DJ et al (1991) Isolation of three novel human cyclins by rescue of G1 cyclin (Cln) function in yeast. Cell 66(6):1197–1206
Lewis JS et al (2005) Selective estrogen receptor modulators (SERMs): mechanisms of anticarcinogenesis and drug resistance. Mutat Res 591(1–2):247–263
Li CI et al (2003) Incidence of invasive breast cancer by hormone receptor status from 1992 to 1998. J Clin Oncol 21(1):28–34
Lim S, Kaldis PJD (2013) Cdks, cyclins and CKIs: roles beyond cell cycle regulation. Development 140(15):3079–3093
Lumachi F et al (2015) Current medical treatment of estrogen receptor-positive breast cancer. World J Biol Chem 6(3):231
Malumbres M (2014) Cyclin-dependent kinases. Genome Biol 15(6):1–10
Malumbres M, Barbacid M (2001) To cycle or not to cycle: a critical decision in cancer. Nat Rev Cancer 1(3):222–231
Malumbres M, Barbacid M (2005) Mammalian cyclin-dependent kinases. Trends Biochem Sci 30(11):630–641
Malumbres M, Barbacid M (2007) Cell cycle kinases in cancer. Curr Opin Genet Dev 17(1):60–65
Malumbres M, Barbacid MJNRC (2009) Cell cycle, CDKs and cancer: a changing paradigm. 9(3): 153–166
Massagué J (2004) G1 cell-cycle control and cancer. Nature 432(7015):298–306
Matsushime H et al (1991) Colony-stimulating factor 1 regulates novel cyclins during the G1 phase of the cell cycle. 65(4):701–713
McIntosh G et al (1995) Determination of the prognostic value of cyclin D1 overexpression in breast cancer. 11(5):885–891
Mehraj U et al (2021) Tumor microenvironment promotes breast cancer chemoresistance. Cancer Chemother Pharmacol:1–12
Mehraj U et al (2022a) Expression pattern and prognostic significance of baculoviral inhibitor of apoptosis repeat-containing 5 (BIRC5) in breast cancer: a comprehensive analysis. Adv Cancer Biol-Metastasis 100037
Mehraj U et al (2022b) Expression pattern and prognostic significance of chemokines in breast cancer: an integrated bioinformatics analysis. Clin Breast Cancer
Mehraj U et al. (2022c) Cryptolepine targets TOP2A and inhibits tumor cell proliferation in breast cancer cells-an in vitro and in silico study. Anti-cancer Agents in Medicinal Chemistry
Meyerson M et al (1994) Identification of G1 kinase activity for cdk6, a novel cyclin D partner. 14(3):2077–2086
Millar E et al (2009) Cyclin D1b protein expression in breast cancer is independent of cyclin D1a and associated with poor disease outcome. 28(15):1812–1820

Musgrove EA, Sutherland RLJCC (2010) RB in breast cancer: differential effects in estrogen receptor-positive and estrogen receptor-negative disease. 9(23):4607–4607
Niederhuber JE, Armitage JO et al (eds) (2019) Abeloff's clinical oncology, 6th edn. Elsevier. ISBN: 9780323476744. https://www.us.elsevierhealth.com/abeloffs-clinical-oncology-9780323476744.html
Niu Y et al (2019) Cyclin-dependent kinases 4/6 inhibitors in breast cancer: current status, resistance, and combination strategies. J Cancer 10(22):5504
Nurse PJBR (2002) Cyclin dependent kinases and cell cycle control. 22(5–6):487–499
O'leary B et al (2016) Treating cancer with selective CDK4/6 inhibitors. 13(7):417–430
Osborne CK, Schiff RJAROM (2011) Mechanisms of endocrine resistance in breast cancer. 62: 233–247
Pan H et al (2017) 20-year risks of breast-cancer recurrence after stopping endocrine therapy at 5 years. 377(19):1836–1846
Pernas S et al (2018) CDK4/6 inhibition in breast cancer: current practice and future directions. 10: 1758835918786451
Peterlin BM, Price DHJMC (2006) Controlling the elongation phase of transcription with P-TEFb. 23(3):297–305
Piezzo M et al (2020) Targeting cell cycle in breast cancer: CDK4/6 inhibitors. 21(18):6479
Qayoom H et al (2022) Expression patterns and therapeutic implications of CDK4 across multiple carcinomas: a molecular docking and MD simulation study. Med Oncol 39(10):1–13
Quereda V et al (2019) Therapeutic targeting of CDK12/CDK13 in triple-negative breast cancer. 36(5):545–558. e547
Reddy HK et al (2005) Cyclin-dependent kinase 4 expression is essential for neu-induced breast tumorigenesis. 65(22):10174–10178
Ren B et al (2002) E2F integrates cell cycle progression with DNA repair, replication, and G2/M checkpoints. 16(2):245–256
Riggins RB et al (2005) Antiestrogens, aromatase inhibitors, and apoptosis in breast cancer. 71: 201–237
Roberts PJ et al (2012) Multiple roles of cyclin-dependent kinase 4/6 inhibitors in cancer therapy. 104(6):476–487
Robinson TJ et al (2013) RB1 status in triple negative breast cancer cells dictates response to radiation treatment and selective therapeutic drugs. 8(11):e78641
Roskoski RJPR Jr (2019) Cyclin-dependent protein serine/threonine kinase inhibitors as anticancer drugs. 139:471–488
Roy PG, Thompson AMJTB (2006) Cyclin D1 and breast cancer. 15(6):718–727
Rudas M et al (2008) Cyclin D1 expression in breast cancer patients receiving adjuvant tamoxifen-based therapy. 14(6):1767–1774
Santo L et al. (2015) Targeting cyclin-dependent kinases and cell cycle progression in human cancers. Seminars in oncology, Elsevier
Satyanarayana A, Kaldis PJO (2009) Mammalian cell-cycle regulation: several Cdks, numerous cyclins and diverse compensatory mechanisms. 28(33):2925–2939
Shao W, Brown MJBCR (2003) Advances in estrogen receptor biology: prospects for improvements in targeted breast cancer therapy. 6(1):1–14
Shapiro GIJJOCO (2006) Cyclin-dependent kinase pathways as targets for cancer treatment. 24(11):1770–1783
Sherr CJ, McCormick FJCC (2002) The RB and p53 pathways in cancer. 2(2):103–112
Sherr CJJS (1996) Cancer cell cycles. 274(5293):1672–1677
Sobhani N et al (2019) Updates on the CDK4/6 inhibitory strategy and combinations in breast cancer. 8(4):321
Sofi S et al (2022a) Targeting cyclin-dependent kinase 1 (CDK1) in cancer: molecular docking and dynamic simulations of potential CDK1 inhibitors. Med Oncol 39(9):1–15
Sofi S et al (2022b) Cyclin-dependent kinases in breast cancer: expression pattern and therapeutic implications. Med Oncol 39(6):1–16

Sommer S and Fuqua SA (2001) Estrogen receptor and breast cancer. Seminars in cancer biology, Elsevier
Spring LM et al (2017) Clinical management of potential toxicities and drug interactions related to cyclin-dependent kinase 4/6 inhibitors in breast cancer: practical considerations and recommendations. 22(9):1039–1048
Stendahl M et al (2004) Cyclin D1 overexpression is a negative predictive factor for tamoxifen response in postmenopausal breast cancer patients. 90(10):1942–1948
Suh H et al (2013) The C-terminal domain of Rpb1 functions on other RNA polymerase II subunits. 51(6):850–858
Susanti NMP, Tjahjono DH (2021) Cyclin-dependent kinase 4 and 6 inhibitors in cell cycle dysregulation for breast cancer treatment. Molecules 26(15):4462
Sutherland RL et al (1983) Tamoxifen induces accumulation of MCF 7 human mammary carcinoma cells in the G0/G1 phase of the cell cycle. 19(5):615–621
Thu K et al (2018) Targeting the cell cycle in breast cancer: towards the next phase. Cell Cycle 17(15):1871–1885
Tinoco G et al (2013) Treating breast cancer in the 21st century: emerging biological therapies. 4(2):117
Trere D et al (2009) High prevalence of retinoblastoma protein loss in triple-negative breast cancers and its association with a good prognosis in patients treated with adjuvant chemotherapy. 20(11):1818–1823
Tsai SY et al (1988) Molecular interactions of steroid hormone receptor with its enhancer element: evidence for receptor dimer formation. 55(2):361–369
Tsang J, Tse GMJAIAP (2020) Molecular classification of breast cancer. 27(1):27–35
Varma H et al (2007) Functional ablation of pRb activates Cdk2 and causes antiestrogen resistance in human breast cancer cells. 2(12):e1256
Viladevall L et al (2009) TFIIH and P-TEFb coordinate transcription with capping enzyme recruitment at specific genes in fission yeast. 33(6):738–751
Watts CK et al (1994) Antiestrogen regulation of cell cycle progression and cyclin D1 gene expression in MCF-7 human breast cancer cells. 31(1):95–105
Weinberg RAJC (1995) The retinoblastoma protein and cell cycle control. 81(3):323–330
Whittaker SR et al (2017) Inhibitors of cyclin-dependent kinases as cancer therapeutics. 173:83–105
Winston JT et al (1996) Regulation of the cell cycle machinery by oncogenic ras. 12(1):127–134
Witkiewicz AK et al (2014) CDK4/6 inhibition provides a potent adjunct to Her2-targeted therapies in preclinical breast cancer models. 5(7–8):261
Xiong Y et al (1991) Human D-type cyclin. 65(4):691–699
Yin J-W, Wang GJD (2014) The Mediator complex: a master coordinator of transcription and cell lineage development. 141(5):977–987
Yu Q et al (2001) Specific protection against breast cancers by cyclin D1 ablation. 411(6841): 1017–1021
Yu Q et al (2006) Requirement for CDK4 kinase function in breast cancer. 9(1):23–32
Zhang Q et al (2011) Cyclin D3 compensates for the loss of cyclin D1 during ErbB2-induced mammary tumor initiation and progression. 71(24):7513–7524
Zwijsen RM et al (1997) CDK-independent activation of estrogen receptor by cyclin D1. 88(3): 405–415

Therapeutic Implications of CDKs in Breast Cancer

11

Manzoor Ahmad Mir and Burhan Ul Haq

11.1 Introduction

A group of serine/threonine kinases known to modulate the cell cycle is the various cyclins and cyclin-dependent kinases (CDKs). To activate, phosphorylate, and stabilize the cyclin-dependent kinases, these lead to the formation of complexes with cyclins in particular cell cycle phases (Malumbres 2014; Nie et al. 2019). In the case of humans, several loci are known to code for cyclin-dependent kinases and cyclins (Malumbres and Barbacid 2005). Nevertheless, only a few of these encoded CDKs and cyclins are directly participating in cell cycle regulation, and the same includes:

1. CDK2, CDK4, and CDK6 are the three interphases CDKs.
2. CDK1, a mitotic CDK.
3. And the other ten cyclins are from four distinct classes (A, B, D, and E type cyclins).

CDKs require modulatory subunits to bind them, known as cyclins, to exert their effects. The latter are formed and destroyed at various cell cycle stages in a specific and timely manner, thus regulating the cell cycle properly (Fig. 11.1).

M. A. Mir (✉)
Department of Bioresources, School of Biological Sciences, University of Kashmir, Srinagar, J&K, India
e-mail: drmanzoor@kashmiruniversity.ac.in

B. U. Haq
Department of Biotechnology, Central University of Kashmir, Ganderbal, J&K, India

M. A. Mir (ed.), *Therapeutic Potential of Cell Cycle Kinases in Breast Cancer*,
https://doi.org/10.1007/978-981-19-8911-7_11

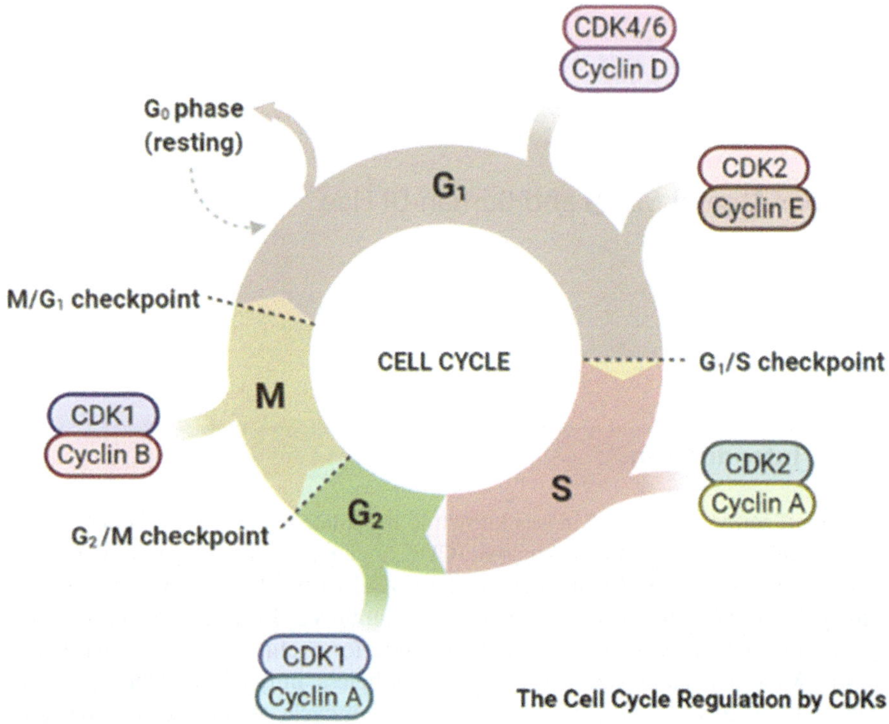

Fig. 11.1 The various CDKs aid in the cell cycle regulation either by inhibiting a phase or activating a particular cycle phase

11.2 Dysregulation of CDKs

Specific complexes of CDK-cyclins are deregulated frequently by mutations associated with tumors, and either unscheduled cell cycle re-entry or continuous proliferation is witnessed due to this deregulation. Furthermore, these two features are seen in most human tumor cells (Malumbres and Barbacid 2001). During the synthesis or formation of DNA and segregation of chromosomes, the checkpoints monitor the proper progression through the cell cycle and sense any defects. Moreover, through the regulation of CDK activity, these activated checkpoints lead to the arrest of the cell cycle (Mir et al. 2022a, b, c, d, e). The purpose of arresting the cell cycle is to provide the time for repairing the defects in the cells properly to hamper their advancement to the daughter cells that will be formed. The endogenous genotoxic agents and the exogenous products could lead to broad changes in the DNA molecule and DNA damage checkpoints help protect the cells from such attacks. Furthermore, when the alterations occur, these by a signaling pathway get sensed leading to CDK inhibition and eventually causing the arrest of cell cycle (Malumbres and Barbacid 2009). The cells may undergo programmed cell death (apoptosis) or enter senescence if the repair process is ineffective due to enormous DNA damage resulting from defects in checkpoints or impaired repair machinery due to genetic defects in the same processes cycle (Mir et al. 2022a, b, c, d, e).

On the flip side, the accumulating changes in DNA may lead to GIN (genomic instability) which leads to the transformation of these cells and thus oncogenesis (Kastan and Bartek 2004). The unscheduled proliferation, the chromosomal instability (CIN), and the genomic instability (GIN) are the primary three defects in the cell cycle. They are mediated by faulty regulation of cyclin-dependent kinesis either directly or indirectly (Mehraj et al. 2021). The chromosomal separation is controlled by SAC (the spindle assembly checkpoint) after the DNA duplication. This signaling process regulates the activity of CDK1 and hinders any defects in the segregation of chromosomes (Kops et al. 2005; Malumbres and Barbacid 2009). On similar grounds, impaired SAC (spindle assembly checkpoint) could lead to an unequal inheritance of DNA. If not repaired, it could aid in tumor progression due to the accumulation of CIN (numerical chromosomal abbreviations). The A-type cyclins are known to activate CDK1 towards the interphase end in order to assist the mitosis onset. After the nuclear envelope is degraded, the A-type cyclins are dissolved to assist in the CDK1-cyclin B complexes formation, which drives the cells through mitosis cycle (Mir et al. 2022a, b, c, d, e).

Given their significant role in the progression of cell cycle, the downregulation of CDKs would end up in defective homeostasis in specific tissues. On the other hand, by initiating the untimely division in cells (progenitor or stem cells), the hyperactivated CDKs could also aid in the development of tumors (Malumbres and Barbacid 2009). Breast cancer can result from many causes, including mutations in DNA repair genes, tumor suppressor genes (p53), and the proto-oncogenes like HER-2, c-myc, as well as cyclin D (Gerger et al. 2007). When BRCA1 and BRCA2 genes are dysregulated, the chance of breast cancer increases. These are known to perform multiple functions, including obstruction of cell cycle progression at the S-phase by halting the rb and possibly CDK2 (Rahman and Stratton 1998; Hashemi et al. 2019) (Fig. 11.2).

The specific CDK inhibitors do not act on multiple CDKs, unlike dual and pan-CDK inhibitors, which are directed against two and more than two types of CDKs, respectively (Mehraj et al. 2022a, b, c, d). The CDK4/6 inhibitors are very significant at preventing the progression of cell cycle and proliferation of tumor cell, and the frequently employed agents include Palbociclib, Abemaciclib, Ribociclib, and Trilaciclib, which are all CDK4/6 inhibitors. These inhibitors are very efficient for BC treatment, and several such drugs are being tested at phases I and II of clinical trials nowadays (Wu et al. 2020). Breast cancer subjects presenting with advanced-stage and metastatic types of breast cancer could be cured using such inhibitors as Abemaciclib (Table 11.1).

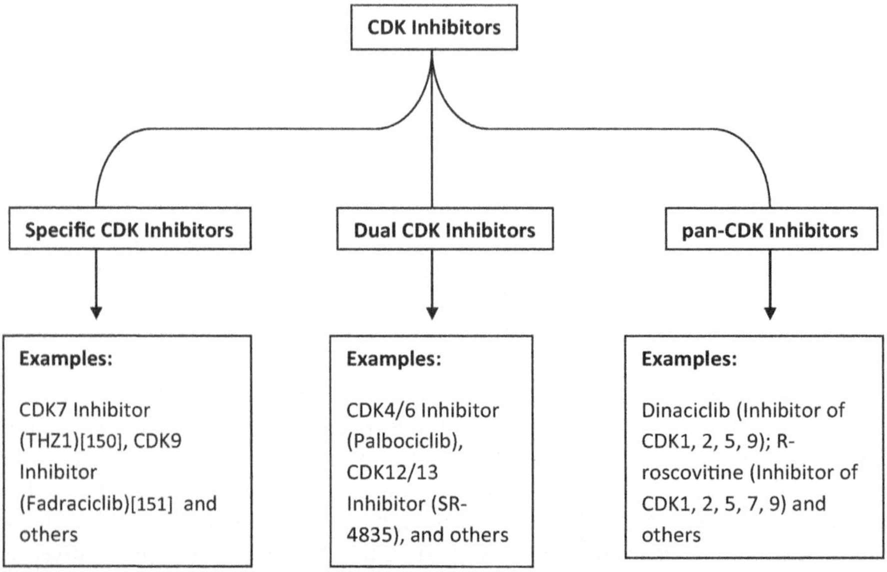

Fig. 11.2 The above flowchart represents various CDK Inhibitors, categorized into three main classes based on their specificity towards the CDKs. (From the above list, only Palbociclib is FDA approved for treating breast cancer)

Table 11.1 Examples of inhibitors (of CDK4/6) used in treatment of BC

S.no.	Drug name	FDA approved	Developed by
1.	Palbociclib	YES	Pfizer
2.	Ribociclib	YES	Novartis
3.	Abemaciclib	YES	Eli Lilly

11.3 Functioning of CDK/Cyclins in the Cell Cycle

Through the target gene phosphorylation like the Rb, a tumor suppressor protein, this complex of cyclin/cyclin-dependent kinases takes control over the cell cycle progression. The active mitogenic signals are known to activate cyclins/cyclin-dependent kinases, and in reaction to the damage of DNA, the Cyclins/CDKs get inhibited via activated cell cycle checkpoints (Otto and Sicinski 2017). The cyclin-dependent kinase inhibitors (CDKIs) modulate the Cyclin/CDKs antagonistically. Examples include the CIP/KIPs as well as the CDK4(INK4) protein inhibitor (Asghar et al. 2015). In addition, Skp1-Cul1-F-box-protein (SCF) complex and anaphase-promoting complex/cyclosome (APC/C) also have a part to play in regulating the mitotic protein expression that impacts the cell cycle transitions, and these are E3 ubiquitin ligases (Sivakumar and Gorbsky 2015; Zhou et al. 2016a, b; Senft et al. 2018).

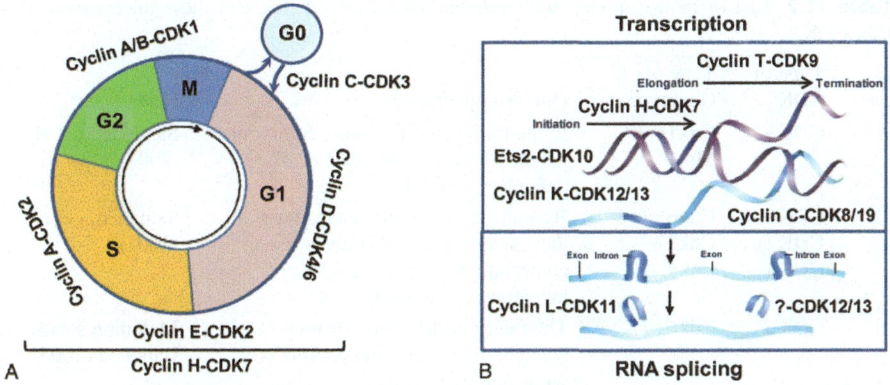

Fig. 11.3 The CDKs can broadly be categorized as: (**a**) CDKs participating in cell cycle and (**b**) CDKs participating in transcription

11.4 Types of CDKs

The cyclin-dependent kinases can be categorized into two types (Fig. 11.3)

1. **CDKs associated with cell cycle** include CDK1, CDK2, CDK4, and CDK6. ***Characteristics:*** These CDKs are known to moderate the progression and cell cycle phases directly.
2. **CDKs associated with transcription** include CDK7, CDK8, CDK9, CDK12, and CDK13. ***Characteristics:*** These CDKs are known to phosphorylate the carboxy-terminal domain (CTD) of the DNA-directed RNA polymerase II subunit (RPB1) of RNA Pol II and another target. The exact mechanism concerning the transcription is not fully elucidated (Asghar et al. 2015).

11.5 Role of CDKs in Breast Cancer

The uncontrolled cellular proliferation manifests as one of the cancer hallmarks (Mehraj et al. 2022a, b, c, d). The same occurs through the cell cycle checkpoint disabling and overriding several safeguards involved with the cyclin/CDKs dysregulation or impairment. Moreover, multiple studies revealed the participation of Cyclin/CDKs dysregulation in various BC phenotypes, as indicated in Table 11.2.

In many solid cancers like BC, uncontrolled cell proliferation is witnessed as a result of cell cycle (Sofi et al. 2022a, b, c) dysregulation and the genetic changes in the proteins involved in cell cycle regulation (Hanahan and Weinberg 2011). The

Table 11.2 List of mouse models with gene-targeted CDK alleles, and these mice strains also lack CDK3

S. no.	Type of CDK	Type of Genotype	Observed phenotype	Reference
1.	CDK1	Cdk1mut/Mut	In the initial cell divisions, embryonic lethality is witnessed because of deficiency of CDK1	Santamaría et al. (2007)
2.	CDK6, CDK2, CDK4	Cdk6−/−; Cdk2−/−; Cdk4−/−	By mid-gestation, the inadequacy of these interphase CDKs gives rise to embryonic lethality as a result of hematopoietic defects.	Santamaría et al. (2007)
3.	CDK6	Cdk6−/−	The faulty proliferation of few hematopoietic cells and anemia is witnessed marginally.	Malumbres and Barbacid (2005)
4.	CDK11	Cdk11−/−	In peri-implantation embryos, lethality complemented by a mitotic anomaly is witnessed.	Malumbres and Barbacid (2005)
5.	CDK4	$Cdk4^{R24C/R24C}$	Development of various tumor types displaying penetrance completely in mice expressing an endogenous Ink4-insensitive $CDK4^{R24C}$ mutant.	Sotillo et al. (2001a, b), Sotillo et al. (2001a, b)

Note: First four represent loss-of-function strains, and the fifth represents gain-of-function strains of the mice. The $Cdk1^{mut}$ allele was made by gene trap vector insertion. A knock-in allele $Cdk4^{R24C}$ wherein Cys takes the place of Arg24 in order to block the Ink4 inhibitor binding

cyclin-dependent kinases, along with their partners and their biological functions in case of breast cancer, are given in Table 11.3.

It has been seen that CDK10, along with its interacting partner Cyclin M typically modulates ETS2 transcription but not via RNAPII phosphorylation, and its attributed roles in breast cancer are the correspondence with metastasis in case of lymph node and resistance witnessed in case of endocrine therapy (Iorns et al. 2008; You et al. 2015; Guen et al. 2017). The CDK11, along with its interacting partner Cyclin L modulates the splicing and transcription of RNA, programmed cell death as well as autophagy, and in breast cancer, this complex partner with growth and angiogenesis, proliferation, and programmed cell death, too (Loyer et al. 2008; Chi et al. 2015; Zhou et al. 2016a, b; dos Santos Paparidis and Canduri 2018; Khan et al. 2022a, b). CDK19 and its interacting partner Cyclin C function as a paralog of CDK8, with akin function to CDK8, although appear to execute some different functions and the attributed role in breast cancer is chemoresistance; furnishes potential targets for enhancing chemotherapy (Galbraith et al. 2013; Zheng et al. 2019). In hepatitis B virus-driven transformation, CDK15/CyclinY takes a part, in the case of breast cancer associated with metastasis and cell invasion (Shiraishi et al. 2014; Li et al. 2019).

Table 11.3 An outline of the biological role of some of the CDKs in BC

Cyclin-dependent kinases	CDK interacting partners (Cyclins)	Attributed role(s) in breast cancer	Reference
Cyclin-dependent kinase **1**	*A/B*	Partners with programmed cell death of MYC-driven triple-negative breast cancer.	Chen et al. (2009), Horiuchi et al. (2012), Sofi et al. (2022a, b, c)
Cyclin-dependent kinase **2**	*A/E*	Corresponds phenotypically with breast cancer or triple-negative BC.	Marais et al. (2010), Nie et al. (2019)
Cyclin-dependent kinase **3**	*C*	Partners with the migration of BC cells including the proliferation, invasion, and the programmed cell death as well.	Cao et al. (2017), Zhang et al. (2017)
Cyclin-dependent kinase **4/6**	*D*	Plays a part in the initiation of breast cancer and tumorigenesis maintenance.	Santo et al. (2015), Malumbres and Barbacid (2001)
Cyclin-dependent kinase **5**	*p35 and p39*	Partners with cell death in breast cancer (ROS-mediated); necessary for TGF-β1-induced EMT.	Pozo et al. (2013), Dorand et al. (2016), NavaneethaKrishnan et al. (2018)
Cyclin-dependent kinase **7**	*H*	Moderates transcriptional fixation in case of triple-negative breast cancer to an essential gene cluster.	Wang et al. (2015), Li et al. (2017)
Cyclin-dependent kinase **8**	*C*	Reacts to breast cancer adjuvant therapy; partners with the progression of the tumor.	Firestein et al. (2008), Nemet et al. (2014), Crown (2017)
Cyclin-dependent kinase **9**	*T*	In subjects presenting with BC acts as a prognostic biomarker, succeeding neoadjuvant chemotherapy.	Schlafstein et al. (2018), Del Re et al. (2019), Mehraj et al. (2022a, b, c, d)

11.6 Need for CDK Inhibitors for Use in BC Treatment

Given the role of CDKs and their dysregulation in BC, several CDK inhibitors have been formed, some of which are FDA approved for their use in clinical settings. Some are being designed and examined as anticancer drugs at various phases of clinical trials (Sánchez-Martínez et al. 2019). Studies on the human BC mouse models indicate that CyclinD1-CDK4/6 axis stimulation ends up with a tumorigenic phenotype and plays a part in the maintenance and initiation of tumorigenesis in HER2+ breast cancer (Dukelow et al. 2015). The ER-negative and PR-negative status in TNBC is indicated by the Cyclin E overexpression and corresponds to a poor prognostic marker in TNBC (Jabbour-Leung et al. 2016; Mintoo et al. 2021). The absence of INK4 and CIP/KIP family proteins, coupled with CDK4/6 overexpression, has been witnessed in the case of breast cancer clinically (Asghar et al. 2015). Research has revealed that different subtypes of breast cancer display different molecular alterations for cell cycle checkpoints (Mir et al. 2022a, b, c, d, e).

11.7 Involvement of Other CDKs

It has also been seen that the overexpression of CDK2 ends up with overexpression of cyclin E and cyclin A, which are its partners—in the case of breast cancer (Santo et al. 2015; Singh et al. 2017). Roles of CDK1 in mitotic progression have been observed along with the overexpression of Cyclin A2 and B1 (Aaltonen et al. 2009). With the help of A- and B-type cyclins, the CDK1 kinase modulates the centrosome cycle and the mitotic onset and represents one of the central modulators of mitosis (Mir et al. 2022a, b, c, d, e). The CDK1 activity is shut down after the successful condensation of chromosomes and their alignment at the metaphase plate in order to permit the segregation of sister chromatid via separase or separin activation. The decondensation of chromosomes, nuclear envelope reformation, and the process of cytokinesis all require this inactivation of CDK1 (Potapova et al. 2006). A study has shown that in the absence of CDK12 protein, there has been much improvement in the triple-negative breast cancer phenotype due to CDK12 loss leading to DNA repair defects (Naidoo et al. 2018). Also, one of the valuable therapies is the inhibition of CDK7 for triple-negative breast cancer patients. It has been witnessed in TNBC that the CDK7 brings about transcriptional addition to a significant gene cluster (Wang et al. 2015; Wadhwa et al. 2020). Due to their participation in sustaining the growth of cancer cells, the CDKs and Cyclins are good targets for treating breast cancers (Mir et al. 2022a, b, c, d, e).

11.8 Types of CDK Inhibitors

The inhibitors directed against CDKs are categorized either as Non-selective or Selective, i.e., either pan-inhibitors or against one single cyclin-dependent kinase, solely based on meticulousness against the CDKs (Ding et al. 2020). Various drugs that are CDK inhibitors in action have entered breast cancer clinical trials and are known to target cell modulators in the cancerous cells, thus furnishing a therapeutic window (Ding et al. 2020; Mir et al. 2022a, b, c, d, e).

11.9 Pan-Inhibitors for BC Treatment

All pan-CDK inhibitors act non-specifically. For example, the targets of Seliciclib (a pan-CDK inhibitor) are CDK1, CDK2, CDK5, CDK7, and CDK9 (53). Several undesirable effects/toxicities have been witnessed due to the use of various pan-CDK inhibitors, including fatigue, myelosuppression, nausea, abnormalities in the liver, vomiting, nerve dysfunction, GIT effects, and for these agents lack of predictive biomarkers for the BC patients (Mehraj et al. 2022a, b, c, d). Thus, these

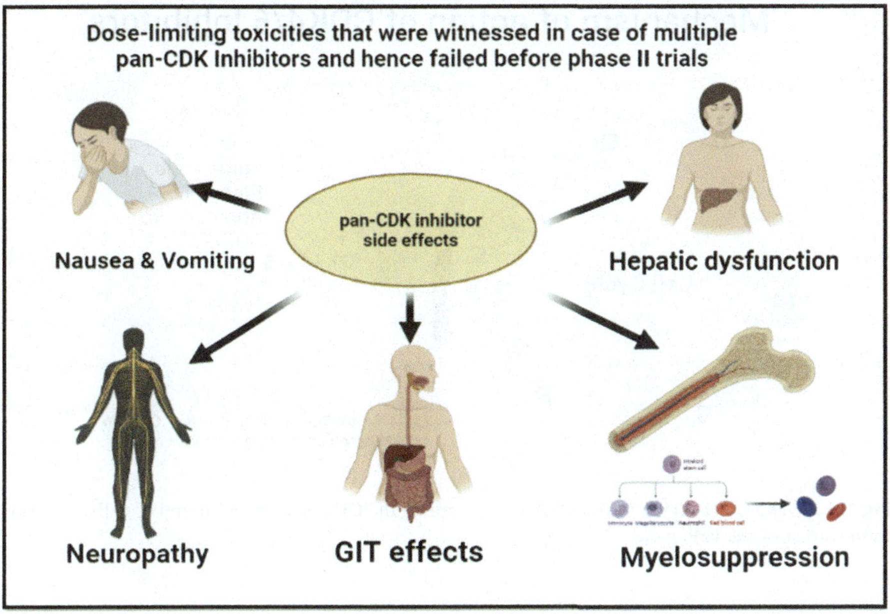

Fig. 11.4 Some side effects observed due to consumption of pan-CDK inhibitors in BC patients

Table 11.4 Examples of some pan-CDK inhibitors that act non-specifically against various CDKs

S. no.	Name of the drug		Developed by
1.	Flavopiridol		Sanofi-Aventis
2.	Dinaciclib		Merck
3.	Seliciclib		Cyclacel
4.	Mitotic kinase inhibitors	4a. AURKB	
		4b. PLK1	

collapsed before phase second trials. The undesirable effects are shown in Fig. 11.4. Some examples of early pan-CDK inhibitors are given in Table 11.4. All pan-CDK inhibitors are non-specific.

Obtained from rohitukine (a chromone alkaloid) is a semi-synthetic flavonoid which represents one of the examples of first-generation inhibitors is flavopiridol. It exerts its anticancer effects by inhibiting CDK1, 2, 4, 6, 7, and 9 (Sedlacek et al. 1996; Shapiro 2006). Flavopiridol in G1 and G2 phases leads to cell cycle arrest and is also known to induce cytotoxicity by blocking CDK7 and CDK9 and c-MYC transcription (Canavese et al. 2012). Another example from the pan-CDK inhibitors is provided by Dinaciclib, which is known to inhibit CDK1, CDK2, CDK5, and CDK9 with excellent Rb phosphorylation inhibitory potency, thus showing a better therapeutic index in comparison with the flavopiridol (Asghar et al. 2015).

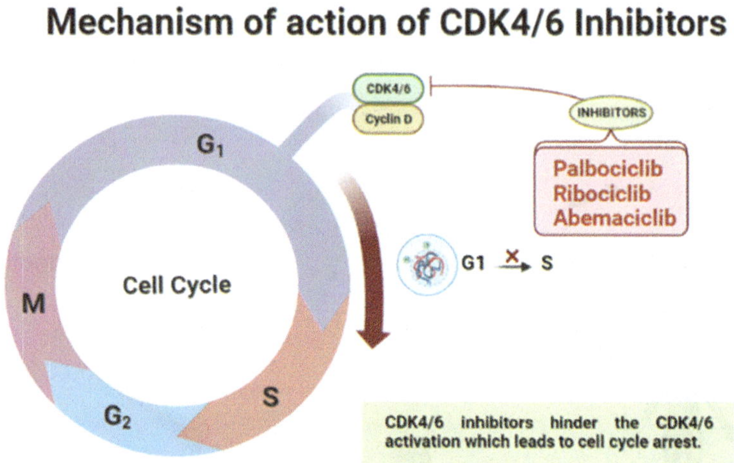

Fig. 11.5 The CDK4/6 inhibitors work by preventing the CDK4/6 activation in BC cells, resulting in the arrest of the cell cycle

11.10 Specific CDK Inhibitors for BC Treatment

For treating ER+ /HER2- advanced and metastatic BC, both FDA and European medicines agency approved CDK4/6 selective inhibitors (Abemaciclib, Palbociclib, and Ribociclib) displaying inhibition of growth in ER+ BC in a dose-dependent manner. The three drugs have the property of binding the CDK4 and CDK6 ATP binding pocket and thus are ATP-competitive drugs and are small molecules. In the ATP binding cleft, these display particular types of interactivity with the residues (Asghar et al. 2015). Due to the resistance in ER+ breast cancer patients induced by the endocrine therapy, the CDK inhibitor development came into focus. The general mechanism of action is shown in Fig. 11.5.

11.11 CDK4/6 Inhibitors and their Mechanism of Action

CDK4/6 inhibitors (palbociclib, ribociclib, and abemaciclib) hinder the CDK4/6 activation, causing cell cycle arrest at the G1 phase of the cell cycle. The complex CDK4/6-CyclinD is responsible for Rb (the tumor suppressor gene) phosphorylation and inhibition of its product. When phosphorylation of Rb occurs, the G1 to S-phase proceeds smoothly, DNA replicates, and mitosis usually occurs. However, when blocked by these inhibitors, the process is hampered, and the cell cycle is arrested. The Palbociclib mechanism of action is shown in Fig. 11.6. In both preclinical and clinical trials for Estrogen Receptor-positive BC, the inhibitors (for CDK4/6) were approved effectively when combined with the anti-estrogen therapies (Sobhani et al.

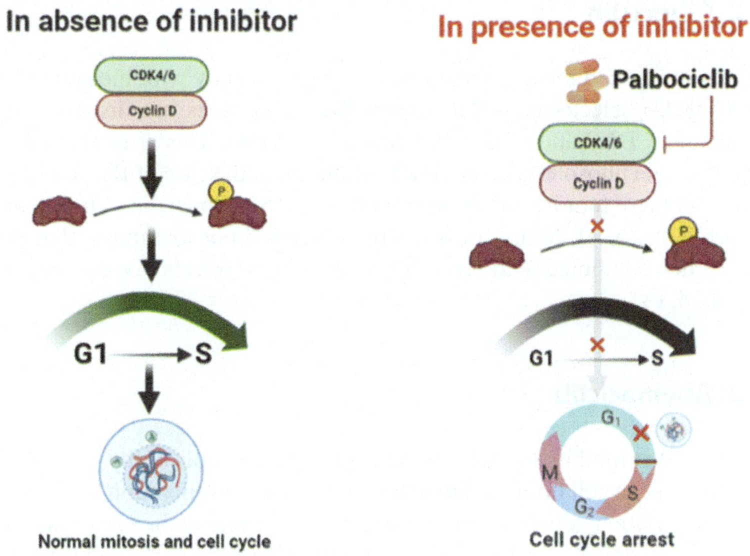

Fig. 11.6 The MOA of Palbociclib, a CDK4/6 inhibitor used to treat BC. [Note that Rb (shown in purple) phosphorylation occurs in the absence of Palbociclib)

2019). Various clinical trials are being carried out to access the inhibitors, which are specific to CDK4/6 in BC (Mir et al. 2022a, b, c, d, e).

11.11.1 Palbociclib

The first CDK4/6 inhibitor that got the approval for treating breast cancer is Palbociclib. This is a potent, small-molecule selective inhibitor of CDK4/6 and is administered orally (Im et al. 2019). It has been observed that the human breast cancer cell line displays varied sensitivity based on its phenotype towards palbociclib. Also, the ER- breast cancer cells with basal-like and triple-negative breast cancer histology show less palbociclib sensitivity than the ER+ breast cancer cell lines with luminal features (Finn et al. 2016a, b; Asghar et al. 2017). Palbociclib and Abemaciclib have been marked for the second-line therapy with Fulvestrant and these display very low efficacy against CDK1, 2, 7, and 9 (Chen et al. 2009). It has also been predicted that the presence of a functional Rb protein is essential for palbociclib. Hampering the Rb phosphorylation causes the arrest of the cell cycle in G_1 phase (Dean et al. 2010). Palbociclib works synergistically in combination with tamoxifen and trastuzumab and efficiently suppresses the ER+ breast cancer cell line proliferation (Finn et al. 2015). In many cancers that occur in humans, the CDK4's (Cyclin-dependent kinase 4) overexpression has been found, including the breast cancer, and palbociclib has been approved by FDA for its treatment due to its specific CDK4 inhibition (Mehraj et al. 2021). This dual inhibitor for CDK4/6 in the case of BC was approved as it specifically shows its inhibitory effect on HER2- breast cancer/ER+ breast cancer (Finn et al. 2016a, b; Finn et al. 2016a, b).

11.11.2 Ribociclib

The drug is orally administered and displays high potency with bioavailability and inhibits CDK4/6 selectivity is Ribociclib. This drug does not display significant activity against CDK2 and CDK1 (Sobhani et al. 2019). This drug also inhibits the Rb + cell lines via phosphorylation of Rb inhibition in the case of BC, leading to cell cycle arrest in these tumor cells, as seen in the case of palbociclib (Chen et al. 2008; Sofi et al. 2022a, b, c). Many trials are being carried out to explore the ribociclib efficiency when combined with other drugs or agents to treat breast cancer (Mir et al. 2022a, b, c, d, e).

11.11.3 Abemaciclib

Another drug that inhibits CDK4/6 is Abemaciclib, which is orally administered. It causes a decrease in cell number and halts cancer cells' proliferation via inhibition of Rb's phosphorylation coupled with the arrest of cell cycle at the G1 phase (Im et al. 2019). A recent study revealed that the inhibitor Abemaciclib could enhance the activation of T-cells and also up-regulated the antigen presentation genes expression in human BC cells (Schaer et al. 2018). In addition, due to a better understanding of CDKs in various subtypes of breast cancer, their mode of inhibition, their side effects, and resistance, many new CDKs are being explored (Mir et al. 2022a, b, c, d, e).

11.11.4 Other CDK Inhibitors

PL (piperlongumine) is a novel CDK inhibitor discovered by Jeong et al. This inhibitor is known to hamper migration and cell proliferation in the case of ER+ breast cancer. The PL is a natural product, and it is obtained from pepper. It hinders the CDK1 and CDK4/6 expression levels and ends up arresting the cell cycle at the G_2/M phase to stop tumorigenesis (Jeong et al. 2019). The highly selective dual inhibitor **SR-4835**: It was revealed by Quereda et al. that this inhibitor acts on CDK12 and CDK13. It can stop cell proliferation in triple-negative breast cancer (Quereda et al. 2019). According to the study's upshot (Mir 2015, Li et al. 2020), in the case of BC, the SR-4835 led to the immunogenic death of cell, thus adding to the antitumor function of PD1-PD-L1 immune checkpoint therapy (Qayoom et al. 2021, Mehraj et al. 2022a, b, c, d). **Panduratin A (PA)** posses several health benefits, including anti-inflammatory, anti-oxidant, and antibacterial, in addition to its anti-cancer activity. It has been seen that PA leads to the block of the cell cycle in G_0/G1 phase by suppressing the expression of CDK4 and cyclin D1 (Liu et al. 2018, Sofi et al. 2022a, b, c). Vanicoside B is phenylpropanoyl sucrose derived from flavonoid glycoside and has been shown to act as a chemopreventive agent (Sofi et al. 2022a, b, c). The vanicoside B could inhibit the CDK8-mediated signaling pathway expression as well as lead to the onset of arrest in the cell cycle in HCC38 and

MDA-MB-231 cells as reported previously (Kim et al. 2019). A role in the modulation of cellular invasion, proliferation as well as migration is played by protein phosphatase Mg^{2+}/Mn^{2+}-dependent 1A (PPM1A) by decreasing retinoblastoma and CDK phosphorylation in the case of TNBC. PPM1A belongs to the Ser/Thr protein phosphatase 2C family (Mazumdar et al. 2019). It has been noted in the study that when the Roscovitine, a pan-CDK inhibitor, is administered sequentially preceding doxorubicin, it is synthetically fatal in triple-negative breast cancer cells (Mehraj et al. 2022a, b, c, d). This inhibitor, when administered, blocks the cell cycle in the G2/M phase, preparing them for DNA damage. It was observed that this combined treatment approach led to an enhancement in DNA double-stranded breaks and lowered the protein recruitment, necessary for homologous recombination compared to the solo treatment by doxorubicin (Mir et al. 2022a, b, c, d, e). It was also witnessed that by employing this combination therapy, tumor volume showed a reduction, and an elevated survival was observed compared to the solo drug or related treatment in the case of xenograft studies (Jabbour-Leung et al. 2016, Mir et al. 2022a, b, c, d, e). This inhibitors of CDK1, 2, 5, 7, and 9 (Zhang et al. 2018, Nie et al. 2019) surfaced as the prime orally available drug from this group to become part of the clinical trials due to its relative success in the preclinical stage, where its success led to the onset of apoptosis in tumor cells (MacCallum et al. 2005, Shapiro 2006, Galons et al. 2010, Mehraj et al. 2022a, b, c, d). CDK1 shows the involvement in the homologous recombination DNA double-stranded break repair pathway. The cyclin-dependent kinase activity is needed for removing the DNA double-stranded breaks to produce single stands during the homologous recombination by recruiting endonucleases Sae2 or CtIP, respectively, in yeast and mammalian cells (Ira et al. 2004, Huertas and Jackson 2009). CDK activity has also been recruited for recruiting and associating the BRCA1 to the MRN [Mre11-Rad50-Nbs1] complex during homologous recombination (Chen et al. 2008; Mir et al. 2022a, b, c, d, e). The inhibition of CDK using the Roscovitine decreased the RPA34, a homologous recombination downstream protein in sarcoma cells that had previously received radiation treatment therapy due to the inefficacy of producing the single stands (Jazayeri et al. 2006; Mir et al. 2020). Therefore weakening homologous recombination by inhibiting the cyclin-dependent kinases could furnish plans to increase the cell sensitivity to chemotherapy in case of TNBC (Jabbour-Leung et al. 2016; Mir et al. 2022a, b, c, d, e). It has also been witnessed that there is high CDK11 expression in the case of triple-negative breast cancer, liposarcoma, and multiple myeloma (Jia et al. 2014; Zhou et al. 2015; Sofi et al. 2022a, b, c). Indirectly the CDK7 may modulate the transcription by phosphorylating and modulating transcription factors like ER and Androgen receptors, both of which have a significant role in breast cancer and prostate cancer, the hormone-driven cancers (Asturias 2004; Compe and Egly 2012; Sainsbury et al. 2015). The co-amplification of CDK12 with ERBB2/HER2 oncogene has been witnessed in subsets of breast cancer (Naidoo et al. 2018). It was revealed through the proteomic analysis that the CDK12 amplification is associated with increased phosphorylation of CDK12, indicating that CDK12 could act as a vital therapeutic target in case of HER2-amplified breast cancers (Mertins et al. 2016; Paculová and Kohoutek 2017). It has already been established via multiple studies and trials that CDKs, particularly CDK4/6inhibitors, enhance the treatment efficiency in combination with hormone therapy than HT alone (Turner et al. 2015; Le Saux et al. 2017).

According to a study on the Indian population (Lakkavalli et al. 2021), hormone therapy was relatively more effective in combination with CDK4/6 inhibitors with tolerable side effects among the HR-positive advanced breast cancer patients (Mir and Mehraj 2019). The study's upshot revealed that the palbociclib, when combined with the hormone therapy, resulted in extended PFS (progression-free survival) compared to the HT alone among those women who presented with the HR-positive metastatic breast cancer (Khan et al. 2022a, b).

11.11.5 Side Effects of CDK4/6 Inhibitors

The CDK4/CDK6 inhibitors Palbociclib, Ribociclib, and Abemaciclib are being used in clinical settings and are employed in the case of BC which is hormone receptor-positive [ER and/or PR expressing]. Improved progression-free survival (PFS) and overall survival (OS) have been witnessed (Finn et al. 2016a, b; Sledge Jr et al. 2017). However, various undesirable effects are also witnessed in BC patients, and developing more specific targets with fewer side effects is the need of the hour. The following side effects have been witnessed in BC patients from CDK4/6 Inhibitors that are approved by FDA (Ettl 2019). The various side effects are listed in Fig. 11.6, and the common side effects observed in the case of all the three approved drugs have been mentioned separately in Fig. 11.7.

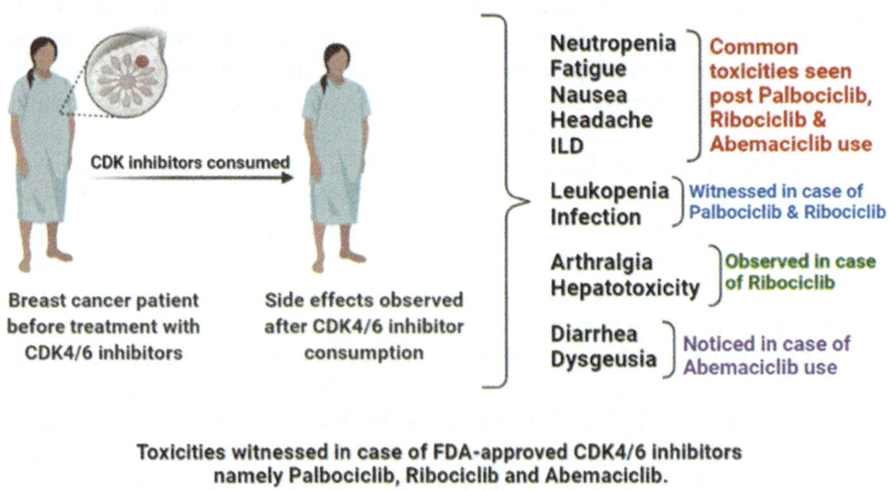

Fig. 11.7 These are various side effects of FDA-approved CDK4/6 inhibitor drugs in the case of BC subjects

11.12 Summary

The active participation of CDKs in coordinating and modulating the cell division provides vast scope for further research to elucidate the process of BC development, especially its metastasis. In this direction, multiple therapeutic implications of CDKs in breast cancer have been explored. Several drugs that help inhibit CDKs, for example, specific CDK4/6 inhibitors, have been successfully developed and used in clinical settings today. Similarly, dual inhibitors, as well as pan-CDK inhibitors, have been explored for treating breast cancer. However, pan-CDK inhibitors have shown antitumor activities in the case of BC patients, given the non-specific mode of action coupled with the undesirable effects on BC patients (Mir et al. 2022a, b, c, d, e). Many such inhibitors failed at different clinical trial phases, and their use in actual clinical practice was not approved. Conclusion: There is a need to reduce the undesirable effects, and the combination of therapy with other drugs may help develop an effective treatment for BC patients. Several such inhibitors are being tested at different phases of clinical trials and can find their uses in treating BC in actual clinical settings (Mir et al. 2022a, b, c, d, e).

11.13 Further Readings

For more insights about the topic, we would suggest detailed findings from the books of Mir MA (2022) https://doi.org/10.1016/C2021-0-02565-7, https://doi.org/10.1016/C2022-0-00074-X and Mir MA (2021) https://doi.org/10.52305/WXJL6770, and from the cancer.net website on the following mentioned below links,

https://www.cancer.net/cancer-types/breast-cancer/types-treatment
https://discovery.ucl.ac.uk/id/eprint/1472740/
https://www.jmedsciences.com/doi/JMEDS/pdf/10.5005/jp-journals-10045-00138

For diagrammatic illustrations, descriptive tables, Lazzeroni (2012) http://www.eurekaselect.com/article/49928

See video links on over all status of cancer, its various types, current new treatment possible options available.

https://www.sciencedirect.com/science/article/pii/S205970292032278X

The readers can have a look upon the following video YouTube links for the better understanding of the chapter:

https://youtu.be/wIsdjfwPUxY
https://youtu.be/SVjJt984PlU
https://www.youtube.com/watch?v=9hgrfXleNsM
https://www.youtube.com/watch?v=ZWqfoBj2bsA
https://www.sciencedirect.com/science/article/pii/S221464741630054X
https://www.ncbi.nlm.nih.gov/pmc/articles/PMC5715522/

References

Aaltonen K et al (2009) High cyclin B1 expression is associated with poor survival in breast cancer. Br J Cancer 100(7):1055–1060
Asghar U et al (2015) The history and future of targeting cyclin-dependent kinases in cancer therapy. Nat Rev Drug Discov 14(2):130–146
Asghar US et al (2017) Single-cell dynamics determines response to CDK4/6 inhibition in triple-negative breast CancerCDK4/6 inhibition in triple-negative breast cancer. Clin Cancer Res 23(18):5561–5572
Asturias FJ (2004) RNA polymerase II structure, and organization of the preinitiation complex. Curr Opin Struct Biol 14(2):121–129
Canavese M et al (2012) Cyclin dependent kinases in cancer: potential for therapeutic intervention. Cancer Biol Ther 13(7):451–457
Cao T et al (2017) CDK3, target of miR-4469, suppresses breast cancer metastasis via inhibiting Wnt/β-catenin pathway. Oncotarget 8(49):84917–84927
Chen L et al (2008) Cell cycle-dependent complex formation of BRCA1· CtIP· MRN is important for DNA double-strand break repair. J Biol Chem 283(12):7713–7720
Chen Y-J et al (2009) A conserved phosphorylation site within the forkhead domain of FoxM1B is required for its activation by cyclin-CDK1. J Biol Chem 284(44):30695–30707
Chi Y et al (2015) Critical role of CDK11p58 in human breast cancer growth and angiogenesis. BMC Cancer 15(1):1–10
Compe E, Egly J-M (2012) TFIIH: when transcription met DNA repair. Nat Rev Mol Cell Biol 13(6):343–354
Crown J (2017) CDK8: a new breast cancer target. Oncotarget 8(9):14269
Dean JL et al (2010) Therapeutic CDK4/6 inhibition in breast cancer: key mechanisms of response and failure. Oncogene 29(28):4018–4032
Del Re M et al (2019) Overexpression of TK1 and CDK9 in plasma-derived exosomes is associated with clinical resistance to CDK4/6 inhibitors in metastatic breast cancer patients. Breast Cancer Res Treat 178(1):57–62
Ding L et al (2020) The roles of cyclin-dependent kinases in cell-cycle progression and therapeutic strategies in human breast cancer. Int J Mol Sci 21(6):1960
Dorand RD et al (2016) Cdk5 disruption attenuates tumor PD-L1 expression and promotes antitumor immunity. Science 353(6297):399–403
Dukelow T et al (2015) CDK4/6 inhibitors in breast cancer. Anti-Cancer Drugs 26(8):797–806
Ettl J (2019) Management of adverse events due to cyclin-dependent kinase 4/6 inhibitors. Breast care 14(2):86–92
Finn RS et al (2015) The cyclin-dependent kinase 4/6 inhibitor palbociclib in combination with letrozole versus letrozole alone as first-line treatment of oestrogen receptor-positive, HER2-negative, advanced breast cancer (PALOMA-1/TRIO-18): a randomised phase 2 study. Lancet Oncol 16(1):25–35
Finn RS et al (2016a) Targeting the cyclin-dependent kinases (CDK) 4/6 in estrogen receptor-positive breast cancers. Breast Cancer Res 18(1):1–11
Finn RS et al (2016b) Palbociclib and letrozole in advanced breast cancer. N Engl J Med 375(20): 1925–1936
Firestein R et al (2008) CDK8 is a colorectal cancer oncogene that regulates β-catenin activity. Nature 455(7212):547–551
Galbraith MD et al (2013) HIF1A employs CDK8-mediator to stimulate RNAPII elongation in response to hypoxia. Cell 153(6):1327–1339
Galons H et al (2010) Cyclin-dependent kinase inhibitors: a survey of recent patent literature. Expert Opin Ther Pat 20(3):377–404
Gerger A et al (2007) A multigenic approach to predict breast cancer risk. Breast Cancer Res Treat 104(2):159–164

Guen VJ et al (2017) The awakening of the CDK10/Cyclin M protein kinase. Oncotarget 8(30): 50174
Hanahan D, Weinberg RA (2011) Hallmarks of cancer: the next generation. Cell 144(5):646–674
Hashemi V et al (2019) The role of DEAD-box RNA helicase p68 (DDX5) in the development and treatment of breast cancer. J Cell Physiol 234(5):5478–5487
Horiuchi D et al (2012) MYC pathway activation in triple-negative breast cancer is synthetic lethal with CDK inhibition. J Exp Med 209(4):679–696
Huertas P, Jackson SP (2009) Human CtIP mediates cell cycle control of DNA end resection and double strand break repair. J Biol Chem 284(14):9558–9565
Im S-A et al (2019) Overall survival with ribociclib plus endocrine therapy in breast cancer. N Engl J Med 381(4):307–316
Iorns E et al (2008) Identification of CDK10 as an important determinant of resistance to endocrine therapy for breast cancer. Cancer Cell 13(2):91–104
Ira G et al (2004) DNA end resection, homologous recombination and DNA damage checkpoint activation require CDK1. Nature 431(7011):1011–1017
Jabbour-Leung NA et al (2016) Sequential combination therapy of CDK inhibition and doxorubicin is synthetically lethal in p53-mutant triple-negative breast CancerCDK inhibition sensitizes p53-mutated cells to doxorubicin. Mol Cancer Ther 15(4):593–607
Jazayeri A et al (2006) ATM-and cell cycle-dependent regulation of ATR in response to DNA double-strand breaks. Nat Cell Biol 8(1):37–45
Jeong CH et al (2019) Piperlongumine induces cell cycle arrest via reactive oxygen species accumulation and IKKβ suppression in human breast cancer cells. Antioxidants 8(11):553
Jia B et al (2014) Cyclin-dependent kinase 11 (CDK11) is crucial in the growth of liposarcoma cells. Cancer Lett 342(1):104–112
Kastan MB, Bartek J (2004) Cell-cycle checkpoints and cancer. Nature 432(7015):316–323
Khan SU et al (2022a) Understanding the cell survival mechanism of anoikis-resistant cancer cells during different steps of metastasis. Clin Exp Metastasis
Khan SU et al (2022b) Activation of lysosomal mediated cell death in the course of autophagy by mTORC1 inhibitor. Sci Rep 12(1):1–13
Kim D et al (2019) Antitumor activity of vanicoside B isolated from Persicaria dissitiflora by targeting CDK8 in triple-negative breast cancer cells. J Nat Prod 82(11):3140–3149
Kops GJPL et al (2005) On the road to cancer: aneuploidy and the mitotic checkpoint. Nat Rev Cancer 5(10):773–785
Lakkavalli RK et al (2021) Metastatic hormone receptor-positive breast cancer in CDK 4/6 era: an outcome audit. J Cancer Res Ther 17(4):994
Le Saux O et al (2017) Assessment of multiple endocrine therapies for metastatic breast cancer in a multicenter national observational study. American Society of Clinical Oncology
Li B et al (2017) Therapeutic rationale to target highly expressed CDK7 conferring poor outcomes in triple-negative breast CancerCDK7 expression and function in triple-negative breast cancer. Cancer Res 77(14):3834–3845
Li S et al (2019) PA28α/β promote breast cancer cell invasion and metastasis via down-regulation of CDK15. Front Oncol 9:1283
Li Y et al (2020) CDK12/13 inhibition induces immunogenic cell death and enhances anti-PD-1 anticancer activity in breast cancer. Cancer Lett 495:12–21
Liu Q et al (2018) Panduratin a inhibits cell proliferation by inducing G0/G1 phase cell cycle arrest and induces apoptosis in breast cancer cells. Biomol Ther 26(3):328
Loyer P et al (2008) Characterization of cyclin L1 and L2 interactions with CDK11 and splicing factors: influence of cyclin L isoforms on splice site selection. J Biol Chem 283(12):7721–7732
MacCallum DE et al (2005) Seliciclib (CYC202, R-Roscovitine) induces cell death in multiple myeloma cells by inhibition of RNA polymerase II-dependent transcription and down-regulation of Mcl-1. Cancer Res 65(12):5399–5407
Malumbres M (2014) Cyclin-dependent kinases. Genome Biol 15(6):1–10

Malumbres M, Barbacid M (2001) To cycle or not to cycle: a critical decision in cancer. Nat Rev Cancer 1(3):222–231

Malumbres M, Barbacid M (2005) Mammalian cyclin-dependent kinases. Trends Biochem Sci 30(11):630–641

Malumbres M, Barbacid M (2009) Cell cycle, CDKs and cancer: a changing paradigm. Nat Rev Cancer 9(3):153–166

Marais A et al (2010) Cell cycle-dependent regulation of the forkhead transcription factor FOXK2 by CDK· cyclin complexes. J Biol Chem 285(46):35728–35739

Mazumdar A et al (2019) The phosphatase PPM1A inhibits triple negative breast cancer growth by blocking cell cycle progression. NPJ Breast Cancer 5(1):1–11

Mehraj U et al (2021) Tumor microenvironment promotes breast cancer chemoresistance. Cancer Chemother Pharmacol 87(2):147–158

Mehraj U et al (2022a) Expression pattern and prognostic significance of baculoviral inhibitor of apoptosis repeat-containing 5 (BIRC5) in breast cancer: a comprehensive analysis. Adv Cancer Biol Metastasis 4:100037

Mehraj U et al (2022b) Expression pattern and prognostic significance of chemokines in breast cancer: an integrated bioinformatics analysis. Clin Breast Cancer

Mehraj U et al (2022c) Adapalene and doxorubicin synergistically promote apoptosis of TNBC cells by hyperactivation of the ERK1/2 pathway through ROS induction. Front Oncol 12

Mehraj U, et al. (2022d) Expression pattern and prognostic significance of CDKs in breast cancer: An integrated bioinformatic study." Cancer Biomarkers(Preprint): 1–15

Mertins P et al (2016) Proteogenomics connects somatic mutations to signalling in breast cancer. Nature 534(7605):55–62

Mintoo M et al (2021) A rohitukine derivative IIIM-290 induces p53 dependent mitochondrial apoptosis in acute lymphoblastic leukemia cells. Mol Carcinog 60(10):671–683

Mir MA (2015) Developing costimulatory molecules for immunotherapy of diseases. Academic Press

Mir MA, Mehraj U (2019) Double-crosser of the immune system: macrophages in tumor progression and metastasis. Curr Immunol Rev 15(2):172–184

Mir MA et al (2020) Targeting different pathways using novel combination therapy in triple negative breast cancer. Curr Cancer Drug Targets 20(8):586–602

Mir MA et al (2022a) Chapter1 - triple-negative breast cancer - an aggressive subtype of breast cancer. In: Mir M (ed) Combinational therapy in triple negative breast cancer. Academic Press, pp 1–28

Mir MA et al (2022b) Chapter 3—current therapeutics and treatment options in TNBC. In: Mir M (ed) Combinational therapy in triple negative breast cancer. Academic Press, pp 61–94

Mir MA et al (2022c) Chapter 8—different drug delivery approaches in combinational therapy in TNBC. In: Mir M (ed) Combinational therapy in triple negative breast cancer. Academic Press, pp 201–230

Mir MA et al. (2022d) Immuno-onco-metabolism and Therapeutic Resistance. Immuno-Oncology Crosstalk and Metabolism, Springer: 45–89

Mir MA et al (2022e) Chapter 4—conventional adjuvant chemotherapy in combination with surgery, radiotherapy, and other specific targets. In: Mir M (ed) Combinational therapy in triple negative breast cancer. Academic Press, pp 95–120

Naidoo K et al (2018) Evaluation of CDK12 protein expression as a potential novel biomarker for DNA damage response–targeted therapies in breast cancer. Mol Cancer Ther 17(1):306–315

NavaneethaKrishnan S et al (2018) Loss of Cdk5 in breast cancer cells promotes ROS-mediated cell death through dysregulation of the mitochondrial permeability transition pore. Oncogene 37(13):1788–1804

Nemet J et al (2014) The two faces of Cdk8, a positive/negative regulator of transcription. Biochimie 97:22–27

Nie L et al (2019) CDK2-mediated site-specific phosphorylation of EZH2 drives and maintains triple-negative breast cancer. Nat Commun 10(1):1–15

Otto T, Sicinski P (2017) Cell cycle proteins as promising targets in cancer therapy. Nat Rev Cancer 17(2):93–115

Paculová H, Kohoutek J (2017) The emerging roles of CDK12 in tumorigenesis. Cell Div 12(1): 1–10

Potapova TA et al (2006) The reversibility of mitotic exit in vertebrate cells. Nature 440(7086): 954–958

Pozo K et al (2013) The role of Cdk5 in neuroendocrine thyroid cancer. Cancer Cell 24(4):499–511

Qayoom H. et al. (2021). Integrating immunotherapy with chemotherapy: a new approach to drug repurposing. Drug Repurposing-Molecular Aspects and Therapeutic Applications, IntechOpen

Quereda V et al (2019) Therapeutic targeting of CDK12/CDK13 in triple-negative breast cancer. Cancer Cell 36(5):545–558

Rahman N, Stratton MR (1998) The genetics of breast cancer susceptibility. Annu Rev Genet 32:95

Sainsbury S et al (2015) Structural basis of transcription initiation by RNA polymerase II. Nat Rev Mol Cell Biol 16(3):129–143

Sánchez-Martínez C et al (2019) Cyclin dependent kinase (CDK) inhibitors as anticancer drugs: recent advances (2015–2019). Bioorg Med Chem Lett 29(20):126637

Santamaría D et al (2007) Cdk1 is sufficient to drive the mammalian cell cycle. Nature 448(7155): 811–815

Santo L, Siu KT, Raje N (2015) Targeting Cyclin-Dependent Kinases and Cell Cycle Progression in Human Cancers. Semin Oncol 42(6):788–800

dos Santos Paparidis NF, Canduri F (2018) The emerging picture of CDK11: genetic, functional and medicinal aspects. Curr Med Chem 25(8):880–888

Schaer DA et al (2018) The CDK4/6 inhibitor Abemaciclib induces a T cell inflamed tumor microenvironment and enhances the efficacy of PD-L1 checkpoint blockade. Cell Rep 22(11): 2978–2994

Schlafstein AJ et al (2018) CDK9 expression shows role as a potential prognostic biomarker in breast cancer patients who fail to achieve pathologic complete response after neoadjuvant chemotherapy. Int J Breast Cancer 2018

Sedlacek H et al (1996) Flavopiridol (L86 8275; NSC 649890), a new kinase inhibitor for tumor therapy. Int J Oncol 9(6):1143–1168

Senft D et al (2018) Ubiquitin ligases in oncogenic transformation and cancer therapy. Nat Rev Cancer 18(2):69–88

Shapiro GI (2006) Cyclin-dependent kinase pathways as targets for cancer treatment. J Clin Oncol 24(11):1770–1783

Shiraishi Y et al (2014) Integrated analysis of whole genome and transcriptome sequencing reveals diverse transcriptomic aberrations driven by somatic genomic changes in liver cancers. PLoS One 9(12):e114263

Singh U et al (2017) Design of novel 3-pyrimidinylazaindole CDK2/9 inhibitors with potent in vitro and in vivo antitumor efficacy in a triple-negative breast cancer model. J Med Chem 60(23): 9470–9489

Sivakumar S, Gorbsky GJ (2015) Spatiotemporal regulation of the anaphase-promoting complex in mitosis. Nat Rev Mol Cell Biol 16(2):82–94

Sledge GW Jr et al (2017) MONARCH 2: abemaciclib in combination with fulvestrant in women with HR+/HER2− advanced breast cancer who had progressed while receiving endocrine therapy. J Clin Oncol 35(25):2875–2884

Sobhani N et al (2019) Updates on the CDK4/6 inhibitory strategy and combinations in breast cancer. Cell 8(4):321

Sofi S et al (2022a) Targeting cyclin-dependent kinase 1 (CDK1) in cancer: molecular docking and dynamic simulations of potential CDK1 inhibitors. Med Oncol 39(9):1–15

Sofi S et al (2022b) Cyclin-dependent kinases in breast cancer: expression pattern and therapeutic implications. Med Oncol 39(6):106

Sofi S et al (2022c) Cyclin-dependent kinases in breast cancer: expression pattern and therapeutic implications. Med Oncol 39(6):1–16

Sotillo R et al (2001a) Wide spectrum of tumors in knock-in mice carrying a Cdk4 protein insensitive to INK4 inhibitors. EMBO J 20(23):6637–6647
Sotillo R et al (2001b) Invasive melanoma in Cdk4-targeted mice. Proc Natl Acad Sci 98(23): 13312–13317
Turner NC et al (2015) Palbociclib in hormone-receptor–positive advanced breast cancer. N Engl J Med 373(3):209–219
Wadhwa B et al (2020) AKT isoforms have discrete expression in triple negative breast cancers and roles in cisplatin sensitivity. Oncotarget 11(45):4178
Wang Y et al (2015) CDK7-dependent transcriptional addiction in triple-negative breast cancer. Cell 163(1):174–186
Wu Y et al (2020) Current therapeutic progress of CDK4/6 inhibitors in breast cancer. Cancer Manag Res 12:3477
You Y et al (2015) Decreased CDK10 expression correlates with lymph node metastasis and predicts poor outcome in breast cancer patients-a short report. Cell Oncol 38(6):485–491
Zhang X et al (2018) Rhomboid domain-containing protein 1 promotes breast cancer progression by regulating the p-Akt and CDK2 levels. Cell Commun Signal 16(1):1–15
Zhang Z et al (2017) HuR promotes breast cancer cell proliferation and survival via binding to CDK3 mRNA. Biomed Pharmacother 91:788–795
Zheng P et al (2019) Long noncoding RNA CASC2 promotes paclitaxel resistance in breast cancer through regulation of miR-18a-5p/CDK19. Histochem Cell Biol 152(4):281–291
Zhou Y et al (2015) Cyclin-dependent kinase 11p110 (CDK11p110) is crucial for human breast cancer cell proliferation and growth. Sci Rep 5(1):1–13
Zhou Y et al (2016a) The emerging roles and therapeutic potential of cyclin-dependent kinase 11 (CDK11) in human cancer. Oncotarget 7(26):40846
Zhou Z et al (2016b) Insights into APC/C: from cellular function to diseases and therapeutics. Cell Div 11(1):1–18

Novel CDK Inhibitors in Breast Cancer 12

Manzoor Ahmad Mir and Tabish Javeed

12.1 Introduction

Cancer is a condition brought on by immortal cells with the capacity to kill people. Ironically, the patient cells must perish for them to survive. Tissue homeostasis is regulated by the division of cells and cell apoptosis (Mir 2021). The development of cancer is significantly influenced by changes in these two physiological systems (Gharbi et al. 2022). New cancer medicines are being developed as a result of extensive research into the components of the cell cycle (Matthews et al. 2022). Due to cyclin overexpression or endogenous CDKI downregulation, CDKs are hyperactive in some cancers (Najm et al. 2019). Based on this knowledge, researchers are investigating whether CDK inhibition might improve cancer therapeutic effectiveness (Mir 2021). Blocking CDKs could be a good way to treat several cancers, including breast cancer (Ettl et al. 2022). Cell cycle control involves three categories of regulatory molecules (cyclins, CDKs, and CDKIs) that indicate how far along in the cell cycle a cell goes (Weinberg 1995; Matthews et al. 2022). Cytokines are the regulatory subunits of an active heterodimer, whereas CDKs are the catalytic subunits (Kciuk et al. 2022). Cyclins do not have any catalytic activity, and CDKs do not do anything if they do not have a cyclin partner. Animal cells have a lot of CDKs. Some of them, such as CDK1, CDK2, and CDK4, have an undeviating mode of action in cell cycle regulation (Satyanarayana and Kaldis 2009). When a binding cyclin activates or inactivates target proteins, CDKs carry out phosphorylation, a common biochemical step that orchestrates organized access into the cell cycle. Later on, in the cell cycle, cyclin-CDK complexes stimulate cyclin-CDK complexes

M. A. Mir (✉) · T. Javeed
Department of Bioresources, School of Biological Sciences, University of Kashmir, Srinagar, J&K, India
e-mail: drmanzoor@kashmiruniversity.ac.in

M. A. Mir (ed.), *Therapeutic Potential of Cell Cycle Kinases in Breast Cancer*,
https://doi.org/10.1007/978-981-19-8911-7_12

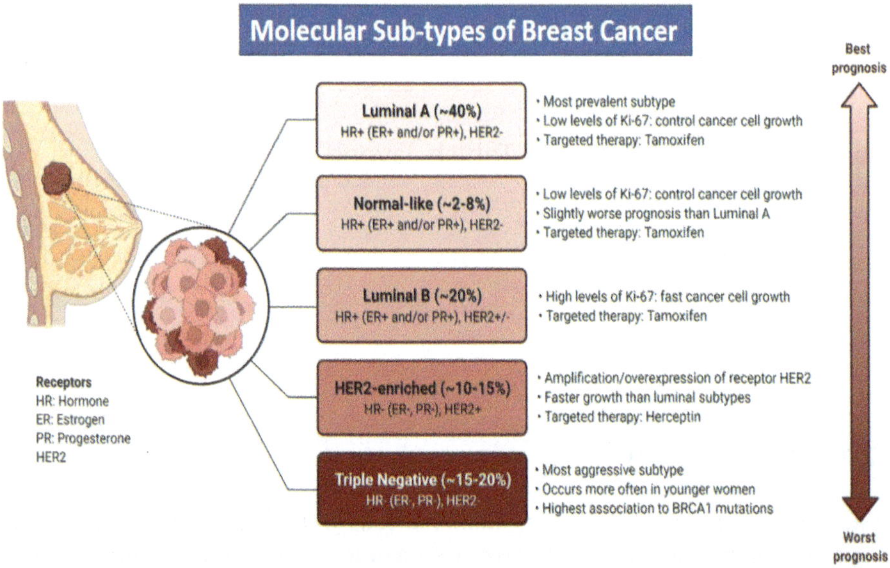

Fig. 12.1 Molecular subtypes of breast cancer

(Akin et al. 2014). Cellular transcription regulation is controlled by a second set of CDKs. They offer a source of nutrition for cancer cells (Mehraj et al. 2021).

A CDKI protein binds to a cyclin-CDK complex and suppresses kinase activity, which happens often during G1 or in response to environmental or damaged DNA signals. The two most important groups of CDKI proteins in the human body are the INK4a/ARF and the Cip/Kip which regulate the cyclin and CDK complexes (Cerqueira et al. 2014). To ensure their survival, all cancers initiate the cell cycle. It is difficult to choose the correct medicine for the right tumor since it is required to first determine which cell cycle regulator controls the cell cycle following an oncogenic event. As a result, animal models have been employed to test whether cell cycle inhibitors are effective against various cancer types. CDKs that are hyperactive or CDK-inhibiting proteins that are defective are seen in a variety of malignancies (Mehraj et al. 2022). CDK4 upregulation or downregulation of p16INK4A causes a loss of cell proliferative control due to increased CDK4 activity, which leads to hyperphosphorylation of Rb proteins and cancer (Willobee et al. 2021). Targeting CDK function to prevent cancer cell overproliferation and employing CDKIs to treat human malignancies makes sense, as per this study. Similarly, CDK2 is downregulated in cancer, which may be caused by oncogenes. One of the substrates that the CDK2 phosphorylates is the retinoblastoma protein (PrB), which triggers the genes essential for S phase via E2F-assisted transcription (Mir 2021).

Breast cancer with different subtypes (Fig. 12.1) being most prevalent across the globe with 276,480 new cases are predicted in 2020 and 42,170 deaths (Santiago-Montero et al. 2020). Patients having metastasis at the time of diagnosis make up about 3–10% of all cases, and they were traditionally regarded to be incurable. The objective of drafting this chapter is to describe how CDKIs are used to treat breast cancer (Cardoso et al. 2020).

12.2 The Cyclin D–CDK4/6-Retinoblastoma Mechanism

Cyclin D1 and CDK4 are necessary for the proliferation of luminal epithelial cells, and these are frequently associated with luminal breast cancers (Fig. 12.2) (Jeselsohn et al. 2010). Cyclin D1 attaches to CDK4, and proteins like p21 maintain the protein complex, making it an active holoenzyme (Sherr and Roberts 1999). The RB protein is then monophosphorylated by the holoenzyme. After CDK2 phosphorylates RB, several genes that aid G1 transition to S phase may be expressed because E2F transcription factors are freed (Joseph et al. 2020). RB phosphorylation, which results in G1 cell cycle arrest and cellular senescence, is blocked by small molecule CDK4/6 inhibitors, which thus stop this process (Choi et al. 2012). There are several approaches to manage CDK4/6's enzyme activity (Knudsen and Witkiewicz 2017). First, certain mitogenic signalling pathways that are involved in breast tumors boost CCND1 expression and/or cyclin D1 protein stability. The activity of CDK4/6 is

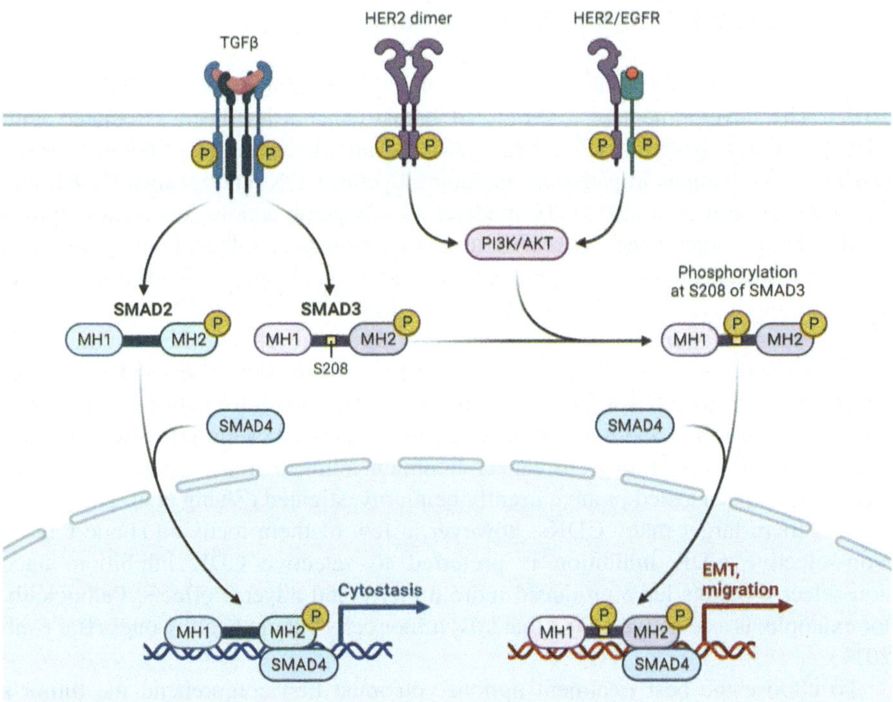

Fig. 12.2 HER2/EGFR signaling pathway in breast cancer

enhanced as a result of this (Vora et al. 2014). Cyclin D1 is required for the growth of breast adenocarcinomas brought on by ERBB2, according to research utilizing animal models, and cyclin D1/CDK4 is likewise connected to the evolution of ERBB2-driven malignancies (Goel et al. 2016). The G1 cell cycle is stopped by the anti-estrogen medications tamoxifen, aromatase inhibitors, and fulvestrant, which block the expression of cyclin D1 (Musgrove et al. 2011). Notably, cyclin D1 may also promote estrogen-independent activation of ER target genes (Zwijsen et al. 1997). Finally, roughly 15% of breast cancers exhibit CCND1 amplification, and these tumors have greater amounts of cyclin D1 protein (Gao et al. 2013). Another strategy to boost CDK4 activity in breast tumors is by this method (Mir and Mehraj 2019).

In spite of the various ways that CDK4 activity may be increased in breast tumors, various cell proteins obstruct CDK4/6 action directly. The INK4 proteins are the main CDK4/6 inhibitors in the body. These proteins bind to the catalytic subunits of CDK4 and CDK6 to inhibit them (Sherr and Roberts 1999). The CDKN2A gene, which produces the protein p16, is deeply deleted in a small subset of breast tumors. Theoretically, these tumors should have higher CDK4/6 activity from the beginning and be more susceptible to CDK4/6 inhibitors. But this issue is still in debate (Finn et al. 2015).

12.3 CDK Inhibitors and Breast Cancer

Breast cancer has been related to a variety of cell cycle abnormalities (Harbeck et al. 2019). The development of a variety of breast cancers has been associated with checkpoint dysregulation (Fig. 12.3). Almost all malignancies, including breast cancer, show changes in pathways including cyclin, CDK, endogenous CDKI, and Rb protein (Akin et al. 2014). To mediate its mitogenic actions, estrogen employs cyclin D1 as target gene. Increased tumor manifestation of cyclin D1 has been connected with decreased recurrence-free endurance and tamoxifen retort in patients having high HER2 expression in multiple studies (Montalto and De Amicis 2020). Tamoxifen's antagonistic action becomes an agonistic effect when cyclin D1 is overexpressed. As a result, cyclin D1 overexpression might be used to predict tamoxifen resistance (Osborne and Schiff 2003). But this information has not been proven to be true yet, and it is still not clear how important cyclin D1 overexpression is for predicting the future. In cancer treatment clinical trials, a vast number of CDKIs have been tested or are currently being investigated (Zhang et al. 2021). The bulk of them target many CDKs; however, a few of them focus on single CDKs. Non-selective CDK inhibition is preferred to selective CDK inhibition since non-selective drugs have produced more toxicity and adverse effects. Palbociclib, for example, is a selective CDKI that kills tumor cells but not healthy ones (Bai et al. 2017).

To choose the best treatment option, you must first comprehend the tumor's biology. CDK4/6 binds cyclin D1 and promotes cell cycle progression by phosphorylating Rb protein and activating E2F transcription factors. The cell cycle

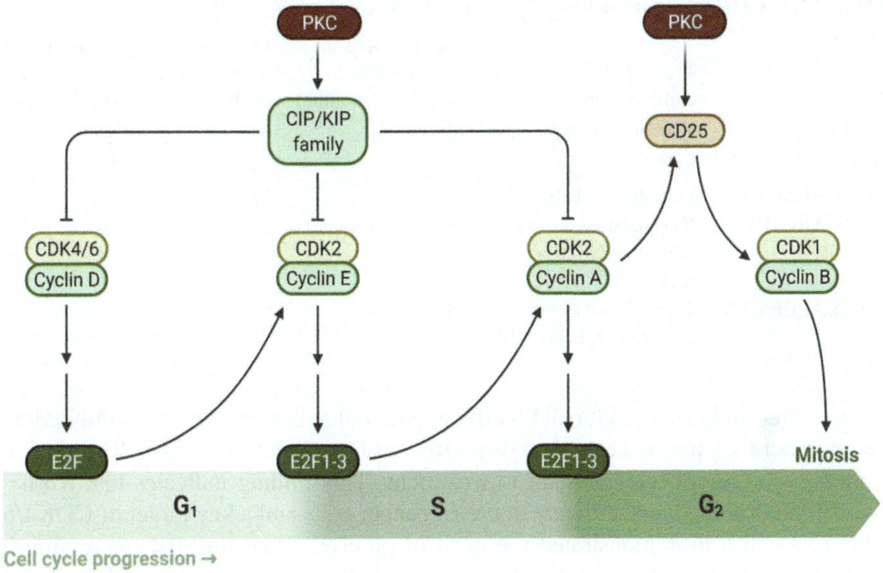

Fig. 12.3 CDK regulation of cell cycle progression

will come to a stop if this well-known route is disturbed in any manner (Narasimha et al. 2014). This technique, however, is insufficient to show that cyclin D1-overexpressing cancers react to any systemic limitation. Cyclin D1 may be required for mantle cell lymphoma cell growth but not survival, and additional resistance mechanisms may exist (Leonard et al. 2012). The research shows that CDK4 and cyclin D1 are needed for the development and persistence of tumor cells in HER2-induced mammary cancer models in mice (Mir et al. 2020). CDK4/6 inhibitors may help tumors that are low-grade and make too much cyclin D1, like ER-positive breast cancer. Cytotoxic or targeted treatments do not shrink tumors. Instead, they stop them from growing (Mir 2021). Palbociclib, an oral CDK4/6 inhibitor, reduces Rb phosphorylation and causes sensitive cell lines to stop growing in the G0/G1 phase by stopping the above process. Palbociclib was first tested on animals to see how well it worked. In vitro, it was shown that Palbociclib alone worked and stopped cell growth (Table 12.1) (Yu et al. 2022). Palbociclib was more effective against non-luminal/basal subtypes, notably those with HER2 amplification in ER-positive cell lines. Rb protein and cyclin D1 levels were greater in sensitive and resistant cells, according to an analysis of variance (Kamdje et al. 2014). Tamoxifen and trastuzumab work better in an in vitro study on breast tumors having HER2-amplified genes, ER-positive luminal subtype is the group of people who are most likely to get better from Palbociclib (Xu et al. 2017). They also found that some drugs, like tamoxifen and trastuzumab, may work better when used together. In this study, it was found that Palbociclib response was linked to higher amounts of cyclin D1 and Rb and lower levels of p16. An experiment was conducted

Table 12.1 CDK inhibitors as first-line treatment options in breast cancer

Study	Treatment combinations	Median free progression survival (months)	Objective response rate (ORR) (%)	Clinical benefit rate (CBR) (%)
PALOMI -1	Letrozole ± Palbociclib	20.2	55	81
PALOMI -2	Letrozole ± Palbociclib	24.8	55	85
MONALEESA-2	Letrozole ± Ribociclib	25.3	53	80
MONARCH-3	Nonsteroidal aromatase inhibitor (NSAI) ± Abemaciclib	Not reached (NR)	59	78
MONALEESA-7	Tamoxifen/NSAI+ goserelin ± Ribociclib	23.8	51	–

out to better understand Palbociclib effectiveness, that indicated it only inhibited the development of tumor cells that were Rb-positive (Ji et al. 2020). Rb-deficient tumors were completely resistant to treatment. This finding indicates that Rb is a crucial factor to enhance efficacy in breast cancer cells and a key target of CDK4/6. This research also demonstrated the need of precisely detecting RB expression in tumors in order to identify individuals who would respond favorably to Palbociclib therapy (Mir 2021).

Palbociclib was used with carboplatin to see if it was helpful when taken with other chemotherapy drugs (Finn et al. 2020). Although it is not the first therapy for breast cancer, carboplatin is used to treat illnesses that have progressed to other regions of the body (Qayoom et al. 2020). In mouse models of metastatic breast cancer, Palbociclib coupled with carboplatin exhibited greater results in comparison to carboplatin alone. Furthermore, when chemotherapy and Palbociclib were combined, there was no increased myelosuppression compared to chemotherapy alone. Another research looked at the effectiveness of Palbociclib in triple-negative breast cancer cell lines when it was combined with doxorubicin (Teo et al. 2017). Rb expression predicted patient reactions to Palbociclib monotherapy or combination therapy. Cancer cells that lack Rb showed no antitumor activity when CDK4/6 was inhibited. Even while Palbociclib prevented doxorubicin from killing cancer cells, the two drugs effectively combined to destroy cancer cells that expressed Rb. Palbociclib and letrozole were combined to check their efficacy and safety (Xu et al. 2017). Twelve postmenopausal women with breast cancer that was ER positive but not HER2 positive responded favorably to the combination (Sofi et al. 2022a, b). In this research, antitumor activity was found. Based on the results of this phase 1 investigation, Palbociclib was developed for a phase 2 experimental test (Cristofanilli et al. 2018). In a phase 2 research, letrozole alone or in combination with Palbociclib was administered to 165 patients with advanced breast cancer. When Palbociclib was supplemented to letrozole, the time it took for the disease to get worse went from 7.5 months to 26.1 months. When Palbociclib and letrozole were used together, progression-free survival in this group of people got a lot better. Only 31% of the people who got letrozole had a reaction that could be seen, but 45% of the people who got the combo had a severe reaction. After 6 months of follow-up, the combined group had a 70% tumor reduction rate, whereas the letrozole alone group had a 44% tumor shrinking rate. Palbociclib enhanced median progression-free survival by roughly 18 months in the long run (Finn et al. 2012).

Ribociclib is another CDK4/6 inhibitor that has in vitro IC50s of around 10 and 40 nM for CDKs 4 and 6, respectively (Mita et al. 2014). In large studies, the most common side effects of grade 3/4 severity were neutropenia and thrombocytopenia, which are also common with Palbociclib (Hortobagyi et al. 2016). However, when Ribociclib is used with endocrine treatment, transaminitis has been seen. It is unclear what caused this hepatotoxicity. Additionally, as shown by electrocardiography, Ribociclib treatment may result in a lengthening of the QT interval. In the clinic, this metric has to be carefully monitored (Mir 2021). The FDA authorized Ribociclib in conjunction with an aromatase inhibitor as the first-line therapy for postmenopausal women with advanced breast cancer that is ER-positive but HER2-negative based on the findings of a phase III study (Hortobagyi et al. 2016). The results of the phase III MONALEESA-7 study, which looked at how well CDK4/6 works in women before and during menopause, were just released (Tripathy et al. 2018). Ovarian function suppression and the oral endocrine medication Ribociclib administered into the participants showed outcomes that were strikingly comparable to those of studies conducted on women who had previously experienced menopause (Finn et al. 2016a, b). This was the first substantial research to combine CDK4/6 inhibition with tamoxifen as an endocrine treatment, and both endocrine regimens significantly increased PFS. Abemaciclib, a potent CDK4/6 inhibitor (Spring et al. 2016). Abemaciclib has a limited effect on CNS cancers, including metastatic lesions (Pernas et al. 2018). Although Abemaciclib CNS penetration is not unique, it is more successful in terms of drug delivery than Palbociclib (Raub et al. 2015). Abemaciclib is less likely to cause hematopoietic toxicity than Palbociclib or Ribociclib; however, the causes underlying this remain unknown (Patnaik et al. 2016). As a result, it may be dosed continuously, with monotherapy commencing at 200 mg bid and endocrine treatment at 150 mg bid. The most frequent Abemaciclib toxicity is Diarrhoea, which usually starts during the first 7 days of treatment and may be treated with loperamide as required. Most people who use Abemaciclib also have an asymptomatic increase in serum creatinine (Mir 2021).

Abemaciclib, like the other CDK4/6 inhibitors, has been shown to be beneficial when combined with hormone treatment in randomized phase III studies. Both ORR and PFS significantly increased when fulvestrant was administered to patients having cancer of advanced type. This was true for individuals who had previously had hormone treatment as well as postmenopausal women (MONARCH 3) (MON-ARCH 2) (Bagegni et al. 2017). Preclinical data, on the other hand, show that Abemaciclib may cause apoptosis in certain breast tumors (Gong et al. 2017) and/or that Abemaciclib may cause an immune response against tumors, which would explain tumor regression and contribute to some of its effectiveness (Deng et al. 2018a, b).

12.4 Combinations of Novel CDK4/6 Inhibitors

12.4.1 Inhibitors of PI3K/AKT/mTOR in Conjunction

As the inhibitors interact, thereby the combining ability to block tumor formation is a smart concept (Vora et al. 2014). It has been shown that ER-positive breast cancer cells that are resistant to CDK4/6 inhibitors as a single agent may be eliminated by combining PI3K and CDK4/6 inhibitors (Herrera-Abreu et al. 2016). This is because suppressing cyclin D1 causes cell death and terminates the cell cycle (Vora et al. 2014). Consequently, downregulating cyclin D1 creates not only a considerable quantity of energy but also a substantial degree of heat both in vitro and in vivo (Herrera-Abreu et al. 2016).

By inhibiting PI3K signalling, CDK4/6 inhibition became more sensitive. This was partially accomplished by decreasing CDK2 activity after mitosis, which caused the cell to enter a "quiescent" state where CDK4/6 activity was required to initiate the cell cycle (Asghar et al. 2017).

12.4.2 Combinations with Immune Checkpoint Inhibitors

Several recent preclinical studies confirm that these inhibitors not only stop tumor cells from dividing but also cause the immune system to attack the tumor (Fig. 12.4). Some of the reasons for this are an increase in the expression of endogenous retroviral sequences in tumor cells, which leads to the production of interferon43 by tumor cells, a decrease in the growth of immune-suppressing regulatory T cells (Goel et al. 2016), and an impact on effector T lymphocytes directly (Deng et al. 2018a, b; Sofi et al. 2022a, b). Immune checkpoint blockade, which specifically targets pathways like the programmed cell death protein-1 (PD-1) axis, further augmented the antitumor immune response induced by CDK4/6 inhibition. This had an additive effect on tumor growth.

In fact, the potential of immunotherapy is that it can help patients with advanced illnesses to have long-lasting responses. It would be significant if CDK4/6 immunotherapy could do this for breast cancer patients. The early research was all based on mouse studies, it is vital to remember that. Nobody is certain how well these simulations reflect the biology of actual cancer. No one is certain that this approach will be successful on people as a result. There were no fresh safety signals throughout the 16-week interim analysis, and an ORR of 14.3% was confirmed (Cardoso et al. 2018). The potential benefits of this method will only be conclusively shown by more developed data and larger, randomized research.

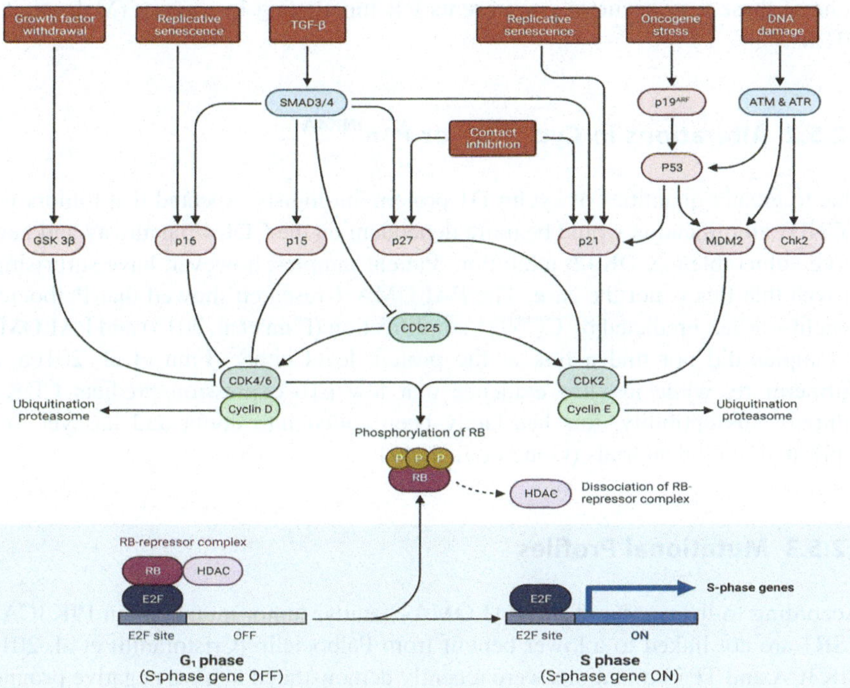

Fig. 12.4 Immune checkpoint inhibitors (G_1 phase/S phase)

12.5 Potential Molecular Biomarkers of CDK4/6 Inhibition Responsiveness and Resistance

Despite the fact that the CDK4/6 pathway is well understood, efforts to develop molecular biomarkers that envisage CDK4/6 inhibitor retort or resistance in human breast tumors have yielded no convincing candidates (Garrido-Castro and Goel 2017). Here is a quick rundown of potential biomarkers.

12.5.1 RB Expression

It is very important that RB is involved in how CDK4/6 inhibitors affect antitumor responses (Finn et al. 2015). So, cancers that do not have a working RB are not likely to respond to CDK4/6 inhibition. Large randomized studies have failed to establish a clear link between RB levels (as evaluated by immunohistochemistry or gene expression) and CDK4/6 inhibitor benefit (Turner et al. 2018). The reasons for this are unknown, although it is possible that it has something to do with the fact that none of these tests accurately represents RB functioning in a tumor (Malorni et al. 2016).

12.5.2 Alterations in Cyclin D1 or $P16^{INK4A}$

Due to greater quantities of cyclin D1 protein, biologists expected that tumors with CCND1 amplification would be more dependent on the CDK4/6 pathway and hence more vulnerable to CDK4/6 inhibition. Patient samples, however, have surprisingly proven that this is not the case. The PALOMA-1 research showed that Palbociclib benefit was not predicted by CCND1 amplification (Finn et al. 2015) and PALOMA-2 samples did not find a link at the protein level either (Finn et al. 2016a, b). Furthermore, while in vitro evidence that low p16 expression predicts CDK4/6 inhibitor susceptibility, this has lately been called into doubt and has yet to be replicated in clinical trials (Gong et al. 2017).

12.5.3 Mutational Profiles

According to the results of the PALOMA-3 study, tumor mutations in PIK3CA or ESR1 are not linked to a lower benefit from Palbociclib (Cristofanilli et al. 2016). PIK3CA and TP53 mutations were recently demonstrated to have negative prognostic effects in patients with breast cancer of advanced type (Hortobagyi et al. 2016).

12.5.4 Other Gene Expression Profiles

To uncover features that predict Palbociclib efficiency, a comprehensive gene expression research of PALOMA-2 and PALOMA-3 tumor tissues is being done (Mir 2021). These outcomes advocate that CDK2 could be a good target to overcome Palbociclib resistance. However, a review of the bigger, first-line PALOMA-2 research did not support this finding, and further trial cohorts are undoubtedly required to support it (Finn et al. 2016a, b). Finally, Palbociclib was shown to be helpful for both luminal A and luminal B malignancies in the PALOMA-2 and PALOMA-3 studies (Finn et al. 2016a, b; Turner et al. 2018).

12.6 Summary

Contemporary advances in the field of CDK inhibitors have enhanced anticancer potencies in vitro and in vivo. However, more research into such field is needed for further development (Fig. 12.5).

Approaches based on structure and bioinformatics should be able to assist tackle challenges that exist in such fields. However, it is not clear yet as to which CDK or combination of CDKs should be utilized in a therapeutic situation. It may be possible

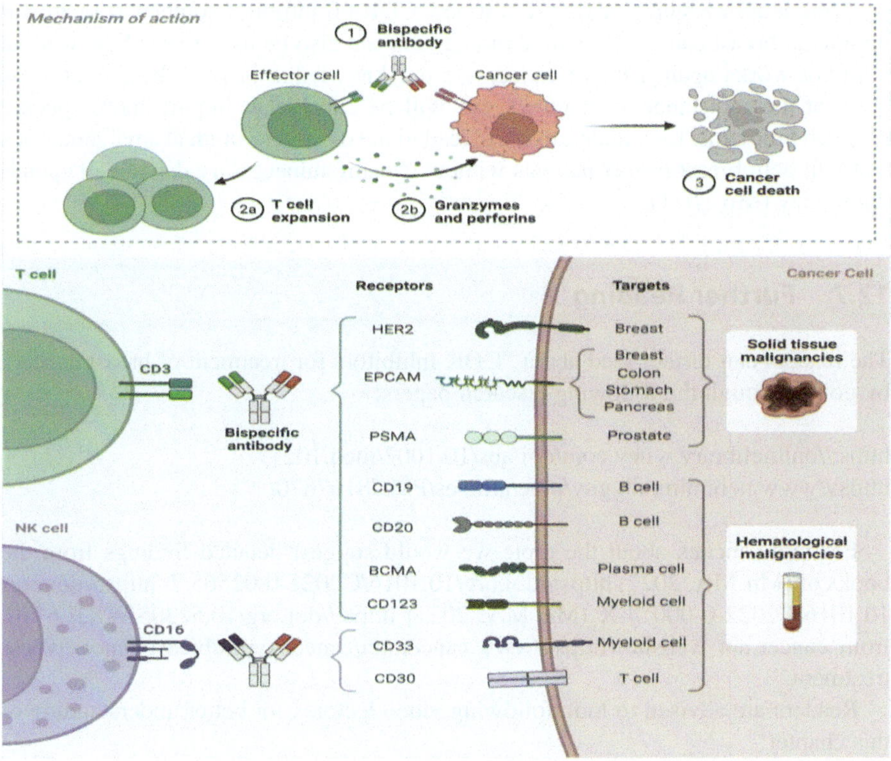

Fig. 12.5 Targeting effector cells with bio-specific antibodies for cancer therapy

to define whether exceedingly selective or broad-range CDK inhibitors are more efficient at treating cancer through clinical (and even preclinical) research (Mir 2021).

Other chemotherapeutic drugs should be researched in addition to CDK inhibitors because they appear to be more effective when used in combination with other chemotherapeutic medications than when used alone. Other targeted medications seem to perform well in combination, thus further clinical research is required in this area (Sofi et al. 2022a, b).

In oncology, targeting CDK or CDKI is a hot topic. Although defining CDKI's role in a particular malignancy is difficult, current clinical trials on these drugs seem to provide sufficient information in this area. Palbociclib, Ribociclib, and Abemaciclib have been proposed as a possible breast cancer treatment. After exhibiting encouraging results in phase 2 trials in terms of progression-free survival, Palbociclib is presently being explored in a phase 3 research for ER-positive breast cancer. If the findings of research on the medication are verified in a large phase 3 trial, Palbociclib along with Abemaciclib will become a new significant targeted therapy for treating breast cancer (Mir 2021).

There is also a sturdy reason to test new CDK4/6 inhibitor combos as treatment options in breast cancers. Potential biomarkers can also be used to see how well an inhibitor works against breast cancer. As our knowledge of how these treatments work at the molecular level grows, we will be capable to inquire more specific research problems. Eventually, this will lead to the discovery of ideal amalgamations that will help breast cancer patients without causing unnecessary damage or spending money (Mir 2021).

12.7 Further Reading

The readers can further read about "CDK Inhibitors for treatment of breast cancer" by going through the following research papers:

https://onlinelibrary.wiley.com/doi/abs/10.1002/med.1021
https://www.ncbi.nlm.nih.gov/pmc/articles/PMC8167670/

For more incites about the topic we would suggest detailed findings from the books of (Mir MA, 2022) https://doi.org/10.1016/C2021-0-02565-7, https://doi.org/10.1016/C2022-0-00074-X (Mir MA, 2021) https://doi.org/10.52305/WXJL6770, from cancer.net website, https://www.cancer.net/cancer-types/breast-cancer/types-treatment.

Readers are advised to look following video lectures for better understanding of this chapter:

https://www.youtube.com/watch?v=BB9jjK7BHkg
https://www.youtube.com/watch?v=vEe3lBduckE

References

Akin S et al (2014) A novel targeted therapy in breast cancer: cyclin dependent kinase inhibitors. J BUON 19(1):42–46

Asghar US et al (2017) Single-cell dynamics determines response to CDK4/6 inhibition in triple-negative breast CancerCDK4/6 inhibition in triple-negative breast cancer. Clin Cancer Res 23(18):5561–5572

Bagegni N et al (2017) Serum thymidine kinase 1 activity as a pharmacodynamic marker of cyclin-dependent kinase 4/6 inhibition in patients with early-stage breast cancer receiving neoadjuvant palbociclib. Breast Cancer Res 19(1):1–11

Bai J et al (2017) Cell cycle regulation and anticancer drug discovery. Cancer Biol Med 14(4):348

Cardoso F et al (2018) 4th ESO–ESMO international consensus guidelines for advanced breast cancer (ABC 4). Ann Oncol 29(8):1634–1657

Cardoso F et al (2020) 5th ESO-ESMO international consensus guidelines for advanced breast cancer (ABC 5). Ann Oncol 31(12):1623–1649

Cerqueira A et al (2014) Genetic characterization of the role of the Cip/kip family of proteins as cyclin-dependent kinase inhibitors and assembly factors. Mol Cell Biol 34(8):1452–1459

Choi S et al (2012) Correlation of the apparent diffusion coefficiency values on diffusion-weighted imaging with prognostic factors for breast cancer. Br J Radiol 85(1016):e474–e479
Cristofanilli M et al (2016) Fulvestrant plus palbociclib versus fulvestrant plus placebo for treatment of hormone-receptor-positive, HER2-negative metastatic breast cancer that progressed on previous endocrine therapy (PALOMA-3): final analysis of the multicentre, double-blind, phase 3 randomised controlled trial. Lancet Oncol 17(4):425–439
Cristofanilli M et al (2018) Predictors of prolonged benefit from palbociclib plus fulvestrant in women with endocrine-resistant hormone receptor–positive/human epidermal growth factor receptor 2–negative metastatic breast cancer in PALOMA-3. Eur J Cancer 104:21–31
Deng J et al (2018a) CDK4/6 inhibition augments antitumor immunity by enhancing T-cell ActivationCDK4/6 inhibition enhances the antitumor T-cell response. Cancer Discov 8(2): 216–233
Deng Y et al (2018b) CDK4/6 inhibitors in combination with hormone therapy for HR+/HER2– advanced breast cancer: a systematic review and meta-analysis of randomized controlled trials. Clin Breast Cancer 18(5):e943–e953
Ettl T et al (2022) The renaissance of cyclin dependent kinase inhibitors. Cancers 14(2):293
Finn R et al (2012) 1000 results of a randomized phase 2 study of Pd 0332991, a cyclin-dependent kinase (Cdk) 4/6 inhibitor, in combination with letrozole vs letrozole alone for first-line treatment of ER+/Her2-advanced breast cancer (Bc). Ann Oncol 23:ii43
Finn RS et al (2015) The cyclin-dependent kinase 4/6 inhibitor palbociclib in combination with letrozole versus letrozole alone as first-line treatment of oestrogen receptor-positive, HER2-negative, advanced breast cancer (PALOMA-1/TRIO-18): a randomised phase 2 study. Lancet Oncol 16(1):25–35
Finn RS et al (2016a) Targeting the cyclin-dependent kinases (CDK) 4/6 in estrogen receptor-positive breast cancers. Breast Cancer Res 18(1):1–11
Finn RS et al (2016b) Palbociclib and letrozole in advanced breast cancer. N Engl J Med 375(20): 1925–1936
Finn RS et al (2020) Biomarker analyses of response to cyclin-dependent kinase 4/6 inhibition and endocrine therapy in women with treatment-Naïve metastatic breast CancerBiomarker analyses of Palbociclib in breast cancer. Clin Cancer Res 26(1):110–121
Gao J et al (2013) Integrative analysis of complex cancer genomics and clinical profiles using the cBioPortal. Sci Signal 6(269):pl1-pl1
Garrido-Castro AC, Goel S (2017) CDK4/6 inhibition in breast cancer: mechanisms of response and treatment failure. Curr Breast Cancer Rep 9(1):26–33
Gharbi S et al (2022) Characterization of the first microRNA in human CDH1 that affects cell cycle and apoptosis and indicates breast cancers progression. J Cell Biochem
Goel S et al (2016) Overcoming therapeutic resistance in HER2-positive breast cancers with CDK4/ 6 inhibitors. Cancer Cell 29(3):255–269
Gong X et al (2017) Genomic aberrations that activate D-type cyclins are associated with enhanced sensitivity to the CDK4 and CDK6 inhibitor abemaciclib. Cancer Cell 32(6):761–776. e766
Harbeck N et al (2019) Breast cancer. Nat Rev Dis Primers 5(1):1–31
Herrera-Abreu MT et al (2016) Early adaptation and acquired resistance to CDK4/6 inhibition in estrogen receptor–positive breast CancerEarly adaption and acquired Palbociclib resistance. Cancer Res 76(8):2301–2313
Hortobagyi GN et al (2016) Ribociclib as first-line therapy for HR-positive, advanced breast cancer. N Engl J Med 375(18):1738–1748
Jeselsohn R et al (2010) Cyclin D1 kinase activity is required for the self-renewal of mammary stem and progenitor cells that are targets of MMTV-ErbB2 tumorigenesis. Cancer Cell 17(1):65–76
Ji W et al (2020) C-myc regulates the sensitivity of breast cancer cells to palbociclib via c-myc/miR-29b-3p/CDK6 axis. Cell Death Dis 11(9):1–13
Joseph C et al (2020) The ITIM-containing receptor: leukocyte-associated immunoglobulin-like receptor-1 (LAIR-1) modulates immune response and confers poor prognosis in invasive breast carcinoma. Cancers 13(1):80

Kamdje AHN et al (2014) Signaling pathways in breast cancer: therapeutic targeting of the microenvironment. Cell Signal 26(12):2843–2856

Kciuk M et al (2022) Metastasis and MAPK pathways. Int J Mol Sci 23(7):3847

Knudsen ES, Witkiewicz AK (2017) The strange case of CDK4/6 inhibitors: mechanisms, resistance, and combination strategies. Trends Cancer 3(1):39–55

Leonard JP et al (2012) Selective CDK4/6 inhibition with tumor responses by PD0332991 in patients with mantle cell lymphoma. Blood 119(20):4597–4607

Malorni L et al (2016) A gene expression signature of retinoblastoma loss-of-function is a predictive biomarker of resistance to palbociclib in breast cancer cell lines and is prognostic in patients with ER positive early breast cancer. Oncotarget 7(42):68012

Matthews HK et al (2022) Cell cycle control in cancer. Nat Rev Mol Cell Biol 23(1):74–88

Mehraj U et al (2021) Tumor microenvironment promotes breast cancer chemoresistance. Cancer Chemother Pharmacol 87(2):147–158

Mehraj U et al (2022) Expression pattern and prognostic significance of CDKs in breast cancer: an integrated bioinformatic study. Cancer Biomarkers(Preprint):1–15

Mir MA (2021) Combination therapies and their effectiveness in breast cancer treatment. NOVA Biomedical Sciences USA, P-1-4010. ISBN: 978-1-68507-195-0. https://doi.org/10.52305/WXJL6770

Mir MA, Mehraj U (2019) Double-crosser of the immune system: macrophages in tumor progression and metastasis. Curr Immunol Rev 15(2):172–184

Mir MA et al (2020) Targeting different pathways using novel combination therapy in triple negative breast cancer. Curr Cancer Drug Targets 20(8):586–602

Mita MM et al (2014) Randomized phase II trial of the cyclin-dependent kinase inhibitor dinaciclib (MK-7965) versus capecitabine in patients with advanced breast cancer. Clin Breast Cancer 14(3):169–176

Montalto FI, De Amicis F (2020) Cyclin D1 in cancer: a molecular connection for cell cycle control, adhesion and invasion in tumor and stroma. Cell 9(12):2648

Musgrove EA et al (2011) Cyclin D as a therapeutic target in cancer. Nat Rev Cancer 11(8):558–572

Najm MA et al (2019) Cyclin-dependent kinase inhibitors as targeted therapy in breast cancer. Biochem Cell Arch 19(1):7–12

Narasimha AM et al (2014) Cyclin D activates the Rb tumor suppressor by mono-phosphorylation. elife 3:e02872

Osborne CK, Schiff R (2003) Growth factor receptor cross-talk with estrogen receptor as a mechanism for tamoxifen resistance in breast cancer. Breast 12(6):362–367

Patnaik A et al (2016) Efficacy and safety of abemaciclib, an inhibitor of CDK4 and CDK6, for patients with breast cancer, non–small cell lung cancer, and other solid tumors. Cancer Discov 6(7):740–753

Pernas S et al (2018) CDK4/6 inhibition in breast cancer: current practice and future directions. Ther Adv Med Oncol 10:1758835918786451

Qayoom H, Bhat BA, Mehraj U, Mir MA (2020) Rising trends of cancers in Kashmir valley: distribution pattern, incidence and causes. J Oncol Res Treat 5(150):2

Raub TJ et al (2015) Brain exposure of two selective dual CDK4 and CDK6 inhibitors and the antitumor activity of CDK4 and CDK6 inhibition in combination with temozolomide in an intracranial glioblastoma xenograft. Drug Metab Dispos 43(9):1360–1371

Santiago-Montero R et al (2020) Novel mathematical model of breast cancer diagnostics using an associative pattern classification. Diagnostics 10(3):136

Satyanarayana A, Kaldis P (2009) Mammalian cell-cycle regulation: several Cdks, numerous cyclins and diverse compensatory mechanisms. Oncogene 28(33):2925–2939

Sherr CJ, Roberts JM (1999) CDK inhibitors: positive and negative regulators of G1-phase progression. Genes Dev 13(12):1501–1512

Sofi S et al (2022a) Targeting cyclin-dependent kinase 1 (CDK1) in cancer: molecular docking and dynamic simulations of potential CDK1 inhibitors. Med Oncol 39(9):1–15

Sofi S et al (2022b) Cyclin-dependent kinases in breast cancer: expression pattern and therapeutic implications. Med Oncol 39(6):1–16

Spring L et al (2016) Targeting the cyclin D–cyclin-dependent kinase (CDK) 4/6–retinoblastoma pathway with selective CDK 4/6 inhibitors in hormone receptor-positive breast cancer: rationale, current status, and future directions. Discov Med 21(113):65

Teo ZL et al (2017) Combined CDK4/6 and PI3Kα inhibition is synergistic and immunogenic in triple-negative breast cancer. Cancer Res 77(22):6340–6352

Tripathy D et al (2018) Ribociclib plus endocrine therapy for premenopausal women with hormone-receptor-positive, advanced breast cancer (MONALEESA-7): a randomised phase 3 trial. Lancet Oncol 19(7):904–915

Turner NC et al (2018) Abstract CT039: cyclin E1 (CCNE1) expression associates with benefit from palbociclib in metastatic breast cancer (MBC) in the PALOMA3 trial. Cancer Res 78-(13_Supplement):CT039-CT039

Vora SR et al (2014) CDK 4/6 inhibitors sensitize PIK3CA mutant breast cancer to PI3K inhibitors. Cancer Cell 26(1):136–149

Weinberg RA (1995) The retinoblastoma protein and cell cycle control. Cell 81(3):323–330

Willobee BA et al (2021) Combined blockade of MEK and CDK4/6 pathways induces senescence to improve survival in pancreatic ductal AdenocarcinomaTargeting MEK and CDK4/6 Signaling improves survival. Mol Cancer Ther 20(7):1246–1256

Xu H et al (2017) Recent advances of highly selective CDK4/6 inhibitors in breast cancer. J Hematol Oncol 10(1):1–12

Yu J et al (2022) Challenges and opportunities in metastatic breast cancer treatments: Nano-drug combinations delivered preferentially to metastatic cells may enhance therapeutic response. Pharmacol Ther 108108

Zhang M et al (2021) CDK inhibitors in cancer therapy, an overview of recent development. Am J Cancer Res 11(5):1913

Zwijsen RM et al (1997) CDK-independent activation of estrogen receptor by cyclin D1. Cell 88(3): 405–415

Targeting CDKs with Other Chemotherapeutic Drugs: A Combinatorial Approach

13

Manzoor Ahmad Mir and Umar Y. Mir

13.1 An Introduction to Cell Cycle

Cell division is an important aspect and fundamental process of various biological activities including tissue regeneration, homeostasis, pathological processes, various physiological activities, and tumorigenesis. The cell cycle is a succession of events by which a cell duplicates its genome and culminates with the formation of two daughter cells. It has two important phases, viz, interphase and M-phase (period of actual division). Interphase is the time during which cell prepares for division by experiencing cell growth and DNA replication. Interphase is divided into 3 phases—G1, S (synthesis phase), and G2 (Fig. 13.1). In the G1 phase the cell prepares for the DNA replication that is regulated by a "restriction point" in mammals. The cell can gain entrance into the cell cycle depending upon various extrinsic (such as growth factors) as well as intrinsic factors (like protein synthesis). Unavailability of any of these factors will lead to cell cycle arrest and entry into Go phase, a dormancy phase. Once the restriction point is passed, the cell is then committed to divide. Cell cycle regulation includes three "check-points"—G1/S, G2/M, and mitotic spindle checkpoints. Several studies have revealed that cyclin-dependent kinases (CDKs) are key drivers of eukaryotic cell cycle that promote the synthesis of DNA and chromosome segregation by phosphorylation of their substrate (Arellano and Moreno 1997, Swaffer et al. 2016). Forfeiture of control over cell cycle is a trademark of cancer. Progression from one phase to next phase of cell cycle requires cyclins and CDKs. Replenishing cell cycle control via CDK inhibition might open some windows for development of targeted cancer therapy.

M. A. Mir (✉) · U. Y. Mir
Department of Bioresources, School of Biological Sciences, University of Kashmir, Srinagar, J&K, India
e-mail: drmanzoor@kashmiruniversity.ac.in

M. A. Mir (ed.), *Therapeutic Potential of Cell Cycle Kinases in Breast Cancer*,
https://doi.org/10.1007/978-981-19-8911-7_13

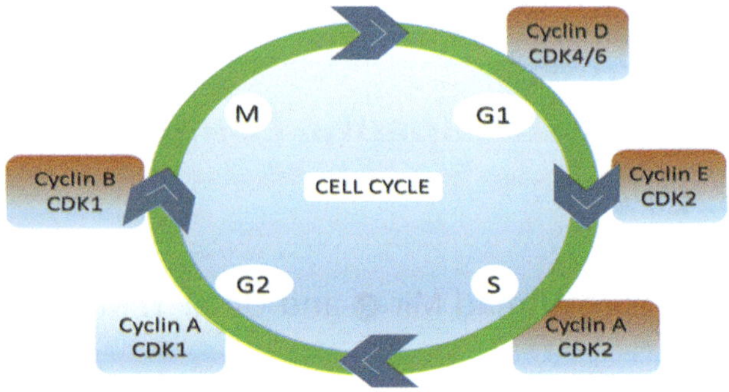

Fig. 13.1 Cyclin and their CDK partners in cell cycle

13.2 CDKs (Cyclin-Dependent Kinases)

CDKs are not just involved in the cell cycle but also have other important functions, such as insulin production, transcription, neuronal functions, and glycogen synthesis (Lim and Kaldis 2013). Until today, based on the homologous sequences in the human genome nearly 21 CDKs and 5 CDK-like genes have been recognized (Malumbres and Barbacid 2009). CDK1 is a critical β determinant of mitotic progression whereas CDK2 is more associated with DNA replication in higher eukaryotes. CDK4/6 promote entry into cell cycle especially G1-S transition in response to various growth signals (Morgan 1997, Mir et al. 2020). Besides cell cycle regulation other cyclin-dependent kinases (CDKs) also participate in cell cycle regulation including CDK7, CDK8, CDK9, and CDK11 (Sofi et al. 2022a, b). CDK7 phosphorylates RNA polymerase II to initiate transcription of genes. CDK8 is a component of the mediator complex which regulates enormous number of genes. CDK9 phosphorylates RNA pol. II and hence stimulates elongation of transcription. CDK11 mostly acts on the splicing machinery. Many evidences suggested that these transcription-associated CDKs can possibly act as probable therapeutic targets for cancer remedy (Qayoom et al. 2022, Sofi et al. 2022a, b). Additional type of CDKs called atypical CDKs are involved in various signaling pathways and post-mitotic functions. These include CDKs 5, 14, 15, 16, 17, and 18. CDK5 plays multiple post-mitotic functions (Mehraj et al. 2022a, b, c, d), while amalgam of CDK14 and Y cyclin is believed to play an important role in β-catenin signaling pathway (Davidson et al. 2009, Jiang et al. 2009). The cell cycle functions in a systematic and sequential manner that is attained by a careful sequence of phases and checkpoints that are regulated by a number of specific proteins that interact with specific cyclin-dependent kinases (CDKs) and maintain the progression of the cell cycle properly (Mehraj et al. n.d., Hanahan and Weinberg 2011). Anomalies in cyclin–CDK–Rb pathway are often associated with breast cancers. Therefore, targeting cancers by inhibiting CDKs will bring into control the cell cycle progression and prove beneficial in the development of targeted cancer therapy (Dickson 2014, Mehraj et al. 2022a, b, c, d). Further the CDKs are not degraded during the course of cell cycle unlike their regulatory counterparts, i.e. cyclins.

As CDKs are critical regulators of transcription, cell cycle, and other important biological functions, inhibitors against these CDKs have come into existence to treat multiple malformations associated with CDK dysfunctions. For over 20 years, many compounds have been developed against the abnormally activated CDKs and some are under clinical trials. Here we will discuss some of CDK inhibitors in synergism with other therapeutics.

13.3 Pan-CDK Inhibitors

CDK inhibitors have been studied since the onset of 1990s, and pan-CDK inhibitors are the first generation of CDK inhibitors including Roscovitine and Flavopiridol, etc. The primary role of the pan-CDK inhibitors is to induce cell cycle arrest and hamper cell proliferation by obstructing CDKs activity. Because of their low specificity and toxicity most of the first-generation pan-CDK inhibitors failed in the clinical trials (Meijer et al. 1997, Mehraj et al. 2021a, b, c). Therefore, second-generation pan-CDK inhibitors such as RGB-286638, AT519, Dinaciclib, TG02, Roniciclib, P276-00, etc., were developed with improved selectivity and reduced cytotoxicity, and have shown good activity in pre-clinical trials with further verification needed. In addition, several pan-CDK inhibitors have gained entry into multiple phase trials and many others have shown promising anti-cancer activity in pre-clinical studies. To improve the efficacy of pan-CDK inhibitors, a number of studies have been performed on drug delivery, most importantly on combination therapy. Despite showing some significant efficacy pan-CDK inhibitors have shown some side effects and safety concerns. However, recent advancements in combination therapeutic approaches are giving positive results in lowering side effects and toxicity of pan-CDK inhibitors. Presently >40 pan-CDK inhibitors are rooming in various clinical stages of development (Mehraj et al. 2021a, b, c, Zhang et al. 2021). In Table 13.1, some pan-CDK inhibitors are represented having their respective targets, phases of clinical trials, and biomarkers of disease. Given below are some pan-CDK inhibitors.

13.3.1 Flavopiridol (Alvocidib)

Flavopiridol is a semisynthetic flavonoid. It is a first-generation pan-CDKI that has been studied on large scale and used in clinical trials. It quenches the activity of multiple CDKs, namely, CDK1, 2, 4, 6, 7, and 9 (Fig. 13.2) (Lin et al. 2010). It also targets positive transcription elongation factor P-TEb (Chao et al. 2000). Pre-clinical study sentenced that flavopiridol has anti-cancer activity in case of prostate cancer, thereby lowers 85% tumor size and improves survival by 30 days (Zeidner and Karp 2015). Despite pre-clinical research, flavopiridol has shown depreciation in efficacy throughout clinical trials of solid tumors and showed some side effects like cardiac dysfunction in AML patients and gastrointestinal toxicity in rodents. This limited the clinical trials of flavopiridol. Therefore, scientists are trying to use it with other drugs, especially paclitaxel, to refine its clinical efficacy (Zeidner and Karp 2015).

Table 13.1 Pan-CDK inhibitors

Name	Alternative name	CDK Target	Structure
Flavopiridol	Alvocidib, L868275, HMR-1275	1, 2, 4, 6, 7, 9	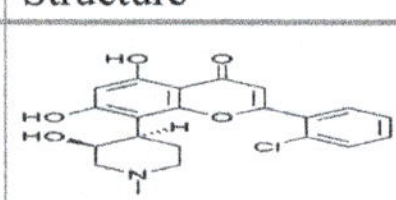
TGO2	SB1317, zotiraciclib	1, 2, 7, 9	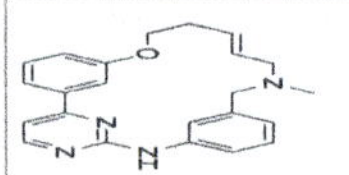
P276-00	Roniciclib hydrochloride, P276	1, 4, 9	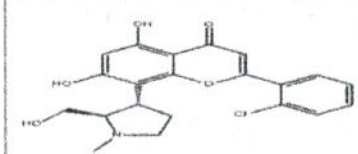
Dinaciclib	SCH 727965	1, 2, 5, 9, 12	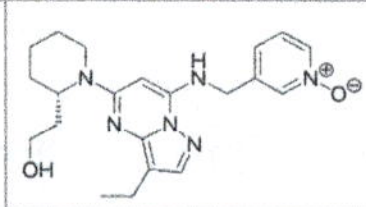
Seliciclib	CYC202, Roscovitine	1, 2, 7, 9	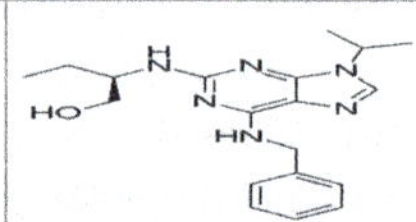
Roniciclib	BAY1000394	1, 2, 4, 7, 9	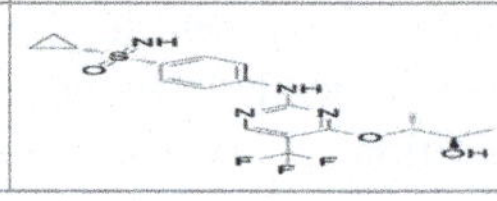

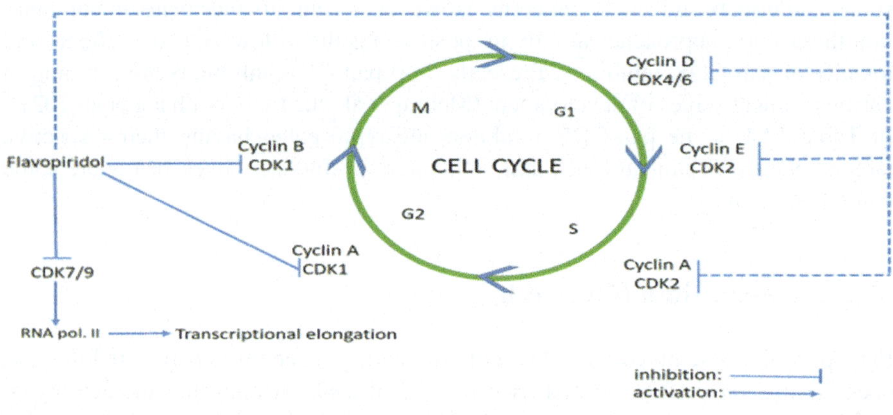

Fig. 13.2 Flavopiridol acting on its target CDK/cyclin complexes

13.3.2 TGO2 (SB1317)

TGO2 is an oral CDK inhibitor which inhibits CDKs including CDK7, CDK2, CDK1, and CDK9 (Fig. 13.3). In pre-clinical studies it has been reported to inhibit glioblastoma cells both alone and in combination (Su et al. 2018). Some studies has

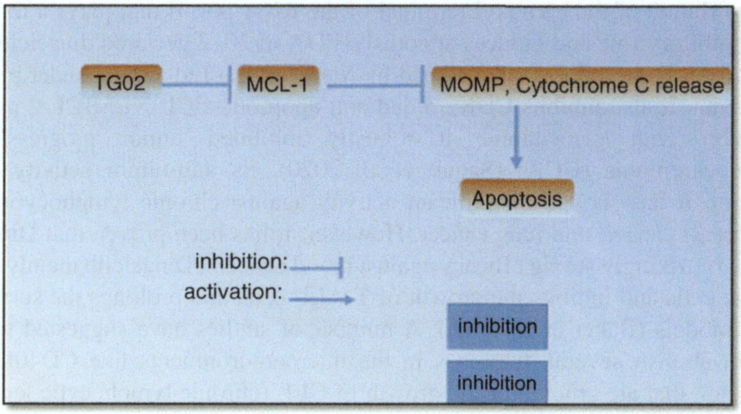

Fig. 13.3 TG02 inducing apoptosis

revealed its activity against hematological diseases and targets a broad range of CDKs and also inhibits JAK/Fit3 activity (Goh et al. 2012). Cultures of cells from AML patients and patients with polycythemia vera are very responsive to TGO2. It has shown positive results against various models of leukemia (Goh et al. 2012, Mehraj et al. 2021a, b, c). Study on patients with AML has shown that TGO2 has inhibitory effect on MCL-1 (inhibitor of apoptosis) and therefore induces apoptosis (Pallis et al. 2017). In addition, in combination TGO2 has shown greater clinical efficacy; for instance, a study has shown that TGO2 when used in combination with carfilzomib (a second-generation proteasome inhibitor) improved the efficacy of refractory multiple myeloma (Ponder et al. 2016).

13.3.3 P276-00

P276-00 is known to inhibit CDK1, CDK4, and CDK9 and has shown activity against cell cycle arrest mostly in the G1 phase and induces apoptosis in the head and neck cancers (Mishra et al. 2013). It has shown anti-tumor activity against mantle cell lymphoma (MCL) cells, with some significant anti-tumor and drug resistant effects in clinical phase II trials. It can repress expression of MCL-1 (anti-apoptotic protein), thereby inducing apoptosis (Shirsath et al. 2012). Research suggests that it has anti-angiogenic activity by inhibiting HIF-1 and lock cancer cells in prostate cancer at G2/M transition (Manohar et al. 2012, Mehraj et al. 2022a, b, c, d). However, the molecular mechanism of P276-00 is not known (Cassaday et al. 2015). Thereby, further verification in this scenario needs further to be evaluated.

13.3.4 Dinaciclib

Dinaciclib is a multi-specific inhibitor inhibiting multiple CDKs such as CDK1, 2, 5, 9, and 12. Dinaciclib mostly obstructs CDK9 activity by preventing the phosphorylation of the carboxyl terminal of the RNA pol. II that plays a transcriptional inhibitory role and induces apoptosis. FDA in 2012 declared dinaciclib as an orphan drug. Dinaciclib was developed by Merck & Co Ltd. and is under phase III clinical trial. It also inhibits CDK2/5 and anti-apoptotic BCL-XL, BCL-2 proteins. In combo with gemcitabine it robustly inhibited tumor progression of cholangiocarcinoma (CCA) (Saqub et al. 2020). Its anti-tumor activity is p53 dependent. It has shown a significant activity against chronic lymphocytic carcinoma, breast cancer, and lung cancer. However, it has been proven that Dinaciclib has shown strikingly strong efficacy against blood cancers. Dinaciclib mainly acts on leukemic cells and inhibits the growth of T-ALL cells and prolongs the survival of clinical models (Baker et al. 2016). A number of studies have suggested that this drug can abolish several cytokines in the microenvironments like CD40L, IL-4, BAFF, etc., that are crucial for the growth of CLL (chronic lymphocytic leukemia) cells (Mehraj et al. n.d., Johnson et al. 2012). Dinaciclib in combination with panobinostat induces MLL-AF9 tumor cell apoptosis with an increase in the survival rate nearly from 33 days to 52 days (Baker et al. 2016).

13.3.5 Seliciclib (Roscovitine/CYC202)

Seliciclib is actually a multipotent pan-cdk inhibitor highly effective against CDK2, CDK7, and CDK9. Seliciclib is being developed by Cyclacel. It has shown antiproliferative activity against lung cancer, multiple myeloma, and hepatocellular carcinoma. It has been shown in vitro to induce by downregulating anti-apoptotic MCL-1 in lung cancers and others. It mainly inhibits transcription depicting its potency against CDK7/9 (Appleyard et al. 2009).

13.3.6 Roniciclib

Roniciclib is another pan-CDK inhibitor that showed its anti-cancer property on thyroid cancer cell lines (Lin et al. 2018). In phase I clinical trials the effectiveness and tolerability have been tested in patients with advanced malignancy. In addition, it was tested in combination with cisplatin on which it has shown synergistic antitumor activity (Syn et al. 2018). Further combination of sorafenib with Roniciclib has reported to have better growth inhibitory effect on xenograft models than alone (Lin et al. 2018). However, in phase II of clinical trials it has shown serious side effects and cytotoxicity, so its usage was terminated (Cho et al. 2018). Therefore, there is a need to re-optimize Roniciclib for its dosage and administration strategy.

13.3.7 PHA-793887

PHA-793887 has been known to hinder the phosphorylation of Rb protein just to obstruct the cell cycle progression (Locatelli et al. 2010). The in vivo results have shown promising effects on leukemic cells, xenograft models, and primary leukemic cell dissemination (Alzani et al. 2010). However, in some patients hepatotoxicity was noticed in phase I clinical trials, due to which the clinical application of it is still under development (Massard et al. 2011).

13.4 Specific CDK Inhibitors

The main drawback associated with pan-CDK inhibitors is their little specificity and increased cytotoxicity on normal cell lines. Pan-CDKs were also associated with pharmacokinetic and administration issues. To improve on these things researchers have successfully developed some specific CDK inhibitors that involve CDK4/6, CDK 7, CDK9, CDK12/13 inhibitors, etc. The discovery of flavopiridol opened a raceway for the evolution of molecules with potent and selective CDK inhibition (Sofi et al. 2022a, b). Each cancer type is known to have a specific landscape of CDK expression, therefore specific inhibitors are expected to specifically provide the therapeutic effect. Multiple specific CDK inhibitors have been elucidated in clinical and pre-clinical studies having anti-tumor effects (Table 13.2). Here we will reveal features of some important specific CDK inhibitors and their role in preventing cancer progression.

13.5 CDK4/6 Inhibitors

CDK4/6 inhibitors are earliest specific inhibitors that are FDA approved for clinical use. These selectively target CDK4/6 with narrow toxicity to normal cells. They arrest G1/S transition and subsequent cell cycle arrest. These inhibitors mainly target cyclin D-CDK4/6-Rb pathway preventing phosphorylation of Rb (tumor suppressor protein), which is otherwise important for cell cycle progression. CDK4/6 inhibition has improved prognosis in various cancers especially breast cancer subtypes; HR-positive and HER-negative breast cancers. Currently three oral agents are approved by FDA as CDK4/6 inhibitors, namely, abemaciclib, palbociclib, Ribociclib (Kwapisz 2017). These three have been approved to use against breast cancer subtypes especially ER+/HER− in combination with anti-hormone therapy (Mehraj et al. 2022a, b, c, d). Besides similarity as CDK inhibitors, these subtly differ in substrate specificity and pharmacodynamics. Palbociclib almost equally inhibits CDK6 and 4, while ribociclib and abemaciclib are more effective to CDK4 and less to CDK6. Ribociclib and palbociclib have shown prolonged half-life as compared to abemaciclib. Research studies have shown that CDK4/6 inhibitors also suppress cancer growth by bringing on anti-tumor immune responses and senescence regulation of cell metabolism (Deng et al. 2018).

Table 13.2 Specific CDK inhibitors

Name	Alternative name	Target CDK	Structure
Palbociclib	PD-0332991	4, 6	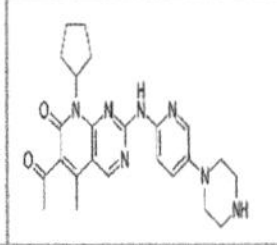
Abemaciclib	Verzenio	4, 6, 1, 2, 9, 14, 16	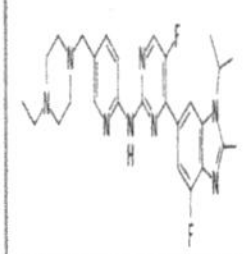
Ribociclib	LEE O11	4, 6	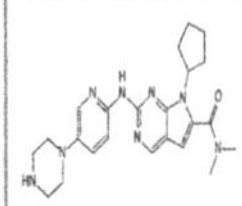
BS-181		7	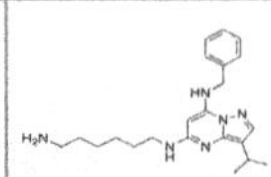
THZ1		7, 12, 13	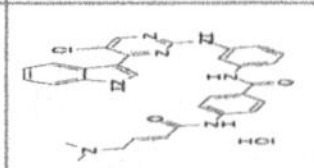
YKL-5-124		7	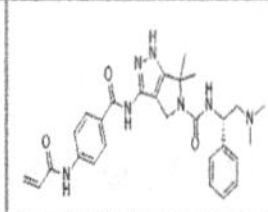
SY-1365	Mevociclib	7	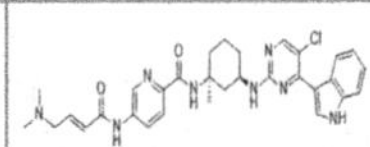
SY-5609		7	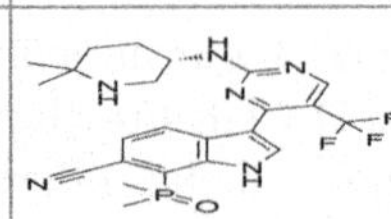
Fadraciclib	CYC065	9	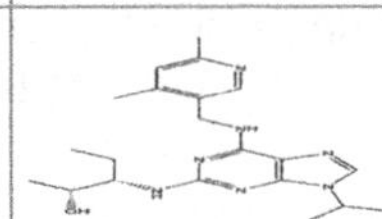
AZD4573		9	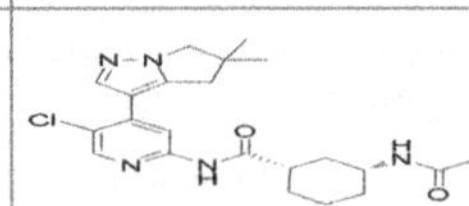
CDKI-73	LS-007	9	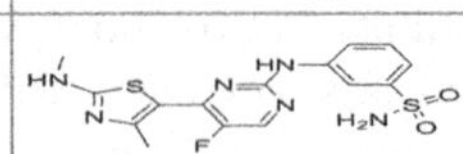
BAY1143572	Atuveciclib	9	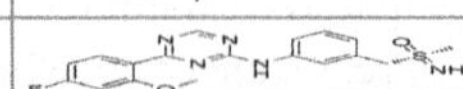
MC180295		9	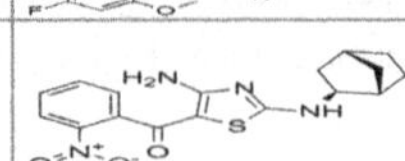

(continued)

Table 13.2 (continued)

Name	Alternative name	Target CDK	Structure
AT7519		1, 2, 4, 6, 7, 9	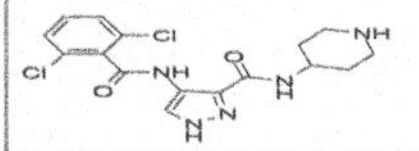
RGB-286638		1, 2, 3, 4, 5, 6, 7, 9	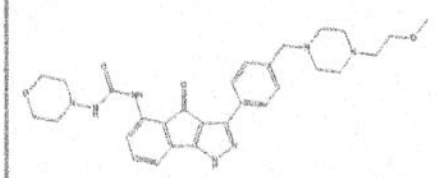
THZ531		12	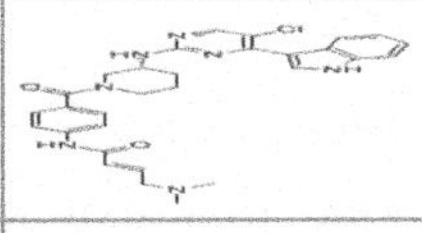
SR-4835		12	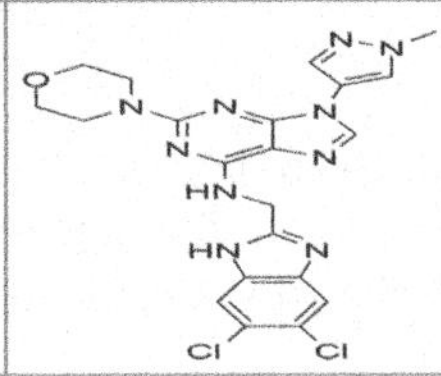

13.5.1 Palbociclib (PD-0332991)

Palbociclib was the primary CDK4/6 inhibitor to be approved as a cancer therapy. It is a popular CDK4/6 inhibitor sold under brand name Ibrance, developed by Pfizer. In vitro studies have shown that palbociclib is effective against colon cancer, lung cancer, myeloma, and particularly breast cancers (Fry et al. 2004, Qayoom et al. 2021). Rb mutant patients are insensitive to palbociclib treatment, depicting that it has role in blocking phosphorylation of Rb to sequester E2F and thus blocks cell from bypassing restriction checkpoint (Sofi et al. 2022a, b). However, patients taking this drug experience neutropenia (low number of neutrophils). Combination therapeutic approach has always yielded better results. In this case, palbociclib has been used in combination with conventional drugs like tamoxifen, trastuzumab, fulvestrant, paclitaxel, and letrozole against breast cancer subtypes. These combinations and a combination anti-hormone therapy work in a synergistic way to inhibit tumor growth (Malorni et al. 2018) .

13.5.2 Abemaciclib

Abemaciclib, sold under brand name Verzenio, is a medication produced by Eli Lilly and acts as an inhibitor of CDK4/6. It is more potent against CDK4 and has shorter half-life as compared to other CDK4/6 inhibitors (Fig. 13.4). It is a quite different one and inhibits other multiple kinases as well like CDK1/2/5/9/14/16 (Fry et al. 2004). Therefore, it has the ability to stop cell cycle in G1 and G2 as well. Various studies have demonstrated that abemaciclib can also induce tumor regression and tumor cell apoptosis as well. However, abemaciclib is associated with neutropenia

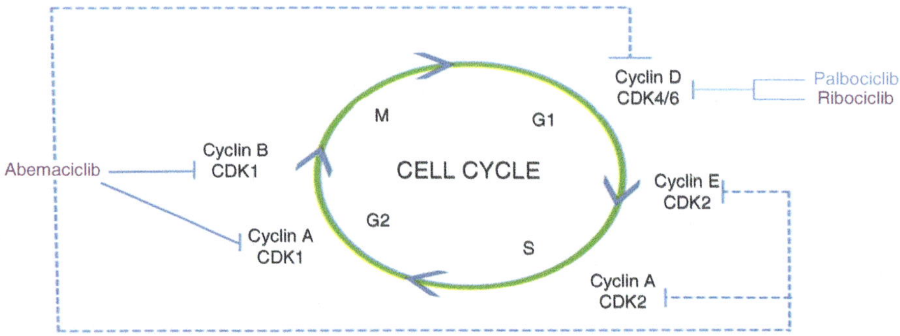

Fig. 13.4 Abemaciclib inhibition on its targets

and GI-toxicities (Rugo et al. 2021). Abemaciclib has shown promising results against cell line deficient in Rb which were resistant to palbociclib and Ribociclib. It is the only specific inhibitor that has been approved for monotherapy against breast cancers.

13.5.3 Ribociclib (LEE 011)

Ribociclib, a CDK4/6 inhibitor, developed by Novartis. It is structurally similar to palbociclib and is used for the treatment of certain breast cancers. Like abemaciclib and palbociclib, it is cytostatic (any agent that slows or stops the growth of cells). It is more powerful against CDK4 than CDK6 like abemaciclib. Neutropenia is associated with its administration. It is used in combination with aromatase inhibitor, letrozole, to antagonize HR+ and HER− breast cancers. Thus, the approved CDK4/6 inhibitors currently trending improved overall survival of breast cancer patients when combined with conventional therapies. Although they all inhibit CDK4/6 but they do differ in substrate specificity, pharmacokinetic properties. Using these drugs in combination therapies might help in reducing side effects (Hortobagyi 2018).

13.6 Specific CDK7 Inhibitors

CDK7 has a bimodal role in progression of cell cycle and transcriptional activation. It is an important component of TFIIH, a general transcription factor, and mediates RNA pol. II phosphorylation at gene regulatory sequences to permit transcription. CDK7 levels are elevated in certain cancers. These findings suggest CDK7 as a potential cancer therapeutic target. Clinical work throughout the world has elucidated number of selective CDK7 inhibitors till date that are in multiple phase trials including LDC4297, BS-181, QS1189, ICEC0942, THZ1, YKL-5-124, THZ2 (Olson et al. 2019).

BS-181 is first specific CDK7 inhibitor with ability to reduce phosphorylation of CDK7 targets thereby inhibiting cell proliferation. But because of its poor cell permeability it has been ruled out of clinical candidature (Ali et al. 2009). Searching for alternatives leads to development of highly selective CDK7 inhibitor ICEC0942. It is an oral cdk7 inhibitor with improved drug properties. Clinically the drug is going through multiple phase trials as monotherapy and in combinatorial approach against prostate cancer, breast cancers, and triple negative breast cancer (TNBC). Its promising pharmacokinetic properties and metabolism make it a good clinical candidate (Patel et al. 2018).

THZ1 is another CDK7 inhibitor with strong anti-tumor and anti-transcriptional activity. It also inhibits activity of CDK12/13 in addition to CDK7. Even research suggests that its anti-tumor activity relies on inhibition of CDK12 and CDK13 (Olson et al. 2019). YKL-5-124 is a potent, highly selective and covalent inhibitor and does not have inhibitory effect on CDK12/13 unlike THZ1. Pre-clinical research studies have demonstrated that YKL-5-124 generates anti-cancer immune response and induces genomic instability (Olson et al. 2019). SY-1365 is another CDK7 selective inhibitor developed by Syros pharmaceuticals derived from THZ. It has entered in phase trials for treatment of ovarian and breast cancers. SY-5609 is another selective CDK7 inhibitor with strong anti-tumor activity, developed by Syros pharmaceuticals. This drug in association with fulvestrant has been approved for clinical trials against TNBC, HR+ and HER2 negative breast cancers. Another CDK7 inhibitor LY3405105 has entered into phase trials against multiple solid advanced tumors.

13.7 CDK9 Inhibitors

CDK9 plays a crucial role in controlling transcription and ensures transcriptional homeostasis. It is normally associated with P-TEFb, a transcriptional elongation factor, modulating gene transcription by phosphorylating C-terminal domain of RNA pol. II. CDK9 dependent phosphorylation is important for processing and maturation of mRNA. CDK9 has been recognized as a valuable target for sorting cancers in which transcription is dysregulated. Moreover, dysregulation of CDK9 pathway has been observed to be associated with various solid and hematological malignancies, making it a valuable anti-cancer target. Functioning of CDK9 has advanced since its discovery and lead to the development of potential molecules as anti-malignants. Numerous selective CDK9 inhibitors have been recognized with strong anti-proliferative property such as AZD-4573, MC180295, CDKI-73, Fadraciclib, etc., some of which have entered clinical development (Borowczak et al. 2022).

Fadraciclib (CYC065), a CDK9 inhibitor with anti-proliferative activity and inhibition of CDK9 mediated transcription (Fig. 13.5). It actually reduces RNA pol. II phosphorylation via CDK9 inhibition and has shown potent anti-tumor activity in multiple cases of malignancies especially neuroblastoma. In addition to CDK9 inhibition, it also inhibits CDK2. Thus, it can also block cell cycle

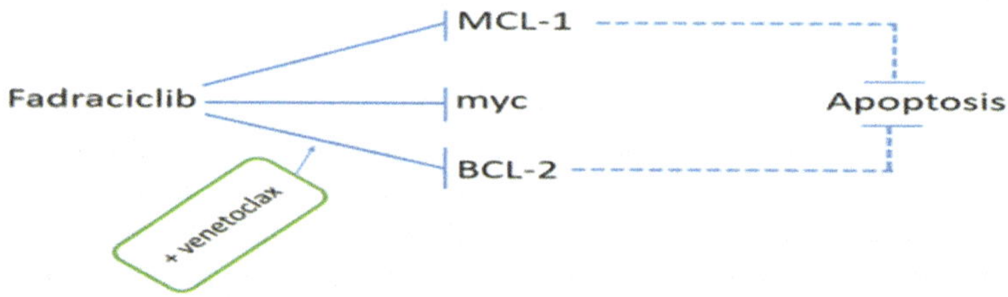

Fig. 13.5 Fadraciclib inducing apoptosis

progression. Fadraciclib represses transcription of Myc oncogene and anti-apoptotic protein Mcl-1. Thus, it may prove beneficial against Triple Negative Breast Cancer (TNBC) in which Myc is over amplified (Poon et al. 2020). Regarding combination therapy, fadraciclib has been used as a combo with other drugs resulting in a more robust anti-tumor response. Like, it has been used with temozolomide against neuroblastoma. Its combination with BCL2 inhibitors including Venetolax has shown promising efficacy against hematological malignancies especially acute myeloid leukemia (AML) (Frame et al. 2020).

AZD4573 is a highly potent selective inhibitor of CDK9 which downregulates anti-apoptotic MCL-1, demonstrating a rapid induction of apoptosis in tumor cells. Although it can block other CDKs as well but its inhibitory effect is many times selective for CDK9. Currently its anti-cancer property is being demonstrated against hematological malignancies (Cidado et al. 2020) .

CDKI-73 is a synthetic CDK9 inhibitor, indirectly represses transcription of E2F via cdk9 inhibition. Thus, blocking G1/S transition. Further research suggests that CDKI-73 exerts pro-apoptotic effect by downregulating anti-apoptotic BCL-2 and MCL-1 via CDK9 inhibition (Borowczak et al. 2022). BAY1143572 (atuveciclib, a benzyl sulfoximine) is another CDK9 inhibitor inhibiting proliferation of cancerous cells even at micromolar concentrations. It reduces RNA pol. II phosphorylation via cdk9 inhibition, lowering transcription of BCL2 and MCL1. Thus, inducing apoptosis also. It has exerted anti-tumor activity against TNBC, lymphomas, esophageal cancers and is currently being investigated in patients with advanced hematological malignancies. NVP-2 is an amino pyrimidine based selective CDK9 inhibitor. It has displayed anti-proliferative activity against certain leukemia's by downregulating MCL-1 and induce apoptosis. 6-bromoindirubin-3'-monoxime is another synthetic CDK9 inhibitor and CDK2 inhibitor. MC180295 is another potent selective CDK9 inhibitor that can reverse the silenced tumor suppressor genes normally silenced during malignancies (Lücking et al. 2017). It also can downregulate myc and MCL1 demonstrating significant anti-cancer activity.

Sangivamycin analogs, also called sangivamycin like molecules (SLM), are nucleoside analogs of anti-tumor and anti-retroviral compound Sangivamycin originally derived from *Streptomyces rimosus*. These analogs possess the same anti-tumor property as demonstrated in pre-clinical study of colon cancer (Cavins et al. 1967). SLM6 is one of the analogs which inhibits CDK9 dependent phosphorylation of RNA pol. II. It can also inhibit CDK1/2, but is more potent against cdk9 as its inhibition on it results in apoptosis of cancer cells. Another analog SM3 appears to inhibit proliferation of multiple myeloma cells but is neutral toward in malignancies (Borowczak et al. 2022). AAP1742 is another CDK9 inhibitor which decreases RNA pol. II phosphorylation and induces cell apoptosis by downregulating XIAP, BCL-2, and MCL-1. AT7519 is a versatile substance that can inhibit CDK1, CDK2, CDK4, CDK6, CDK7, and CDK9. RGB-286638 is a non-selective inhibitor of CDK1, 2, 3, 4, 5, 6, 7, 9 but increasing potency toward CDK9. It causes robust inhibition of transcription and induction of apoptosis. But its clinical development needs further investigation (Borowczak et al. 2022). SNS-032 is another CDk inhibitor which strongly inhibits CDK9 and weakly inhibits other kinases like CDK2 and CDK7.

Thus, CDK9 inhibitors are broad spectrum inhibitors showing anti-cancer activity by repressing number of anti-apoptotic proteins and downregulating transcription. CDK9 inhibitors may complement current treatment protocols to improve efficacy. Like, these inhibitors have been used in combination therapy with other drugs mainly Doxorubicin, Cyclophosphamide, Rituximab, Bortezomib, BET-inhibitors, etc., just to enhance strength of current treatment regimes. Further most of first-generation CDK9 inhibitors are multipotent targeting multiple CDKs with toxicity. Novel selective CDK9 inhibitors are designed to enhance efficacy and compliance of current treatments. Although there is improvement in treatment efficacy still there are certain pharmacodynamic obstacles that need resolution.

13.8 CDK12 Inhibitors

CDK12 is a transcription related CDK regulating RNA splicing, translation, transcription, and DNA Damage Response (DDR). In association with cyclin k CDK12 regulates transcriptional elongation by phosphorylating RNA pol. II (Blazek et al. 2011). Recently, growing evidences demonstrate involvement of CDK12 in cancers. This may be attributed to its functional role that it does in transcription and other roles. Mutation or anomalic expression of CDK12 has been detected in certain cancers, such as prostate cancer, breast cancer, ovarian cancer, etc. Research suggests that it shows both tumorigenic and tumor suppressive effect. Overexpression of CDK12 in certain cancers lead to malignancy as seen in HER2+ breast cancer, while loss of CDK12 has tumor suppressive effect because of downregulation of DDR genes in this case. Recently, it has been validated as a potential therapeutic target of cancer. In recent years various inhibitors have been developed such as Dinaciclib, THZ1, THZ531, SR-4835, etc.

THZ1 is CDK12/13 and CDK7 inhibitor with strong anti-malignant activity. It inhibits myc oncogene expression and has shown promising results in Ewing Sarcoma, ovarian cancer, and neuroblastoma. THZ531 is a THZ1 based CDK12 inhibitor that damages DNA damage repair pathway (Iniguez et al. 2018). Collaboration of THZ531 with Sorafenib has showed positive results in hepatocellular carcinoma. Dinaciclib is a multi-specific inhibitor inhibiting multiple CDKs including CDK12. It inhibits phosphorylation of RNA pol. II. SR-4835 is another CDK12 inhibitor that downpours DNA Damage repair genes and induces apoptosis. Studies have shown that loss of CDK12 enhances anti-tumor effect of PARP (Poly-ADP Ribose Polymerase) inhibitors and cell cycle inhibitors checkpoint like Chk1 inhibitor. CDK12 inhibitors in collaboration with PARP inhibitors and Chk1 inhibitor could collaborate to target cancers. Like, combination of PARP inhibitors and SR-4835 has shown synergistic effect in TNBC (Quereda et al. 2019).

13.9 Natural Compounds Acting as CDK Inhibitors

Natural compounds from marine species, plants, and microorganisms have been recognized as non-specific CDK inhibitors. They lead to discovery of novel compounds with favorable biomedical applications like anti-cancer treatment. Several organisms have been studied as sources of these anti-cancer drugs and have proven effective for destroying tumors in various clinical settings. Here we will reveal some of naturally derived substances from marine species, plants, and microorganisms with CDK inhibitory properties (Table 13.3).

Olomoucine, a derivative of plant cytokinin, is one of the first natural CDK inhibitors obtained from cotyledons of Radish (*Raphanus sativus*). Primarily it was shown to inhibit CDK1 then it was discovered to inhibit CDK2, 5, 7, and 9 also. Olomoucine discovery resulted in the preparation of variety of analogs such as Roscovitine. Roscovitine (Seliciclib) is a synthetic drug inhibiting CDK2/7/9 exhibiting strong anti-proliferative effects (Vermeulen et al. 2002). Butyrolactone-1 is known to inhibit CDK1 and CDK2. It was originally isolated from *Aspergillus terreus*. Butyrolactone-1 arrests both G1 and G2/M along with apoptosis induction (Kitagawa et al. 1994). UCN-01 (7-Hydroxystaurosporine) is a natural CDK1/2 inhibitor isolated from *Streptomyces species*. It arrests cell cycle in G2 by inhibiting Cyclin B-CDK1 (MPF) (Diaz-Padilla et al. 2009). Hymenialdisine lactam is another natural CDK inhibitor derived from *Axinella sp,* a marine sponge. It is an inhibitor of GSK-3B, CDK1/5, MEK-1, and Casein Kinase 1 (Wan et al. 2004).

Chromone alkaloids and flavoalkaloids are natural compounds known to have anti-cancer properties. One of the alkaloid Rohitukine isolated from a flowering plant *Dysoxylum binectariferum* led to evolution of Flavopiridol, first pan-CDK inhibitor known to inhibit multiple CDKs [66]. Indirubins are the first natural compounds used by humans as CDK inhibitors. These are derived from gastropod mollusks, indigo-producing plants, and bacteria. Indirubins were principal ingredients of traditional Chinese medication to cure Chronic myelogenous leukemia (CML). Indirubins are renowned inhibitors of CDK1/2 (Kumar et al. 2014). Based on indirubins, synthetic substituted indirubins have been prepared including 6-bromoindirubin and 6-bromoindirubin-3'-monooxime (CDK2/9 inhibitor). Wogonin is a mono-flavonoid natural CDK inhibitor derived from Scutellaria radix. It is known to inhibit CDK2/4 and preferentially CDK9. It reduces transcription of anti-apoptotic MCL-1 by inhibiting CDK9 dependent phosphorylation of RNA Pol. II. Wogonin has shown anti-proliferative properties in experiments on Hodgkin's lymphoma, melanoma, hepatocellular and pancreatic carcinoma. Fangchinoline is an alkaloid based natural CDK inhibitor derived from roots of

Table 13.3 Natural compounds acting as CDK inhibitors

Name	Source	CDK Target	Structure
Olomoucine	*Raphanus*	, 2, 5, 7, 9	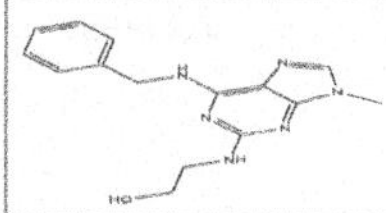
Butyrolactone-1	*Aspergillus terreus*	1, 2	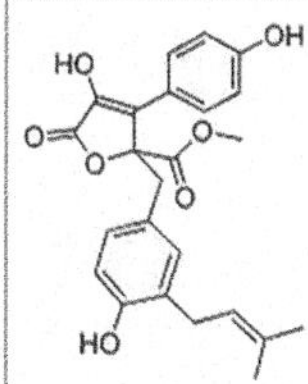
UCN-01	*Streptomyces species*	1, 2	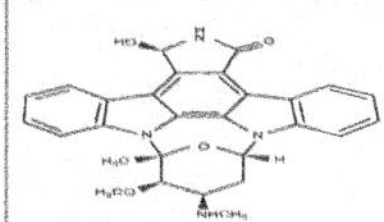
Hymenialdisine	*Axinella sp.*	1, 5, GSK-3B, CK1, MEK	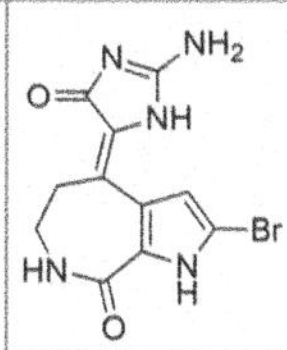
Rohitukine	*Dysoxylum binectariferum*	2, 9	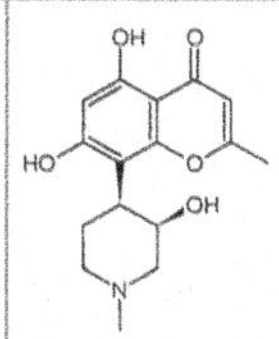
Wogonin	*Scutellaria radix*	*2, 4, 9*	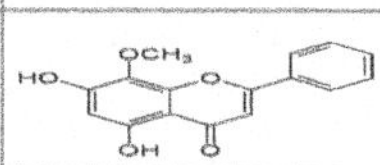
Fangchinoline	*Stephania tetrandra*	4, 6	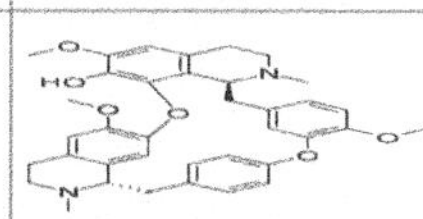
Berberine	*Berberis spp.*	2, 4	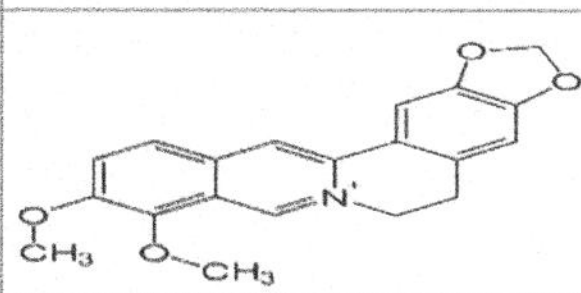
Indirubin	*Bacteria*	*1, 2, 4, 5*	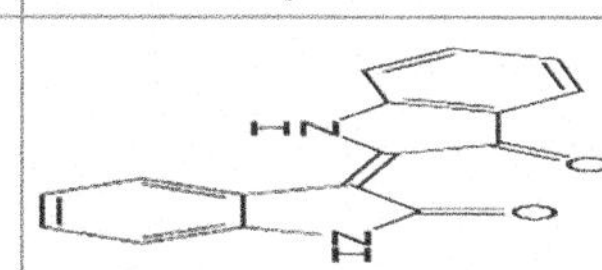

(continued)

Table 13.3 (continued)

Name	Source	CDK Target	Structure
EGCG	Green tea	2, 4, 6	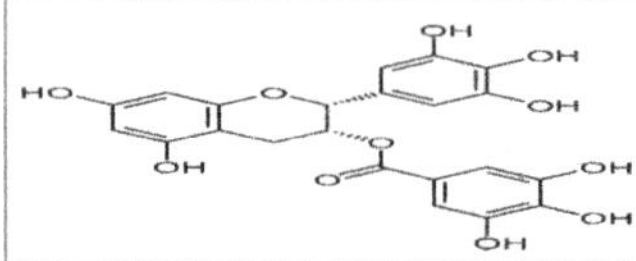
Proanthocyanidin	*Vitis vinifera*	2, 4, 6	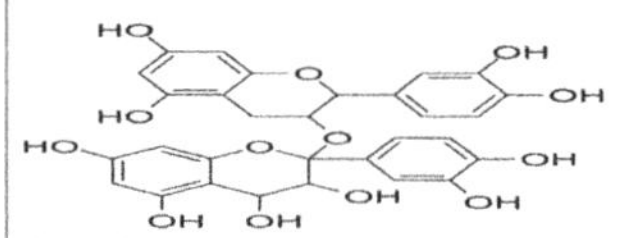
Indole-3-carbinol	Brassicaceae	2, 4, 6	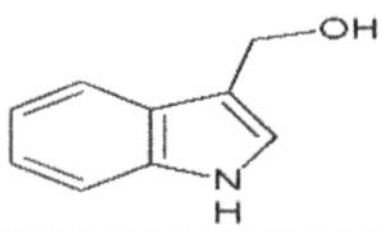
Apigenin	Fruits and vegetables	2, 4, 6	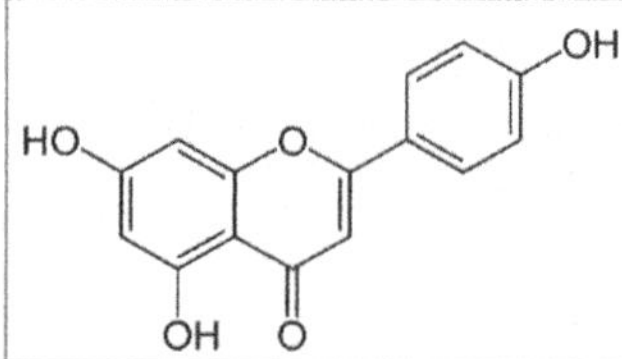
Resveratrol	Grapes	2, 4, 6	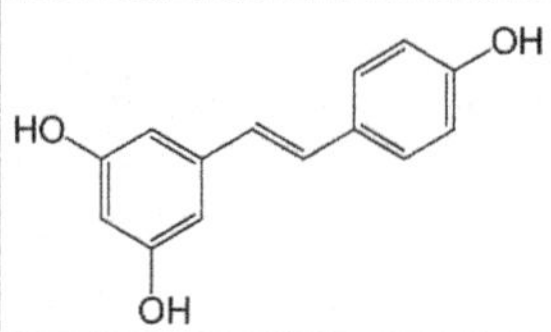
Quercetin	Buckwheat	2, 7	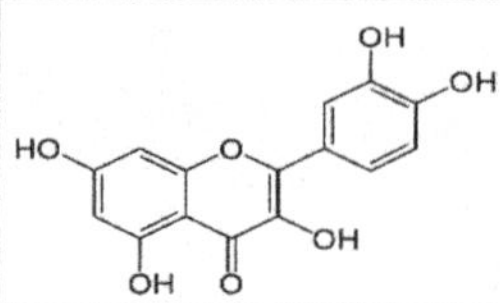

vine plant, *Stephania tetrandra.* It is known to inhibit kinase activities of CDK2, 4, and 6 (Birdsall 1997). Berberine is another alkaloid CDK inhibitor obtained from berberis species. It decreases the levels of CDK2/4.

Triterpenoid compounds called Limonoids including isoobacunoic acid, limonexic acid, isolimonexic acid, and methyl nomilinate derived from citrus species found to have anti-proliferative activity. Methyl nomilinate particularly inhibits CDK4/6 and blocks cell in G1 (Kim et al. 2012). Polyphenols derived from leaf extract of mulberry plant is known to have anti-tumor activity by decreasing activities of CDK4/6/ (Chan et al. 2010). Acetylbritannilactone, a naturally occurring CDK 2/4/6 inhibitor derived from British yellowhead, *Inula britannica.* Flavonoids such as quercetin, isoquercetin, and rutin, from buckwheat seeds and bran downregulate CDK2/7. Polyphenol epigallocatechin-3-gallate (EGCG), a component of green tea causes cell cycle arrest by inhibiting CDK2/4/6 (Shankar et al.

2007). Another polyphenol, namely, Resveratrol derived from grapes well known for its inhibition on cell cycle progression by reducing the levels of CDK2/4 and 6. Proanthocyanidins, from *Vitis vinifera*, have shown to inhibit cell proliferation of epidermoid carcinomas decreasing kinase activities of CDK2/4 and 6. Indole-3-carbinol from Brassicaceae family has shown anti-proliferative activity against colon cancer, endometrial cancer, prostate cancer, and breast cancer. It induces cell cycle arrest by inhibiting CDK2/4/6. Propolis or bee glue is a resinous mixture derived from *Apis mellifera* has shown to inhibit tumor growth by downregulating CDK4. Apigenin, a dietary flavonoid abundant in fruits and vegetables has shown to inhibit tumor growth in mice models. It inhibits CDK2/4/6 (Bailon-Moscoso et al. 2017).

Thus, it is fair to say that naturally occurring compounds derived from plants, microorganisms, and other organisms have tendency to inhibit tumor growth and are modulators of cell cycle regulation. Moreover, there is a certain need of scrutiny of possible targets of these naturally derived compounds and their analogs to increase selectivity and specificity of treatment. To address resistance of tumor cells against these natural compounds a combinatorial approach should be taken. Thus, trials and studies are definite requirements to authenticate clinical approach of these natural compounds either singly or in various combination therapeutic approaches.

13.10 Summary

Cell division is an important aspect and fundamental process of various biological activities including tissue regeneration, homeostasis, pathological processes, various physiological activities, and tumorigenesis. The cell cycle is a succession of events by which a cell duplicates its genome and culminates with the formation of two daughter cells. Cyclin-dependent kinases (CDKs) are key drivers of eukaryotic cell cycle that promote the synthesis of DNA and chromosome segregation by phosphorylation of their substrates. Forfeiture of control over cell cycle is a trademark of cancer. Progression from one phase to next phase of cell cycle requires cyclins and CDKs. Replenishing cell cycle control via CDK inhibition might open some windows for development of targeted cancer therapy. CDKs are not just involved in the cell cycle but also have other important functions, such as insulin production, transcription, neuronal functions, and glycogen synthesis. As CDKs are critical regulators of transcription, cell cycle, and other important biological functions, inhibitors against these CDKs have come into existence to treat multiple malformations associated with CDK dysfunctions. For over 20 years, many compounds have been developed against the abnormally activated CDKs and some are under clinical trials. In this chapter, we have summarized various categories of CDK inhibitors that can be used to treat various malignancies.

Pan-CDK inhibitors are the first generation of CDK inhibitors including Roscovitine and Flavopiridol. The primary role of the pan-CDK inhibitors is to induce cell cycle arrest and hamper cell proliferation by obstructing CDKs activity. Because of their low specificity and toxicity most of the first-generation pan-CDK inhibitors failed in the clinical trials [16–18]. Therefore, second-generation pan-CDK inhibitors such as RGB-286638, AT519, Dinaciclib, TG02, Roniciclib, P276-00, etc., were developed with improved selectivity and reduced cytotoxicity, and have shown good activity in pre-clinical trials. CDK4/6 inhibitors are earliest specific inhibitors that are FDA approved for clinical use. These selectively target CDK4/6 with narrow toxicity to normal cells. These inhibitors mainly target cyclin D-CDK4/6-Rb pathway preventing phosphorylation of Rb (tumor suppressor protein), which is otherwise important for cell cycle progression. Currently three oral agents are approved by FDA as CDK4/6 inhibitors, namely; abemaciclib, palbociclib, Ribociclib. Specific CDK7 inhibitors like LDC4297, BS-181, QS1189, ICEC0942, THZ1, YKL-5-124, THZ2 are currently recognized as potential inhibitors of CDK7 that inhibit CDK7 induced transcriptional activation. Numerous selective CDK9 inhibitors have been recognized with strong anti-proliferative property such as AZD-4573, MC180295, CDKI-73, and Fadraciclib. Naturally derived substances from marine species, plants, and microorganisms with CDK inhibitory properties such as triterpenoids, olomoucine, proanthocyanidins, and many more.

Thus, numerous synthetic and naturally derived CDK inhibitors have been recognized as anti-tumor drugs that can inhibit the malignancies arising from dysregulation of normal CDKS. Moreover, there is need of scrutinizing more such compounds which will prove effective drugs against different malignancies.

13.11 Further Reading

The readers can further read about "CDK Inhibitors for treatment of breast cancer" by going through the following research papers:

https://onlinelibrary.wiley.com/doi/abs/10.1002/med.1021

https://www.ncbi.nlm.nih.gov/pmc/articles/PMC8167670/

For more insights about the topic, we would suggest detailed findings from the books of

(Mir MA, 2022) https://doi.org/10.1016/C2021-0-02565-7, https://doi.org/10.1016/C2022-0-00074-X

(Mir MA, 2021) https://doi.org/10.52305/WXJL6770,

from cancer.net website, https://www.cancer.net/cancer-types/breast-cancer/types-treatment

Readers are advised to look following video lectures for better understanding of this chapter:

https://www.youtube.com/watch?v=BB9jjK7BHkg

https://www.youtube.com/watch?v=vEe3lBduckE

References

Ali S et al (2009) The development of a selective cyclin-dependent kinase inhibitor that shows antitumor activity. Cancer Res 69(15):6208–6215

Alzani R et al (2010) Therapeutic efficacy of the pan-cdk inhibitor PHA-793887 in vitro and in vivo in engraftment and high-burden leukemia models. Exp Hematol 38(4):259–269
Appleyard MVCL et al (2009) Seliciclib (CYC202, R-roscovitine) enhances the antitumor effect of doxorubicin in vivo in a breast cancer xenograft model. Int J Cancer 124(2):465–472
Arellano M, Moreno S (1997) Regulation of CDK/cyclin complexes during the cell cycle. Int J Biochem Cell Biol 29(4):559–573
Bailon-Moscoso N et al (2017) Natural compounds as modulators of cell cycle arrest: application for anticancer chemotherapies. Curr Genomics 18(2):106–131
Baker A et al (2016) The CDK9 inhibitor dinaciclib exerts potent apoptotic and antitumor effects in preclinical models of MLL-rearranged acute myeloid leukemia. Cancer Res 76(5):1158–1169
Birdsall TC (1997) Berberine: therapeutic potential of alkaloid found in several medicinal plants. Altern Med Rev 2:94–103
Blazek D et al (2011) The cyclin K/Cdk12 complex maintains genomic stability via regulation of expression of DNA damage response genes. Genes Dev 25(20):2158–2172
Borowczak J et al (2022) CDK9 inhibitors in multiple myeloma: a review of progress and perspectives. Med Oncol 39(4):1–15
Cassaday RD et al (2015) A phase II, single-arm, open-label, multicenter study to evaluate the efficacy and safety of P276-00, a cyclin-dependent kinase inhibitor, in patients with relapsed or refractory mantle cell lymphoma. Clinical Lymphoma Myeloma and Leukemia 15(7):392–397
Cavins JA et al (1967) Initial toxicity study of sangivamycin (NSC-65346). Cancer Chemother Rep 51(4):197–200
Chan K-C et al (2010) Polyphenol-rich extract from mulberry leaf inhibits vascular smooth muscle cell proliferation involving upregulation of p53 and inhibition of cyclin-dependent kinase. J Agric Food Chem 58(4):2536–2542
Chao S-H et al (2000) Flavopiridol inhibits P-TEFb and blocks HIV-1 replication. J Biol Chem 275(37):28345–28348
Cho BC et al (2018) Phase Ib/II study of the pan-cyclin-dependent kinase inhibitor roniciclib in combination with chemotherapy in patients with extensive-disease small-cell lung cancer. Lung Cancer 123:14–21
Cidado J et al (2020) AZD4573 is a highly selective CDK9 inhibitor that suppresses MCL-1 and induces apoptosis in hematologic cancer CellsAZD4573, a selective CDK9 inhibitor, targets MCL-1. Clin Cancer Res 26(4):922–934
Davidson G et al (2009) Cell cycle control of wnt receptor activation. Dev Cell 17(6):788–799
Deng J et al (2018) CDK4/6 inhibition augments antitumor immunity by enhancing T-cell activation. Cancer Discov 8(2):216–233
Diaz-Padilla I et al (2009) Cyclin-dependent kinase inhibitors as potential targeted anticancer agents. Investig New Drugs 27(6):586–594
Dickson MA (2014) Molecular pathways: CDK4 inhibitors for cancer TherapyCDK4 inhibitors. Clin Cancer Res 20(13):3379–3383
Frame S et al (2020) Fadraciclib (CYC065), a novel CDK inhibitor, targets key pro-survival and oncogenic pathways in cancer. PLoS One 15(7):e0234103
Fry DW et al (2004) Specific inhibition of cyclin-dependent kinase 4/6 by PD 0332991 and associated antitumor activity in human tumor xenografts. Mol Cancer Ther 3(11):1427–1438
Goh KC et al (2012) TG02, a novel oral multi-kinase inhibitor of CDKs, JAK2 and FLT3 with potent anti-leukemic properties. Leukemia 26(2):236–243
Hanahan D, Weinberg RA (2011) Hallmarks of cancer: the next generation. Cell 144(5):646–674
Hortobagyi GN (2018) Ribociclib for the first-line treatment of advanced hormone receptor-positive breast cancer: a review of subgroup analyses from the MONALEESA-2 trial. Breast Cancer Res 20(1):1–11
Iniguez AB et al (2018) EWS/FLI confers tumor cell synthetic lethality to CDK12 inhibition in Ewing sarcoma. Cancer Cell 33(2):202–216
Jiang M et al (2009) Cyclin Y, a novel membrane-associated cyclin, interacts with PFTK1. FEBS Lett 583(13):2171–2178

Johnson AJ et al (2012) The novel cyclin-dependent kinase inhibitor dinaciclib (SCH727965) promotes apoptosis and abrogates microenvironmental cytokine protection in chronic lymphocytic leukemia cells. Leukemia 26(12):2554–2557

Kim J et al (2012) Methyl nomilinate from citrus can modulate cell cycle regulators to induce cytotoxicity in human colon cancer (SW480) cells in vitro. Toxicol In Vitro 26(7):1216–1223

Kitagawa M et al (1994) A cyclin-dependent kinase inhibitor, butyrolactone I, inhibits phosphorylation of RB protein and cell cycle progression. Oncogene 9(9):2549–2557

Kumar N et al (2014) Chebulagic acid from Terminalia chebula causes G1 arrest, inhibits NFκB and induces apoptosis in retinoblastoma cells. BMC Complement Altern Med 14(1):1–10

Kwapisz D (2017) Cyclin-dependent kinase 4/6 inhibitors in breast cancer: palbociclib, ribociclib, and abemaciclib. Breast Cancer Res Treat 166(1):41–54

Lim S, Kaldis P (2013) Cdks, cyclins and CKIs: roles beyond cell cycle regulation. Development 140(15):3079–3093

Lin S-F et al (2018) Activity of roniciclib in medullary thyroid cancer. Oncotarget 9(46):28030

Lin TS et al (2010) Flavopiridol, fludarabine, and rituximab in mantle cell lymphoma and indolent B-cell lymphoproliferative disorders. J Clin Oncol 28(3):418

Locatelli G et al (2010) Transcriptional analysis of an E2F gene signature as a biomarker of activity of the cyclin-dependent kinase inhibitor PHA-793887 in tumor and skin biopsies from a phase I clinical study. Mol Cancer Ther 9(5):1265–1273

Lücking U et al (2017) Identification of atuveciclib (BAY 1143572), the first highly selective, clinical PTEFb/CDK9 inhibitor for the treatment of cancer. ChemMedChem 12(21):1776–1793

Malorni L et al (2018) Palbociclib as single agent or in combination with the endocrine therapy received before disease progression for estrogen receptor-positive, HER2-negative metastatic breast cancer: TREnd trial. Ann Oncol 29(8):1748–1754

Malumbres M, Barbacid M (2009) Cell cycle, CDKs and cancer: a changing paradigm. Nat Rev Cancer 9(3):153–166

Manohar SM et al (2012) Cyclin-dependent kinase inhibitor, P276-00, inhibits HIF-1α and induces G2/M arrest under hypoxia in prostate cancer cells. Prostate Cancer Prostatic Dis 15(1):15–27

Massard C et al (2011) A first in man, phase I dose-escalation study of PHA-793887, an inhibitor of multiple cyclin-dependent kinases (CDK2, 1 and 4) reveals unexpected hepatotoxicity in patients with solid tumors. Cell Cycle 10(6):963–970

Mehraj U et al (2022a) Expression pattern and prognostic significance of baculoviral inhibitor of apoptosis repeat-containing 5 (BIRC5) in breast cancer: a comprehensive analysis. Advances in Cancer Biology-Metastasis 100037:100037

Mehraj U et al (2021a) Tumor microenvironment promotes breast cancer chemoresistance. Cancer Chemother Pharmacol 87:1–12

Mehraj U et al (2021b) The tumor microenvironment as driver of stemness and therapeutic resistance in breast cancer: new challenges and therapeutic opportunities. Cell Oncol 44:1–21

Mehraj U et al (2022b) Adapalene synergistically with doxorubicin promotes apoptosis of TNBC cells by hyperactivation of the ERK1/2 pathway through ROS induction, vol 12

Mehraj U et al (n.d.) Chemokines in triple-negative breast cancer heterogeneity: new challenges for clinical implications. Elsevier

Mehraj U et al (2021c) Prognostic significance and targeting tumor-associated macrophages in cancer: new insights and future perspectives. Breast Cancer 28(3):539–555

Mehraj, U., et al. (2022c). Cryptolepine targets TOP2A and inhibits tumor cell proliferation in breast cancer cells-an in vitro and in silico study. Anti-cancer Agents in Medicinal Chemistry

Mehraj U et al (2022d) Expression pattern and prognostic significance of CDKs in breast cancer: an integrated bioinformatic study. Cancer Biomarkers(Preprint):1–15

Meijer L et al (1997) Biochemical and cellular effects of roscovitine, a potent and selective inhibitor of the cyclin-dependent kinases cdc2, cdk2 and cdk5. Eur J Biochem 243(1–2):527–536

Mir MA et al (2020) Targeting different pathways using novel combination therapy in triple negative breast cancer. Curr Cancer Drug Targets 20(8):586–602

Mishra PB et al (2013) Molecular mechanisms of anti-tumor properties of P276-00 in head and neck squamous cell carcinoma. J Transl Med 11(1):1–11
Morgan DO (1997) Cyclin-dependent kinases: engines, clocks, and microprocessors. Annu Rev Cell Dev Biol 13:261
Olson CM et al (2019) Development of a selective CDK7 covalent inhibitor reveals predominant cell-cycle phenotype. Cell chemical biology 26(6):792–803
Pallis M et al (2017) Complementary dynamic BH3 profiles predict co-operativity between the multi-kinase inhibitor TG02 and the BH3 mimetic ABT-199 in acute myeloid leukaemia cells. Oncotarget 8(10):16220
Patel H et al (2018) ICEC0942, an orally bioavailable selective inhibitor of CDK7 for cancer TreatmentICEC0942, a CDK7 inhibitor for cancer therapy. Mol Cancer Ther 17(6):1156–1166
Ponder KG et al (2016) Dual inhibition of mcl-1 by the combination of carfilzomib and TG02 in multiple myeloma. Cancer Biol Ther 17(7):769–777
Poon E et al (2020) Orally bioavailable CDK9/2 inhibitor shows mechanism-based therapeutic potential in MYCN-driven neuroblastoma. J Clin Invest 130(11):5875–5892
Qayoom H et al (2022) Expression patterns and therapeutic implications of CDK4 across multiple carcinomas: a molecular docking and MD simulation study. Med Oncol 39(10):1–13
Qayoom H et al (2021) An insight into the cancer stem cell survival pathways involved in chemoresistance in triple-negative breast cancer. Future Oncol 17(31):4185–4206
Quereda V et al (2019) Therapeutic targeting of CDK12/CDK13 in triple-negative breast cancer. Cancer Cell 36(5):545–558
Rugo HS et al (2021) Management of abemaciclib-associated adverse events in patients with hormone receptor-positive, human epidermal growth factor receptor 2-negative advanced breast cancer: safety analysis of MONARCH 2 and MONARCH 3. Oncologist 26(1):e53–e65
Saqub H et al (2020) Dinaciclib, a cyclin-dependent kinase inhibitor, suppresses cholangiocarcinoma growth by targeting CDK2/5/9. Sci Rep 10(1):1–13
Shankar S et al (2007) Epigallocatechin-3-gallate inhibits cell cycle and induces apoptosis in pancreatic cancer. Frontiers in Bioscience-Landmark 12(13):5039–5051
Shirsath NP et al (2012) P276-00, a cyclin-dependent kinase inhibitor, modulates cell cycle and induces apoptosis in vitro and in vivo in mantle cell lymphoma cell lines. Mol Cancer 11(1): 1–12
Sofi S et al (2022a) Targeting cyclin-dependent kinase 1 (CDK1) in cancer: molecular docking and dynamic simulations of potential CDK1 inhibitors. Med Oncol 39(9):1–15
Sofi S et al (2022b) Cyclin-dependent kinases in breast cancer: expression pattern and therapeutic implications. Med Oncol 39(6):1–16
Su Y-T et al (2018) Novel targeting of transcription and metabolism in GlioblastomaTG02 in glioblastoma treatment. Clin Cancer Res 24(5):1124–1137
Swaffer MP et al (2016) CDK substrate phosphorylation and ordering the cell cycle. Cell 167(7): 1750–1761
Syn NL et al (2018) Pan-CDK inhibition augments cisplatin lethality in nasopharyngeal carcinoma cell lines and xenograft models. Signal Transduct Target Ther 3(1):1–9
Vermeulen K et al (2002) Antiproliferative effect of plant cytokinin analogues with an inhibitory activity on cyclin-dependent kinases. Leukemia 16(3):299–305
Wan Y et al (2004) Synthesis and target identification of hymenialdisine analogs. Chem Biol 11(2): 247–259
Zeidner JF, Karp JE (2015) Clinical activity of alvocidib (flavopiridol) in acute myeloid leukemia. Leuk Res 39(12):1312–1318
Zhang M et al (2021) CDK inhibitors in cancer therapy, an overview of recent development. Am J Cancer Res 11(5):1913

CDKs in Cell-Cycle Progression and Therapeutic Strategies in Human Breast Cancer

14

Manzoor Ahmad Mir and Ulfat Jan

14.1 Introduction

Over the last 20 years, numerous scientific studies have illustrated that there is a relationship between cancer and dysregulation in the cell cycle (Malumbres and Barbacid 2009). Cancer cells develop genetic instabilities that cause uncontrollable mitogenic signaling and faulty anti-mitogenic signaling, resulting ultimately in uncontrolled growth (Malumbres and Barbacid 2001, Massagué 2004). Furthermore, many cancers develop epigenetic changes, which contribute to further abnormalities, and also chromosome instabilities, which cause numeric alterations in chromatin, i.e., resulting in GIN and CIN, respectively (Kastan and Bartek 2004, Kops et al. 2005, Mehraj et al. 2021a, b). Such changes, when combined, result in higher sensitivity toward the aggregation of further gene modifications which lead to tumorigenesis and the development of far more malignant morphologies, as well as higher proliferative benefits. Unplanned growth, genomic instability, and chromosomal instability are the three cell cycle abnormalities that have been controlled, overtly or covertly, via CDKs dysregulation (Malumbres and Barbacid 2005). Attachment of governing components called cyclins is required for CDK activation. These proteins are generated and eliminated at specified moments throughout the cell growth, allowing for precise control of autophosphorylation activity. Both cyclin-dependent kinases and cyclins are encoded by numerous human genes (13 and 25 loci) (Malumbres and Barbacid 2005). But, only a few of these protein complexes are engaged throughout to drive the whole process of the cell cycle. These are CDK1,2,4,6 and ten cyclins of 4 distinct categories. Cancer abnormalities typically counter distinct CDK–cyclin complexes, leading to either prolonged

M. A. Mir (✉) · U. Jan
Department of Bioresources, School of Biological Sciences, University of Kashmir, Srinagar, J&K, India
e-mail: drmanzoor@kashmiruniversity.ac.in

M. A. Mir (ed.), *Therapeutic Potential of Cell Cycle Kinases in Breast Cancer*,
https://doi.org/10.1007/978-981-19-8911-7_14

multiplication or unplanned re-merge into the cell cycle, 2 characteristics shared by the majority of human cancer cells (Malumbres and Barbacid 2001). Procedures that detect probable abnormalities in DNA synthesis and chromosomal segregation regulate normal cell cycle continuance. These procedures are called cell cycle checkpoints. Via regulation of CDK function, activation of such checkpoints causes cell cycle arrest. This restriction permits each and every cell to correct such damage errors correctly and prevent them from being passed into the new cells. This DNA damage checkpoint shields tissues against environmental and genetic genotoxic agents. Such agents prompt numerous defects in the genetic material of an organism and lead to abnormal growth. Such changes are detected via signaling pathway, which results in CDK suppression and cycle arrest (Bartek et al. 2004). Cells might undergo senescence or experience death if correction is failed due to severe DNA damage or gene mutations in the checkpoint or DNA repair machinery. Genomic instability, on the other hand, can originate from a build-up of Gene mutations, resulting in cell mutation and tumorigenesis (Kastan and Bartek 2004). The spindle assembly checkpoint (SAC), a pathway primarily governs CDK1 activities as well as prevents chromosomal segregation errors, is in charge of appropriate chromosome segregation once the DNA has been replicated (Kops et al. 2005) (Perez de Castro, I., de Carcer, G. & Malumbres, M. A census of mitotic cancer genes: new insights into tumor cell biology and cancer therapy. Carcinogenesis 28, 899–912 (2007); Musacchio, A. & Salmon, E. D. The spindle-assembly checkpoints in space and time. Nature Rev. Mol. Cell Biol. 8, 379–393 (2007)). This faulty checkpoint may result in inconsistent transmission of the genetic information, which, if left uncorrected, might aid cancer growth via amassing numeric chromosomal instabilities (Malumbres and Barbacid 2009).

14.2 CDKs and Cell Cycle Progression

One of the fundamental pathological features of tumor is the dysregulated cell cycle, which results in uncontrolled cell expansion, so discovering treatment strategies to stop cell growth is a commonly used strategy for treating tumors. To multiply, a cell should first go via a sequence of processes; all of these processes are governed by a large number of regulatory proteins, and this sequence of processes is called the cell cycle – a mechanism that is largely consistent throughout eukaryotic organisms (Hartwell 1974, Sofi 2021). All the phases of a cell division cycle should always be completed in order, including high precision across all critical steps exerted via signaling gates, called cell cycle checkpoints, thus preventing the progression in the event of, say, DNA damages to the cells (Kastan and Bartek 2004). Passage from one phase of the cell cycle into the other phase is mainly managed by the family of proteins called cyclin-dependent kinases (CDKs), and these CDKs are in turn activated by the other family of proteins called cyclins (Sofi et al. 2022a, b). Both CDKs and cyclins form complexes so as to progress the cell cycle from one stage to another stage in a precise manner. Having such importance in the continuance of the cell division, these CDKs were used as promising drug targets for cancer drugs, but such first-generation CDK inhibitor drugs do not provide beneficial clinical results (Lapenna and Giordano 2009, Malumbres and Barbacid 2009, Qayoom et al. 2021). Non-specific pan-CDK suppression was revealed to be harmful to non-cancer cells, at least to some extent (Asghar et al. 2015, Mehraj et al. 2022a, b, c). The process of the cell cycle comprises of four distinct phases; these phases are G1 phase/G0 phase, S phase, G2 phase, and M phase, respectively. G1 and G2, called gap phases, are the basic regulatory checkpoints, controlled by various CDKs and their cyclin partners, and these checkpoints determine whether the cell enters the S phase and proceeds toward the mitosis stage of the cell cycle. CDK4 and CDK6, two kinases both these CDKs govern the G1 phase. D1, D2, and D3 type cyclins bind to these CDKs and modulate their catalytic activities (Massagué 2004, Cardoso et al. 2018, Qayoom et al. 2022). D1, D2, and D3 type cyclins bind and stimulate CDK4 and CDK6 during the initial G1 phase in reaction to mitogenic stimuli, and the complex cyclin D-CDK4/6 then preferentially phosphorylates as well as deactivates the members of the Retinoblastoma-associated proteins (pRb), like p110 (encoded by RB1), the linked pocket protein p107 (encoded by RBL1), and p130 (encoded by RB2), so as to permit the expression of E1 and E2 of the class E cyclins (Malumbres and Barbacid 2009, Malumbres 2014). Additionally, the above pocket proteins are phosphorylated by CDK2–cyclin E complexes, resulting in subsequent deactivation (Lundberg et al. 1998, Harbour et al. 1999, Mehraj et al. 2022a, b, c). E-type cyclins are only available in the early phases of DNA synthesis, yet their distribution is closely regulated throughout the whole process of the cell growth. Such finding, including the capacity of predominate negative mutants plus anti-CDK2 antibodies effectively stop the cellular proliferation in certain human cancer cell lines, suggests that CDK2 is indeed a basic controller of the cell division cycle (Pagano et al. 1993, van den Heuvel and Harlow 1993). Throughout the later stages of the gap1 stage, E1 and E2 are stimulated, that in turn activates CDK2 via binding with it, which was previously sequestered via 2 inhibitors of the CDKs, i.e., members of the CIP and KIP families. CDK2 is also activated by CDC25A (cell division cycle 25 A), that removes phosphorylation from the CDK2 (Watanabe et al. 1995). Moreover, when CDK2 is activated it is having the capacity to fully phosphorylate a considerable array of proteins that are necessary for the cell division cycle continuance, genome duplication, histone synthesis as well as centrosome replication (Ma et al. 2000, Okuda et al. 2000, Sever-Chroneos et al. 2001). CDK2/E functional complex also has the capability to modulate retinoblastoma to reverse the G1- S checkpoint at the late stages of G1-S phase of the cell cycle that in turn results in the commencement of the S-phase of cell cycle, via forming a + ve feedback complex. CDK2 and CDK4/6 in combination activity regulate the cell division cycle transition toward Synthetic phase, also known as the "restriction barrier," where mitogenic proteins are no longer needed to pass the ongoing cell cycle. Toward its end, cyclin A displaces cyclin E and generates a novel product, CDK2/cyclin A, in which cyclin E is promptly destroyed via ubiquitylation facilitated by the F-box/WD string polypeptide 7 (FBXW7) (Ma et al. 2000, Okuda et al. 2000). Through phosphorylation E2F1 and CDC6, the cyclin A/CDK2 pair ends the S phase and pushes the cell cycle transition from S to G2. Cyclin A then stimulates CDK1, causing the cell to undergo

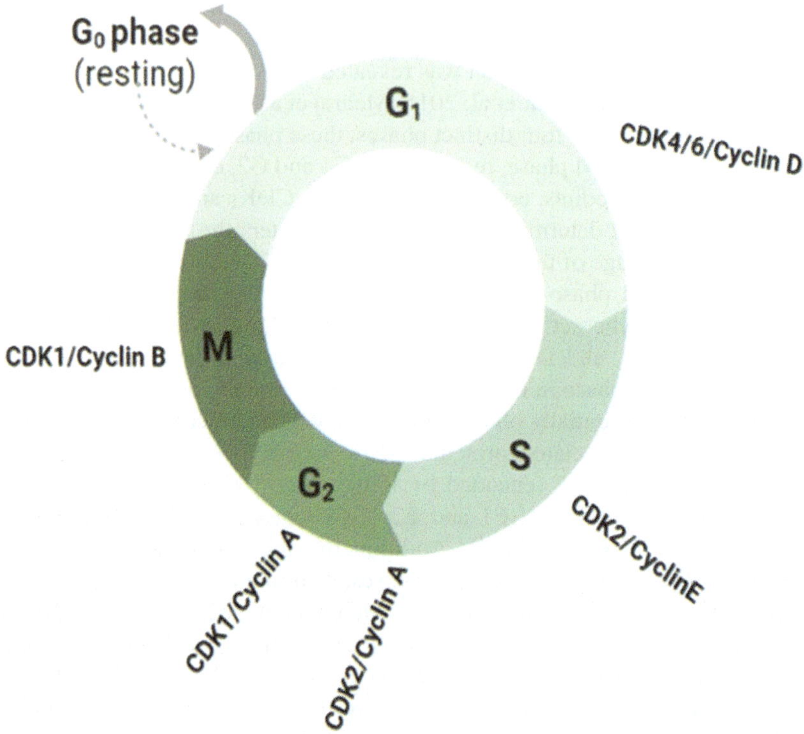

Fig. 14.1 Different phases of the cell cycle and CDKs along with their cyclin members that control the whole of the cell cycle and its progression

the M phase transition. The complex cyclin B/CDK1 maintains CDK1 activity throughout mitosis (Fig. 14.1). Such phosphorylation of stimulated CDK1 causes the nuclear envelope to disintegrate, its chromosomes to condense, as well as the bipolar spindle to assemble. Several spindle assembly checkpoints regulate the transition from mitotic metaphase to anaphase, and anaphase is triggered by a reduction in CDK1 activity caused by the breakdown of cyclin B via Anaphase-promoting Complex/Cyclosome (Gavet and Pines 2010, Sofi et al. 2022a, b). Dysregulation in the CDK1 expression levels allows chromosomal segregation as well as the accomplishment of cytokinesis and karyokinesis. Out of all the CDKs, CDK1 seems to be the only CDK, that is needed for cell cycle advancement, as it commences mitosis and assures that all the important steps in cell duplication proceed in the correct order with higher precision (Santamaría et al. 2007). In living organisms, Gap1 to Synthetic phase transition is said to be operated by the two main mechanisms, i.e., CDK4/6/D and CDK2/E (Satyanarayana and Kaldis 2009, Aguilar and Fajas 2010). Along with it, these two CDKs phosphorylate RB, via releasing its inhibition of elongation factor 2 (E2F), thus allowing the cell cycle to enter into the S phase (Kato et al. 1993, Vermeulen et al. 2003, Lim and Kaldis 2013, Mir et al.

2020a, b). Recent data, meanwhile, suggest that Cdk4/D is not required for G1/S transition in typical developmental stages. 3-D-type cyclins are found in animals (D1, D2, and D3). Ignoring the fact that almost all organs grow properly and cells multiply, CycD1/D2/D3 triple-knockout mice survived until mid-gestation, when most died due to cardiac defects (Finn et al. 2016a, b, Mehraj et al. 2021a, b). Among all of these D-type cyclins, cyclin D1 is the most well-studied of them. The production of cyclin D1 is characterized like a "delayed-early" reaction to mitogenic signal, and also its expression is controlled by complicated promoter and enhancer interaction (Matsushime et al. 1991). Despite being quite extensively explored, cyclin D3 follows the same pattern of cyclin D1, whereas the programming of cyclin D2 is less clear, in spite of the fact that cyclin D2 can induce multiplication in some circumstances (Sicinski et al. 1996, Kushner et al. 2005, Yu et al. 2005, Cooper et al. 2006, Spofford et al. 2006). Diverse D-type cyclin paralogs were probably synthesized to increase CDK6 or CDK4 activity, reflecting tissue-specific elements of cellular homeostasis (Ciemerych et al. 2002, Lim and Kaldis 2013).

14.3 Dysregulation of CDKs in Breast Cancer

Several of the important principles in CDK biology were found more than two decades back when studying fungus and the simultaneous divisions observed in fetal samples; in fact, the discoveries out of these investigations contributed to the awarding of a Noble Peace prize to all these scientists (Hartwell 2002, Nurse 2002). CDK1 has been observed as a critical controller of mitosis, while CDK2 was found to be additionally more important for DNA synthesis in eukaryotic organisms. CDK6 and CDK4 are susceptible to a variety of major physiological cues and are responsible for almost all of the management, controlling of the cell cycle entrance in multicellular organisms. Following that, in contrast to CDKs that effectively drive cell cycle advancement (including CDK1, CDK2, CDK4, and CDK6), and a new family of CDKs which control transcription (CDK9, CDK7, and CDK8) has been discovered (Drapkin et al. 1996, Bregman et al. 2000, Lim and Kaldis 2013, Nemet et al. 2014, Qayoom et al. 2021). Such kinase proteins exhibiting post-mitotic roles in specialized tissue environments, like CDK5, have been discovered. Because CDKs play such an important part in cell differentiation monitoring, it is not unexpected that almost all malignancies do have characteristics that disrupt the cell cycle's typical regulation (Malumbres and Barbacid 2001). And along with it, from about last two decades several medications specifically targeting the CDKs functioning have evolved and are also being tried in various clinical settings (Asghar et al. 2015) (Table 14.1).

Because the CDK4/6–RB pathway is so important for cell cycle entrance, there is no surprise that a large number of malignancies use it to enhance multiplication (Sherr 1996, Bartkova et al. 2006, Burkhart and Sage 2008, Knudsen and Knudsen 2008). Upregulation of p16INK4A by many of the oncogenes serves as just an inherent checkpoint against abnormal development (Serrano et al. 1997, Michaloglou et al. 2005, Bartkova et al. 2006, Burd et al. 2013). Upregulation of

Table 14.1 Cyclin/CDK complexes and their function in the cell cycle

S. No	Stage of the cell cycle	Cyclin/ CDK complex	Major roles
01.	Gap 1 stage	Cyc D/Cdk4/6	This complex controls the cell growth as well as continuation of the cell cycle Possible target of this particular complex is the Rb. This Rb acts as an inhibitor of the cell cycle continuation and via CycD/Cdk4/6 complex this Rb is phosphorylated and in turn helps in cell cycle growth and progression.
02.	Gap1 to S stage	Cyc E/Cdk2	Main function of this complex is the G1 to S transition This complex also phosphorylates Rb, thus promoting cell cycle progression as well as directing the commencement of the genome duplication It promotes the expression of Cyc A, thus allowing the progression toward the S phase of the cell cycles
03.	Synthetic stage	Cyc A/Cdk2	Main function of this complex is the commencement of the DNA duplication This complex also inhibits the production of new complexes that aid in replication, assuring that one cell contains only one copy of DNA. This whole inhibition process is carried out through phosphorylation of the CDC6, main component of the replication process
04.	Gap2 to mitotic stage	CycA/Cdk1	It controls the whole of the M-phase of the cell cycle and regulates and maintains the CycB/Cdk1 complex
05.	Mitotic stage	Cyc B/Cdk1	A crucial regulator of the mitotic stage transition as well as takes part in the phosphorylation and stimulation of numerous downstream protein kinases

p16INK4A causes Retinoblastoma (RB) to inhibit proliferation and cellular proliferation, resulting in oncogene-induced cell death. To allow continued carcinogenic growth, such cell death should be reversed, that happens in cancers via two main mechanisms: deletion of p16INK4A or lack of Retinoblastoma (Witkiewicz et al. 2011, LaPak and Burd 2014). Deletion of p16INK4A separates carcinogenic burden with CDK4 or CDK6 expression inhibition, while absence of Retinoblastoma disrupts cell division cycle regulation downwards. The cell growth arrest caused by p16INK4A requires RB, which is comparable with just this hypothesis (Lukas et al. 1995). Furthermore, because RB-negative cancers release supraphysiological amounts of p16INK4A, they remain immune to increased p16INK4A production (Witkiewicz et al. 2011). Targeted carcinogenic stimulation of CDK6 or CDK4 expression is a different way to disrupt the CDK6/4–RB pathway. Several cancers exhibit dysregulated cyclin D1 expression levels, genetic displacement, and gene expansion (Motokura et al. 1991, Jiang et al. 1992, Buckley et al. 1993, Bartkova et al. 1994, Knudsen et al. 2006). The particular carcinogenic potential of cyclin D1 is supported by a multitude of diverse data sets (Sherr 1995, Diehl and therapy 2002, Knudsen et al. 2006). Moreover, it has been seen that there are increased expression levels of both CDK4 and CDK6 in various types of cancer (Asghar et al. 2015).

Furthermore, the various methods of the CDK4/6/D complex dysfunction are generally restrictive and often cancer-type-specific. Lack of Retinoblastoma protein, for instance, is a characteristic of lung cancer, cyc D1 dysregulation is prevalent throughout BC, whereas p161NK4A deletion is fairly prevalent in GBM (Glioblastoma) (Asghar et al. 2015). May interfere in the cell multiplication via increasing growth of the cells, that is a biochemical pathway, in contrast to phosphorylating Retinoblastoma (Romero-Pozuelo et al. 2020). Certain cancers overexpress cyclin D1, suggesting that deregulation of CDK4/6:Cyc D1 axis plays a role in BC (Arnold and Papanikolaou 2005). Recently, CDK2 dysregulation has been discovered to be common in several cancers (Scaltriti et al. 2011). Several malignancies, especially uterus and ovary tumors, have cyc E1 or cyc E2 augmentations as a significant carcinogenic activity (Etemadmoghadam et al. 2013, Karst et al. 2014, Kuhn et al. 2014). Synthetic phase is initiated without the necessity for CDK6 or CDK4 expression when cyc E is expressed ectopically (Lukas et al. 1997, Knudsen et al. 1998, Caldon et al. 2012). As a result, it is thought that Cyc E amplification could be cancer-causing in the same way (specifically, eliminating the biological necessity for CDK6/4 activity to start E-Cyc E production). p27KIP1, a CDKI, is dysregulated in several malignancies; however, chromosomal deletion of p27KIP1 is very uncommon (Chu et al. 2008, Hershko 2010).

14.4 CDKs as Therapeutic Targets of Breast Cancer

BC is now highly prevalent but also the utmost common element of mortality and morbidity in women, preceded by pulmonary and colorectal cancer in terms of deaths and conversely in prevalence rate. Breast cancer instances and fatalities in women are predicted to reach 2,800,000 and 627,000, correspondingly, over the globe (Bray et al. 2018). Hormone receptors, including ER and PR and HER2, represent key predictive and therapeutic indicators for endocrine therapy and anti-HER2 targeted therapy. Progesterone and Estrogen represent key markers of hormone treatment sensitivity, so they are found in about 75% of all BC manifestations (Hammond et al. 2010, Tsang and Tse 2020). Estrogen-positive tumors are almost always Progesterone-positive, with a tiny percentage expressing single hormone receptor positives, which means they are more malignant but less receptive to hormone treatment (Cui et al. 2005, Ethier et al. 2018). Increased expressions of the HER2 and associated proteins are seen in 15% of BCs and are linked toward meager diagnosis, a severe clinical history, as well as a predicted responsiveness to anti-HER2 therapeutic strategies (Bedard et al. 2009). The remaining 10–15% of BC patients, on the other hand, include triple-negative breast cancer (TNBC). Such a form of BC has a significant risk of recurrence and has a poor prognosis. Currently available medicines are ineffective in treating it (Yam et al. 2017). For individuals with BC, targeted treatment therapy extends their lives. Furthermore, pre-clinical research has revealed that stimulation of the CDK6/CDK4–cyclin D group results in excess cell growth, rendering pharmacologic suppression of the unit a promising treatment option (Pernas et al. 2018, Sobhani et al. 2019). In case of the

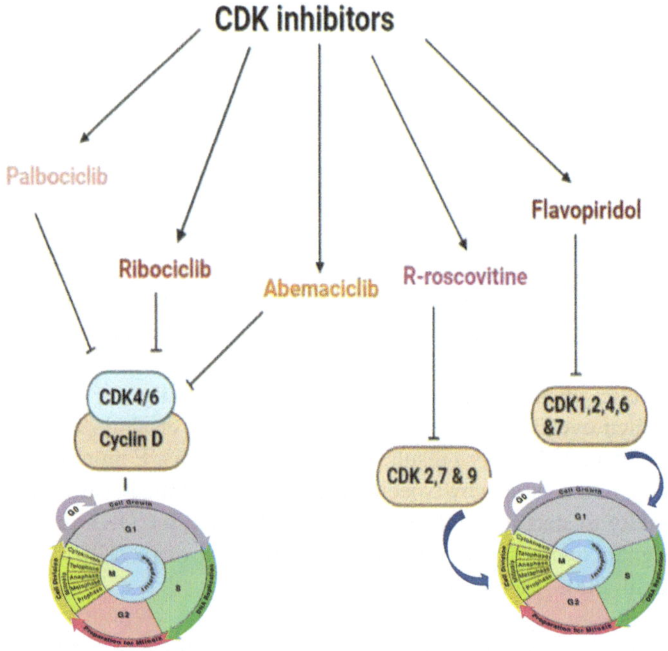

Fig. 14.2 Various types of CDK inhibitors and the CDK that is specifically inhibited by these inhibitors

estrogen-positive type of BC, the basic approved drug is the CDK inhibitors such as CDK6/CDK4 inhibitors (Fig. 14.2). Such subtype of BC maintains the functioning of the retinoblastoma, specifying that the function of inhibitor of CDK4/CDK6 remains as an integral part (Finn et al. 2009). Furthermore, another of the Estrogen receptors direct substrates is CCND1 (encoding cyclin D1), which is typically overexpressed in estrogen-positive breast cancers. Several randomized controlled trials have proven that combining CDK6/CDK4 inhibitors with traditional anti-estrogen treatments seems to have a synergistic impact (Bedard et al. 2009, Finn et al. 2009, Finn et al. 2015, Finn et al. 2016a, b, Hortobagyi et al. 2016, Goetz et al. 2017, Yam et al. 2017, Pernas et al. 2018, Sobhani et al. 2019). Three main CDKIs (CDK6/4 inhibitors) that are accepted by the US FDA include Palbociclib, Abemaciclib, and Ribociclib. All of the three drugs are used against hormone receptor-positive or human epidermal receptor 2-negative BCs. When these inhibitors are employed in combination with hormone therapy, the best result in case of survival rates is that decreased progression levels are seen in women with hormone receptor-positive and HER2-negative breast cancers (Finn et al. 2015, Finn et al. 2016a, b, Hortobagyi et al. 2016, Dickler et al. 2017, Goetz et al. 2017, Sledge et al. 2017). Unfortunately, such first-generation CDKIs (Flavopiridol or UCN-01) has not shown substantial therapeutic benefits (Shapiro 2006). CDKIs are basically classified into two categories including first-generation CDKIs and second-generation CDKIs. R-Roscovitine, Flavopiridol, and UCN-01 belong to first-generation CDKIs that are generally least sensitive as well as possess a wide range of specificity against a number of CDKs. Second-generation CDKIs are more precise against various CDKs and are also safer as compared to first-generation inhibitors (Finn et al. 2016a, b).

14.5 CDK4/6 Inhibitors as a Monotherapeutic Approaches

14.5.1 Palbociclib

Palbociclib, a Cdk inhibitor, has been tested in 41 patients having metastatic tumors during stage-I dosage progression trial, that had been pre-screened to see if they have Retinoblastoma activity (Rb+) (Flaherty et al. 2012). It is the primary CDK4/6 inhibitors approved for use in humans. It is not even an extremely specific Cdk4/6 blocker, and it also inhibits Cdk4 or Ckd6 similarly well. It enters a steady condition in 8 days after reaching the maximal level around 6 and 12 h (Morikawa and Henry 2015). This drug is typically prescribed as a 125-milligram capsule administered as 1 tablet daily with meals for 3 weeks, preceded by just a 1-week break from medication (https://www.fda.gov/ accessed on 13 June 2021). It has been demonstrated in the preclinical trials that Palbociclib possesses the capacity to decrease the proliferation of ER+ BC cells and can function effectively in a synergistic way with anti-estrogens as well as reverts hormone resistance (Finn et al. 2009). Such outcomes prompted the development as well as application of PALOMA-1, a transparent, randomized, solid evidence trial comparing Palbociclib with Letrozole vs Letrozole solely as the primary-line treatment for postmenopausal women with ER+, HER2-advanced BC (Finn et al. 2015). This substituent revealed that Palbociclib in combination with Letrozole resulted in a considerably progression-free life than Letrozole solely; these results enabled the FDA's rapid authorization of Palbociclib with Letrozole for such an application within USA (Beaver et al. 2015, Mir 2015). Palbociclib has been derived out of a series of pyridopyrimidine chemicals, owing because of its advantageous physical and pharmacological features (Fry et al. 2004). It specially and in equal manner inhibits the activities of Cdk4 as well as Cdk6 Cyc-D1 kinase activities (Fry et al. 2004). Also, it is extremely selective toward Cdk4/6, having little efficacy versus a variety of all other kinases. In a variety of RB-competent carcinoma cells, especially BCs, Palbociclib reduces cell proliferation and decreases replication at low nM levels. Rb-negative BC cells show little response as predicted (Fry et al. 2004). The selectivity of this drug in addressing Cdk4/6/Cyc D seems critical because it allows cancerous processes to be inhibited whereas healthy cells remained fairly quiescent and non-cycling phase (Dean et al. 2010). It is hardly remarkable, from what we have seen about different features of BC intrinsic subgroups (Perou et al. 2000), as well as the modifications within Retinoblastoma axis that they cause (Brigham et al. 2012), suggesting Cdk4/6 knockdown susceptibility is determined by biological profile. In an experiment, it was found that the susceptibility of Palbociclib in several

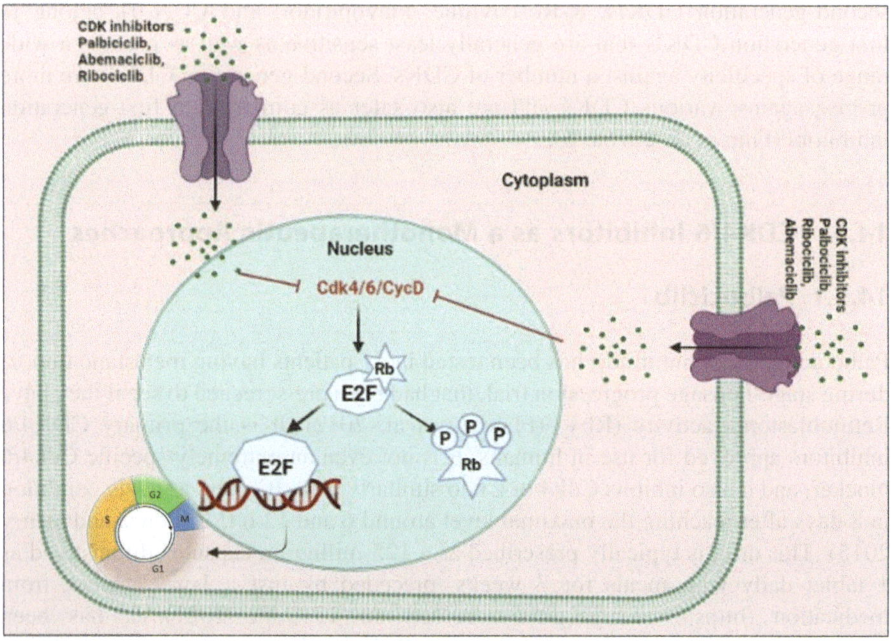

Fig. 14.3 Three main CDK inhibitors and their mechanism of action

BC subgroups, that there is a strong relation between the subgroup of BC and the susceptibility toward this particular drug (Finn et al. 2009), and in this experiment, it was found that the ER+ BC cells having luminal characters are more susceptible versus the basal type cancer cells that are resistant toward this drug (Mir and Agrewala 2008, Finn et al. 2009). In tamoxifen-resistant BC disease models, Palbociclib efficiently dephosphorylates Rb and suppresses cellular proliferation (Fig. 14.3) (Finn et al. 2009, Thangavel et al. 2011).

Palbociclib can possibly prevent multiplication as well as promote senescence in BC models which are resistant toward hormone treatment, culminating in a sustained cell-cycle arrest that differs with that of Estrogen inhibitors (Thangavel et al. 2011). Cdk4/6 suppression is a valuable and realistic treatment option for cancers that have developed resistance to hormone treatment (Thangavel et al. 2011). The use of Palbociclib has been well explored in a range of human cancers (Fig. 14.4).

Specific research has been carried out on 17 patients out of which five individuals possessing recurrent MCL show metastasis-free life for 1 year or more when given Palbociclib medication, having 1 CR and 2 PRs (Leonard et al. 2012). Two individuals having aggressive Retinoblastoma-positive BC achieved partial response in a phase II evaluation of Palbociclib (DeMichele et al. 2013). Apart from the Palbociclib, there are other Cdk4/6 blockers such as Ribociclib, Abemaciclib that have reached stage 3 clinical trials depending just on the results of the Palbociclib (Finn et al. 2016a, b, Mir and Mehraj 2019).

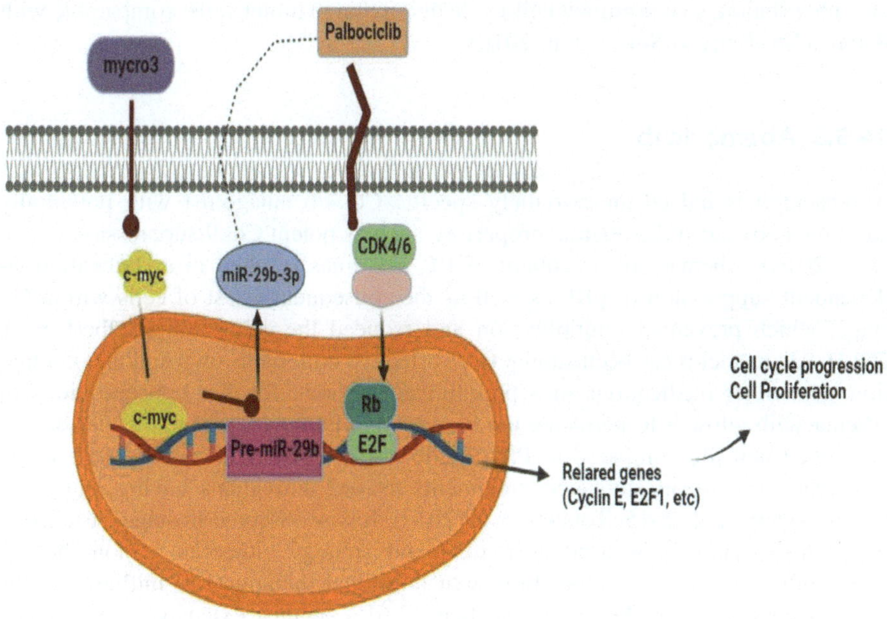

Fig. 14.4 Pathway utilized by the approved drug Palbociclib to specifically inhibit the cell cycle progression as well as cellular proliferation

14.5.2 Ribociclib

After Palbociclib, another drug approved by the US FDA is named Ribociclib. It belongs to the category of the first line of CDKIs and it especially inhibits Cdk4/6 (Goel et al. 2016, Hortobagyi et al. 2016). This drug is administered orally with a beginning dose of 600 milligrams once a day for about 21 continuous days and ensures a 7 days off medication (https://www.fda.gov/ accessed on 13 June 2021). It has IC50s around 10 nanometers and 40 nanometers, correspondingly. Food and Drug Administration has increased overall authorization of the inhibitor to encompass pre/perimenopausal women with Estrogen-positive/Human epidermal receptor 2-negative metastatic BC (Vora et al. 2014). One hundred twenty-eight cancer patients having Retinoblastoma-positive malignant solid tumors and malignancies took part in the first stage I dosage progression trial of separate Ribociclib. On a 21 of 28 day pattern, the maximum tolerated dose and recommended phase second dose was set at 900 milligrams and 600 milligrams, correspondingly (Infante et al. 2014). The main dose-limiting toxicities include elevated creatinine, tiredness, vomiting, electrolyte imbalance, DVT, oral thrush, neutropenia, thrombocytopenia etc. (Infante et al. 2014, Qayoom and Bhat 2020). It is an ATP-competitive tiny medication which binds to an ATP-binding groove of Cdk6 and Cdk4, forming strong interactions with proteins inside the ATP-binding groove (Asghar et al. 2015, Mir et al. 2020a, b). It suppresses RB-positive Breast cancer cells by blocking RB phosphorylation, so it promotes cell-cycle disruption in tumor cells, comparable with Palbociclib (Barroso-Sousa et al. 2016).

14.5.3 Abemaciclib

Abemaciclib is indeed an extremely specific Cdk4/6 antagonist with potentially additional advanced therapeutic properties, such as potent Cdk9 suppression (Chen et al. 2016). Abemaciclib treatment of BC cell lines resulted in a concentration-dependent suppression of pRb as well as the consequent arrest of cells within G1 stage, which prevented multiplication and reduced the cell count (Gelbert et al. 2014). Abemaciclib has been shown to be effective both on its own and in combination with other medications in a pre-clinical research. Smaller concentrations of Abemaciclib allow it to permeate the blood–brain boundary and that could have a prolonged half-life compared to Palbociclib. As a result, such qualities are being researched for use in anticancer treatments toward individuals having malignant tumors (Raub et al. 2015, Tolaney et al. 2017). It was observed in stage-I trials that when Abemaciclib was used as a treatment strategy either as a monotherapy (200 milligrams) or in combination with hormonal therapy (150 milligrams) the total success rate was 31% but also that of 61% against extensively pre-treated hormone positive breast tumors, it showed either a response or steady illness that can end up to 6 months, and this indicates that this drug could be useful as a monotherapeutic approach (Spring et al. 2016, Mir et al. 2022a, b). The suggested beginning dosages include 150 milligrams two times in a day when combined with Fulvestrant or 200 milligrams two times daily when used as monotherapy (https://www.fda.gov/ accessed on 31 May 2021). Frequent side effects of all grades of therapy include Diarrhoea (52%), nausea (33%), tiredness (21%), vomiting (21%), and neutropenia (19%) (Shapiro et al. 2013). Such findings led the FDA to designate Abemaciclib like a "Innovative Treatment" having "Green Card" certification in Oct 2015 for the treatment of metastatic BC (Lilly 2015).

14.6 CDK4/6 Inhibitors as a Combinational Approach

Numerous stage 3 trials are now being conducted, but it is anticipated that Cdk4/6 blockers are most beneficial when used in conjunction with other treatments (ClinicalTrials.gov. From: <https://clinicaltrials.gov/>. Accessed January 2016). The ultimate level of treatments for hormone receptor-positive BC is hormone therapies. Approximately 70% of all mammary tumors are hormone receptor-positive, meaning they exhibit one or both of the progesterone receptors (PgR) or the estrogen receptor (ER). Hormonal therapies prevent HR-positive BC cells from using hormonal signaling to proliferate and multiply, although up to 50% of hormone-positive BCs develop resistance to it over time (acquired resistance) or show it for the first time (de novo) (Lange and Yee 2011). Primary luminal B BCs and hormone therapy-resistant BC cell lines both show stimulation of the cyc

D-CDK4/6-INK4-Rb axis but might be responsive to Cdk4/6 suppression (Thangavel et al. 2011). Along with it, there are up to 35% of BCs in which CCND1 overexpression has been recognized as well as in some cases Cyc D amplification is also recognized (Gillett et al. 1994, Dickson et al. 1995, Musgrove et al. 2011, Brigham et al. 2012). In clinical studies on animal models, Palbociclib has been found to prevent the multiplication of BC cells that are sensitive to hormonal therapies (Thangavel et al. 2011, Mehraj et al. 2022a, b, c, Mir et al. 2022a, b). Additionally, it has been seen that when Ribociclib is combined with Fulvestrant or Letrozole, it inhibits the cancer progression in Estrogen-positive xenograft animal models (O'Brien et al. 2014). ALOMA-1/TRIO-18, a first randomized stage 2 preclinical trial testing Palbociclib, a Cdk4/6 suppressor in combination with the Letrozole, aromatase blocker, as a standard therapy (Stage II trial) against postmenopausal female having metastatic ER-positive BC, HER2-ve metastatic BC (Total patients = 165). After this trial, it has been seen that this combination therapy nearly doubles the progression-free survival 10.2 months to 20.2 months as compared to Letrozole alone possessing low progression-free survival in such breast cancer patients (Finn et al. 2015). None of the biomarkers was found which can predict sensitivity toward Palbociclib; enhanced progression-free survival was seen despite the CCND1 overexpression and lack of p^{16INK4A} (Finn et al. 2015). On the basis of various beneficial findings of the PALOMA-1, FDA expedited the authorization of the Palbociclib + Letrozole in combination as a primary line of therapy against the advanced hormone receptor-positive BC (U.S. Food and Drug Administration. Palbociclib accelerated approval. From: <http://www.fda.gov/Drugs/Information On Drugs/Approved Drugs/ucm432886.htm>. Accessed July 2015). Furthermore, the use of Palbociclib in prophylactic and therapeutic situations is being studied. For instance, a stage III evaluation of adjuvant Palbociclib combined with conventional hormonal therapy is currently being conducted in individuals having HR-positive, HER2-negative BC who still have illness following neoadjuvant chemotherapy and surgeries (ClinicalTrials.gov. From: <https://clinicaltrials.gov/> Accessed January 2016). Like the Palbociclib, all other Cdk4/6 inhibitors like Abemaciclib, Ribociclib etc. are used in combination with various hormonal therapies and are under various trial stages to be used against hormone receptor-positive or human epidermal receptor factor 2-negative BCs.

Pre-clinical research yields inconsistent results when it comes to the advantages of mixing CDKIs especially Cdk4/6 with conventional cytotoxic chemotherapeutic treatment. In a lung cancer mouse xenograft, Abemaciclib improved gemcitabine's anticancer effects (Gelbert et al. 2014). Furthermore, combining Palbociclib with the mTOR complex 1 blocker Everolimus prevented the development of NSCLC cells in such a highly efficient manner (Gopalan et al. 2013). When Letrozole or Fulvestrant were added, monotherapeutic agent such as Ribociclib's ability to prevent tumorigenesis in Estrogen receptor-positive xenograft models significantly increased. The three-drug combo of Letrozole or Fulvestrant, Ribociclib, as well as a PI3K blocker (buparlisib [BKM120] or Alpelisib [BYL719]), showed the best tumor suppression (O'Brien et al. 2014). On the basis of such studies, a growing rise of using CDKIs in three-drug combo with hormonal therapies as well as with pi3K/AKT/mTOR pathway inhibitor. The effectiveness of first-line three-drug combo with Ribociclib is being investigated in a randomized stage 1b research in patients having metastatic Estrogen receptor-positive, HER2-negative BC (NCT01872260) (ClinicalTrials.gov. From: <https://clinicaltrials.gov/>. Accessed January 2016 (Juric et al. 2016). Several other combination therapeutic strategies are used against various types of BCs; some are under study and some have been approved by the FDA, but still, more research is needed in order to discover new medications to tackle such a devastating disease.

14.7 Novel CDKIs

A greater focus in finding specific unique CDKIs has been sparked by the current study of new CDK antagonists targeting CDKs as well as an improved knowledge of different Breast cancer subtypes and associated negative effects (Table 14.2). Keeping this in view, numerous scientists have examined various inhibitors versus CDKs. Inhibitive role played by piper longumine in case of metastatic estrogen receptor-positive breast cancer thus inhibits the cancer growth (Jeong et al. 2019). In case of triple-negative breast cancer, SR-4835 can specifically suppress Cdk 12 and 13 and repress cell growth, according to research carried out by Quereda et al. (2019). It has been seen that the pandurate A possesses several attributes including anti-tumor, G0/G1 arrest via inhibiting Cdk4/CycD1 complex (Liu et al. 2018a, b). Vanicoside B, a natural product, plays a main role as an inhibiting agent but can specifically inhibit Cdk8 as well as arrest the cell cycle in HCC38 and MDA-MB-321 cells (Kim et al. 2019). Current findings reveal that the MTH-3, a water-soluble bis(hydroxymethyl) alkanoate analog of curcumin, inhibits Cdk1 activity so as to stop the G2 to M stage from progressing in MDAMB-231 cell lines (Chang et al. 2018). Along with it, Galangi a compound obtained from genus Alpinia possesses

Table 14.2 Various types of Novel CDKIs and their function against breast cancer

S. No	Novel CDK inhibitor	Obtained from	Function
01.	Piper longuimine	Plants	It specifically suppresses the proliferation as well as metastasis of the ER-positive BC
02.	SR-4835		Specifically inhibits both CDK12 and CDK13, in turn inhibiting the cell growth
03.	Pandurate A		Anticancerous via inhibiting the CDK4/CycD1
04.	Vanicoside B	Plants	Anticancer against TNBC Plays main role as an inhibiting agent but can specifically inhibit Cdk8 as well as arrest the cell cycle in HCC38 and MDA-MB-321 cells.
05.	Galangi	Plants	It possesses anti-tumor attributes as well as opposes the growth of MCF-7 cells, that in turn results in cell death via inhibition of Cdk2,1 and, that results in arrest of the cell division cycle

anti-tumor attributes as well as opposes the growth of MCF-7 cells, which in turn results in cell death via inhibition of Cdk2,1 and arrest of the cell division cycle (Liu et al. 2018a, b).

In conclusion, breast cancer is a difficult condition that could be addressed with the CDKIs indicated in the upper text as combined with several medications, producing a positive therapeutic outcome. For example, research have shown that the combination of Palbociclib with paclitaxel has a substantially greater impact than either drug alone at hampering cell proliferation and amplifying apoptosis (Cretella et al. 2019). At last, we can conclude that various CDKIs can be combined with various novel therapies that can result in a beneficial and breakthrough discovery in the medical field that can specifically stop cell cycle progression as well as can amplify apoptosis and can slow down the process of proliferation and metastasis of various types of BCs.

14.8 Summary

Breast cancer is one of the deadly diseases associated with several characteristics that usually benefit this disease to progress as well spread to other tissues. One of the basic hallmarks associated with this disease is the dysregulated cellular cycle due to dysfunctional cyclin-dependent kinases. So, from the very past, this cell cycle is targeted as a therapeutic target, especially CDKs and in CDKs usually CDK6/4 is the main target of anticancer medications as this complex is the key controller of cellular proliferation. Nowadays, there are numerous drugs that can suppress or inhibit uncontrollable cellular proliferation of cancerous cells via targeting various CDKs, but none of the drugs is promising. So, there is a continuous need for novel drugs as well as therapies that can specifically amplify the process of apoptosis of cancerous cells as well as can slow down the uncontrolled process of cellular proliferation. In the near future, we are expecting some promising drugs that can have both the characteristics of amplifying as well as slowing down the processes of apoptosis and proliferation, respectively.

14.9 Further Readings

Modern textbooks on cell cycle and its regulation and CDKs as therapeutic strategies in various human breast cancers include those by (Giordano and Soprano 2002), (Sledge and Baselga 2012, Choudhary 2018), (Kaldis 2006), (Kaldis 2010), and (Afroze et al. 2020). Also, go through Chap. 3 of Giordano and Soprano (2002); it informs us about the CDK inhibitors and therapeutic strategies in breast cancer, and Chaps. 5 and 6 of Choudhary (2018) gives a brief idea about natural and synthetic CDK inhibitors that can result as potentially helpful medicinal strategies to curb this fatal disease. Video lectures by EMSD Open Cancer Horizons

https://www.sciencedirect.com/science/article/pii/S205970292032278X

Video lecture by HMP Education · 25-Feb-2017

https://www.youtube.com/watch?v=RXsWAvdWG0s

Video lecture by European Medical Journal · EMJ 24-Apr-2015

https://www.emjreviews.com/oncology/video/cdk-4-6-inhibitor-palbociclib-for-the-treatment-of-metastatic-breast-cancer/

References

Afroze D et al (2020) Breast cancer biology. IntechOpen

Aguilar V, Fajas L (2010) Cycling through metabolism. EMBO Mol Med 2(9):338–348

Arnold A, Papanikolaou A (2005) Cyclin D1 in breast cancer pathogenesis. J Clin Oncol 23 (18):4215–4224

Asghar U et al (2015) The history and future of targeting cyclin-dependent kinases in cancer therapy. Nat Rev Drug Discov 14(2):130–146

Barroso-Sousa R et al (2016) Clinical development of the CDK4/6 inhibitors ribociclib and abemaciclib in breast cancer. Breast Care (Basel) 11(3):167–173

Bartek J et al (2004) Checking on DNA damage in S phase. Nat Rev Mol Cell Biol 5(10):792–804

Bartkova J et al (1994) Cyclin D1 protein expression and function in human breast cancer. Int J Cancer 57(3):353–361

Bartkova J et al (2006) Oncogene-induced senescence is part of the tumorigenesis barrier imposed by DNA damage checkpoints. Nature 444(7119):633–637

Beaver JA et al (2015) FDA Approval: Palbociclib for the Treatment of Postmenopausal Patients with Estrogen Receptor–Positive, HER2-Negative Metastatic Breast CancerFDA Approval Summary for Palbociclib for ER+, HER2− MBC. Clin Cancer Res 21(21):4760–4766

Bedard PL et al (2009) Stemming resistance to HER-2 targeted therapy. J Mammary Gland Biol Neoplasia 14(1):55–66

Bray F et al (2018) Global cancer statistics 2018: GLOBOCAN estimates of incidence and mortality worldwide for 36 cancers in 185 countries. CA Cancer J Clin 68(6):394–424

Bregman DB et al (2000) Cell cycle regulation and RNA polymerase II. Front Biosci 5(5):D244–D257

Brigham et al (2012) Comprehensive molecular portraits of human breast tumours. Nature 490 (7418):61–70

Buckley MF et al (1993) Expression and amplification of cyclin genes in human breast cancer. Oncogene 8(8):2127–2133

Burd CE et al (2013) Monitoring tumorigenesis and senescence in vivo with a p16INK4a-luciferase model. Cell 152(1–2):340–351

Burkhart DL, Sage J (2008) Cellular mechanisms of tumour suppression by the retinoblastoma gene. Nat Rev Cancer 8(9):671–682

Caldon CE et al (2012) Cyclin E2 overexpression is associated with endocrine resistance but not insensitivity to CDK2 inhibition in human breast cancer cells. Mol Cancer Ther 11(7):1488–1499

Cardoso F et al (2018) 4th ESO–ESMO international consensus guidelines for advanced breast cancer (ABC 4). Ann Oncol 29(8):1634–1657

Chang L-C et al (2018) Effect of bis (hydroxymethyl) alkanoate curcuminoid derivative MTH-3 on cell cycle arrest, apoptotic and autophagic pathway in triple-negative breast adenocarcinoma MDA-MB-231 cells: an in vitro study. Int J Oncol 52(1):67–76

Chen P et al (2016) Spectrum and degree of CDK drug interactions predicts clinical performance spectrum and degree of CDK drug engagement. Mol Cancer Ther 15(10):2273–2281

Choudhary MI (2018) Frontiers in anti-cancer drug discovery. Bentham Science Publishers

Chu IM et al (2008) The Cdk inhibitor p27 in human cancer: prognostic potential and relevance to anticancer therapy. Nat Rev Cancer 8(4):253–267

Ciemerych MA et al (2002) Development of mice expressing a single D-type cyclin. Genes Dev 16 (24):3277–3289
Cooper AB et al (2006) A unique function for cyclin D3 in early B cell development. Nat Immunol 7(5):489–497
Cretella D et al (2019) Pre-treatment with the CDK4/6 inhibitor palbociclib improves the efficacy of paclitaxel in TNBC cells. Sci Rep 9(1):1–11
Cui X et al (2005) Biology of progesterone receptor loss in breast cancer and its implications for endocrine therapy. J Clin Oncol 23(30):7721–7735
Dean J et al (2010) Therapeutic CDK4/6 inhibition in breast cancer: key mechanisms of response and failure. Oncogene 29(28):4018–4032
DeMichele A et al (2013) A phase II trial of an oral CDK 4/6 inhibitor, PD0332991, in advanced breast cancer. Am Soc Clin Oncol 31:519
Dickler MN et al (2017) MONARCH 1, a phase II study of abemaciclib, a CDK4 and CDK6 inhibitor, as a single agent, in patients with refractory HR+/HER2− metastatic breast cancer. Clin Cancer Res 23(17):5218–5224
Dickson C et al (1995) Amplification of chromosome band 11q13 and a role for cyclin D1 in human breast cancer. Cancer Lett 90(1):43–50
Diehl JA (2002) Cycling to cancer with cyclin D1. Cancer Biol Ther 1(3):226–231
Drapkin R et al (1996) Human cyclin-dependent kinase-activating kinase exists in three distinct complexes. Proc Natl Acad Sci U S A 93(13):6488–6493
Etemadmoghadam D et al (2013) Synthetic lethality between CCNE1 amplification and loss of BRCA1. Proc Natl Acad Sci U S A 110(48):19489–19494
Ethier J et al (2018) Outcomes of single versus double hormone receptor–positive breast cancer A GEICAM/9906 sub-study. Eur J Cancer 94:199–205
Finn RS et al (2016a) Targeting the cyclin-dependent kinases (CDK) 4/6 in estrogen receptor-positive breast cancers. Breast Cancer Res 18(1):1–11
Finn RS et al (2015) The cyclin-dependent kinase 4/6 inhibitor palbociclib in combination with letrozole versus letrozole alone as first-line treatment of oestrogen receptor-positive, HER2-negative, advanced breast cancer (PALOMA-1/TRIO-18): a randomised phase 2 study. Lancet Oncol 16(1):25–35
Finn RS et al (2009) PD 0332991, a selective cyclin D kinase 4/6 inhibitor, preferentially inhibits proliferation of luminal estrogen receptor-positive human breast cancer cell lines in vitro. Breast Cancer Res 11(5):1–13
Finn RS et al (2016b) Palbociclib and letrozole in advanced breast cancer. N Engl J Med 375 (20):1925–1936
Flaherty KT et al (2012) Phase I, dose-escalation trial of the oral cyclin-dependent kinase 4/6 inhibitor PD 0332991, administered using a 21-day schedule in patients with advanced CancerPhase I, dose-escalation trial of PD 0332991, 4-week cycle. Clin Cancer Res 18 (2):568–576
Fry DW et al (2004) Specific inhibition of cyclin-dependent kinase 4/6 by PD 0332991 and associated antitumor activity in human tumor xenografts. Mol Cancer Ther 3(11):1427–1438
Gavet O, Pines J (2010) Progressive activation of CyclinB1-Cdk1 coordinates entry to mitosis. Dev Cell 18(4):533–543
Gelbert LM et al (2014) Preclinical characterization of the CDK4/6 inhibitor LY2835219: in-vivo cell cycledependent/independent anti-tumor activities alone/in combination with gemcitabine. Investig New Drugs 32(5):825–837
Gillett C et al (1994) Amplification and overexpression of cyclin D1 in breast cancer detected by immunohistochemical staining. Cancer Res 54(7):1812–1817
Giordano A, Soprano KJ (2002) Cell cycle inhibitors in cancer therapy: current strategies. Humana Press
Goel S et al (2016) Overcoming therapeutic resistance in HER2-positive breast cancers with CDK4/ 6 inhibitors. Cancer Cell 29(3):255–269

Goetz MP et al (2017) MONARCH 3: abemaciclib as initial therapy for advanced breast cancer. J Clin Oncol 35(32)
Gopalan PK et al (2013) Inhibitory effect of the CDK4/6 inhibitor, PD 0332991, is enhanced by mTOR inhibition in Non-Small Cell Lung Cancer (NSCLC). Oncotarget 73(8_Supplement):693
Hammond MEH et al (2010) American Society of Clinical Oncology/College of American Pathologists guideline recommendations for immunohistochemical testing of estrogen and progesterone receptors in breast cancer (unabridged version). Arch Pathol Lab Med 134(7): e48–e72
Harbour JW et al (1999) Cdk phosphorylation triggers sequential intramolecular interactions that progressively block Rb functions as cells move through G1. Cell 98(6):859–869
Hartwell LH (1974) Saccharomyces cerevisiae cell cycle. Bacteriol Rev 38(2):164–198
Hartwell LH (2002) Nobel lecture: yeast and cancer. Biosci Rep 22(3):373–394
Hershko DDJFO (2010) Cyclin-dependent kinase inhibitor p27 as a prognostic biomarker and potential cancer therapeutic target. Future Oncol 6(12):1837–1847
Hortobagyi GN et al (2016) Ribociclib as first-line therapy for HR-positive, advanced breast cancer. N Engl J Med 375(18):1738–1748
Infante JR et al (2014) A phase I study of the single-agent CDK4/6 inhibitor LEE011 in pts with advanced solid tumors and lymphomas. Am Soc Clin Oncol 32:2528
Jeong CH et al (2019) Piperlongumine induces cell cycle arrest via reactive oxygen species accumulation and IKKβ suppression in human breast cancer cells. Antioxidants (Basel) 8 (11):553
Jiang W et al (1992) Amplification and expression of the human cyclin D gene in esophageal cancer. Cancer Res 52(10):2980–2983
Juric D et al (2016) Abstract P3–14-01: phase Ib/II study of ribociclib and alpelisib and letrozole in ER+, HER2–breast cancer: safety, preliminary efficacy and molecular analysis. Clin Cancer Res 76(4_Supplement).:P3-14-01-P13-14-01
Kaldis P (2006) Cell Cycle Regulation. Springer, Berlin Heidelberg
Kaldis P (2010) Cell Cycle Regulation. Springer, Berlin Heidelberg
Karst AM et al (2014) Cyclin E1 deregulation occurs early in secretory cell transformation to promote formation of fallopian tube–derived high-grade serous ovarian cancers. Cancer Res 74 (4):1141–1152
Kastan MB, Bartek JJN (2004) Cell-cycle checkpoints and cancer. Nature 432(7015):316–323
Kato J-Y et al (1993) Direct binding of cyclin D to the retinoblastoma gene product (pRb) and pRb phosphorylation by the cyclin D-dependent kinase CDK4. Genes Dev 7(3):331–342
Kim D et al (2019) Antitumor activity of vanicoside B isolated from Persicaria dissitiflora by targeting CDK8 in triple-negative breast cancer cells. J Nat Prod 82(11):3140–3149
Knudsen ES et al (1998) Inhibition of DNA synthesis by RB: effects on G1/S transition and S-phase progression. Genes Dev 12(15):2278–2292
Knudsen ES, Knudsen KE (2008) Tailoring to RB: tumour suppressor status and therapeutic response. Nat Rev Cancer 8(9):714–724
Knudsen K et al (2006) Cyclin D1: polymorphism, aberrant splicing and cancer risk. Oncogene 25 (11):1620–1628
Kops GJ et al (2005) On the road to cancer: aneuploidy and the mitotic checkpoint. Nat Rev Cancer 5(10):773–785
Kuhn E et al (2014) Frequent CCNE1 amplification in endometrial intraepithelial carcinoma and uterine serous carcinoma. Mod Pathol 27(7):1014–1019
Kushner JA et al (2005) Cyclins D2 and D1 are essential for postnatal pancreatic β-cell growth. Mol Cell Biol 25(9):3752–3762
Lange CA, Yee D (2011) Killing the second messenger: targeting loss of cell cycle control in endocrine-resistant breast cancer. Endocr Relat Cancer 18(4):C19
LaPak KM, Burd CEJMCR (2014) The molecular balancing act of p16INK4a in cancer and aging. Mol Cancer Res 12(2):167–183

Lapenna S, Giordano A (2009) Cell cycle kinases as therapeutic targets for cancer. Nat Rev Drug Discov 8(7):547–566

Leonard JP et al (2012) Selective CDK4/6 inhibition with tumor responses by PD0332991 in patients with mantle cell lymphoma. Blood 119(20):4597–4607

Lilly E (2015) Company. Lilly receives FDA breakthrough therapy designation for abemaciclib–a CDK 4 and 6 inhibitor–in advanced breast cancer. https://investor.lilly.com/news-releases/news-release-details/lilly-receives-fda-breakthrough-therapy-designation-abemaciclib.

Lim S, Kaldis PJD (2013) Cdks, cyclins and CKIs: roles beyond cell cycle regulation. Development 140(15):3079–3093

Liu D et al (2018a) Galangin induces apoptosis in MCF-7 human breast cancer cells through mitochondrial pathway and phosphatidylinositol 3-kinase/Akt inhibition. Pharmacology 102(1–2):58–66

Liu Q et al (2018b) Panduratin A inhibits cell proliferation by inducing G0/G1 phase cell cycle arrest and induces apoptosis in breast cancer cells. Biomol Ther (Seoul) 26(3):328

Lukas J et al (1995) Cyclin D1 is dispensable for G1 control in retinoblastoma gene-deficient cells independently of cdk4 activity. Mol Cell Biol 15(5):2600–2611

Lukas J et al (1997) Cyclin E-induced S phase without activation of the pRb/E2F pathway. Genes Dev 11(11):1479–1492

Lundberg AS et al (1998) Functional inactivation of the retinoblastoma protein requires sequential modification by at least two distinct cyclin-cdk complexes. Mol Cell Biol 18(2):753–761

Ma T et al (2000) Cell cycle–regulated phosphorylation of p220NPAT by cyclin E/Cdk2 in Cajal bodies promotes histone gene transcription. Genes Dev 14(18):2298–2313

Malumbres M, Barbacid M (2001) To cycle or not to cycle: a critical decision in cancer. Nat Rev Cancer 1(3):222–231

Malumbres M, Barbacid M (2009) Cell cycle, CDKs and cancer: a changing paradigm. Nat Rev Cancer 9(3):153–166

Malumbres M, Barbacid M (2005) Mammalian cyclin-dependent kinases. Trends Biochem Sci 30 (11):630–641

Malumbres MJG (2014) Cyclin-dependent kinases. Genome Biol 15(6):1–10

Massagué JJN (2004) G1 cell-cycle control and cancer. Nature 432(7015):298–306

Matsushime H et al (1991) Colony-stimulating factor 1 regulates novel cyclins during the G1 phase of the cell cycle. Cell 65(4):701–713

Mehraj U et al (2022a) Expression pattern and prognostic significance of baculoviral inhibitor of apoptosis repeat-containing 5 (BIRC5) in breast cancer: a comprehensive analysis. Adv Cancer Biol Metastasis 4:100037

Mehraj U et al (2021a) Tumor microenvironment promotes breast cancer chemoresistance. Cancer Chemother Pharmacol 87(2):147–158

Mehraj U et al (2021b) Prognostic significance and targeting tumor-associated macrophages in cancer: new insights and future perspectives. Breast Cancer 28(3):539–555

Mehraj U et al (2022b) Cryptolepine targets TOP2A and inhibits tumor cell proliferation in breast cancer cells-an in vitro and in silico study. Anti Cancer Agents Med Chem 22(17):3025–3037

Mehraj U et al (2022c) Expression pattern and prognostic significance of CDKs in breast cancer: an integrated bioinformatic study. Cancer Biomark 34(3):505–519(Preprint):1–15

Michaloglou C et al (2005) BRAFE600-associated senescence-like cell cycle arrest of human naevi. Nature 436(7051):720–724

Mir MA (2015) Developing costimulatory molecules for immunotherapy of diseases. Academic Press, pp 1–299, ISBN: 9780128025857. https://doi.org/10.1016/C2014-0-02898-5

Mir MA, Agrewala JN (2008) Signaling through CD80: an approach for treating lymphomas. Expert Opin Ther Targets 12(8):969–979

Mir MA, Mehraj U (2019) Double-crosser of the immune system: macrophages in tumor progression and metastasis. Curr Immunol Rev 15(2):172–184

Mir MA et al (2020a) Nanobodies: the "magic bullets" in therapeutics, drug delivery and diagnostics. Hum Antibodies 28(1):29–51

Mir MA et al (2020b) Targeting different pathways using novel combination therapy in triple negative breast cancer. Curr Cancer Drug Targets 20(8):586–602

Mir MA, Sofi S, Qayoom H (2022a) Conventional adjuvant chemotherapy in combination with surgery, radiotherapy, and other specific targets. Elsevier, pp 95–120

Mir MA, Sofi S, Qayoom H (2022b) Different drug delivery approaches in combinational therapy in TNBC. Elsevier, pp 201–230

Morikawa A, Henry NL (2015) Palbociclib for the Treatment of Estrogen Receptor–Positive, HER2-Negative Metastatic Breast CancerPalbociclib for Metastatic Breast Cancer. 21(16): 3591–3596

Motokura T et al (1991) A novel cyclin encoded by a bcl1-linked candidate oncogene. Nature 350 (6318):512–515

Musgrove EA et al (2011) Cyclin D as a therapeutic target in cancer. Nat Rev Cancer 11(8):558–572

Nemet J et al (2014) The two faces of Cdk8, a positive/negative regulator of transcription. Biochimie 97:22–27

Nurse PJBR (2002) Cyclin dependent kinases and cell cycle control. Biosci Rep 22(5–6):487–499

O'Brien NA et al (2014) In vivo efficacy of combined targeting of CDK4/6, ER and PI3K signaling in ER+ breast cancer. Cancer Res 74(19_Supplement):4756–4756

Okuda M et al (2000) Nucleophosmin/B23 is a target of CDK2/cyclin E in centrosome duplication. Cell 103(1):127–140

Pagano M et al (1993) Regulation of the cell cycle by the cdk2 protein kinase in cultured human fibroblasts. J Cell Biol 121(1):101–111

Pernas S et al (2018) CDK4/6 inhibition in breast cancer: current practice and future directions. Ther Adv Med Oncol 10:1758835918786451

Perou CM et al (2000) Molecular portraits of human breast tumours. Nature 406(6797):747–752

Qayoom H, Bhat BA (2020) U Mehraj U, Mir MA (2020) rising trends of cancers in Kashmir valley: distribution pattern, incidence and causes. J Oncol Res Treat 5(150):2

Qayoom H et al (2022) Expression patterns and therapeutic implications of CDK4 across multiple carcinomas: a molecular docking and MD simulation study. Med Oncol 39(10):1–13

Qayoom H et al (2021) An insight into the cancer stem cell survival pathways involved in chemoresistance in triple-negative breast cancer. Future Oncol 17(31):4185–4206

Quereda V et al (2019) Therapeutic targeting of CDK12/CDK13 in triple-negative breast cancer. Cancer Cell 36(5):545–558. e547

Raub TJ et al (2015) Brain exposure of two selective dual CDK4 and CDK6 inhibitors and the antitumor activity of CDK4 and CDK6 inhibition in combination with temozolomide in an intracranial glioblastoma xenograft. Drug Metab Dispos 43(9):1360–1371

Romero-Pozuelo J et al (2020) Cdk4 and Cdk6 couple the cell-cycle machinery to cell growth via mTORC1. Cell Rep 31(2):107504

Santamaría D et al (2007) Cdk1 is sufficient to drive the mammalian cell cycle. Nature 448 (7155):811–815

Satyanarayana A, Kaldis PJO (2009) Mammalian cell-cycle regulation: several Cdks, numerous cyclins and diverse compensatory mechanisms. Oncogene 28(33):2925–2939

Scaltriti M et al (2011) Cyclin E amplification/overexpression is a mechanism of trastuzumab resistance in HER2+ breast cancer patients. Proc Natl Acad Sci U S A 108(9):3761–3766

Serrano M et al (1997) Oncogenic ras provokes premature cell senescence associated with accumulation of p53 and p16INK4a. Cell 88(5):593–602

Sever-Chroneos Z et al (2001) Retinoblastoma tumor suppressor protein signals through inhibition of cyclin-dependent kinase 2 activity to disrupt PCNA function in S phase. Mol Cell Biol 21 (12):4032–4045

Shapiro G et al (2013) A first-in-human phase I study of the CDK4/6 inhibitor, LY2835219, for patients with advanced cancer. Am Soc Clin Oncol 31:2500

Shapiro GI (2006) Cyclin-dependent kinase pathways as targets for cancer treatment. J Clin Oncol 24(11):1770–1783

Sherr CJ (1996) Cancer cell cycles. Science 274(5293):1672–1677
Sherr CJ (1995) D-type cyclins. Trends Biochem Sci 20(5):187–190
Sicinski P et al (1996) Cyclin D2 is an FSH-responsive gene involved in gonadal cell proliferation and oncogenesis. Nature 384(6608):470–474
Sledge GW, Baselga J (2012) Targeted therapies in breast cancer. Clinical Publishing
Sledge GW et al (2017) MONARCH 2: Abemaciclib in combination with fulvestrant in patients with HR+/HER2-advanced breast cancer who progressed on endocrine therapy. Am Soc Clin Oncol 35:1000
Sobhani N et al (2019) Updates on the CDK4/6 inhibitory strategy and combinations in breast cancer. Cells 8(4):321
Sofi S et al (2022a) Targeting cyclin-dependent kinase 1 (CDK1) in cancer: molecular docking and dynamic simulations of potential CDK1 inhibitors. Med Oncol 39(9):1–15
Sofi S et al (2022b) Cyclin-dependent kinases in breast cancer: expression pattern and therapeutic implications. Med Oncol 39(6):1–16
Sofi S (2021) Novel biomarkers in breast cancer. Elsevier, pp 47–92
Spofford LS et al (2006) Cyclin D3 expression in melanoma cells is regulated by adhesion-dependent phosphatidylinositol 3-kinase signaling and contributes to G1-S progression. J Biol Chem 281(35):25644–25651
Spring L et al (2016) Targeting the cyclin D–cyclin-dependent kinase (CDK) 4/6–retinoblastoma pathway with selective CDK 4/6 inhibitors in hormone receptor-positive breast cancer: rationale, current status, and future directions. J Biol Chem 21(113):65
Thangavel C et al (2011) Therapeutically activating RB: reestablishing cell cycle control in endocrine therapy-resistant breast cancer. Endocr Relat Cancer 18(3):333–345
Tolaney SM et al (2017) Abemaciclib for the treatment of brain metastases (BM) secondary to hormone receptor positive (HR+), HER2 negative breast cancer. Am Soc Clin Oncol 35:1019
Tsang J, Tse GM (2020) Molecular classification of breast cancer. Adv Anat Pathol 27(1):27–35
van den Heuvel S, Harlow EJS (1993) Distinct roles for cyclin-dependent kinases in cell cycle control. Science 262(5142):2050–2054
Vermeulen K et al (2003) The cell cycle: a review of regulation, deregulation and therapeutic targets in cancer. Cell Prolif 36(3):131–149
Vora SR et al (2014) CDK 4/6 inhibitors sensitize PIK3CA mutant breast cancer to PI3K inhibitors. Cancer Cell 26(1):136–149
Watanabe N et al (1995) Regulation of the human WEE1Hu CDK tyrosine 15-kinase during the cell cycle. EMBO J 14(9):1878–1891
Witkiewicz AK et al (2011) The meaning of p16ink4a expression in tumors: functional significance, clinical associations and future developments. Cell Cycle 10(15):2497–2503
Yam C et al (2017) Targeting the molecular subtypes of triple negative breast cancer: understanding the diversity to progress the field. Oncologist 22(9):1086–1093
Yu Q et al (2005) Ras and Myc can drive oncogenic cell proliferation through individual D-cyclins. Oncogene 24(47):7114–7119

CDk Inhibitor for Treatment of Breast Cancer

15

Manzoor Ahmad Mir, Aabida Gul, Shazia Sofi, and M. Sultan Khan

15.1 Introduction

Breast cancer, defined as the uncontrolled growth of mammary glands, is a global disease that influences over 1.3 million people every year and accounts for around 23% of all cancers (Aguirre-Ghiso et al. 2001). The average survival time for women with metastatic breast cancer (MBC) is slightly over 2 years (Mehraj et al. 2021). Among different types of breast cancers, hormone receptor-positive (HR+) breast cancer is the most common subtype, and it affects thousands of people each year. CDKs are the important regulators of cell cycle and cell division (Altucci et al. 1997). Cell undergoes divisions in a cyclic manner (cell cycle) where it passes through different stages and at phase transitions different check points play an important role to allow or block the cell division progression. Retinoblastoma (Rb) the tumor suppressor protein controls the cell cycle at a crucial and early stage by binding to E2 transcription factor (E2F) and inhibiting the G1/S transition and its (Rb) inactivation allows cell division to continue (Asghar et al. 2015). However, upon binding of cyclin D with CDK4/6, the Rb gets phosphorylated and allows the E2F to be released and thereby resulting in cell cycle progression. The involvement of hormone estrogen in the CDK4/6–Rb axis develops estrogen receptor-positive (ER+) breast cancer has also been reported by accelerating the G1/S transition (Balduzzi et al. 2014, Qayoom et al. 2021). The estrogen binds to ER-alpha triggers cyclin-D1transcription, followed by CDK4/6 activation and Rb phosphorylation, which leads to cell cycle progression. The untimely binding of

M. A. Mir (✉) · A. Gul · S. Sofi
Department of Bioresources, School of Biological Sciences, Srinagar, J&K, India
e-mail: drmanzoor@kashmiruniversity.ac.in

M. S. Khan
Neurobiology and Molecular Chronobiology Laboratory, Department of Animal Biology, School of Life Sciences, University of Hyderabad, Hyderabad, India

M. A. Mir (ed.), *Therapeutic Potential of Cell Cycle Kinases in Breast Cancer*,
https://doi.org/10.1007/978-981-19-8911-7_15

cyclin and CDKs in the G1 phase has a potential to cause uncontrolled cellular growth and the introduction of CDK (4/6) inhibitors has been a revolutionary area of research for the treatment of HR + MBC. Due to their specificity, selectivity, and reversible CDK4/6 inhibitory activity, drugs like Palbociclib, Ribociclib, and Abemaciclib have recently got Food and Drug Administration (FDA) approved for HR+ MBC in combination with particular endocrine therapy (Mir 2022a, b, Qayoom et al. 2022). Targeting tumors with CDK inhibition is an attempt to resume cell cycle regulation has proven to be an appealing alternative in treating breast cancers and has helped to increase the survival of cancer patients (Austreid et al. 2014).

Cancer cells show resistance to the targeted therapy which can be of two types; de-novo or in-vivo referring to built-in mechanisms of resistance and resistance developed over time to targeted therapies, respectively (Mir 2022a, b). Need of ER/PR expression is the most commonly identified mechanism of de novo endocrine therapy resistance. This route may be exploited by cancer cells through a variety of ways, including changes in CDK inhibitor proteins (Asghar et al. 2015);).

15.2 The Cell Cycle and its Function

Many malignancies use the cell cycle as a target for tumor development (Asghar et al. 2015). The two most important physiological processes that govern tissue homeostasis are cell division and cell death, and any abnormalities in two important processes lead to the development of immortal cancerous cells. During the division of a cell, it undergoes four different phases/stages including G1, S, G2, and M phase. G1 represents the first resting phase during which the majority of cell advancement and maturation take place. The cell cycle then either enters the DNA synthesis phase or S phase. After synthetic phase, cell can undergo another phase of rest (G2) where it grows and synthesizes proteins and organelles to be used in the next mitotic phase (M phase) which represents the actual divisional phase of cell cycle, leading to the formation of two daughter cells. The newly formed daughter cells may start a fresh cycle of division by continuing G1/S transition and continue the process or may undergo a temporary resting phase (G0) depending on the type of cells and involving special regulatory checkpoints (Asghar et al. 2015). Critical regulatory checkpoints involved in the gap stages (G1 and G2) are shown in Fig. 15.1 (Sofi et al. 2022a, b).

15.3 Cell-Cycle Control and Cyclins

The primary requirement for cell cycle involves the succeeding activation and inactivation of the serine or threonine CDKs in eukaryotes (Crasta et al. 2006). These enzymes contain a CDK catalytic subunit whose expression is sustained all over the cell cycle, as well as a regulatory subunit whose activity is at different stages is regulated by transcriptional regulation, subcellular localization, and protein degradation etc. (Sherr 1996, Sofi et al. 2022a, b). The activation of D-type cyclins (cyclins D1, D2, and D3) by mitogenic stimulation of growth-arrested cells is

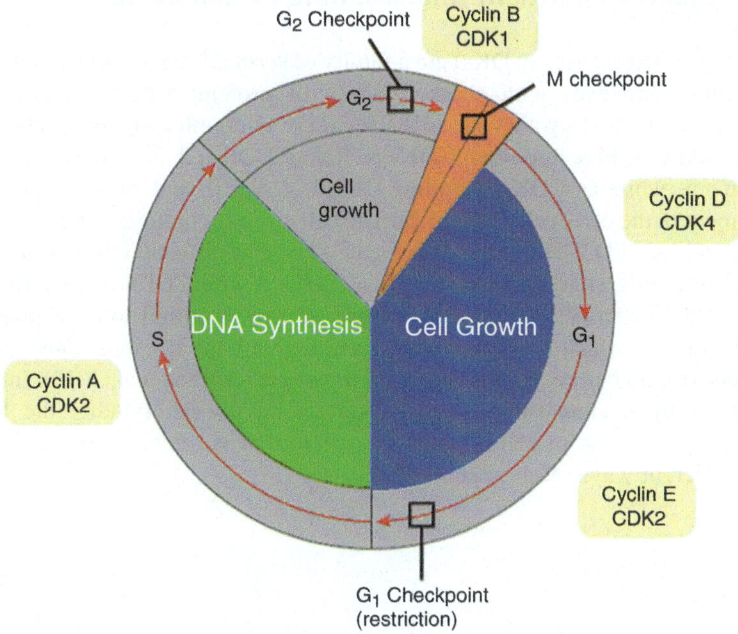

Fig. 15.1 Stages of typical Cell cycles and its phases. Between the G1 and S phases of interphase, the restriction point occurs. Between the G2 and M phases, the G2-M checkpoint occurs. During the M phase, there is a spindle checkpoint. Each phase's key cyclins are highlighted here in this figure

thought to be the major activity of these molecules, which link extracellular messages to the cell cycle and its mechanisms. The D-type cyclins bind to CDK (4/6) preferentially and phosphorylates critical downstream substrates, primarily pRb and other proteins including p107 and p130 (Dyson 1998). The partial phosphorylation of Rb allows bound transcription factors, particularly those of the E2F family, to be released, which then bind to the upstream regulatory elements of genes whose transcription and function are required for S phase advancement. Because cyclin E1 is an E2F target gene, partial phosphorylation of retinoblastoma protein (pRb) causes cyclin-E protein expression and the creation of active cyclin E-Cdk2 complexes in the mid-to late-G1 phase. The inhibitory effect of pRb on G1 to S phase development is eliminated when it is completely phosphorylated. The D cyclins titrate the balance of the Kip inhibitors, p21 Waf1 = Cip1 andp27Kip1, between cyclin E–Cdk2 complexes, in which they forbid kinase activity, and cyclin D–Cdk4/6 complexes, in which they act as stabilizing assembly factors, in addition to this primarily transcriptional mechanism control G1 phase cyclin–Cdk complexes (Mehraj et al. 2022a, b). These processes ensure that the D-type and E cyclins work together in order to control cell cycle progression after mitogenic activation. As a result, genetic alterations that disrupt this homeostatic process at any level are likely to result in the deprivation of normal growth regulation leading to oncogenesis.

15.4 Cyclin-Dependent Kinases: Role in Cell Cycle

Cyclin-dependent kinases (CDKs) are a family of serine/threonine kinases that work by forming complexes with another family of proteins called cyclins. In this heterodimer complex, cyclins act as the regulatory subunits whereas CDKs act as catalytic subunits. Five important CDKs including CDK1, 2, 3, 4, and 6 show direct involvement, while as CDKs including CDK7, 8, 9, 10, and 11 show an indirect involvement in the cell cycle (Fig. 15.2). The cell cycle is controlled and managed by specific cyclins and CDKs which regulate cell cycle at different checkpoints during the cell cycle progression. Cyclins-D are associated with CDK4 and CDK6 where (D1, D2, D3) is encouraging the cell cycle to advance from G1 to the S phase. The E2F gene family, which is dormant when coupled with the Rb, must finish essential gene transcription before undergoing G1-S transition (Asghar et al. 2015). Important role played by cyclins and dependent kinases in phosphorylating the Rb and releasing E2F leading to cell division is a route that is often exploited by cancer cells in different ways (Mehraj et al. 2022a, b).

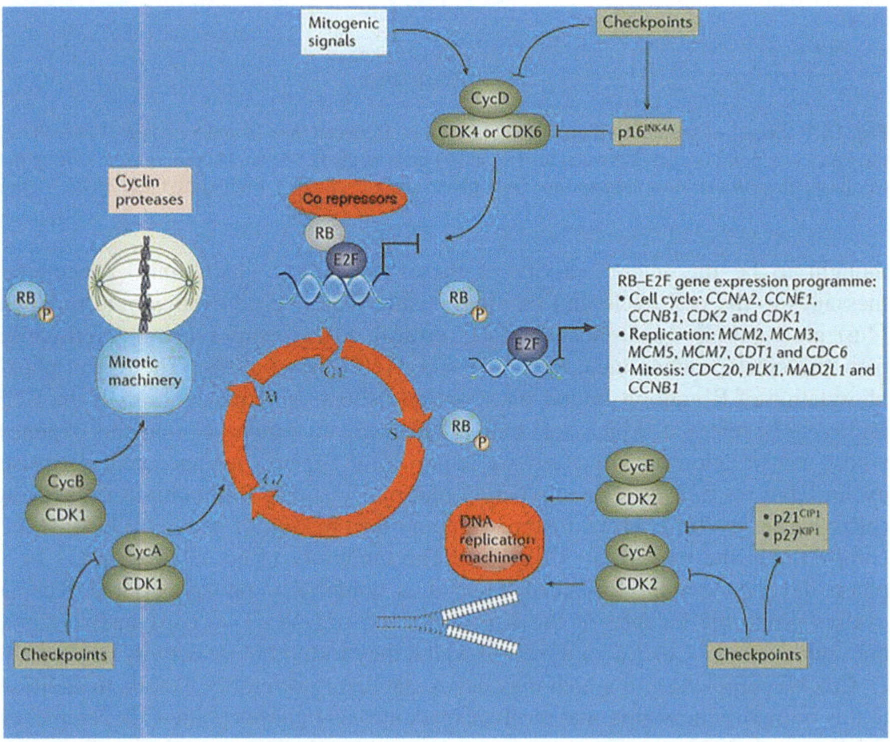

Fig. 15.2 CDKs are accountable for cell cycle progression. Anti-proliferative checkpoints decrease CDK4 and CD6 activity or elevate the expression of the CDK4 and CDK6 inhibitor p16INK4A, whereas mitogenic signals activate CDK4 and CDK6 and encourage entrance into the cell cycle

Two important cyclin-dependent kinases 4 and 6, (CDK4 and CDK6) phosphorylates numerous critical substrates, including the tumor suppressor retinoblastoma protein, permit the E2F transcription factors to coordinate a gene expression program. Both of these kinases are involved in transcriptionally activating type-E and type-A cyclins, where CDK4 activates cyclin E and CDK6 activates cyclin A. It also activates CDK2, which phosphorylates RB and also the starts replication of its DNA. Checkpoints can reduce CDK2 activity directly or by involving inhibitor proteins like p21CIP1 and p27KIP1 also known as CDK interacting protein/kinase inhibitory proteins (CIP/KIP). CDK1–Cyc A and CDK1–Cyc B complexes arise after DNA replication is completed, and then phosphorylates the target in the G2 phase. The default mechanism of cell division involves activation of CDK1–CycB complex and moves toward mitotic phase in the absence of any DNA damage and after proper chromosomal segregation. There are, however, effective checkpoints to regulate CDK1 and control or stop mitotic progression. Entering into the anaphase and generating two daughter cells in the new G1 phase of the cell cycle necessitate the subsequent breakdown of CycB. RB is dephosphorylated during M-G1 transition, making cell cycle once again susceptible to mitogenic and anti-proliferative signals (Mehraj et al. 2022a, b).

15.5 Proliferation of Breast Epithelial Cells and Cyclins

G1-cyclin expression and regulation in mammary epithelial cells were first studied in cultured, normal, and cancerous human breast epithelial. The fact that "Normal" cells were produced from basal epithelial cells, which were not known to give rise to breast carcinomas at the time, and the breast carcinoma cells emerged from breast luminal epithelial cells, confused the interpretation of these researches. Both D1 and D3 cyclins were expressed in the later cells, but not D2, which appeared to be restricted to cultures of "normal" basal epithelial cells (Bartek and Lukas 2003). The expected transcriptional activation of cyclin-D1, D3, and E in early, middle, and late phases of G1, respectively, with concurrent formation of active Cdk complexes and progression into S phase was observed when cancer cell lines (breast) were stimulated with numerous growth factors, including EGF, IGF, and heregulin families (Sherr and Roberts 2004).

Studies of actions of estrogens and progestins, which were previously thought to regulate cell cycle progression through impacts on G1 phase progression, revealed some new cell cycle regulatory mechanisms. The essential part of estrogen in the enhancement of breast cancer has sparked research into the relationship between estrogen and the cell cycle machinery (Sherr and Roberts 1999). Estrogen-induced transcriptional activation of c-myc, cyclin D1, cyclinD1–Cdk4, cyclin E–Cdk2 complexes, pRb, and cell cycle progression in breast cancer cells arrested in G0/G1 by prior treatment with estrogen antagonists (Poon et al. 1994). The fact that these processes could be blocked by antisense oligomers or antibodies to cyclin D1 and that inducible production of cyclin D1 or c-myc could recapitulate the effects of estrogen in this paradigm suggested that cyclin D1 played a key role (Prall et al.

1998, Mir 2015). The activation of cyclin E–Cdk2 occurred in the mid-G1 phase of the response to estrogen and was not accompanied by significant changes in cyclin E protein levels. Rather, cyclin is a protein that takes part in cell division.

15.6 Breast Cancer and Cyclins

Depletion of p21Waf1 from various complexes appears to be the primary activator of E–Cdk2. This effect was caused by p21Waf1 sequestration into recently established cyclin D1–Cdk4 complexes at the expense of cyclinE–p21Waf1–Cdk2 inhibitory complexes induced by anti-estrogen, as well as estrogen-mediated inhibition of p21Waf1 transcription, which allowed newly synthesized cyclin E forming E-Cdk2 complex in the absence of any inhibitor (Prall et al. 1998). Since c-myc is elevated within the first hour of estrogen stimulation, the latter process could be the result of c-myc-mediated transcriptional repression of p21. Role of estrogen is shown in Fig. 15.3.

Progestins are growth inhibitory and cause the arrest of the cell cycle in G1 phase in some model systems. Down-regulated cyclin D1 and E, along with activation of Cdk inhibitor p18Ink4c, are associated with growth arrest. This INK4 inhibitor prevents the development of complex between cyclin D and Cdk4, resulting in the cyclin–Cdk–inhibitor complexes being relegated and the availability ofp27Kip1 being increased, allowing inhibitory cyclin E–Cdk2-p27Kip1complexes to form (Swarbrick et al. 2000). Following progestin administration, both cyclin D–Cdk4 complex and cyclin E–Cdk2 activities are suppressed, causing decreased pRb and G1 arrest (Musgrove et al. 2001). Hence steroids such as estrogen and progesterone act as important regulators of cell cycle and can promote or inhibit cell cycle development by interconnecting with various targets in the pRb pathway. As a

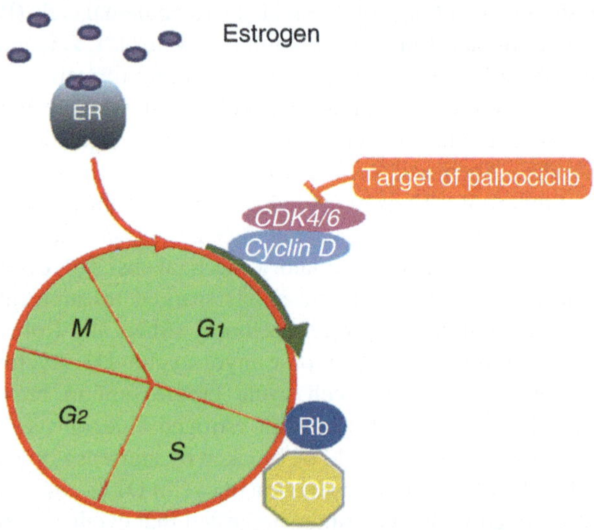

Fig. 15.3 Role of Palbociclib in preventing the progression of cell cycle from G1 to the S phase by inhibiting CDK4/6. ER = receptor for estrogen. (Adapted and modified from Pfizer inc. 2016)

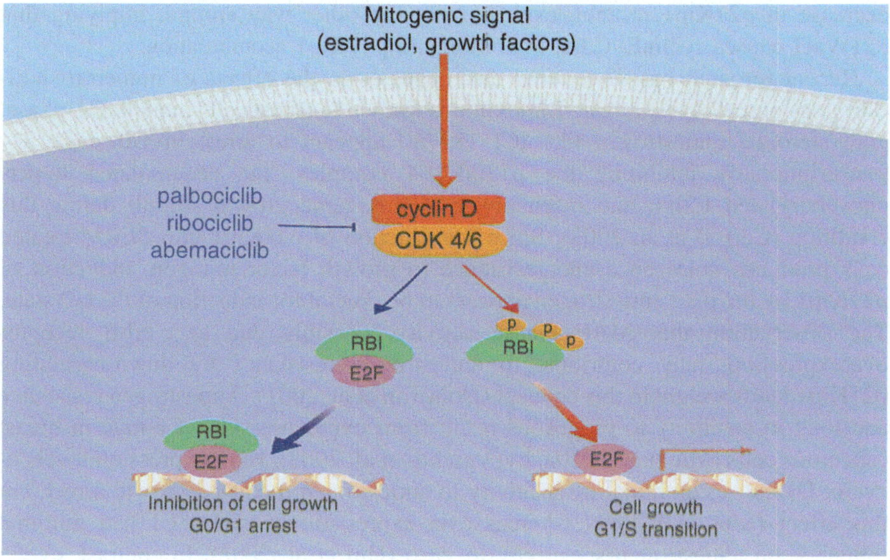

Fig. 15.4 Diagrammatic representation of the function of CDK4/6 inhibitors in cancer cells. External mitogenic signals encourage CDK4/6 and cyclin D complex formation. The release of the E2F, G1 to S phase transition, and the hyperphosphorylation of RB1 are all made possible by the CDK4/6-cyclin D complex, which promotes cell development. The phosphorylation of RB1, which is still coupled to the E2F transcription factor, is inhibited by CDK4/6 inhibitors such as Palbociclib, Ribociclib, or abemaciclib, which shows to cell G1/S cycle arrest and the reduction of cell growth

result, disruptions in this system significantly regulate these mechanisms, as well as steroid sensitivity and responsiveness. Such changes could have far-reaching implications in developing a novel breast cancer therapy.

Anti-estrogens known to cease cell cycle in G0/G1 phase are still used for treating breast cancer (Sutherland et al. 1983). Their effects on certain molecules in the pRb pathway have piqued people's interest. Reduced expression of D1 cyclin, inactivation of its complex with Cdk4 (cyclinD1–Cdk4), and reduced phosphorylation levels of Rb are all linked to anti-estrogen-induced cell cycle arrest (Agarwal et al. 1995) (Fig. 15.4). Inhibition of cyclinE–Cdk2 in Synthetic phase also dependents on the recruitment of p21Waf1 to cyclinE–Cdk2 complexes, as indicated by the studies of Carroll et al. 2000 (Carroll et al. 2003). Less expression levels of cyclin D1 gene, which is relying on anti-estrogen suppression of c-myc gene expression, is required for recruiting p21Waf1 to cyclin E–Cdk2 complexes. Indeed, antisense oligonucleotide inhibition of c-myc expression to levels that resemble anti-estrogen-induced decreased expression is sufficient to trigger the same cascade of events described above for anti-estrogen inhibition of breast cancer cell proliferation (Casimiro and Pestell 2012). Since p27Kip1 abrogates antiestrogen-induced cell cycle arrest in MCF–7 cells, it is also required for anti-estrogen-induced cell cycle arrest (Mir and Mehraj 2019)((Chen et al. 1997). However, antisense therapy of p21Waf1 causes a decrease in p27Kip1 protein level, but not the other way around, implying that p21Waf1 causes cyclinE–Cdk2 inhibition and p27Kip1 accumulation.

Recent research has discovered discrepancies in the effects of numerous anti-estrogens on cell cycle arrest. Tamoxifen is known to arrest cells in early G1 phase, pure steroidal antiestrogen like ICI 182780 appears to arrest in G0 stage. By transcriptionally inhibiting the p130/E2F4 complex, the accumulated hyper-phosphorylated E2F4, and insensitivity to mitogenic growth stimuli define this condition (Carroll et al. 2003). Since transduction of p27Kip1 into SERM-treated cells produces quiescence and resistance to growth factor mitogen, induction of p27Kip1 by the pure antiestrogen appears to be crucial for induction of the G0 state. The observation that MAP kinase activity, possibly due to c-erbB receptor overexpression, may contribute to antiestrogen resistance by down-regulating p27Kip1 lends weight to this concept (Donovan et al. 2001). Antiestrogen resistance and total insensitivity to progestins result from overexpression of c-myc in breast cancerous cells (Mir et al. 2020a, b) (Venditti et al. 2002). High expression levels of cyclin D1 causes an initial insensitivity to antiestrogen-mediated growth arrest, but this effect fades with time, whereas over-expression of cyclin E1 had minimal influence on antiestrogen sensitivity in vitro (Hui et al. 2002). In contrast, cyclin D1 confers practically total resistance to progestin-induced growth inhibition, whereas cyclin E1 has a strong but less pronounced effect (Musgrove et al. 2001). Hence, overexpression of cyclin D1 and E1 and downregulation of p21Waf1 and p27Kip1 modify the sensitivity of breast cancer cells to therapeutically relevant hormone-responsive breast cancer therapy in vitro.

15.7 Role of Cyclins in Carcinogenesis of Mammary Glands

In vivo research using genetically altered mice have provided more intuition into the involvement of cyclins in breast cancer. Several labs have found that mammary gland development in transgenic mice producing D- and E-type cyclins is controlled by promoters found in epithelial cells of mammary glands, especially the mouse mammary tumor virus long terminal repeat promoter (MMTV-LTR). Mammary gland development is disrupted in cyclin D1 transgenics, with enhanced proliferation and precocious lobuloalveolar development, which is typical of early pregnancy followed by adenocarcinoma (Fu et al. 2004, Mir et al. 2020a, b). Cancers occur in roughly 75% of mice after an 18-month latency period, indicating that cyclin D1 is a feeble oncogene in contrast to activated c-neu, Ha-ras, and c-myc, which produce tumors at three, six, and eleven months, respectively, when overexpressed under the control of MMTV-LTR (Muller et al. 1988). These findings also show that for cyclinD1 to exert its carcinogenic potential, other genetic processes may be required. More recently, it was discovered that MMTV-cyclin D1transgenics lack the p16 Ink4A expression pulse associated with normal mammary gland involution, implying that this defect may result in an enlargement of the stem cell population responsible for long-term proliferation (Gartel et al. 2001).

In contrast to the precocious lobuloalveolar development produced by cyclin D1, MMTV-driven cyclin D2 overexpression in mammary tissues results in enhanced cell proliferation in the pregnant gland but partial or full inhibition of alveolar differentiation. This was accompanied by a decrease in cyclinD1 isoform abundance and an increase in p27Kip1 expression, which could explain the phenotype given the significance of cyclin D1 function in normal alveologenesis (Kornberg 2005). Overactivation of cyclinD2 resulted in a modest tumor frequency, with only 19% of mice getting tumors (Kong et al. 2002) and because the cyclinD2 gene is generally methylated in breast cancers, the significance of these findings to breast cancer deserves additional exploration (Egloff and Murphy 2008). These findings could imply that when cyclin D2 is overexpressed in the luminal epithelium, it can mimic the actions of cyclin D1 in causing carcinogenesis in this experimental paradigm. To know about the similarity index in functionality of D1 and D2, more studies are needed. Data from mice that only express one D-type cyclin suggest that there is a large amount of redundancy (Cowling and Cole 2010). Cyclin D2, on the other hand, exhibits a distinct selectivity for Cdk activation, preferentially binding and activating Cdk2 in human breast epithelial cells and being ineffectual in connecting with transcription factors and stimulating gene expression, a characteristic that appears to be unique (Swarbrick et al. 2000). As a result, more research into the involvement of cyclin D2 in mammary cancer is needed. Studies of cyclin D3's effects on mammary carcinogenesis have been anxiously sought because it is frequently overexpressed in cancer and is linked to high-grade breast tumors (Russell and Nurse 1987). Compared to their cyclin-D1 counterparts, MMTV-cyclin-D3 mice showed normal mammary gland growth and involution after breastfeeding, according to a recent study.

However, after many pregnancies, these mice developed mammary carcinoma at a significant rate, with 73% of animals developing mammary carcinoma. Surprisingly, these mice developed squamous cell carcinoma rather than the more common adenocarcinoma seen in the other two models (Pirkmaier et al. 2003). These findings show that overexpression of any of D-type cyclins in the mammary gland can lead to cancer formation, but there are gene-specific variations. These differences can be seen in the varied impacts on normal breast development and the cancer phenotypes, such as D1 and D2 cyclins causing adenocarcinoma whereas cyclin D3 causes a mainly squamous phenotype. These three Cyclin-D type cyclin genes effect epithelial cell development and differentiation of mammary glands.

In transgenic mice, the effects of cyclin E1 overexpression were studied in which the human gene was produced under the control of the ovine beta-globulin promoter, resulting in mammary-specific expression during pregnancy and lactation (Bodrug et al. 1994, Mir 2022a, b). During the first pregnancy, this expression resulted in hyperplastic papillary projections, the bulk of which was removed during subsequent mammary gland involution after weaning. After 8–13 months, 10% of female mice developed adenocarcinomas, which exhibited considerably higher expression of cyclin E at both mRNA and protein levels, additionally cyclin E-associated kinase activity. Cyclin E1 is a "weak" oncogene in mammary epithelium, similar to D-type cyclins. The capacity of cyclin D1 overexpression to cause

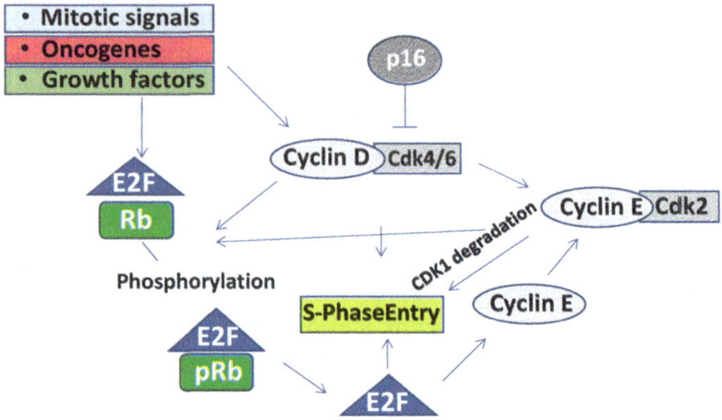

Fig. 15.5 Diagrammatic representation showing cyclin and cyclin-dependent kinase regulated cell cycle programming; A mitotic signal activating cyclin-dependent kinases, CDK4 and CDK6 which phosphorylates numerous critical substrates, including the tumor suppressor retinoblastoma (Rb), permit the E2F transcription factors to coordinate a gene expression program and also activates CDK2, which phosphorylates RB and also the starts replication of its DNA

breast cancer, as well as the requirement of cyclin D1 activity for cell cycle development raises the question of whether cyclin D1 is required for tumor formation (Fig. 15.5). The basic ductal structure grows correctly at puberty in cyclin D1 deficient mice without the development of alveoli during pregnancy and failure of lactation (Fan et al. 1997)). This issue does not appear to be due to a requirement for cyclin D1, as epithelium lacking both cyclin D1 and p27Kip1 may help in normal mammary gland development, so does the epithelium with cyclin D1 replaced by cyclin E1 (Holstege et al. 1998)) (Gegonne et al. 2008); instead, it appears that the necessity is for epithelial cell growth to occur in a timely manner. Surprisingly, cyclin D2 and D3 are not necessary for proper mammary gland development, as gland development is normal in cyclin D2 and D3 null mice, despite the fact that overexpression of these genes causes breast cancer, as previously stated (Cahill et al. 1998). Crossing cyclin D1 null mice with animals expressing distinct mammary oncogenes under the control of the MMTV promoter has revealed new information on cyclin D1's role in various oncogenic pathways. The discovery that mammary tumorigenesis is hampered in absence of cyclin D1 in c-neu and Ha-ras transgenics, but not in c-myc and Wnt-1 MMTV transgenics, identified cyclin D1 as a crucial component of several mammary carcinogenesis pathways (Yu et al. 2001). Further research into the molecular basis of these effects will provide a better understanding of how distinct oncogenes interact in a tissue-specific setting.

Because the deficit in lobuloalveolar development occurs in the presence of potentially non-essential D2 and D3 cyclins, previous investigations in cyclin-D1 null mice clearly show that cyclin D1 serves a highly particular role in breast development. This effect is considered to be caused by the failure of other cyclins in the mammary gland to u-regulate, rather than by the cyclins functional variety. In cyclin D2 and D3 deficient mice, carcinogens produced by MMTV-neu and

Table 15.1 Potential and functional roles for G1 cyclins

Cyclins	Binding partner	FUNCTIONS
D1/D2/ D3	Cdk4/6	Cell cycle progression
D2	Cdk2	Cell cycle progression
E	Cdk2	Cell cycle progression
D1	ER,C/EBPβ	Transcriptional activation
D1	AR,beta2/neuro D,DMP1,Myb,MyoD,SP1, STAT3,TR	Transcriptional activation
D1	AIB-1.GRIP-1,SRC1a	Co-activation
D1	CBP/p300,P/CAF	Chromatin remodeling
D1	TAFπ250	It helps in recruiting the enzyme, RNA polymerase II
E	AR	Co-activation

MMTV-ras remain unaltered, and cyclinE1 could replace cyclin-D1 in breast progression and its tumorigenesis (Mir and Mehraj 2019). These results recommend that tissue-specific cyclin-D1 expression regulation and timing are crucial for these processes. In MMTV-neu-induced and MMTV-ras-induced breast tumors, upregulated expression of cyclin D1 without the involvement of cyclins D2 and D3 was detected, whereas Wnt-1 and c-myc-induced tumors expressed both cyclins D1 and D2 (Yu et al. 2001). Given the evidence that these two oncogenes can activate cyclinE–Cdk2 via different routes in breast cancer cells andt hat c-myc can stimulate cyclin D2 expression and sequestration ofp27Kip1 into Cdk 4/6 complexes, the absence of dependence of c-myc-induced malignancies on cyclin D1 is likely not surprising (Hartgrink et al. 2009).

More recent results from transcript profiles of distinct oncogene-induced mammary malignancies in mice corroborate that cyclin D1 is upregulated in MMTV-neu and –ras induced tumors, whereas cyclins D2, E1, and E2 are upregulated in MMTV-myc-induced tumors (Sofi et al. 2022a, b)). The ability of Wnt-1 to cause premature and early development of mammary gland and cancer in cyclin D1 null animals suggests that Wnt-1 signaling via beta-catenin and enhanced cyclinD1 gene expression is not the primary mechanism of activation, as has been shown in other cell types.

While this pathway is most likely intact in breast cancer development, these findings point to Wnt-1 activation of cyclin D2 as a key downstream effect on mammary gland epithelium (Qayoom and Bhat 2020). Taken together, these findings show that proper regulation and expression of cyclin-D1gene is needed for mammary gland development and the generation of any particular type of mammary cancer in mammals (here, mice). Cyclins D2, D3, E1, and E2 are not required for proper mammary gland/breast development; however, it is unknown whether they are required downstream of some mammary oncogenes, such as c-myc. (Table 15.1).

15.8 In Breast Cancer; Cyclin Overexpression

Since the breakthrough of cyclin D1, it is known as the most frequently commonly expressed gene (onco) in breast cancer development, but a lot of research has been undertaken to understand the link between this cyclin and various characteristics of breast cancer (Buckleyetal.1993). It was reported that Cyclin-D1 is upregulated by almost 30–60% in breast cancers, particularly in the early phases (Oyama et al. 1998).

Various reports have suggested a link between Cyclin-D1 and estrogen where this cyclin acts at a downstream position of estrogen receptors (Prall et al. 1998). However, the role of cyclin-D1 and estrogen expression in the development of cancer is still a debate (Gillett et al. 1996). Given the confusing concerns of the link with ER state, the interrelationships with other molecules in the pRb pathway, most of which were not tested concurrently, this debate is somewhat unsurprising. Subgroup studies within tiny patient cohorts have also been a key roadblock to reaching any firm results. Despite these severe limitations, a link between CCND1 gene amplification and poor clinical prognosis in ER+ patients appears to exist (Buckley et al. 1993). Other researches have not been able to corroborate this link. The possibility that cyclin D1 overexpression can contribute to more problematic results by making the target cells resistant to endocrine therapies further complicates the interpretation (Hwang and Clurman 2005). One small clinical trial found that ERC patients with low cyclin D1 had a considerably prolonged duration of response to tamoxifen than those with high cyclinD1 (Kenny et al. 1999), which supports this theory. More extensive researches are required to resolve these concerns. In humans, Cyclin-D2 is present in normal mammary gland cells (epithelial) but is rarely found in breast cancer tissues. Although numerous important functions are hypothesized about cyclinD2, this observation is owing to promote hyper-methylation in the majority of malignancies (Keyomarsi et al. 2002) but the functional importance, if any, in breast oncogenesis has yet to be explained. Overexpression of Cyclin D3 is found in breast tumors;however, there is little information on its link to phenotypic and disease outcome (Russell and Nurse 1987). According to studies Buckley et al. 1993 and Keyomarsi and Pardee 1993, cyclin E1 is inappropriately expressed in 40% of breast tumors. The level of expression rises with the stages and grades of tumor. Unlike cyclinD1, cyclin E is largely over-expressed in the ER-phenotype ((Keyomarsi et al. 2002); (Span et al. 2003). Furthermore, cyclin E over expression is associated with elevated p16Ink4A levels in roughly 40% of breast tumors, showing that cyclin E expression is linked to pRb pathway dysregulation (Loden et al. 2002). Because of the strong relationship between cyclin E and p27Kip1 in cell cycle control, cyclin E over-expression in combination with low p27Kip1 expression is more prognostic than cyclin E alone (Porter et al. 1997).

The relative ratios of wild type and truncated versions of the protein also have an impact on cyclin E1 over-expression (Keyomarsi and Pardee 1993). The fact that these low molecular weight, N-terminally shortened versions are tumor cell specific and more effective at facilitating the G1toS phase transition are possible explanations for this impact (Porter et al. 1997). Since the parameters tested include cyclin-E protein levels by Immunohistochemistry (IHC) or Western blot and mRNA related studies, more research is needed to confirm these findings. The latter method failed to find a link between cyclin E mRNA levels and relapse-free or overall survival in a recent investigation. High cyclin E mRNA levels, on the other hand, were only related with poor relapse-free survival of patients undergoing adjuvant endocrine therapies, supporting the theory that cyclinE confers endocrine resistance (Span et al. 2003). Hence, cyclins D1 and E1 are involved in the progression of ERC and ER breast cancers, respectively. They also present preliminary evidence that these genes could be valuable markers of disease progression and therapy responsiveness, but further research is needed to fully understand these concerns.

15.9 The CDK4/6 Targeted Preclinical Research

Flavopiridol, a "pan-CDK" inhibitor targeting CDKs including CDK1/2/4/6/7 and9, was the focus of early efforts to target CDKs (Bose et al. 2013, Dickson 2014). Although activity has been seen in blood related malignancies, the drug progression has been marred by naxiousness as well as complex pharmacokinetics and management difficulties. Pan-CDK inhibitors of the next generation, such as dinaciclib, are being tried in clinical trials for a range of cancers (Criscitiello et al. 2014). Following that, researchers concentrated onCDK4/6 inhibitors and their anti-breast cancer effectiveness. These drugs block the cyclin D1—CDK4/6 interaction, preventing the cell cycle from progressing through the G1/S stages and resulting in cell cycle arrest. Preclinical studies with CDK4/6 inhibitors have shown that they are effective in tumors with cyclin D1 over expression, common in ER+ breast cancer (Konecny et al. 2011) With the administration of CDK4/6 inhibitors, single agent pharmacodynamic reduction of phospho-Rb and decrease in the proliferative marker Ki-67 has been reported in a range of benign tumors as well as mantle cell lymphoma (Fry et al. 2004). Finn et al. 2014 tested the CDK inhibitor Palbociclib's in vitro sensitivity in a panel of molecularly defined human breast cancer cell lines.

In ER-resistant cell lines, CDK4/6 inhibition also increased tamoxifen responsiveness (Finn et al. 2009). Palbociclib (PD0332991), Abemaciclib (LY2835219), and LEE011 are the three highly selective oral medicines that are now in active clinical development (Table 15.2).

15.10 CDK4/6Inhibitors'MechanismsofAction

Because of the small structural variations between the ATP-binding pockets of distinct CDKs, highly selective CDK4/6 inhibitors can be designed (Choi and Anders 2014). CDK inhibitors are able to take part in binding despite chemical differences via hydrogen bonds and other hydrophobic interactions (Knockaert et al. 2002). The backbone carbonyl works as an acceptor for hydrogen bonds, while the amino acid side chains act as hydrogen bond donors, according to various chemical assays. CDKs have a modest ATP binding site that can adopt a (Knockaert et al.

Table 15.2 Phase Ib/II and phase III efficacy results using cyclin dependent kinase 4/6 inhibitors to treat advanced breast cancer

Drug	Treatment	No. of patients evaluated	Response rate %[a]	Clinical benefits %[b]	Progression free survival, mo	Overall surviva, mo
Palbociclib	Palbociclib	33	6.8	21	4.5	NA
	Placebo+ Letrozole	81	33.3	58.0	10.2	33.3
	Palbociclib+ Letrozole	84	42.9	81.0	20.2	37.5
	Placebo+ Fulvestrant	174	6.3	19.0	3.8	NA
	Palbociclib +Fulvestrant	347	10.4	34.0	9.2	NA
Ribociclib	Ribociclib+ Exemestane+ Everolimus	13	7.7	46.2	NA	NA
	Ribociclib+ Letrozole	10	10.0	30	NA	NA
	Robococlib+ BYL719	7	0	28.6	NA	NA
Abemaciclib	Abemaciclib	36	25.0	61.1	9.1	NA
	Abemaciclib+ Letrozole or Anastrozole	36	5.6	67.0	NA	NA
	Abemaciclib+ tamoxifen	16	NA	75.0	NA	NA

Table 15.3 Clinical trials on Palbociclib for breast cancer

Study	DeMichele et al	Salmon et al	Finn et al
Meeting	ASCO 2013 Annual meeting	ASCO 2013 Annual meeting	AACR annual meeting 2014
Phase	II	Ib	II
N	37	12	165
Primary endpoint	Safety and efficacy	Safety and tolerability	PFS
Therapy	Palbociclib	Palbociclib+ letrozole	Palbociclib+ letrozole vs letrozole
Breast cancer subtype	ER+ HER2–29/37 ER+ HER2–2/37 TNBC 6/37	ER+ HER2-	ER+ HER2-
Prior chemotherapy for advanced disease	34/37 (92%)	8(67%)	34(40%)
Response rate	2/36 (1%) PR 18/36(50%) SD	3/12 (25%) PR 9/12(75%) SD	NR NR
PFS	ER+ 3.8 months TNBC 1.9 months	NR	20.2 months 10.2 months

2002) variety of configurations, resulting in both selective and non-selective inhibitors.

However, because the amino acids in the CDK ATP-binding region are generally maintained, there are currently no CDK inhibitors that target just one kinase. Preclinical investigations with Palbociclib are the main source of current knowledge about the mechanism of action of CDK4/6Inhibitors. Palbociclib, unlike pan-CDK inhibitors, fits closely into the ATP-binding pocket. Palbociclib has a broader binding interface with its target as a result of this, which could lead to enhanced efficacy (Choi and Anders 2014). In tumor cells, Palbociclib suppresses Rb phosphorylation at two different CDK4/6 sites, Ser780 and Ser795 (Shapiro 2006). This inhibition causes a concentration-dependent arrest of Rb-positive tumors in G1, as well as decreased production of E2F-dependent genes and Ki-67 staining. Palbociclib was discovered to prevent the incorporation of thymidine and prevent DNA replication. However, no action was observed against Rb-deficient cells, implying that there are no other targets outside CDK4/6 (Shapiro 2006). Palbociclib and other CDK4/6 inhibitors like Ribociclib and Abemaciclib have shown a reversible blockage of the cell cycle progression in BC cell lines with selectivity in preclinical trials (Choi and Anders 2014) (Table 15.3).

15.11 Prevention and Treatment

15.11.1 Chemo-prevention

It involves the use of pharmacological and natural agents capable of inhibiting the tumor/cancer development either by protecting/repairing the DNA damage that induces carcinogenesis or reverse the development of premalignant cells (Sporn 1976). Chemotherapy targets estrogen receptor because more than 70% breast cancers are reported to have its origin in estrogen receptor. Therefore, estrogen inhibitors such as selective estrogen receptor modulators (SERMs) and the aromatase inhibitors (AIs) are the large class of receptors acts as either antagonists or agonists against the estrogen receptor (Mehraj et al. 2021). Tamoxifen (TAM) is a most often used SERM, used for the treating breast cancer over decades (Bozovic-Spasojevic et al. 2012)). Most importantly the TAM has been used to cure the all stages of breast cancers (Nagini 2017). Many large scale trials such as the Royal Marsden Prevention Trial, the Italian Prevention Trial, the International Breast Cancer Intervention Study, and the Breast Cancer Prevention Trial (NSABP-1) have time and again proved that TAM could significantly reduce the risk of mammary gland tumorigenesis (Qayoom et al. 2021). However, the TAM has shown effectiveness only against the breast cancers that originated from estrogen receptors (estrogen-positive) and no significant ameliorating effect has been reported in ER-negative cancers (Cuzick et al. 2003). Raloxifene, another kind of SERMs reportedly having fewer side effects, is being used against breast cancer in postmenopausal women and also those suffering from osteoporosis and heart diseases (Barrett-Connor et al. 2006). Due to its less risk in thromboembolic complications and endometrial cancers, raloxifene is believed to be a good therapeutic against invasive breast cancers. Although various other SERMs including arzoxifene, ospemifen, lasofoxifene (LFX), and bazedoxifene (BZA) were discovered and tried for its anticancerous properties, only BZA successfully reached up to the clinical trials(Marty et al. 2005). Recently, aromatase inhibitors (AIs) were used as a first choice of therapy instead of TAM against postmenopausal breast cancer patients. It has been studied that AIs cause the inhibition of aromatase that catalyzes the conversion of estrogen from androgen (Hiscox et al. 2009). The two classes of AIs such as steroidal inhibitors and non-steroidal inhibitors have also been used for the treatment of breast cancers. Non-steroidal inhibitors like letrozole and anastrozole can bind irreversibly to the enzymes active site. However, there is no significant difference in their efficiency to prevent the breast cancer (Dowsett et al. 2010). However, there are also limitations of AIs. As they inhibit the biosynthesis of estrogen, therefore they are given only to postmenopausal women. Furthermore, the main side effect that AIs carries is the possibility of osteoporosis, which poses a remarkable health threat to elderly women. It has been observed that a prolonged treatment of SERMs results into Acquired resistance to AIs. Moreover, crosstalks between estrogen receptor pathways and various signaling pathways including PI3K/Akt/mTOR and Ras/Raf/MEK/MAPK leads to resistance of cancer cells to

AIs (Lønning and Eikesdal 2013). Therefore, AIs in combination with inhibitors of various cancer signaling pathways can provide a hopeful alternative.

15.11.2 Biological Prevention

With advancing technology, monoclonal antibodies were developed targeted towards HER2, a gene responsible for about 20–30% of all breast cancers (Mir and Mehraj 2019; (Elizalde et al. 2016). The Trastuzumab (Herceptin) is a first kind of HER2 targeted drug approved by FDA (Mir et al. 2015). It directly interacts with the C-terminal portion of domain IV of HER2 (Choi and Anders 2014). The trastuzumab helps in degrading HER2 by proteasomal complexes (Junttila et al. 2009). Trastuzumab was initially tested against MBC and showed significant results with 26% objective response rate (ORR). Furthermore, invitro studies have demonstrated the synergetic effect of trastuzumab when used alongside other anti-cancer drugs like 4-hydroxycyclophosphamide, docetaxel, nimotuzumab, and carboplatin (Vogel et al. 2002). For example, a random trial conducted have shown trastuzumab in combination with docetaxel showed significantly higher efficiency with 50% ORR (Marty et al. 2005) More importantly, various heart related issues were reported as a side effect in trastuzumab treated patients (Balduzzi et al. 2014). Similarly, pertuzumab, an another monoclonal antibody based drug approved for the treatment of breast cancer shows various side effects as well (Loibl et al. 2017).

15.12 Summary

Breast cancer is one of the most commonly detected cancers in females across globe and 1 in every 8 females have a risk of developing a breast cancer during their lifetime (Stewart et al. 1984). It is a complicated disease with no single origin, rather is a multistep process and the pathogenesis is yet to be elucidated completely. Moreover, genetic and environmental factors influence the breast cancer. Therefore, the prevention strategies must be targeted before risk factors develop the disease. It has been reported that developed countries have high rate of breast cancer incidences as compared to developing countries. However, the recovery rate is very low in developing countries than developed countries may be due to lack of research investment and technical know-how. There are ample medical resources such as chemo-preventive drugs available in developed countries and a global cooperation on health issues could significantly reduce the rate of the breast cancer across the globe. Most of the breast cancers occur through estrogen receptor led mechanism, and various therapies like, endocrine therapies that target the estrogen receptor has shown some promising results, however, the resistance to this kind of therapy is inevitable in advanced stages. Therefore, research studies must focus on the factors that drive the breast cancer through estrogen-receptor-positive mechanism. In fact, this progress has reaped some research outcomes and thereby, several endocrine therapies have been developed aimed at targeting the drivers of breast cancer like mTOR, cyclin-dependent kinases CDK4 and CDK6, besides other pathways like PI3K, AKT, and HER2 which significantly have been dissolved through various kinds of inhibitors. However, still challenges are there to cure the disease completely.

15.13 Further Reading

The readers can further read about "CDK Inhibitors for treatment of breast cancer" by going through the following research papers:

https://onlinelibrary.wiley.com/doi/abs/10.1002/med.1021

https://www.ncbi.nlm.nih.gov/pmc/articles/PMC8167670/

For more incites about the topic we would suggest detailed findings from the books of (Mir MA, 2022) https://doi.org/10.1016/C2021-0-02565-7, (Mir MA, 2021) https://doi.org/10.52305/WXJL6770, from cancer.net website, https://www.cancer.net/cancer-types/breast-cancer/types-treatment

Also, readers are advised to look following video lectures for better understanding of this chapter:

https://www.youtube.com/watch?v=BB9jjK7BHkg

https://www.youtube.com/watch?v=vEe3lBduckE

References

Agarwal ML et al (1995) p53 controls both the G2/M and the G1 cell cycle checkpoints and mediates reversible growth arrest in human fibroblasts. Proc Natl Acad Sci 92(18):8493–8497

Aguirre-Ghiso JA et al (2001) Urokinase receptor and fibronectin regulate the ERKMAPK to p38MAPK activity ratios that determine carcinoma cell proliferation or dormancy in vivo. Mol Biol Cell 12(4):863–879

Altucci L et al (1997) Estrogen induces early and timed activation of cyclin-dependent kinases 4, 5, and 6 and increases cyclin messenger ribonucleic acid expression in rat uterus. Endocrinology 138(3):978–984

Asghar U et al (2015) The history and future of targeting cyclin-dependent kinases in cancer therapy. Nat Rev Drug Discov 14(2):130–146

Austreid E et al (2014) The emergence of targeted drugs in breast cancer to prevent resistance to endocrine treatment and chemotherapy. Expert Opin Pharmacother 15(5):681–700

Balduzzi S et al (2014) Trastuzumab-containing regimens for metastatic breast cancer. Cochrane Database of Systematic Reviews (6)

Barrett-Connor E et al (2006) Effects of raloxifene on cardiovascular events and breast cancer in postmenopausal women. N Engl J Med 355(2):125–137

Bartek J, Lukas J (2003) Chk1 and Chk2 kinases in checkpoint control and cancer. Cancer Cell 3(5):421–429

Bodrug SE et al (1994) Cyclin D1 transgene impedes lymphocyte maturation and collaborates in lymphomagenesis with the myc gene. EMBO J 13(9):2124–2130

Bose P et al (2013) Cyclin-dependent kinase inhibitor therapy for hematologic malignancies. Expert Opin Investig Drugs 22(6):723–738

Bozovic-Spasojevic I et al (2012) Chemoprevention for breast cancer. Cancer Treat Rev 38(5): 329–339

Buckley MF et al (1993) Expression and amplification of cyclin genes in human breast cancer. Oncogene 8(8):2127–2133
Cahill DP et al (1998) Mutations of mitotic checkpoint genes in human cancers. Nature 392(6673): 300–303
Carroll JS et al (2003) p27Kip1 induces quiescence and growth factor insensitivity in tamoxifen-treated breast cancer cells. Cancer Res 63(15):4322–4326
Casimiro MC, Pestell RG (2012) Cyclin d1 induces chromosomal instability. Oncotarget 3(3):224
Chen MD et al (1997) Phase I trial of taxol as a radiation sensitizer with cisplatin in advanced cervical cancer. Gynecol Oncol 67(2):131–136
Choi YJ, Anders L (2014) Signaling through cyclin D-dependent kinases. Oncogene 33(15): 1890–1903
Cowling VH, Cole MD (2010) Myc regulation of mRNA cap methylation. Genes Cancer 1(6): 576–579
Crasta K et al (2006) Cdk1 regulates centrosome separation by restraining proteolysis of microtubule-associated proteins. EMBO J 25(11):2551–2563
Criscitiello C et al (2014) Dinaciclib for the treatment of breast cancer. Expert Opin Investig Drugs 23(9):1305–1312
Cuzick J et al (2003) Overview of the main outcomes in breast-cancer prevention trials. Lancet 361(9354):296–300
Dickson MA (2014) Molecular pathways: CDK4 inhibitors for cancer TherapyCDK4 inhibitors. Clin Cancer Res 20(13):3379–3383
Dowsett M et al (2010) Meta-analysis of breast cancer outcomes in adjuvant trials of aromatase inhibitors versus tamoxifen. J Clin Oncol 28(3):509–518
Dyson N (1998) The regulation of E2F by pRB-family proteins. Genes Dev 12(15):2245–2262
Egloff S, Murphy S (2008) Cracking the RNA polymerase II CTD code. Trends Genet 24(6): 280–288
Elizalde PV et al (2016) ErbB-2 nuclear function in breast cancer growth, metastasis and resistance to therapy. Endocr Relat Cancer 23(12):T243–T257
Fan S et al (1997) Cells lacking CIP1/WAF1 genes exhibit preferential sensitivity to cisplatin and nitrogen mustard. Oncogene 14(18):2127–2136
Fry DW et al (2004) Specific inhibition of cyclin-dependent kinase 4/6 by PD 0332991 and associated antitumor activity in human tumor xenografts. Mol Cancer Ther 3(11):1427–1438
Fu M et al (2004) Minireview: cyclin D1: normal and abnormal functions. Endocrinology 145(12): 5439–5447
Gartel AL et al (2001) Myc represses the p21 (WAF1/CIP1) promoter and interacts with Sp1/Sp3. Proc Natl Acad Sci 98(8):4510–4515
Gegonne A et al (2008) TFIID component TAF7 functionally interacts with both TFIIH and P-TEFb. Proc Natl Acad Sci 105(14):5367–5372
Gillett C et al (1996) Cyclin D1 and prognosis in human breast cancer. Int J Cancer 69(2):92–99
Hartgrink HH et al (2009) Gastric cancer. Lancet 374(9688):477–490
Hiscox S et al (2009) Aromatase inhibitors in breast cancer. Maturitas 63(4):275–279
Holstege FCP et al (1998) Dissecting the regulatory circuitry of a eukaryotic genome. Cell 95(5): 717–728
Hui R et al (2002) Constitutive overexpression of cyclin D1 but not cyclin E confers acute resistance to antiestrogens in T-47D breast cancer cells. Cancer Res 62(23):6916–6923
Hwang HC, Clurman BE (2005) Cyclin E in normal and neoplastic cell cycles. Oncogene 24(17): 2776–2786
Junttila TT et al (2009) Ligand-independent HER2/HER3/PI3K complex is disrupted by trastuzumab and is effectively inhibited by the PI3K inhibitor GDC-0941. Cancer Cell 15(5): 429–440
Kenny FS et al (1999) Overexpression of cyclin D1 messenger RNA predicts for poor prognosis in estrogen receptor-positive breast cancer. Clin Cancer Res 5(8):2069–2076

Keyomarsi K, Pardee AB (1993) Redundant cyclin overexpression and gene amplification in breast cancer cells. Proc Natl Acad Sci 90(3):1112–1116
Keyomarsi K et al (2002) Cyclin E and survival in patients with breast cancer. N Engl J Med 347(20):1566–1575
Knockaert M et al (2002) Pharmacological inhibitors of cyclin-dependent kinases. Trends Pharmacol Sci 23(9):417–425
Konecny GE et al (2011) Expression of p16 and retinoblastoma determines response to CDK4/6 inhibition in ovarian CancerCDK4/6 inhibition in ovarian cancer. Clin Cancer Res 17(6): 1591–1602
Kornberg RD (2005) Mediator and the mechanism of transcriptional activation. Trends Biochem Sci 30(5):235–239
Loden M et al (2002) The cyclin D1 high and cyclin E high subgroups of breast cancer: separate pathways in tumorogenesis based on pattern of genetic aberrations and inactivation of the pRb node. Oncogene 21(30):4680–4690
Loibl S et al (2017) Dual HER2-blockade with pertuzumab and trastuzumab in HER2-positive early breast cancer: a subanalysis of data from the randomized phase III GeparSepto trial. Ann Oncol 28(3):497–504
Lønning PE, Eikesdal HP (2013) Aromatase inhibition 2013: clinical state of the art and questions that remain to be solved. Endocr Relat Cancer 20(4):R183
Marty M et al (2005) Randomized phase II trial of the efficacy and safety of trastuzumab combined with docetaxel in patients with human epidermal growth factor receptor 2–positive metastatic breast cancer administered as first-line treatment: the M77001 study group. J Clin Oncol 23(19): 4265–4274
Mehraj U et al (2021) Tumor microenvironment promotes breast cancer chemoresistance. Cancer Chemother Pharmacol 87(2):147–158
Mehraj U et al (2022a) Expression pattern and prognostic significance of baculoviral inhibitor of apoptosis repeat-containing 5 (BIRC5) in breast cancer: a comprehensive analysis. Advances in Cancer Biology-Metastasis 4:100037
Mehraj U et al (2022b) Expression pattern and prognostic significance of CDKs in breast cancer: an integrated bioinformatic study. Cancer Biomark 34(3):505–519. Cancer Biomarkers (Preprint):1–15
Mir MA (2015) Developing costimulatory molecules for immunotherapy of diseases. Academic Press. https://doi.org/10.1016/C2014-0-02898-5
Mir MA, Mehraj U (2019) Double-crosser of the immune system: macrophages in tumor progression and metastasis. Curr Immunol Rev 15(2):172–184
Mir MA et al (2020a) Nanobodies: the "magic bullets" in therapeutics, drug delivery and diagnostics. Hum Antibodies 28(1):29–51
Mir MA et al (2020b) Targeting different pathways using novel combination therapy in triple negative breast cancer. Curr Cancer Drug Targets 20(8):586–602
Mir MA, Sofi S, Qayoom H (2022a) Conventional adjuvant chemotherapy in combination with surgery, radiotherapy, and other specific targets. Elsevier, pp 95–120. https://doi.org/10.1016/B978-0-323-96136-3.00008-X. ISBN 9780323961363
Mir MA, Sofi S, Qayoom H (2022b) Different drug delivery approaches in combinational therapy in TNBC. Elsevier, pp 201–230. https://doi.org/10.1016/B978-0-323-96136-3.00012-1
Muller WJ et al (1988) Single-step induction of mammary adenocarcinoma in transgenic mice bearing the activated c-neu oncogene. Cell 54(1):105–115
Musgrove EA et al (2001) Cyclin D1 overexpression induces progestin resistance in T-47D breast cancer cells despite p27Kip1 association with cyclin E-Cdk2. J Biol Chem 276(50): 47675–47683
Nagini S (2017) Breast cancer: current molecular therapeutic targets and new players. Anti-Cancer Agents in Medicinal Chemistry (Formerly Current Medicinal Chemistry-Anti-Cancer Agents) 17(2):152–163

Oyama T et al (1998) Frequent overexpression of the cyclin D1 oncogene in invasive lobular carcinoma of the breast. Cancer Res 58(13):2876–2880
Pirkmaier A et al (2003) Alternative mammary oncogenic pathways are induced by D-type cyclins; MMTV-cyclin D3 transgenic mice develop squamous cell carcinoma. Oncogene 22(28): 4425–4433
Poon RY et al (1994) Cell cycle regulation of the p34cdc2/p33cdk2-activating kinase p40MO15. J Cell Sci 107(10):2789–2799
Porter PL et al (1997) Expression of cell-cycle regulators p27Kip1 and cyclin E, alone and in combination, correlate with survival in young breast cancer patients. Nat Med 3(2):222–225
Prall OWJ et al (1998) Estrogen regulation of cell cycle progression in breast cancer cells. J Steroid Biochem Mol Biol 65(1–6):169–174
Qayoom H, Bhat BA (2020) U Mehraj U, Mir MA (2020) rising trends of cancers in Kashmir valley: distribution pattern, incidence and causes. J Oncol Res Treat 5(150):2
Qayoom H et al (2022) Expression patterns and therapeutic implications of CDK4 across multiple carcinomas: a molecular docking and MD simulation study. Med Oncol 39(10):1–13
Qayoom H et al (2021) An insight into the cancer stem cell survival pathways involved in chemoresistance in triple-negative breast cancer. Future Oncol 17(31):4185–4206
Russell P, Nurse P (1987) Negative regulation of mitosis by wee1+, a gene encoding a protein kinase homolog. Cell 49(4):559–567
Shapiro GI (2006) Cyclin-dependent kinase pathways as targets for cancer treatment. J Clin Oncol 24(11):1770–1783
Sherr CJ (1996) Cancer cell cycles. Science 274(5293):1672–1677
Sherr CJ, Roberts JM (1999) CDK inhibitors: positive and negative regulators of G1-phase progression. Genes Dev 13(12):1501–1512
Sherr CJ, Roberts JM (2004) Living with or without cyclins and cyclin-dependent kinases. Genes Dev 18(22):2699–2711
Sofi S et al (2022a) Targeting cyclin-dependent kinase 1 (CDK1) in cancer: molecular docking and dynamic simulations of potential CDK1 inhibitors. Med Oncol 39(9):1–15
Sofi S et al (2022b) Cyclin-dependent kinases in breast cancer: expression pattern and therapeutic implications. Med Oncol 39(6):1–16
Span PN et al (2003) Cyclin-E is a strong predictor of endocrine therapy failure in human breast cancer. Oncogene 22(31):4898–4904
Sporn, M. B. (1976). Approaches to prevention of epithelial cancer during the preneoplastic period., AACR
Stewart TA et al (1984) Spontaneous mammary adenocarcinomas in transgenic mice that carry and express MTV/myc fusion genes. Cell 38(3):627–637
Swarbrick A et al (2000) Cooperation of p27Kip1 and p18INK4c in progestin-mediated cell cycle arrest in T-47D breast cancer cells. Mol Cell Biol 20(7):2581–2591
Venditti M et al (2002) C-myc gene expression alone is sufficient to confer resistance to antiestrogen in human breast cancer cells. Int J Cancer 99(1):35–42
Vogel CL et al (2002) Efficacy and safety of trastuzumab as a single agent in first-line treatment of HER2-overexpressing metastatic breast cancer. J Clin Oncol 20(3):719–726
Yu Q et al (2001) Specific protection against breast cancers by cyclin D1 ablation. Nature 411(6841):1017–1021

Response of Therapy in Cell-Cycle Regulatory Genes in Breast Cancer

16

Manzoor Ahmad Mir, Shariqa Aisha, Kaneez Fatima, and Fayaz Ahmad Malik

16.1 Introduction

Cancerous cells bypass the cell cycle's multiple shielding and checkpoints, allowing them to multiply indefinitely despite aneuploidy as well as other abnormalities that can stop non-malignant cells from proliferating. This behavior is acquired by accumulating a variety of genetic or epigenetic genomic changes that hyper-activate or deactivate key elements of the cell cycle that put unique cellular demands on cancerous cells in order to maintain abnormal growth. Diverse subtypes of breast cancer have various molecular changes and reliance on the cell cycle as well as its checkpoints. Tumor cells in ER (+) /HER2 (−) "luminal" breast tumors still rely on estrogen for oncogenic actions. When estrogen binds to hormonal receptors, it triggers a signaling cascade that results in receptor-driven gene expression which promotes cellular growth, survival, and multiplication. Cyc D1 is an ER targeting gene which promotes cell cycle progress across the limitation point by permitting CDK4/6/cyclin D interactions (Platet et al. 2004). ESR1, the gene that encodes the estrogen receptors, and its related protein, are overexpressed in ER+ tumors. Augmentation of the CCND1 gene is also common in luminal tumors. In ER+ breast tumors, activating alterations in PIK3CA are prevalent, and they lead to the progression of cell cycle via oncogenic AKT/mTOR signaling. ER+/luminal cancers, unlike TNBC and HER2+ types, usually have functional p53 and Rb tumor suppressor mechanisms as well as are gnomically stable due to their major reliance on estrogen signaling. TNBCs, on the other hand, have RB1 alterations or deletions that disrupt

M. A. Mir (✉) · S. Aisha
Department of Bioresources, School of Biological Sciences, Srinagar, J&K, India
e-mail: drmanzoor@kashmiruniversity.ac.in

K. Fatima · F. A. Malik
Pharmacology Division, CSIR-Indian Institute of Integrative Medicine, Sanat Nagar, Srinagar, J&K, India

M. A. Mir (ed.), *Therapeutic Potential of Cell Cycle Kinases in Breast Cancer*,
https://doi.org/10.1007/978-981-19-8911-7_16

the stability of the Rb/E2F/CDK4/6 pathway's cell cycle regulation, and also frequent changes in DNA damage reaction genes like BRCA1. Such tumors are also extremely aneuploid, with almost ubiquitous TP53 loss, recurring CCNE1 DNA multiplications, or decreased PTEN function (Curtis et al. 2012; Xu et al. 2014; Bianchini et al. 2016). TNBC tumors are dependent on the spindle assembly checkpoints and express elevated amount of mitotic checkpoint proteins (e.g., BUB1, TKK, AURKB, and MAD2) and DNA repairing proteins, according to numerous cellular dependency investigations (Yuan et al. 2006; Brough et al. 2011; Daniel et al. 2011; Curtis et al. 2012; Mehraj et al. 2021a). This is apparently due to their significant levels of genomic instabilities. CDK4 copy increases are widespread in all forms of breast cancers, although they are most common in HER2+ tumors. Furthermore, PIK3CA, TP53, and PTEN alterations, as well as CCND1 DNA duplication, are all common in this subtype.

Dysregulation of a cell cycle caused by tumor suppressor deactivation and abnormal stimulation of cyclins and CDKs is a hallmark of breast carcinoma. Given how important this biological mechanism is for cancer cell growth, it is no surprise that it is a prospective treatment target (Ignatiadis and Sotiriou 2013; Dominguez-Brauer et al. 2015; Mir et al. 2020). Nevertheless, because of the loss of target selectivity and dose-limiting toxicity, previously cell cycle targeted medicines performed badly in the clinic. Despite earlier medications' poor clinical performance, improvement of the treatment approach to increase therapeutic efficacy and the introduction of novel potent and specific inhibitors have reawakened interest in using the cell cycle as an antitumor therapy approach.

16.2 Treatment Response of the HER-2 Oncogene in Breast Cancer

HER-2, commonly called as HER-2/neu or erbB-2, is a 185-kDa transmembrane tyr kinase growth regulator receptor that is found on chromosome 17q (Yarden and Sliwkowski 2001; Mir et al. 2022a; b; c; d). Growth factor receptors are activated by attaching to targeting ligands or, if expressed in adequate receptors concentration on the cell membrane, by themselves, following by dimerization or receptors autophosphorylation, that results in various transduction pathways functioning via a number of routes. Angiogenesis, proliferation, abnormal cell interactions, enhanced cell mobility, metastasis, and apoptosis inhibition are all induced by the MAP kinase or 3-kinase (PI3K)/Akt routes (Oved and Yarden 2002; Wadhwa et al. 2020). The finding of HER-2 gene duplication and higher expression in early human breast carcinoma, as well as its link to more severe therapeutic behavior (Slamon et al. 1987), sparked initial interest in diagnosis and treatment applications. HER-2 gene is infrequently elevated in benign breast carcinoma, and its expression differs by histological subtype, as it is nearly exclusively detected in ductal vs. lobular initial breast malignancies. The HER-2 gene is increased and highly expressed in 20%–30% of aggressive cases of BC, as well as in the most of elevated Ductal carcinoma in situ patients (van de Vijver et al. 1988). Numerous researches have

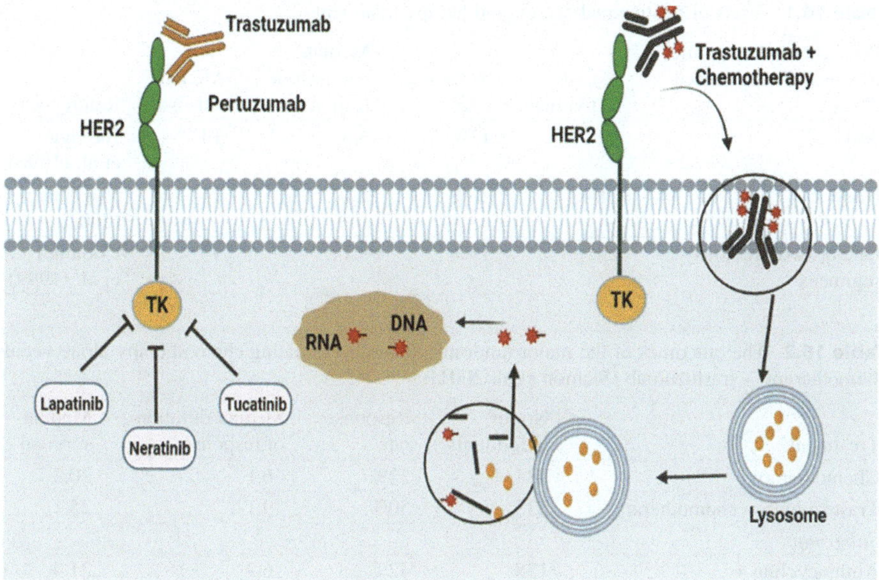

Fig. 16.1 Mechanism of action of anti-HER2 therapies

linked it to an elevated relapse in early breast malignancy, as well as greater resistance to endocrine treatment (possibly more with tamoxifen than it is with aromatase antagonists), resistance to non-anthracycline treatment, improved responsiveness to doxorubicin, as well as, in some cases, taxane-based treatment (Ross et al. 2003; Mehraj et al. 2021b). Nevertheless, apart from identifying individuals for trastuzumab treatment, HER-2 expression is not widely advised for decision-making at this time. It raises the likelihood of tumor relapse in early breast carcinoma and may thus affect adjuvant therapy selection.

Antibodies targeting growth hormone receptors have been proven in experimental animals to suppress growth. Trastuzumab is a humanized synthetic monoclonal antibody that targets the HER-2 protein's extracellular region (Carter et al. 1992). Trastuzumab's mode of activity is thought to involve modulatory impacts on cell signaling in animal systems, although there are additional indications of an immunological impact (Clynes et al. 2000) (Fig. 16.1). The findings of the earliest trastuzumab studies are summarized in Tables 16.1 and 16.2 (Baselga et al. 1996; Cobleigh et al. 1999; Slamon et al. 2001; Vogel et al. 2002). When trastuzumab was given as a single drug, response percentages ranged from 11% to 26%, and this effectiveness was greater (35%) in individuals who, in retrospect, had really HER-2+ tumors based on improved immunohistochemistry (IHC) or gene amplification standards. When trastuzumab was coupled with chemotherapy, it showed increased effectiveness, including increases in reaction rates, time to progression of the disease, durability of response, and survival in the major randomized study. With using trastuzumab, particularly in conjunction with anthracycline treatment,

Table 16.1 Trials of trastuzumab as a monotherapy treatment

Prior chemotherapeutic treatment for advanced illness	No. of individuals	Response rate	Median duration of response	Median survival	References
Any	43	12%	6.6	14	Baselga et al. (1996)
None	114	26%	>12	24	Vogel et al. (2002)
One or two previous regimens	222	15%	9.1	13	Cobleigh et al. (1999)

Table 16.2 The outcomes of the major randomized study evaluating chemotherapy alone versus chemotherapy + trastuzumab (Slamon et al. 2001)

Treatment	No. of individuals	Response rate	Median duration of response	Median survival
Chemotherapy	234	32%	6.1	20.3
Trastuzumab + chemotherapy	235	50%	9.1	25.1
Subgroups				
Anthracycline + cyclophosphamide	138	42%	6.7	21.4
Anthracycline + cyclophosphamide + trastuzumab	143	56%	9.1	26.8
Paclitaxel	96	17	4.5	18.4
Trastuzumab + paclitaxel	92	41	10.5	22.1

cardiomyopathy that is generally temporary and resolves over time has also observed. This is an example of how difficult it is to forecast the effects of tailored medicine. While HER-2 expression in mature myocytes is modest, the HER signaling cascade is known to have a role in embryonic cardiac and neural growth, as well as stress reactions and remodeling in the mature heart (Schneider et al. 2002)

Growth hormone receptor systems interact with additional routes, including those implicated in hormone-receptor signaling and DNA repairing, indicating that trastuzumab as well as other traditional breast carcinoma treatments could be additive or synergistic in some cases. Platinum medicines, vinorelbine, and docetaxel were reported to have the highest levels of synergy in preclinical studies, although differing outcomes were found by other scientists using various cell lines (Pegram et al. 1999). Trastuzumab had showed the most efficacies when combined with docetaxel, vinorelbine, and, to a smaller extent, gemcitabine in phase II experimental studies, although broader comparative studies are needed to enhance these combinations (O'Shaughnessy 2003; Esteva et al. 2002; Burstein et al. 2003). The combination of carboplatin with chemotherapy has been demonstrated to improve responsiveness and duration to progression in one research (Robert 2002). Trastuzumab as well as related HER-family-targeted medicines is also being utilized to overcome resistance to hormonal therapies or increase their efficacy (Johnston et al. 2003a).

Pertuzumab (2C4), a novel anti-HER-2 Mab, binds the external region of HER-2 as well, but it produces steric interference and inhibits receptors dimerization. This antibody was found in preclinical studies to suppress the development of cells that express reduced amounts of HER-2, likely by interacting with the production of HER family heterodimers (Agus et al. 2002). In solid tumors, phase I screening revealed efficacy (3/21 patients, 15%) (Agus et al. 2003), and research in breast carcinoma, including HER-2– or trastuzumab-refractory HER-2+ breast malignancy, are currently ongoing.

16.3 Endocrine Resistance

Because antiestrogens, a commonly utilized and successful treatment for hormone-responsive BCs, quickly downregulate cyc D1, unregulated transcription of this cyclin may be supposed to impact susceptibility to these drugs. Several clinical evidence supports the concept that individuals with elevated cyclin D1 have a shorter length of responsiveness to antiestrogen treatment, and that higher expression of cyc D1 in breast tumor cultured cells leads to transient antiestrogen resistance (Butt et al. 2005; Qayoom et al. 2021). More evidence from laboratory models and primary breast tumors would be needed to answer the issue of whether cyclin D1 overexpression impacts antiestrogen responsiveness. Despite the fact that cyclin E upregulation in breast carcinoma cells has only a minor impact on antiestrogen responsiveness in vitro, one research discovered that elevated cyc E expression has been linked with worse RFS in patient populations treated with hormonal therapies (Sutherland and Musgrove 2004).

p27 mediates the cell-cycle stop of breast carcinoma cells by therapeutically effective pharmaceutical drugs that impede estrogen activity (Sutherland and Musgrove 2004). Administration of MCF-7 BC cells with synthetic steroidal antiestrogen ICI 182780 (Mir 2015) (Faslodex), for instance, resulted in enhanced p27 expression, increased p27–cyclin E–Cdk2 interaction, or cell-cycle halt. Moreover, this antiestrogen's high stimulation of p27 helps to induce a quiescent, growth factor-insensitive condition. Antiestrogen sensitivity is conferred in breast tumor cells in vitro when p21 or p27 is downregulated by antisense oligonucleotides or when Skp2 is overexpressed. MEK inhibition recovers p27 suppression of cyclin E–Cdk2 complex and treatment response in antiestrogen sensitive breast tumor cells, demonstrating that antiestrogen responsiveness may be regained in resistant cells by treating with specific signal cascade inhibitors. In a therapeutic setting, tumor p21 and p27 status could be prognostic of antiestrogen response. Elevated p27 expression has been linked with enhanced relapse-free as well as overall survival in a research of premenopausal females with initial phase breast tumor obtaining combinational endocrine treatment of tamoxifen and goserelin (Pohl et al. 2003), and p21 levels had also been linked with reaction to antiestrogens in several, but not every, clinical research (Butt et al. 2005).

16.4 Early Generation Cell Cycle/CDK Inhibitors and Microtubule Binding Drugs

Microtubule binding agents (MTBAs) are a cornerstone in cancer treatment that work by stabilizing (taxanes, such as docetaxel and paclitaxel) or disrupting (eribulin, vinca alkaloids) microtubules during mitosis of multiplying cancerous cells (Dumontet and Jordan 2010). These drugs trigger the spindle assembly/mitotic checkpoints (SAC) that inhibit anaphase and mitotic escape till all chromosomes had established bipolar adhesion to the spindle. MTBAs cause cell cycle stop in mitosis as a result of this, and persistent mitotic pause results in apoptosis (Dumontet and Jordan 2010; Mehraj et al. 2022a). Numerous MTBAs have been licensed for breast carcinoma treatment, either individually or in conjunction with additional chemotherapy drugs (or HER2-targeted MABs). When utilized as adjuvant therapy for early-stage breast carcinoma, taxanes were shown to boost treatment efficacy, and substantial increase in tumor responsiveness can be attained in initial lines of treatment for metastatic cancer (De Laurentiis et al. 2008; Gradishar 2012; Mir et al. 2022a; b; c; d) (Fig. 16.2).

The cell cycle's first-generation targeted inhibitors, in comparison to MTBAs, showed minimal effectiveness in the management of solid tumors (Dumontet and Jordan 2010). Dinaciclib, flavopiridol, and seliciclib (Finn et al. 2016) are non-selective multi-CDK antagonists, as are many similar mitotic kinase blockers like those inhibiting AURKB and PLK1. The surprising insufficiency of such medicines had been ascribed in large portion to dose-limiting effects produced by unwanted target suppression in non-malignant tissues, such as neurotoxicity, myelosuppression, and gastrointestinal issues (Dumontet and Jordan 2010; Finn et al. 2016). The absence of prognostic biomarkers to guide patient choice for these medications may possibly had led to their failure (Finn et al. 2016; Mir et al. 2022a; b; c; d). Although preclinical research suggests that some of these medicines,

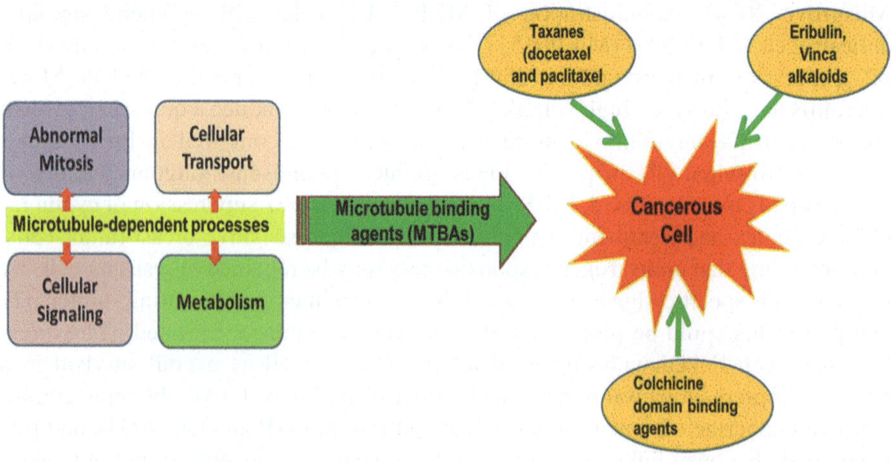

Fig. 16.2 Different microtubule binding drugs

like dinaciclib, could be repurposed, the processes are dependent on transcriptional suppression instead of cell cycle-related actions (Johnson et al. 2016).

In order to accomplish a treatment efficacy for breast cancer specificity, these experiences formalized the necessity of utilizing cancer- and subtype-specific vulnerabilities (such as genetic instability and anomalies in cell-cycle proteins which are not crucial in non-malignant cells). This need has directed the advancement of the latest generation of agents that target cell cycle.

16.4.1 Inhibitors of TTK

TTK protein kinases (TTK), also called as MPS1 (monopolar spindle 1), is an important modulator of the SAC (Spindle checkpoint assembly) (Lara-Gonzalez et al. 2012; Liu and Winey 2012; Musacchio 2015) (Fig. 16.3). TTK establishes and maintains the mitotic checkpoint by phosphorylating its substrate and recruiting checkpoint protein to kinetochores throughout mitosis. Because it delays anaphase and the commencement of mitotic escape till all chromosomes had attained bipolar linkage to the mitotic spindle, the checkpoint is crucial for preserving genomic integrity throughout mitosis (Lara-Gonzalez et al. 2012). TTK promotes proper chromosomal segregation and genetic stability by regulating the SAC. TTK is frequently abundantly expressed in cancerous cells, that also might be due to its role in mitosis, as aggressive tumors with elevated mitotic indicators naturally have elevated expression rates of cell-cycle genes; nevertheless, it may also be due to tumor cells' reliance on the SAC to facilitate feasible separation of their aneuploid as well as unreliable genomes into new cells (Yuan et al. 2006; Daniel et al. 2011; Curtis et al. 2012; Patel et al. 2018a). TNBCs were shown to have a gene expression profile linked with "aggressiveness," which was elevated for genes associated with

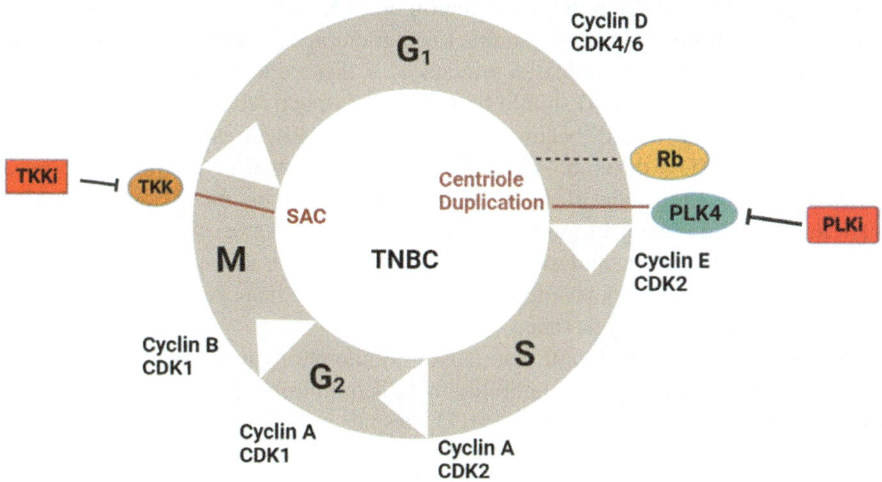

Fig. 16.3 Cell cycle vulnerabilities of TNBC tumors

Table 16.3 TTK inhibitors in clinical trials

Drug	Phase	Patient Cohort	Trail ID
BOS172722 (CCT289346) ± paclitaxel	Phase I	Advanced solid tumors	NCT03328494
BAY1161909 + paclitaxel	Phase I	Advances cancers	NCT021138812
BAY 1217389 + paclitaxel	Phase I	Advanced cancers	NCT02366949
CFI-402257	Phase I	Advanced cancers	NCT02792465
S81694 + paclitaxel	Phase I/II	TNBC	NCT03411161

genomic instability, notably TTK (Liu and Winey 2012). TNBC cells' reliance on TTK was also validated in this investigation. TTKs expression differentiates aneuploid cancerous cells from non-malignant diploid cells which have supplemental competent paths protecting integrity of the genome, and offers a treatment efficacy to target breast tumor cells, irrespective of whether its upregulation reflects a correlation with extremely aggressive tumor physiology or a functional reliance of cancerous cells on the SAC. The idea behind using TTK inhibitors to cure genetically unstable tumors like TNBC is to bypass the SAC and force cells to undergo mitosis when the chromosomes could be properly segregated. In susceptible cancerous cells, this leads to mitotic segregation faults and unbearable levels of genetic instability, eventually resulting in cell death. Numerous TTK inhibitors, like CFI-402257 (Table 16.3), are now being examined in initial stage clinical studies as individual medicines or in conjunction with taxane therapy. AZ3146, CCT271850, NMS-P715, CCT251455, MPI-0479605, and MPS1-IN-3 are some of the other TTK inhibitors in experimental research, highlighting the interest and therapeutic promise of this family of anti-cancer drugs (Naud et al. 2013; Tannous et al. 2013; Kusakabe et al. 2015; Faisal et al. 2017).

Gatekeeper alterations in the active region of TTK were shown to give resistance in in vitro to TTK inhibitors, comparable to other therapeutic kinase inhibitors (Koch et al. 2016), though the clinical significance of these alterations in patients who acquire resistance (Fig. 16.4) should be verified prospectively.

Patient biopsies would become accessible for genetic sequencing as TTK inhibitors precede through clinical studies, allowing pharmacogenomics investigations to find molecular correlations linked with treatment responsiveness. Such data will be crucial in guiding patient treatment categorization. In the meanwhile, once tumor genetic and medication reaction data are accessible, possible TTK inhibitor biomarkers responsiveness could be developed in a preclinical context and examined for therapeutic relevance.

Zaman and coworkers found that activation alterations in CTNNB1 (β-catenin) were related with increased susceptibility to TTK inhibitors when contrasted to wild-type CTNNB1 mice (Zaman et al. 2017), despite only a basic assessment of the working effect of CTNNB1 mutant on TTK inhibitor sensitivity was carried in an in vitro. Moreover, the frequency of CTNNB1 alterations in BC restricts the use of this potential biomarker in this illness. p53−/− HCT-116 colon cancer cells were

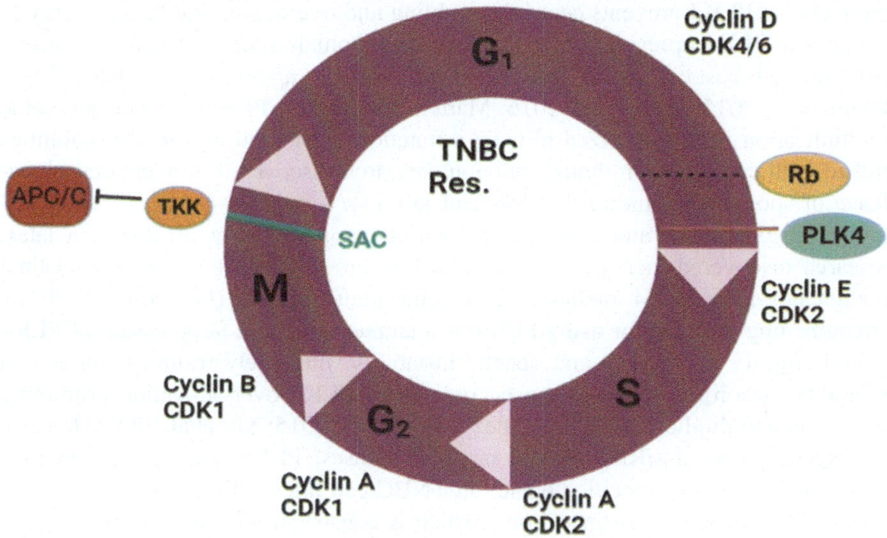

Fig. 16.4 Potent resistance mechanisms to TTK inhibition in TNBC

shown to be highly susceptible to the wide ranging S/T kinase inhibitors, SP600125, that suppress TTK with an IC50 = 1.95 μM (Jemaa et al. 2012), suggesting that p53 status might be linked to responsiveness to TTK suppression. The lack of specificity of SP600125, on the other hand, makes it difficult to understand the significance of TTK suppression in triggering selective p53-deficient cell death in this research. Furthermore, utilizing siRNA screenings to compare genetic dependence in PTEN-mutant versus wild-type BC cell lines, it was discovered that TTK inhibition is selectively detrimental in PTEN-deficient cancerous cells, implying that PTEN may be a biomarker for TTK inhibitor effectiveness (Dumontet and Jordan 2010; Mendes-Pereira et al. 2012).

16.4.2 PLK4 Inhibitors

Centrosomes, which are made up of 2 barrel-like structures termed centrioles, are primary microtubule organizing centers in cells. In G1, the centrosome's centrioles detach from each other. Centriole replication begins at the same time as the G1/S shift. In the S and G2 stages, procentriole synthesis and extension occur, and centrosome development and division occur, resulting in 2 centrosomes that constitute the mitotic spindle (Wang et al. 2014; Nigg and Holland 2018). PLK4 (Polo-like kinase 4) is the controller of centriole replication. The activation of centrosome biology proteins by PLK4 is an upstream step in the centriole formation signaling cascades, which is necessary for centrosome synthesis (Wang et al. 2014; Maniswami et al. 2018; Nigg and Holland 2018). PLK4 transcription is strictly controlled to maintain the numeric integrity of centrioles and centrosomes, as reduction of PLK4 prevents centriole doubling and overexpression causes centriole expansion. PLK4 upregulation and centrosome anomalies are prevalent in cancers, particularly breast carcinoma, and are linked to disease aggressiveness (Chan 2011; Pannu et al. 2015; Denu et al. 2016; Marteil et al. 2018). PLK4-driven centrosome multiplication is hypothesized to have a function in the chromosomal instabilities linked with cancer. Even though previous research discovered no increased prevalence of spontaneous tumor development in p53+/− or p53−/− mice after PLK4 increased expression and centrosome multiplication (Vitre et al. 2015), a latest research discovered increased tumor growth in an APCMin/+ model of intestinal malignancy after PLK4-mediated centrosome multiplication (Levine et al. 2017), strengthening the case for using PLK4 as a targeted therapy. Suppression of PLK4 would amplify aneuploidy and genetic instability; ultimately result in cancer cell mortality, according to the proposed process of PLK4 over-activation promoting centrosome multiplication (Dominguez-Brauer et al. 2015; Mir et al. 2022a; b; c; d).

Depending on a siRNA screen targeted kinases, PLK4 was considered as a potential option with specific potency in TNBC cell lines. CFI-400945 has distinct phenotypic impacts at distinct doses, which is consistent with the inverse impacts mentioned for PLK4 removal and over-activation in the literature, with depletion of centrosome at elevated doses as well as centriole over-duplication (and subsequent centrosome overexpression) at small doses. PLK4 was partially inhibited by CFI-400945 at low doses, blocking auto-regulation through trans-autophosphorylation of its degron, resulting in enhanced PLK4 proteins expression (Cunha-Ferreira et al. 2013; Mason et al. 2014; Bedard et al. 2016). CFI400945 administration causes chromosomal segregation abnormalities due to the production of abnormal mitotic spindles (for example, multi-polar spindles), elevated DNA content associated with genetic instability, or finally death of cells in several cancer types (Mason et al. 2014; Lohse et al. 2017; Kawakami et al. 2018).

Employing in vitro models of cancer, researchers have yet to find significant pathways of developed resistance to CFI-400945, and this remained an important field of research. In diploid non-malignant RPE1 cells, cell resistance to centrosome loss generated by the tool chemical centrinone, a relatively specific inhibitor of PLK4, was revealed to be controlled by a p53-dependent 53BP1-USP28 pathway. In two classical investigations that created models of RPE1 in which natural PLK4 was substituted by an analog-sensitive variant that is inhibited following chemical stimulation to promote centrosome removal, this result of 53BP1 and USP28 deactivation generating resistance toward centrosome loss was similarly seen (Fong et al. 2016; Lambrus et al. 2016). Although this mechanism of resistance may exist in cancers with good p53 signaling, its therapeutic significance is unknown. TP53-independent processes would likely promote resistance in tumors with elevated amounts of genetic instability, centrosomal abnormalities, and common TP53 alterations, such as TNBC (Marteil et al. 2018).

Early functional assessment of CFI400945 impacts in breast tumor cell lines suggested a possible link among PTEN status and drug responsiveness, with PTEN loss being linked to response (Mason et al. 2014). This result was in line with a study that found inactivation of PLK4 was synthetically fatal in breast cancers with PTEN mutation (Brough et al. 2011), implying that PTEN status might be used as a marker for CFI-400945. PTEN as a possible biomarker in BC PDX with characterized reactions to CFI-400945, as well as other postulated biomarkers dependent on PLK4 biology, like instability of chromosomes or copy numbers of centrosomes, is now being investigated. Surprisingly, a recent thorough evaluation of centrosome multiplication in the NCI-60 tumor cell line panel revealed richness of centrosome multiplication in aggressive TNBC and microsatellite stable colorectal malignancy (vs MSI CRC) that are 2 diseases for which CFI-400945 has shown substantial preclinical and clinical antitumor action (Marteil et al. 2018). Clinically tumor samples from existing clinical studies will be used to examine the possible connections among PTEN inactivation, centrosome duplication, and CFI-400945 responsiveness.

16.5 Modulators for Downstream Signal Transduction

Signaling pathway controlled by growth factors stimulates several major kinases which operate as master regulators, controlling multiple routes. Because alternative receptors can potentially initiate signals, targeting downstream messengers may produce therapeutically favorable benefits which growth factor channel inhibition cannot, but it can also lead to extra toxicities. The mTOR is a critical downstream regulator that regulates the cell cycle by coupling growth impulses from receptor or cytosolic kinases. Rapamycin as well as its equivalents suppress mTOR activation, preventing downstream stimulation of S6 kinase or 4E binding protein-1, and thereby reducing translation of essential protein biosynthesis machinery elements and cell cycle transcription factors (like c-Myc or cyclin D1) (Mita et al. 2003; Khan et al. 2022b; Mir et al. 2022a; b; c; d). CCI-779 is a rapamycin analog which had been through stage I research and has shown toxicities such as dermatitis, myelosuppression, and elevated liver enzymes. Improvements were observed in numerous tumor types, particularly breast, on a weekly frequency, which looked to be the most tolerated (Hidalgo et al. 2000). A phase-2 trial comparing 75 mg CCI779 intravenously weekly versus 250 mg CCI779 intravenously once a week for DOXO and/or taxane-refractory BC had shown preliminary cumulative outcomes of 9 responses from 106 patient populations (8.5%), with a 10% occurrence of level III/IV hepatocellular, skin, as well as hematological toxicities (Chan et al. 2003). More research is required to establish its efficacy when administered early in the development of the illness or in combination treatment, as well as whether biological subtypes of individuals are somewhat more likely to benefit. The MAP kinase and PI3K/Akt cascades are activated by Ras, a downstream main signaling protein. FTIs (farnesyl transferase inhibitors) block Ras from reaching the inner layer, where it is triggered. Even though certain oncogenic variants of Ras are poorly suppressed by FTIs, and RAS alterations are rarely shown in breast carcinoma, FTIs could still have a role in breast cancer because Ras is involved in growth factor receptors and other processes. In a stage II study of 76 individuals, the FTI tipifarnib produced a 12% reaction rate as well as a 24% therapeutic improvement rate, with thrombocytopenia, neurotoxicity, and granulocytopenia as adverse effects (Johnston et al. 2003b). Small molecules or antisense inhibitor of Ras downstream elements (for example, Raf, MEK kinase) are being studied, however no outcomes in breast cancer have been reported yet.

16.6 Cell Cycle Modulators and Cyclins

The process of entering cell cycle and active multiplication is strongly regulated. CDKs are a set of proteins that are appropriately distributed throughout the cell cycle. When CDKs are triggered, they enhance the activation of other proteins, particularly pRb, a key gatekeeper which enables the cell to transition from G0 to dynamic cycling and mitosis. Cyclins control CDKs favorably, while CDK inhibitors inhibit them negatively (CKIs). Cyc D1 and cyc E expression rates fluctuate with the cell cycle and both are important in the cell's transition from G1 to S stage (Vermeulen et al. 2003; Sofi et al. 2022a).

The cyclin D1 gene which is located on chromosome 11q13 had been shown to be highly expressed in 40%–50% of aggressive breast tumors and increased in 10%–20% of instances (Steeg and Zhou 1998; Mehraj et al. 2022b). The pRb tumor suppressor factor is activated when cyclin D1 is bonded with its CDK counterpart, liberating the transcription component E2F and stimulating proteins essential for DNA replication. Elevated levels of cyclin D1 expression seem to be linked to ER positivity and a higher proliferation index (Loden et al. 2002). The cyclin E gene is found on 19q12 chromosome and is only slightly increased in breast carcinoma (2%); nevertheless, upregulation and changes in the breakdown pathway leading to the aggregation of limited-molecular-weight variants have been observed in 20%–30% of breast malignancies (Keyomarsi et al. 2002). Occasionally, both cyclin D1 and cyclin E are abundantly expressed at the same time. Overexpression of cyc E, like cyc D1, causes hyper-phosphorylation of pRb and enhanced proliferation. In comparison to elevated cyc D1 cancers, strong cyclin E tumors are also capable of inducing S phase without pRb phosphorylation or E2F stimulation. Overall, there is a large loss in cell cycle regulation as well as a substantial deregulation of multiplication as a consequence of this. Elevated cyc E cancers are much more prone to be of a greater grade than elevated cyc D1 tumors, are HR (−), possess a greater proliferation score, and had a worse prognosis (Keyomarsi et al. 2002; Loden et al. 2002; Mir 2022). Many characteristics linked to elevated cyclin E levels could indicate why this phenotype is highly aggressive. As previously stated, tumors that overexpress cyclin E are likely to skip the pRb node, enabling for more rapid cell cycle. Furthermore, higher cyc E levels had been linked to greater genetic instability, as compared to higher cyclin D1 levels. Moreover, the enzyme, elastase that breaks down cyclin E to its low-molecular-weight variants, has been linked to a higher propensity for infiltration and metastasis, which may help to elucidate the aggressive phenotype (Keyomarsi et al. 2002; Mir and Agrewala 2008; Khan et al. 2022a). Yet, there is no evidence that cyclins or their variants should be used routinely for predictive or therapeutic purposes.

Cell cycle control is an attractive target because it is a critical end point for many signaling cascades (Dai and Grant 2003). Flavopiridol is a nonspecific CDK blocker and a semi-synthetic flavone derivative of rohitukine, an antitumor drug derived from an Indian plant (Tan and Swain 2002). This chemical seems to cause apoptosis by interfering with CDKs for ATP binding and disrupting P-TEFb (the CDK9-cyclin T complexes), perhaps as a result of downregulation of anti-apoptotic proteins. Secretory diarrhea and hypotension were found to be dosage limiting in phase I investigations. In a phase II study in mantle cell lymphoma, that is linked to cyclin D1 amplification, 3 patients (11%) responded, with diarrhea, lethargy, and nausea typical side effects, as well as modest hematologic damage (Kouroukis et al. 2003). In breast carcinoma, trials of Flavopiridol in conjunction with a variety of chemotherapeutic drugs are continuing, and preliminary outcomes from a stage I study of Flavopiridol plus docetaxel showed that this conjunction is well accepted (Patel et al. 2018b; Mir et al. 2022a; b; c; d). Because growth hormone-receptor signaling ultimately leads to cell cycle entry, addressing the distal and proximal elements of this pathway with trastuzumab in conjunction with Flavopiridol was examined in HER-2+ cell lines, and combinatorial cytotoxicity was shown (Nahta et al. 2002). Therefore, experiments that combine early and middle signaling with specific cell cycle modulators could be likely to produce improved cell deaths while causing less host damage.

Ro 31-7453 is a nonspecific oral cell-cycle blocker that has been shown to be effective against a variety of tumor cell lines in vitro. It induces loss of mitotic spindle assembly in dividing cells, resulting to M-phase halt, by slightly inhibiting CDK2, CDK1, and CDK4 and tubulin assembly. Two of 32 (6 percent) individuals in a phase II trial of tax and anthracycline-resistant BC reacted, with diarrhea and nausea being the most common adverse effects (Osborne et al. 2004). UCN-01 (7hydroxy-staurosporine) is similarly a wide blocker of CDKs and PDK1, exhibiting hypotension as an adverse effect in stage I studies and no reactions in renal cell carcinoma (Shaw et al. 2009; Sofi et al. 2022b). Chemotherapeutic combinations are being investigated. CDK4 and CDK6 inhibitors with greater specificity have been designed and are tested in clinical studies.

Proteasomal inhibitors had become of interest (Nalepa and Harper 2003; Qayoom et al. 2022) because numerous CKIs as well as similar negative regulating molecules are generally controlled by ubiquitin–proteasome degradation. They are currently being tested in clinical trials for breast carcinoma. These medications impact not just CKIs but also a variety of many additional short-lived proteins, like the repressor of NF-κB, a critical mediator of stress and immune system response systems, and so may block other signaling routes as well. In anthracycline-pretreated breast carcinoma, a phase-2 trial of the proteasome blocker bortezomib in conjunction with docetaxel generated response in 6 out of 14 patients (43%) (Albanell et al. 2004).

16.7 Summary

Based on the initial clinical achievement of CDK4/6 inhibitors as well as the hopeful anti-cancer properties of TTK inhibitors as well as CFI-400945, the upcoming rational stages in the advancement of these molecules are to better comprehend treatment resistance, recognize biomarkers for patient choice, and advise treatment sequence data. Long-term therapy of terminal metastatic malignancies with these novel drugs will very certainly accelerate tumor evolution and give selection pressure for drug-resistant clones to spread. Knowing how this evolving process alters tumor genomes to produce treatment resistance phenotypes would be crucial in developing approaches to counteract progression of the disease. Understanding resistance processes and drug-induced biological changes could lead to the development of biomarkers that can forecast innate tumor responses. The proportion of persons with developed tolerance to CDK 4 and 6 inhibitors, that are currently conventional of care for the management of metastatic ER +/HER2+ BCs, would continue to rise. Multiple attempts are being made to explain the genetic evolution of therapeutic resistance utilizing both liquid (i.e., ctDNA) and paired tumor tissues, and testing of advancing illness will be critical for discovering resistance pathways.

16.8 Further Reading

The readers can further read about the role of CDKs in breast cancer by going through the following papers

- https://www.nature.com/articles/s41523-017-0009-7
- http://egetipdergisi.com.tr/en/pub/issue/36515/414615

The following visual presentations are also available for the readers to view for a better conceptual grasp of CDKs and their function in breast cancer

- https://www.youtube.com/watch?v=0Sj3rbJPeXQ
- https://www.youtube.com/watch?v=vEe3lBduckE
- https://www.youtube.com/watch?v=9Rd74mqd-jw

For more insights about the topic, we would suggest detailed findings from the books of (Mir MA, 2022) https://doi.org/10.1016/C2021-0-02565-7, https://doi.org/10.1016/C2014-0-02898-5 (Mir MA, 2021) https://doi.org/10.52305/WXJL6770, from cancer.net website, https://www.cancer.net/cancer-types/breast-cancer/types-treatment

For diagrammatic illustrations, descriptive tables, (Lazzeroni, 2012) http://www.eurekaselect.com/article/49928

References

Agus DB et al (2002) Targeting ligand-activated ErbB2 signaling inhibits breast and prostate tumor growth. Cancer Cell 2(2):127–137

Albanell J et al (2004) Phase I study of bortezomib in combination with docetaxel in anthracycline-pretreated advanced breast cancer. In: Proteasome inhibitors in cancer therapy, pp 271–282. https://doi.org/10.1007/978-1-59259-794-9_20

Baselga J et al (1996) Phase II study of weekly intravenous recombinant humanized anti-p185HER2 monoclonal antibody in patients with HER2/neu-overexpressing metastatic breast cancer. J Clin Oncol 14(3):737–744

Bedard PL et al (2016) Abstract CT066: first-in-human phase I trial of the oral PLK4 inhibitor CFI-400945 in patients with advanced solid tumors. AACR

Bianchini G et al (2016) Triple-negative breast cancer: challenges and opportunities of a heterogeneous disease. Nat Rev Clin Oncol 13(11):674–690

Brough R et al (2011) Functional viability profiles of breast cancer. Cancer Discov 1(3):260–273

Burstein HJ et al (2003) Trastuzumab and vinorelbine as first-line therapy for HER2-overexpressing metastatic breast cancer: multicenter phase II trial with clinical outcomes, analysis of serum tumor markers as predictive factors, and cardiac surveillance algorithm. J Clin Oncol 21(15): 2889–2895

Butt AJ et al (2005) Downstream targets of growth factor and oestrogen signalling and endocrine resistance: the potential roles of c-Myc, cyclin D1 and cyclin E. Endocrine Related Cancer 12(1):S47

Carter P et al (1992) Humanization of an anti-p185HER2 antibody for human cancer therapy. Proc Natl Acad Sci 89(10):4285–4289

Chan S et al (2003) Phase 2 study of two dose levels of CCI-779 in locally advanced or metastatic breast cancer (MBC) failing prior anthracycline and/or taxane regimens. J Clin Oncol 23 (23):5314–5322

Chan JY (2011) A clinical overview of centrosome amplification in human cancers. Int J Biol Sci 7(8):1122

Clynes RA et al (2000) Inhibitory Fc receptors modulate in vivo cytotoxicity against tumor targets. Nat Med 6(4):443–446

Cobleigh MA et al (1999) Multinational study of the efficacy and safety of humanized anti-HER2 monoclonal antibody in women who have HER2-overexpressing metastatic breast cancer that has progressed after chemotherapy for metastatic disease. J Clin Oncol 17(9):2639–2639

Cunha-Ferreira I et al (2013) Regulation of autophosphorylation controls PLK4 self-destruction and centriole number. Curr Biol 23(22):2245–2254

Curtis C et al (2012) The genomic and transcriptomic architecture of 2,000 breast tumours reveals novel subgroups. Nature 486(7403):346–352

Dai Y, Grant S (2003) Cyclin-dependent kinase inhibitors. Curr Opin Pharmacol 3(4):362–370

Daniel J et al (2011) High levels of the Mps1 checkpoint protein are protective of aneuploidy in breast cancer cells. Proc Natl Acad Sci 108(13):5384–5389

De Laurentiis M et al (2008) Taxane-based combinations as adjuvant chemotherapy of early breast cancer: a meta-analysis of randomized trials. Database of Abstracts of Reviews of Effects. (DARE): Quality-assessed Reviews [Internet], Centre for Reviews and Dissemination (UK)

Denu RA et al (2016) Centrosome amplification induces high grade features and is prognostic of worse outcomes in breast cancer. BMC Cancer 16(1):1–13

Dominguez-Brauer C et al (2015) Targeting mitosis in cancer: emerging strategies. Mol Cell 60(4): 524–536

Dumontet C, Jordan MA (2010) Microtubule-binding agents: a dynamic field of cancer therapeutics. Nat Rev Drug Discov 9(10):790–803

El-Rayes BF et al (2006) A phase I study of flavopiridol and docetaxel. Investig New Drugs 24 (4):305–310. https://doi.org/10.1007/s10637-005-4343-5

Esteva FJ et al (2002) Phase II study of weekly docetaxel and trastuzumab for patients with HER-2–overexpressing metastatic breast cancer. J Clin Oncol 20(7):1800–1808
Faisal A et al (2017) Characterisation of CCT271850, a selective, oral and potent MPS1 inhibitor, used to directly measure in vivo MPS1 inhibition vs therapeutic efficacy. Br J Cancer 116(9): 1166–1176
Finn RS et al (2016) Targeting the cyclin-dependent kinases (CDK) 4/6 in estrogen receptor-positive breast cancers. Breast Cancer Res 18(1):1–11
Fong CS et al (2016) 53BP1 and USP28 mediate p53-dependent cell cycle arrest in response to centrosome loss and prolonged mitosis. elife 5:E1491
Gradishar WJ (2012) Taxanes for the treatment of metastatic breast cancer. Breast cancer: basic and clinical research 6:BCBCR-S8205
Hidalgo M et al (2000) CCI-779, a rapamycin analog and multifaceted inhibitor of signal transduction: a phase I study. Oncogene 19(56):6680–6686
Ignatiadis M, Sotiriou C (2013) Luminal breast cancer: from biology to treatment. Nat Rev Clin Oncol 10(9):494–506
Jemaa M et al (2012) Selective killing of p53-deficient cancer cells by SP600125. EMBO Mol Med 4(6):500–514
Johnson SF et al (2016) CDK12 inhibition reverses de novo and acquired PARP inhibitor resistance in BRCA wild-type and mutated models of triple-negative breast cancer. Cell Rep 17(9): 2367–2381
Johnston SRD et al (2003a) Integration of signal transduction inhibitors with endocrine therapy: an approach to overcoming hormone resistance in breast cancer. Clin Cancer Res 9(1):524s–532s
Johnston SRD et al (2003b) Phase II study of the efficacy and tolerability of two dosing regimens of the farnesyl transferase inhibitor, R115777, in advanced breast cancer. J Clin Oncol 21(13): 2492–2499
Kawakami M et al (2018) Polo-like kinase 4 inhibition produces polyploidy and apoptotic death of lung cancers. Proc Natl Acad Sci 115(8):1913–1918
Keyomarsi K et al (2002) Cyclin E and survival in patients with breast cancer. N Engl J Med 347(20):1566–1575
Khan SU et al (2022a) Understanding the cell survival mechanism of anoikis-resistant cancer cells during different steps of metastasis. Clin Exp Metastasis 39:715
Khan SU et al (2022b) Activation of lysosomal mediated cell death in the course of autophagy by mTORC1 inhibitor. Sci Rep 12(1):1–13
Koch A et al (2016) Molecular basis underlying resistance to Mps1/TTK inhibitors. Oncogene 35(19):2518–2528
Kouroukis CT et al (2003) Flavopiridol in untreated or relapsed mantle-cell lymphoma: results of a phase II study of the National Cancer Institute of Canada clinical trials group. J Clin Oncol 21(9):1740–1745
Kusakabe K-I et al (2015) Discovery of imidazo [1, 2-b] pyridazine derivatives: selective and orally available Mps1 (TTK) kinase inhibitors exhibiting remarkable antiproliferative activity. J Med Chem 58(4):1760–1775
Lambrus BG et al (2016) A USP28–53BP1–p53–p21 signaling axis arrests growth after centrosome loss or prolonged mitosis. J Cell Biol 214(2):143–153
Lara-Gonzalez P et al (2012) The spindle assembly checkpoint. Curr Biol 22(22):R966–R980
Levine MS et al (2017) Centrosome amplification is sufficient to promote spontaneous tumorigenesis in mammals. Dev Cell 40(3):313–322
Liu X, Winey M (2012) The MPS1 family of protein kinases. Annu Rev Biochem 81:561
Loden M et al (2002) The cyclin D1 high and cyclin E high subgroups of breast cancer: separate pathways in tumorogenesis based on pattern of genetic aberrations and inactivation of the pRb node. Oncogene 21(30):4680–4690
Lohse I et al (2017) Activity of the novel polo-like kinase 4 inhibitor CFI-400945 in pancreatic cancer patient-derived xenografts. Oncotarget 8(2):3064

Maniswami RR et al (2018) PLK4: a link between centriole biogenesis and cancer. Expert Opin Ther Targets 22(1):59–73

Marteil G et al (2018) Over-elongation of centrioles in cancer promotes centriole amplification and chromosome missegregation. Nat Commun 9(1):1–17

Mason JM et al (2014) Functional characterization of CFI-400945, a polo-like kinase 4 inhibitor, as a potential anticancer agent. Cancer Cell 26(2):163–176

Mehraj U et al (2022a) Expression pattern and prognostic significance of baculoviral inhibitor of apoptosis repeat-containing 5 (BIRC5) in breast cancer: a comprehensive analysis. Advances in Cancer Biology - Metastasis 4:100037

Mehraj U et al (2021a) Tumor microenvironment promotes breast cancer chemoresistance. Cancer Chemother Pharmacol 87(2):147–158

Mehraj U et al (2021b) Prognostic significance and targeting tumor-associated macrophages in cancer: new insights and future perspectives. Breast Cancer 28(3):539–555

Mehraj, U., et al. (2022b). "Cryptolepine targets TOP2A and inhibits tumor cell proliferation in breast cancer cells-an in vitro and in silico study." Anti-cancer Agents in Medicinal Chemistry

Mendes-Pereira AM et al (2012) NLK is a novel therapeutic target for PTEN deficient tumour cells. PLoS One 7(10):e47249

Mir MA (2015) Developing costimulatory molecules for immunotherapy of diseases. Academic Press. https://doi.org/10.1016/C2014-0-02898-5. ISBN: 9780128025857

Mir MA, Agrewala JN (2008) Signaling through CD80: an approach for treating lymphomas. Expert Opin Ther Targets 12(8):969–979

Mir MA et al (2020) Targeting different pathways using novel combination therapy in triple negative breast cancer. Curr Cancer Drug Targets 20(8):586–602

Mir MA et al (2022a) Chapter 2 - novel biomarkers in triple-negative breast cancer - role and perspective. In: Mir M (ed) Combinational therapy in triple negative breast cancer. Academic Press, pp 29–60. pp 47–96. https://doi.org/10.52305/DXSK7394

Mir MA et al (2022b) Chapter 3 - current therapeutics and treatment options in TNBC. In: Mir M (ed) Combinational therapy in triple negative breast cancer. Academic Press, pp 61–94. https://doi.org/10.1016/B978-0-323-96136-3.00007-8

Mir MA et al (2022c) Chapter 8 - different drug delivery approaches in combinational therapy in TNBC. In: Mir M (ed) Combinational therapy in triple negative breast cancer. Academic Press, pp 201–230. https://doi.org/10.52305/DHHG6044

Mir MA et al (2022d) Chapter 4 - conventional adjuvant chemotherapy in combination with surgery, radiotherapy, and other specific targets. In: Mir M (ed) Combinational therapy in triple negative breast cancer. Academic Press, pp 95–120. https://doi.org/10.1016/B978-0-323-96136-3.00008-X

Mir MA, Sofi S, Qayoom H (2022) Different drug delivery approaches in combinational therapy in TNBC. Elsevier, pp 201–230. https://doi.org/10.1016/B978-0-323-96136-3.00012-1

Mita MM et al (2003) Mammalian target of rapamycin: a new molecular target for breast cancer. Clin Breast Cancer 4(2):126–137

Musacchio A (2015) The molecular biology of spindle assembly checkpoint signaling dynamics. Curr Biol 25(20):R1002–R1018

Nahta R et al (2002) Rate-limiting effects of cyclin D1 in transformation by ErbB2 predicts synergy between herceptin and flavopiridol. Cancer Res 62(8):2267–2271

Nalepa G, Harper JW (2003) Therapeutic anti-cancer targets upstream of the proteasome. Cancer Treat Rev 29:49–57

Naud S et al (2013) Structure-based design of orally bioavailable 1 H-pyrrolo [3, 2-c] pyridine inhibitors of mitotic kinase monopolar spindle 1 (MPS1). J Med Chem 56(24):10045–10065

Nigg EA, Holland AJ (2018) Once and only once: mechanisms of centriole duplication and their deregulation in disease. Nat Rev Mol Cell Biol 19(5):297–312

O'Shaughnessy J (2003) Gemcitabine and trastuzumab in metastatic breast cancer, Elsevier. Semin Oncol 30(2 Suppl 3):22–26. https://doi.org/10.1053/sonc.2003.37272

Osborne C et al (2004) Oncogenes and tumor suppressor genes in breast cancer: potential diagnostic and therapeutic applications. Oncologist 9(4):361–377
Oved S, Yarden Y (2002) Molecular ticket to enter cells. Nature 416(6877):133–136
Pannu V et al (2015) Rampant centrosome amplification underlies more aggressive disease course of triple negative breast cancers. Oncotarget 6(12):10487
Patel N et al (2018a) Integrated genomics and functional validation identifies malignant cell specific dependencies in triple negative breast cancer. Nat Commun 9(1):1–16
Patel AB, Louder RK, Greber BJ, Grünberg S, Luo J, Fang J, Liu Y, Ranish J, Hahn S, Nogales E (2018b) Structure of human TFIID and mechanism of TBP loading onto promoter DNA. Science 362(6421):eaau8872. https://doi.org/10.1126/science.aau8872
Pegram M et al (1999) Inhibitory effects of combinations of HER-2/neu antibody and chemotherapeutic agents used for treatment of human breast cancers. Oncogene 18(13):2241–2251
Platet N et al (2004) Estrogens and their receptors in breast cancer progression: a dual role in cancer proliferation and invasion. Crit Rev Oncol Hematol 51(1):55–67
Pohl G et al (2003) High p27Kip1 expression predicts superior relapse-free and overall survival for premenopausal women with early-stage breast cancer receiving adjuvant treatment with tamoxifen plus goserelin. J Clin Oncol 21(19):3594–3600
Qayoom H et al (2022) Expression patterns and therapeutic implications of CDK4 across multiple carcinomas: a molecular docking and MD simulation study. Med Oncol 39(10):1–13
Qayoom H et al (2021) An insight into the cancer stem cell survival pathways involved in chemoresistance in triple-negative breast cancer. Future Oncol 17(31):4185–4206
Robert N (2002) Phase III comparative study of trastuzumab and paclitaxel with and without carboplatin in patients with HER-2/neu positive advanced breast cancer. Breast Cancer Res Treat 76(1):S37
Ross JS et al (2003) The her-2/neu gene and protein in breast cancer 2003: biomarker and target of therapy. Oncologist 8(4):307–325
Schneider JW et al (2002) Trastuzumab cardiotoxicity: speculations regarding pathophysiology and targets for further study. Semin Oncol 29(3 Suppl 11):22–28. https://doi.org/10.1053/sonc.2002.34051.Elsevier
Shaw VA et al (2009) A phase 2 study of UCN-01 in advanced renal cell carcinoma. https://adisinsight.springer.com/trials/700002384
Slamon DJ et al (1987) Human breast cancer: correlation of relapse and survival with amplification of the HER-2/neu oncogene. Science 235(4785):177–182
Slamon DJ et al (2001) Use of chemotherapy plus a monoclonal antibody against HER2 for metastatic breast cancer that overexpresses HER2. N Engl J Med 344(11):783–792
Sofi S et al (2022a) Targeting cyclin-dependent kinase 1 (CDK1) in cancer: molecular docking and dynamic simulations of potential CDK1 inhibitors. Med Oncol 39(9):1–15
Sofi S et al (2022b) Cyclin-dependent kinases in breast cancer: expression pattern and therapeutic implications. Med Oncol 39(6):106
Steeg PS, Zhou Q (1998) Cyclins and breast cancer. Breast Cancer Res Treat 52(1):17–28
Sutherland RL, Musgrove EA (2004) Cyclins and breast cancer. J Mammary Gland Biol Neoplasia 9(1):95–104
Tan AR, Swain SM (2002) Review of flavopiridol, a cyclin-dependent kinase inhibitor, as breast cancer therapy. Semin Oncol 29(3 Suppl 11):77–85. https://doi.org/10.1053/sonc.2002.34059.Elsevier
Tannous BA et al (2013) Effects of the selective MPS1 inhibitor MPS1-IN-3 on glioblastoma sensitivity to antimitotic drugs. J Natl Cancer Inst 105(17):1322–1331
van de Vijver MJ et al (1988) Neu-protein overexpression in breast cancer. N Engl J Med 319(19):1239–1245
Vermeulen K et al (2003) The cell cycle: a review of regulation, deregulation and therapeutic targets in cancer. Cell Prolif 36(3):131–149
Vitre B et al (2015) Chronic centrosome amplification without tumorigenesis. Proc Natl Acad Sci 112(46):E6321–E6330

Vogel CL et al (2002) Efficacy and safety of trastuzumab as a single agent in first-line treatment of HER2-overexpressing metastatic breast cancer. J Clin Oncol 20(3):719–726

Wadhwa B et al (2020) AKT isoforms have discrete expression in triple negative breast cancers and roles in cisplatin sensitivity. Oncotarget 11(45):4178

Wang G et al (2014) The role of mitotic kinases in coupling the centrosome cycle with the assembly of the mitotic spindle. J Cell Sci 127(19):4111–4122

Xu H et al (2014) The omics of triple-negative breast cancers. Clin Chem 60(1):122–133

Yarden Y, Sliwkowski MX (2001) Untangling the ErbB signalling network. Nat Rev Mol Cell Biol 2(2):127–137

Yuan B et al (2006) Increased expression of mitotic checkpoint genes in breast cancer cells with chromosomal instability. Clin Cancer Res 12(2):405–410

Zaman GJR et al (2017) TTK inhibitors as a targeted therapy for CTNNB1 (β-catenin) mutant cancers. Mol Cancer Therapeut 16(11):2609–2617

17 Different Cyclins and Their Significance in Breast Cancer

Manzoor Ahmad Mir and Hina Qayoom

17.1 Introduction

The cooperation of cyclins and their particular cyclin-dependent kinases (CDKs) controls the progression of the cell cycle (Malumbres and Barbacid 2009). In order to create the complex and control the progression through the cell cycle, cyclins are the regulatory subunits of CDKs. This cyclin–CDK complex triggers a series of processes that move in a straight line from the resting state (G0), through the growth phase (G1), DNA replication (S), the growth phase gap 2 (G2), and cell division at the end (M). Any irregularity in any cell cycle phase results in arrest, which lasts until the problem is fixed (Malumbres and Barbacid 2005). The cell cycle is orchestrated by cyclins and cyclin-dependent kinases (CDKs), whose expression varies during particular stages (Malumbres and Barbacid 2009; Malumbres 2014; Sofi et al. 2022a, 2022b). By phosphorylating targets like Rb, cyclin–CDK complex formation permits control of cell cycle progression. Cyclin-dependent kinases inhibitors (CKIs) such as INK4s, such as p16INK4A/CDKN2A, p15INK4B/CDKN2B, p18INK4C/CDKN2C, and p19INK4D/CDKN2D, and CDK-interacting protein/kinase inhibitory proteins (CIP/KIPs), such as p21CIP1/WAF Additionally, Skp1–Cul1–F-box protein (SCF) complex and anaphase-promoting complex/cyclosome (APC/C) expression are regulated by E3 ubiquitin ligases in order to modulate cell cycle transitions (Malumbres 2014; Sivakumar and Gorbsky 2015) (Fig. 17.1).

M. A. Mir (✉) · H. Qayoom
Department of Bioresources, School of Biological Sciences, University of Kashmir, Srinagar, J&K, India
e-mail: drmanzoor@kashmiruniversity.ac.in

M. A. Mir (ed.), *Therapeutic Potential of Cell Cycle Kinases in Breast Cancer*, https://doi.org/10.1007/978-981-19-8911-7_17

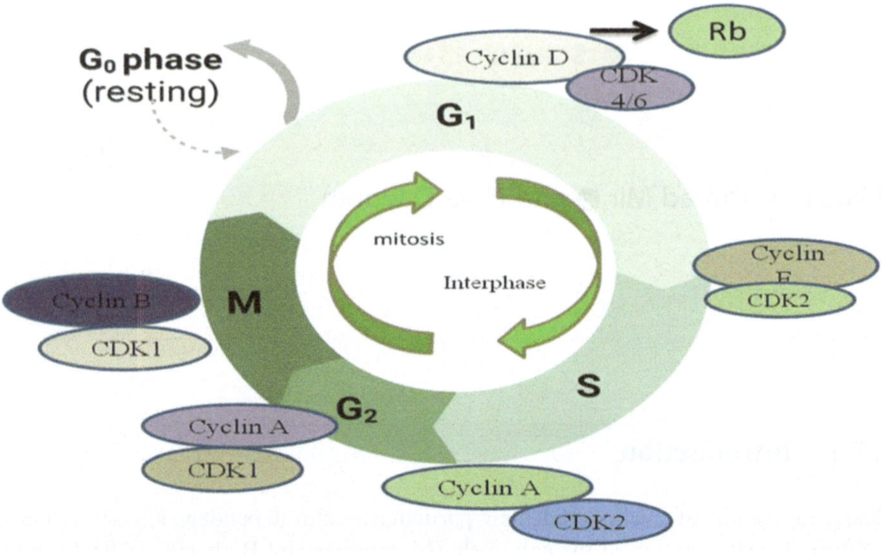

Fig. 17.1 Different phases of cell cycle

17.2 History of Cell Cycle

Leland Hartwell, Tim Hunt, and Paul Nurse collaborated to discover the proteins: cyclins and cyclin-dependent kinases (CDKs), which are in charge of controlling the cell cycle. Their work earned them the 2001 Nobel Prize in Physiology or Medicine in recognition of their significant contributions, and it further revealed the mechanisms underlying cell cycle control. The basic systems that control the cell cycle have been remarkably preserved throughout evolution. Only two stages of the cell cycle were recognized before the 1950s: interphase and mitosis, which could only be seen under a microscope.

However, Stephen Pele and Howard were the ones to make this discovery later on. Using 32P and autoradiography, they showed that DNA synthesis took place within a distinct interphase period and named this phase S (synthesis) phase. Additionally, they showed that a gap that eventually came to be known as G1 and G2, respectively, was seen after the conclusion of mitosis and the beginning of a new S phase (Mehraj et al. 2022b).

Hartwell also introduced the idea of "cell-cycle regulation" in addition to "cell cycle regulation." And that stress or exposure to the environment triggers the activation of the cell cycle checkpoints. Hartwell, however, investigated how sensitive yeast is to ionizing radiations and DNA-damaging effects. He developed the term "checkpoint" to describe the idea that cells arrest, repair the damage, and then move on to the next phase of the cell cycle in light of the discovery that cells are momentarily paused in response to DNA damage. Nurse discovered the cdc2 genes in *S. pombe* in the middle of the 1970s and came to the conclusion that they played a crucial role in the passage from the G2 to the mitotic (M) phase. Nurse also discovered a human gene that encodes a CDK and was eventually given the name CDK1 (cyclin-dependent kinase-1). He proposed that reversible phosphorylation is necessary for CDK1 activity (i.e., its reversible modification by the addition or removal of phosphate groups) (Sofi et al. 2022a, 2022b).

By creating complexes with one of the multiple cyclins, the CDKs are activated. Tim Hunt made the accidental discovery of the cyclins while taking a physiology course at the Marine Biological Laboratories (MBL) in Woods Hole in the early 1980s. He observed that a particular protein was eliminated at each cell division and then resynthesized in the following cycle during synchronous cleavage divisions of the sea urchin *Arbacia punctulata* embryo. Based on the cyclic pattern of this protein's synthesis and breakdown, he gave these proteins the name "cyclin" (Casimiro et al. 2012; Qayoom et al. 2022).

Tim's research was later supported by numerous other studies. For instance, Joan Ruderman found that several cyclins are involved in the embryonic cleavage process and that these cyclins interact specifically with various CDKs at various stages of the cell cycle, becoming activated by phosphorylation and becoming inactive by dephosphorylation.

17.3 Cancer and Cell Cycle

Cell divisions that are out of control are one of cancer's key characteristics. Cancer cells divide abnormally because cell proliferation is unchecked and cell cycle checkpoints are damaged. In addition to cyclins and CDKs, a number of additional proteins, most of which are either oncogenes or tumor-suppressor genes, are also implicated in the abnormal activation of cell proliferation. Mutations that cause excessive and unchecked cell growth, such as those seen in the Ras gene, activate oncogenes. However, in nearly all malignancies, tumor suppressors that act as regulators of cell proliferation are either deleted or altered (via genetic recombination, heterozygosity, or epigenetic alteration) (Mehraj et al. 2022a) (Fig. 17.2).

17.3.1 Cyclins

A family of proteins known as cyclins is essential for controlling the cell cycle. It can be recognized structurally by the "cyclin-box" region, a preserved area. About 150 amino acid residues make up this conserved area, which is arranged into five helical sections and is crucial for cyclins to interact with CDKs. There have been more than 20 cyclins or cyclin-like proteins discovered, many of which have been shown to have no known biological purpose. All 56 kDa-sized cyclins with known functions were essential for the progression of the cell cycle through each phase (D'Angiolella et al. 2013).

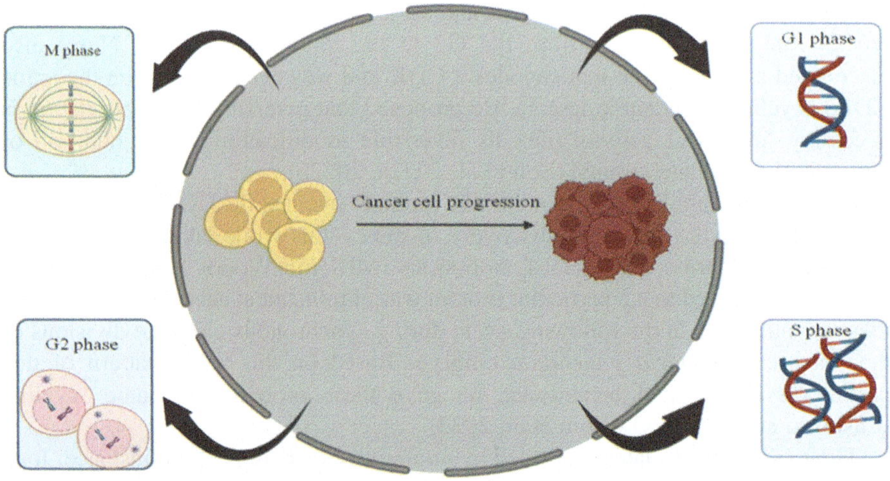

Fig. 17.2 Cancer cell progression via dysregulation of cell cycle phases

By phosphorylating and inactivating their target substrates, cyclins are the regulatory subunits of holoenzyme CDK complexes that control the progression through cell cycle checkpoints (Zhang et al. 2018, Mir et al. 2022).

By binding with various CDKs designated for the cell cycle phase at various time points, the cyclins maintain cell cycle regulation. After a study in which a chromosome breakpoint common to B cell lymphomas that were known to encode cyclin D1 was cloned, the aberrant cyclin expression was linked to malignancy. Consequently, cell cycle deregulation in malignancies has been linked to the overexpression of various cyclins, including cyclin E. Cells that overexpress cyclin E are known to enter the cell cycle's S phase too soon and spend less time in the G1 phase. Cyclin E is typically seen as being crucial for preserving chromosomal stability (Malumbres 2014).

17.3.2 Cyclins and Cell Cycle

The unique complexes of particular CDKs and corresponding cyclins play a vital role in controlling the cell cycle and division. The cyclins function as the complex's regulatory subunit and regulate the cell cycle. The cyclin–CDK complex controls the cell's passage through its several stages, which include resting (G0), growth phase (G1), DNA replication (S phase), and cell division (M). Any issue, including stress and DNA damage, causes the cell cycle to be arrested. Nearly 11 functional cyclins have been identified in humans, including members of the D-type cyclin subfamily D1, D2, and D3 (Fig. 17.3).

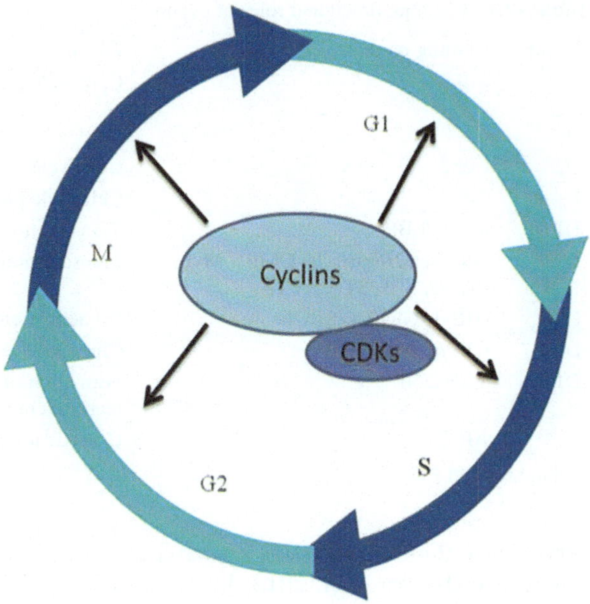

Fig. 17.3 Cyclins direct the progression of different cell cycle phases in complex with CDKs

17.4 Different Types of Cyclins with Their Functional Significance:

17.4.1 Cyclin D

With CDK4 or CDK6, cyclin D is known to generate a variety of heterodimeric complexes that regulate the cell cycle's progression through the G1 phase and the start of DNA replication at the transition from the G1 to the S phase (Malumbres and Barbacid 2009; Sherr et al. 2016). The concentration, activation, subcellular distribution, stabilization, and degradation of the cyclin D and CDK complexes are all firmly maintained in response to mitogenic signals. On the other hand, cancer and frequent chemotherapeutic resistance result from overexpressed cyclin D.

Cyclin D1 is more often deregulated in several solid tumors than other cyclins, such as cyclin D2 and cyclin D3 (Casimiro et al. 2012). The second most often amplified locus in solid tumors is the CCND1 gene, which codes for the protein cyclin D1 Table 17.1 (Beroukhim et al. 2010).

According to research, tumors with elevated levels of cyclin D1 exhibit unchecked growth and proliferation in response to a malfunction of the cell cycle at the restriction point in the G1 phase. It controls the course of the cell cycle by creating a complex with CDK4/6 or other partners such as transcription factors, chromatin-modifying enzymes, or cytosolic proteins. In fact, cyclin D1 is known to control the development and spread of malignancies. These include chromosome duplication and stability (Zeng et al. 2010; Casimiro et al. 2012), immunological

Table 17.1 Potential functional roles of cyclins

Cyclin	Binding partner	Function
D1/ D2/ D3	Cdk4/6	Cell cycle progression
D2	Cdk2	Cell cycle progression
E	Cdk2	Cell cycle progression
D1	ER, C/EBPβ	Transcriptional activation
D1	AR, Beta2/Neuro D, DMP1, Myb, MyoD, SP1, STAT3, TR	Transcriptional activation
D1	AIB-1, GRIP-1, SRCIa	Co-activation
D1	CBP/p300, P/CAF	Chromatin remodeling
D1	TAFn250	Formation of the initiation complex and recruitment of RNA polymerase II
E	AR	Co-activation

Malumbres and Barbacid (2009)

surveillance (Goel et al. 2017; Zhang et al. 2018), migration (Sofi and Mir 2021), senescence (Brown et al. 2012; Laphanuwat et al. 2018), DNA damage response (Li et al. 2010; Jirawatnotai et al. 2011), mitochondrial respiration (Sakamaki et al. 2006; Tchakarska et al. 2011), and migration (Neumeister et al. 2003; Li et al. 2008; Mehraj et al. 2021b). Cyclin D1 is connected to the mitochondria's outer membrane and is known to move back and forth between the cytoplasm and nucleus in healthy cells.

However, in tumor cells or specific tumor locations, it builds up in the cytoplasm and nucleus. Breast cancer and other malignancies including pancreatic, melanoma, endometrial, colorectal, and non-small cell lung carcinoma exhibit overexpression of cyclin D1 (Zhong et al. 2010; Mir et al. 2020a) Fig. 17.4.

17.4.2 Cyclin E

Human cyclin E was first identified in 1991 (Koff et al. 1991) from cDNA libraries of genes that might fix the G1 cyclin mutation in the yeast *S. cerevisiae* (Patel et al. 2016). Cyclin E is produced by a gene on the 19q12–q13 chromosome. This unique gene encodes a large number of polypeptides with a kDa between 32 and 54 (Porter and Keyomarsi 2000). Dysregulation of cyclin E expression, which is also implicated in a number of carcinogenic processes, is the precursor to cell cancer (Geisen and Möröy 2002). Cyclin E is mostly present in some tumor cells and is known to play a role in the growth of cancer. Overexpression of cyclin E is brought on by gene amplification.

Cell cycle dysregulation results from cyclin E overexpression. For instance, the cyclin E gene is amplified by 8 times and its mRNA levels are elevated 64-fold in breast cancer, which has a major effect on the development of the disease (Buckley et al. 1993). The stimulation of S phase in conjunction with CDK2 and activated

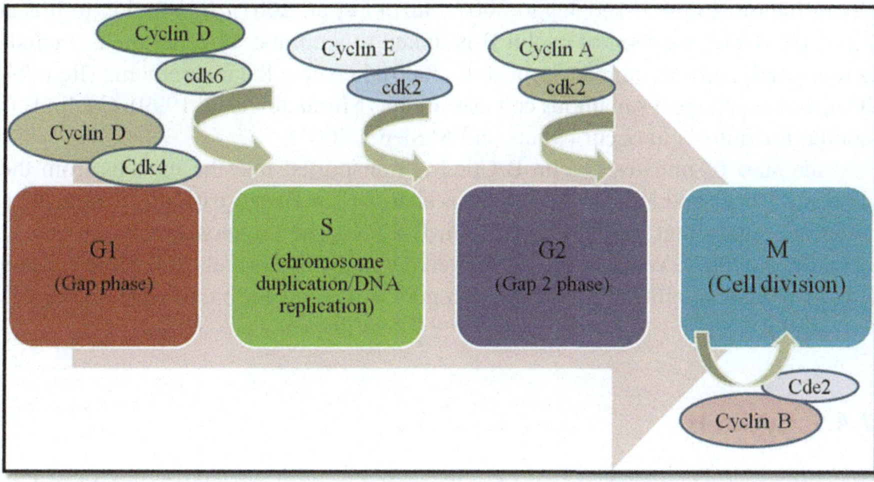

Fig. 17.4 The cyclins in complex with specific CDKs direct the different cell cycle phases

transcriptional regulators including human B-MYB and NPAT, which are vital for cell proliferation, is the usual role of cyclin E (Sala et al. 1997).

17.4.3 Cyclin A

Another form of cyclin is cyclin A, which works in both the S and M phases of the cell cycle and is an important cyclin due to its interaction with CDK1 (also known as CDC2) and CDK2 (Pagano et al. 1992). Cyclin A is known to facilitate the G2-M transition and serve as an MPF when it forms a complex with CDC2 (Furuno et al. 1999). Cancer invasion and metastasis are linked to overexpression of cyclin A, which is also associated with poor survival and prognosis (Li et al. 2002; Baldini et al. 2006; Wegiel et al. 2008). The primary role of the cyclin A–CDK complex during mitosis is to start chromosomal condensation and nuclear membrane disintegration (Pagano and Draetta 1991; Furuno et al. 1999; Gong et al. 2007). Cyclin A is redundant in the proliferation of fibroblast cells but crucial for the development of hematopoietic and embryonic stem cells (Kalaszczynska et al. 2009).

17.4.4 Cyclin B

Cyclin B controls the change from G2 to M phase and is crucial for the start of mitosis in its complex with CDK1 (Pines and Hunter 1990). For a proper G2–M transition to occur, Cyclin B is very important; otherwise, genomic instability and the development of cancer will result. As a result, abnormal cyclin B expression will ultimately cause unchecked cell proliferation and encourage malignant transformation (Löbrich and Jeggo 2007; Suzuki et al. 2007). In addition to breast cancer, the overexpression of cyclin B is linked to a number of carcinomas, including non-small cell carcinoma (Soria et al. 2000), hepatocellular carcinoma (Ito et al. 2000), and esophageal squamous cell carcinoma (Murakami et al. 1999). Cyclin B is essential for mitosis to occur (Smits and Medema 2001).

At the start of mitosis, Cyclin B/Cdc2 is transported into the nucleus from the cytoplasm (Pines and Hunter 1991; Santos et al. 2012). Because of this, Suzuki et al. shown in a study that only nuclear cyclin B affects the prognosis of breast cancer (Suzuki et al. 2007). Winter's et al. discovered that cyclin B levels in the nucleus and cytoplasm were significant predictors of a poor prognosis for breast cancer (Winters et al. 2001).

17.4.5 Cyclin H

Cyclin H, a member of the cyclin family, collaborates with CdK7 and MAT1 to create a trimeric CDK-activating kinase (CAK) complex, which is essential for the control of the cell cycle and healthy proliferation (Patel et al. 2016). Cyclin H is a polypeptide with 323 amino acids that, along with CDK7 and MAT1, forms the CDK-activating kinase (CAK) (Larochelle et al. 2001). The central component of TFIIH, CAK phosphorylates the RNA polymerase II (RNAPII) CTD subunit before taking part in transcription (Shiekhattar et al. 1995; Fuss and Tainer 2011). Additionally, CDK2 is phosphorylated by CAK, which accelerates the cell cycle from G1 to S phase (Desai et al. 1995). Evidence has demonstrated a link between cyclin H genetic variation and tumor growth as well as chemosensitivity (Kayaselcuk et al. 2006; Murali et al. 2014; Palugulla et al. 2018).

17.4.6 Cyclin T

There are various subtypes of cyclin T, including T1, T2, and T2b. It is known to form the PTEFb complex with CDK9, which phosphorylated the RNAPII CTD S2 to control effective transcription elongation (Peng et al. 1998; Price 2000; Gegonne et al. 2008). Similar to cyclin H, cyclin T levels do not fluctuate throughout the cell cycle, indicating that these cyclins have essential roles that are independent of the cell cycle stages (Brown et al. 1994; Poon et al. 1994; Tassan et al. 1994; Moiola et al. 2010).

The expression of cyclin T1 is controlled during T-cell activation (Garriga et al. 1998; Herrmann et al. 1998; Ghose et al. 2001; Marshall et al. 2005). Cyclin T-CDK9 plays a crucial role in controlling a number of cellular functions, including lymphoid development. In NIH 3T3 cells in vitro, overexpression of cyclin T is sufficient to cause foci and colony formation, and in Nu/Nu mice in vivo, it causes tumor growth. Cyclin T probably has a role in lymphomas originating from B- and T-cell lineages, presumably by inhibiting apoptosis (Bellan et al. 2004).

17.4.7 Cyclin K

Cyclin K regulates the phosphorylation of S2 and S5 of the RNAPII CTD by binding CDK12 and CDK13, most likely in two distinct complexes (Bartkowiak et al. 2010; Blazek et al. 2011). The capacity of human cyclin K to restore vitality to yeast cells deficient in all G1-cyclin proteins led to its initial cloning (Edwards et al. 1998). Despite this, its purpose is still not fully understood. The genetic deletion of cyclin K causes mortality at a relatively early stage of mouse development, according to a recent study (Blazek et al. 2011).

Later research hypothesized that cyclin K might control the transcription of a number of genes (Bartkowiak et al. 2010; Blazek et al. 2011; Davidson et al. 2014). But neither can it account for the early embryonic fatal phenotype in mice nor is it apparent whether the transcriptional deficiency is a direct effect. In non-proliferative adult human and mouse tissues, we previously discovered that the cyclin K protein is scarcely detectable (Xiang et al. 2014). However, it is strongly expressed in stem cells that develop quickly (Dai et al. 2012).

17.4.8 Cyclin F

A non-canonical cyclin, cyclin F does not bind to or activate CDKs (D'Angiolella et al. 2013). Instead, cyclin F belongs to the family of around 70 F-box proteins, a group of receptors that recognize substrates and draw them to the Skp1-Cul1-Fbox protein (SCF) E3 ligase (Cardozo and Pagano 2004). By causing the breakdown of cell cycle inhibitors, SCF ligases play an evolutionarily conserved role in driving cell cycle progression. For instance, the CDK inhibitors Sic1 and p27 are both destroyed by yeast SCFCdc4 and human SCFSkp2, respectively (Feldman et al. 1997). Cyclin F has cell cycle-dependent action because both its mRNA and protein levels fluctuate during the cell cycle (Bai et al. 1996).

Cyclin F starts to build up at the G1/S transition, reaches a peak in G2, and then has its protein levels reduced by proteasomal degradation in mitosis and G1 (Mavrommati et al. 2018). where, after being phosphorylated by casein kinase II, cyclin F is ubiquitinated by SCFTRCP (Mavrommati et al. 2018). The cell cycle E3 Anaphase Promoting Complex/Cyclosome (APC/C) also ubiquitinates Cyclin F (Choudhury et al. 2016). Additionally, various cell cycle proteins have been linked to cyclin F via ubiquitination and degradation. Together, cyclin F's dynamic regulation and substrate repertoire underscore the role it plays in controlling the cell cycle (Emanuele et al. 2020).

17.4.9 Cyclin G

In rat fibroblasts, src kinase family screening led to the discovery of Cyclin G, a new addition to the cyclin family. Shortly after, transcriptional targets of the tumor suppressor gene p53 were found by differential screening (Okamoto and Beach

1994; Shimizu et al. 1998). Cyclin G has similarities with human cyclins A and I, B-type cyclins, and Cig1 from fission yeast. Cyclin G, in contrast to other members of the cyclin family, lacks the "destruction box" motif that regulates ubiquitin-dependent degradation but does have an autophosphorylation pattern resembling that of the epidermal growth factor receptor (Horne et al. 1996). Cyclin G is the only known cyclin that the p53 tumor suppressor gene transcriptionally activates, indicating that it may be involved in p53-mediated cell growth control (Zauberman et al. 1995; Smith et al. 1997; Kimura et al. 2001).

Furthermore, Cyclin G's biological role is still unknown despite the fact that it shares a high degree of similarity with other cyclins and has yet to be matched with a cyclin-dependent kinase-binding partner. However, upon its activation by p53, Cyc G forms a compound with the B9 regulatory subunits of protein phosphatase 2A (Okamoto et al. 1996). Cyclin G overexpression has been seen in human osteosarcoma cells, which is consistent with our observation (Wu et al. 1994; Skotzko et al. 1995).

17.5 Role of Cyclins in Regulation of Transcription

Cyclins are known to control gene transcription and mRNA processing in addition to their role in the course of the cell cycle. Cyclin D1 is the most important protein involved in the control of transcription among the cyclins. Cyclin D1 controls the transcriptional activity of the androgen receptor and the estrogen receptor and is known to bind with over 30 additional transcription factors (Fu et al. 2004). Cyclin D1 is bound by the histone acetyltransferases P/CAF, p300, and AIB1 (Zwijsen et al. 1998; Reutens et al. 2001). The interaction of cyclin D1 with target gene promoters was shown by chromatin immunoprecipitation (ChIP), which was associated with H3 lysine 9 deacetylation. The restoration of cyclin D1 resulted in the deacetylation of H3 Lys9 and the concurrent recruitment of HDAC1/HDAC3 (Klein and Assoian 2008). Furthermore, the environment of local chromatin influences the recruitment of cyclin D1 to specific target genes (Bienvenu et al. 2005; Fu et al. 2005; Wang et al. 2018; Mir and Mehraj 2019).

The movement of the histone acetyltransferase p300/CBP to control the genes directing DNA damage repair signals was also linked to Cyclin D1 recruitment to genomic DNA. It has been demonstrated that Cyclin D1 controls p300 activity in a kinase-independent manner. Cyclin D1 was hypothesized as a transcriptional regulator by co-occupancy with p300 at target DNA binding sites because p300 is known as a transcriptional co-integrator (Bienvenu et al. 2005). The transcriptional machinery, a multicomplex protein made up of transcription factors (TFIIB, -D, -E, -F, and -H), the Mediator complex, and RNA polymerase II (RNAPII), is responsible for the transcription of protein-coding genes (Sikorski and Buratowski 2009; Mehraj et al. 2021a).

In eukaryotes, the Mediator complex serves as a link between transcription factors and RNAPII (Kornberg 2005; Malik and Roeder 2010). All yeast RNAPII genes must have a mediator in order to be transcriptionally transcribed (Holstege et al. 1998). The biggest RNAPII subunit is phosphorylated by several cyclins, including cyclin C-CDK8, cyclin H-CDK7, cyclin T-CDK9, and cyclin K-CDK12 or CDK13, to control transcription. The carboxy terminal domain (CTD), also known as the heptapeptide (YS2PTS5PS7) repeats, is where RNAPII is phosphorylated (Phatnani and Greenleaf 2006; Mir and Agrewala 2008). The transcriptional cycle is impacted from its inception through its elongation and termination by a succession of phosphorylation events at the CTD's S2, S5, and S7 (Egloff and Murphy 2008, Cowling and Cole 2010, Mir et al. 2020b, Mehraj et al. 2022a).

By phosphorylating the RNAPII CTD and the cyclin H subunit of TFIIH, Cyclin C-CDK8 forms a subcomplex with Mediator proteins Med12 and Med13 that interacts with the main Mediator complex to suppress active transcription. The main S5 kinase that facilitates promoter clearance and makes the transition from transcription initiation possible is the cyclin H-CDK7-Mat1 (CDK-activating kinase [CAK]) complex, which binds TFIIH (Roy et al. 1994; Devault et al. 1995; Yankulov and Bentley 1997; Mir 2015). This establishes a direct connection between the cell cycle machinery and transcription regulation.

Additionally, the cyclin H-CDK7-Mat1 complex directly interacts with transcription factors to control the activity of these molecules. Phosphorylation of PPAR by CDK7 prevents lipogenesis in 3T3-L1 cells and mouse embryonic fibroblasts (MEFs) (Helenius et al. 2009).

To control productive transcription elongation in humans, cyclin T (T1, T2, and T2b) and CDK9 join forces to create the PTEFb complex, which phosphorylates CTD S2 of RNAPII (Peng et al. 1998; Gegonne et al. 2008). Similar to cyclin H, the amount of cyclin T does not fluctuate throughout the cell cycle, indicating that these cyclins have essential roles that are independent of the cell cycle stage (Brown et al. 1994; Poon et al. 1994; Tassan et al. 1994; Moiola et al. 2010). The expression of cyclin T1 is controlled during T-cell activation (Bartkowiak et al. 2010; Qayoom et al. 2021). Cyclin K regulates the phosphorylation of S2 and S5 of the RNAPII CTD by binding CDK12 and CDK13, most likely in two distinct complexes.

In human cells, CDK12 depletion significantly reduces CTD S2 phosphorylation. The expression of a small number of genes is net downregulated, but these genes—FANCI, FANCD2, ATR, and BRCA1—are crucial for the DNA damage response (DDR) (Blazek et al. 2011). Depletion of cyclin K-CDK12 increases -H2AX foci and makes cells more vulnerable to DNA-damaging substances, both of which are consistent with its function in controlling the DDR. Cyclin L is closely related to cyclin K, cyclin T1, and cyclin T2 (L1 and L2). By interacting with the SR splicing protein family and CDK11, Cyclin L regulates splicing. Cytoclin L (Redon et al. 2002; Blazek et al. 2011) is a possible oncogene for head and neck cancer.

17.6 Role of Cyclins in Chromosomal Instability

Chromosomal instability (CIN) is a common characteristic that cells from solid tumors possess and is regarded as a defining characteristic of cancer (Bakhoum and Compton 2012). Multiple pathways can contribute to CIN, which results in an aberrant chromosomal complement. It is hotly contested whether CIN is a cause or a result of cancer, but it does happen early in the course of cancer development and is linked to a bad prognosis (Pfau and Amon 2012). Most human breast cancers overexpress cyclin D1, according to research. Although cyclin D1 is necessary for carcinogenesis, there are a variety of cyclin D1 kinase-independent tasks that cyclin D1 performs (Qayoom et al. 2021).

Furthermore, a number of pieces of data point to a possible discrepancy between cyclin D1's capacity to phosphorylate pRB and its role in oncogenesis. In this context, cyclin D1 overexpression in human breast cancer is not associated with pRB phosphorylation or the proliferative marker Ki67 (Van Diest et al. 1997; Shoker et al. 2001).

We had postulated an alternate mechanism by which cyclin D1 may drive carcinogenesis by inducing CIN based on a considerable number of papers from our group and others (Casimiro and Pestell 2012). Early research revealed that in rat embryonic fibroblasts, cyclin D1 did not cause aneuploidy (Spruck et al. 1999). Aneuploidy, improper mitosis, the accumulation of extra centrosomes, abnormalities in the mitotic spindle, and irregular mitosis were all caused by transiently expressing cyclin D1, according to a recent investigation on mouse primary hepatocytes (Nelsen et al. 2005). Only CIN-positive bladder cancer samples exhibited Cyclin D1 gene amplification, which was linked with tumor grade. Our research employing cyclin D1 ChIP and sequencing (ChIP-Seq) showed that the genes that control CIN have an enrichment of the gene regulatory elements bound by cyclin D1 (Casimiro et al. 2014).

A number of human cancers, including colorectal cancer, non-small cell lung cancer, and head and neck squamous cell carcinoma, have been found to overexpress cyclin B1, and its overexpression is strongly linked to a poor prognosis in breast cancer (Wang et al. 1997). Additionally, aneuploidy and rapid proliferation of human breast carcinomas are linked to overexpression of cyclin B1 (Suzuki et al. 2007). The maturation/M phase-promoting factor (MPF) in eukaryotes controls the beginning of the M phase (Hunt 1989). In order to trigger the meiotic G2-M phase transition in immature oocytes and the mitosis of somatic cells, the MPF complex, which is made up of cyclin B1 and CDK1, is necessary (Lohka et al. 1988).

Cyclin B1 mediates nuclear envelope dissolution and chromosomal condensation, while cyclin B2 mediates Golgi disassembly. Protein phosphatase 2A/B55 (PP2A-B55) is a key MPF inhibitor (Castilho et al. 2009). Along with lymphomas and leukemias, cyclin E transcript and protein levels are elevated in carcinomas of the lungs, digestive system, and breast (Wingate et al. 2009). A prognostic marker called increased cyclin E expression has been found in 18%–22% of breast tumors (Keyomarsi et al. 1994). Breast cancer exhibits a low molecular weight hyperactive form of cyclin E, which is linked to a very bad prognosis (Porter et al. 2001). According to one theory, Cyclin E is a crucial cell cycle regulatory protein that helps with the G1–S phase transition (Hwang and Clurman 2005).

Following cyclin D1-CDK4/6, cyclin E binds its catalytic subunit CDK2 and phosphorylates Rb to release E2F transcription factors that control genes involved in the advancement of the S phase. However, a cyclin E deletion model disproved the crucial cyclin E requirement for S phase advancement by showing that mitotic cell division does not necessitate cyclins E1 and E2. Additionally, CDK2-null mice underwent normal development, and null fibroblasts displayed a typical cell cycle profile. These models do, however, highlight a number of cyclin E prerequisites. Mice lacking cyclin E have substantially defective placental trophoblast large cells and megakaryocyte endoreplication (Geng et al. 2003).

Additionally, null MEFs exhibit a deficiency in MCM loading onto replication forks, are resistant to oncogenic transformation, and are unable to enter the S phase. The redundancy provided by cyclin E–CDK1 complexes may explain why these important abnormalities are not present in CDK2-null mice. It's interesting to note that cyclin E's kinase-independent G0-S phase and replication licensing functions. Gene amplification, flaws in the p16-Rb-cyclin D1 signalling axis, or flaws in the ubiquitin-mediated degradation pathway can all lead to abnormal cyclin E production (Keyomarsi and Pardee 1993). Given that it frequently correlates with proliferation indices, higher cyclin E abundance may result from increased proliferation rates (Rudolph et al. 2003).

It might, however, function as a molecular catalyst for transformation by using CIN. Aneuploidy was produced when cyclin E was induced in rat fibroblasts or human epithelial cells (Spruck et al. 1999). Defective p53 and dysregulated cyclin E expression resulted in increased ploidy and genomic instability in basic human cells. High expression of a lung-specific, degradation-resistant cyclin E commonly resulted in dysplasia, numerous lung adenocarcinomas, and tumors with CIN in a mouse model (Ma et al. 2007).

17.7 Summary

Cancer cells divide abnormally because cell proliferation is unchecked and cell cycle checkpoints are damaged. In addition to cyclins and CDKs, a number of additional proteins, most of which are either oncogenes or tumor-suppressor genes, are also implicated in the abnormal activation of cell proliferation. The cyclins function as the complex's regulatory subunit and regulate the cell cycle. There are various types of cyclins that function with respect to the different cell cycle phases such as cyclin A, cyclin B, cyclin E, cyclin T, cyclin H, cyclin F, and cyclin G. These cyclins along with their partners (CDKs) play an important role in the regulation of cell cycle. The deregulation in the expression of either cyclin or CDKs or both play an important role in the progression of cancer, including breast cancer.

17.8 Further Readings

The readers can have a look at the following articles for a better understanding of the given topic:

i) https://www.ncbi.nlm.nih.gov/pmc/articles/PMC3636749/
ii) https://ascopubs.org/doi/10.1200/JCO.2005.05.064

The readers can also take a look at the following visual presentations:

i) https://youtu.be/foR2tZHj5Eo
ii) https://youtu.be/tBoG9d0tGCE

For more incites about the topic we would suggest detailed findings from the books of

(Mir 2022) https://doi.org/10.1016/C2021-0-02565-7,

(Mir 2021) https://doi.org/10.52305/WXJL6770, from cancer.net website, https://www.cancer.net/cancer-types/breast-cancer/types-treatment

References

Bai C et al (1996) SKP1 connects cell cycle regulators to the ubiquitin proteolysis machinery through a novel motif, the F-box. Cell 86(2):263–274

Bakhoum SF, Compton DA (2012) Chromosomal instability and cancer: a complex relationship with therapeutic potential. J Clin Investig 122(4):1138–1143

Baldini E et al (2006) Cyclin A and E2F1 overexpression correlate with reduced disease-free survival in node-negative breast cancer patients. Anticancer Res 26(6B):4415–4421

Bartkowiak B et al (2010) CDK12 is a transcription elongation-associated CTD kinase, the metazoan ortholog of yeast Ctk1. Genes Dev 24(20):2303–2316

Bellan C et al (2004) CDK9/CYCLIN T1 expression during normal lymphoid differentiation and malignant transformation. J Pathol 203(4):946–952

Beroukhim R et al (2010) The landscape of somatic copy-number alteration across human cancers. Nature 463(7283):899–905

Bienvenu F et al (2005) Transcriptional regulation by a DNA-associated form of cyclin D1. Mol Biol Cell 16(4):1850–1858

Blazek D et al (2011) The Cyclin K/Cdk12 complex maintains genomic stability via regulation of expression of DNA damage response genes. Genes Dev 25(20):2158–2172

Brown AJ et al (1994) Expression and activity of p40MO15, the catalytic subunit of cdk-activating kinase, during Xenopus oogenesis and embryogenesis. Mol Biol Cell 5(8):921–932

Brown NE et al (2012) Cyclin D1 activity regulates autophagy and senescence in the mammary epithelium cyclin D1 activity and autophagy. Cancer Res 72(24):6477–6489

Buckley MF et al (1993) Expression and amplification of cyclin genes in human breast cancer. Oncogene 8(8):2127–2133

Cardozo T, Pagano M (2004) The SCF ubiquitin ligase: insights into a molecular machine. Nat Rev Mol Cell Biol 5(9):739–751

Casimiro MC et al (2012) Cyclins and cell cycle control in cancer and disease. Genes Cancer 3(11-12):649–657

Casimiro MC, Pestell RG (2012) Cyclin d1 induces chromosomal instability. Oncotarget 3(3):224

Casimiro MC et al (2014) Overview of cyclins D1 function in cancer and the CDK inhibitor landscape: past and present. Expert Opin Investig Drugs 23(3):295–304

Castilho PV et al (2009) The M phase kinase Greatwall (Gwl) promotes inactivation of PP2A/B55δ, a phosphatase directed against CDK phosphosites. Mol Biol Cell 20(22):4777–4789

Choudhury R et al (2016) APC/C and SCF cyclin F constitute a reciprocal feedback circuit controlling S-phase entry. Cell Rep 16(12):3359–3372

Cowling VH, Cole MD (2010) Myc regulation of mRNA cap methylation. Genes Cancer 1(6): 576–579
D'Angiolella V et al (2013) A cyclin without cyclin-dependent kinases: cyclin F controls genome stability through ubiquitin-mediated proteolysis. Trends Cell Biol 23(3):135–140
Dai Q et al (2012) Cyclin K-containing kinase complexes maintain self-renewal in murine embryonic stem cells. J Biol Chem 287(30):25344–25352
Davidson L et al (2014) 3′ end formation of pre-mRNA and phosphorylation of Ser2 on the RNA polymerase II CTD are reciprocally coupled in human cells. Genes Dev 28(4):342–356
Desai D et al (1995) Effects of phosphorylation by CAK on cyclin binding by CDC2 and CDK2. Mol Cell Biol 15(1):345–350
Devault A et al (1995) MAT1 ('menage à trois') a new RING finger protein subunit stabilizing cyclin H-cdk7 complexes in starfish and Xenopus CAK. EMBO J 14(20):5027–5036
Edwards MC et al (1998) Human cyclin K, a novel RNA polymerase II-associated cyclin possessing both carboxy-terminal domain kinase and Cdk-activating kinase activity. Mol Cell Biol 18(7):4291–4300
Egloff S, Murphy S (2008) Cracking the RNA polymerase II CTD code. Trends Genet 24(6): 280–288
Emanuele MJ et al (2020) Complex cartography: regulation of E2F transcription factors by cyclin F and ubiquitin. Trends Cell Biol 30(8):640–652
Feldman RMR et al (1997) A complex of Cdc4p, Skp1p, and Cdc53p/cullin catalyzes ubiquitination of the phosphorylated CDK inhibitor Sic1p. Cell 91(2):221–230
Fu M et al (2004) Minireview: Cyclin D1: normal and abnormal functions. Endocrinology 145(12): 5439–5447
Fu M et al (2005) Cyclin D1 represses p300 transactivation through a cyclin-dependent kinase-independent mechanism. J Biol Chem 280(33):29728–29742
Furuno N et al (1999) Human cyclin A is required for mitosis until mid prophase. J Cell Biol 147(2): 295–306
Fuss JO, Tainer JA (2011) XPB and XPD helicases in TFIIH orchestrate DNA duplex opening and damage verification to coordinate repair with transcription and cell cycle via CAK kinase. DNA Repair 10(7):697–713
Garriga J et al (1998) Upregulation of cyclin T1/CDK9 complexes during T cell activation. Oncogene 17(24):3093–3102
Gegonne A et al (2008) TFIID component TAF7 functionally interacts with both TFIIH and P-TEFb. Proc Natl Acad Sci 105(14):5367–5372
Geisen C, Möröy T (2002) The oncogenic activity of cyclin E is not confined to Cdk2 activation alone but relies on several other, distinct functions of the protein. J Biol Chem 277(42): 39909–39918
Geng Y et al (2003) Cyclin E ablation in the mouse. Cell 114(4):431–443
Ghose R et al (2001) Induction of TAK (cyclin T1/P-TEFb) in purified resting CD4+ T lymphocytes by combination of cytokines. J Virol 75(23):11336–11343
Goel S et al (2017) CDK4/6 inhibition triggers anti-tumour immunity. Nature 548(7668):471–475
Gong D et al (2007) Cyclin A2 regulates nuclear-envelope breakdown and the nuclear accumulation of cyclin B1. Curr Biol 17(1):85–91
Helenius K et al (2009) Mat1 inhibits peroxisome proliferator-activated receptor γ-mediated adipocyte differentiation. Mol Cell Biol 29(2):315–323
Herrmann CH et al (1998) Tat-associated kinase, TAK, activity is regulated by distinct mechanisms in peripheral blood lymphocytes and promonocytic cell lines. J Virol 72(12):9881–9888
Holstege FCP et al (1998) Dissecting the regulatory circuitry of a eukaryotic genome. Cell 95(5): 717–728
Horne MC et al (1996) Cyclin G1 and cyclin G2 comprise a new family of cyclins with contrasting tissue-specific and cell cycle-regulated expression (∗). J Biol Chem 271(11):6050–6061
Hunt T (1989) Maturation promoting factor, cyclin and the control of M-phase. Curr Opin Cell Biol 1(2):268–274

Hwang HC, Clurman BE (2005) Cyclin E in normal and neoplastic cell cycles. Oncogene 24(17): 2776–2786

Ito Y et al (2000) Expression and prognostic role of cyclin-dependent kinase 1 (cdc2) in hepatocellular carcinoma. Oncology 59(1):68–74

Jirawatnotai S et al (2011) A function for cyclin D1 in DNA repair uncovered by protein interactome analyses in human cancers. Nature 474(7350):230–234

Kalaszczynska I et al (2009) Cyclin A is redundant in fibroblasts but essential in hematopoietic and embryonic stem cells. Cell 138(2):352–365

Kayaselcuk F et al (2006) Expression of cyclin H in normal and cancerous endometrium, its correlation with other cyclins, and association with clinicopathologic parameters. Int J Gynecol Cancer 16(1):402

Keyomarsi K et al (1994) Cyclin E, a potential prognostic marker for breast cancer. Cancer Res 54(2):380–385

Keyomarsi K, Pardee AB (1993) Redundant cyclin overexpression and gene amplification in breast cancer cells. Proc Natl Acad Sci 90(3):1112–1116

Kimura SH et al (2001) Cyclin G1 is involved in G2/M arrest in response to DNA damage and in growth control after damage recovery. Oncogene 20(25):3290–3300

Klein EA, Assoian RK (2008) Transcriptional regulation of the cyclin D1 gene at a glance. J Cell Sci 121(23):3853–3857

Koff A et al (1991) Human cyclin E, a new cyclin that interacts with two members of the CDC2 gene family. Cell 66(6):1217–1228

Kornberg RD (2005) Mediator and the mechanism of transcriptional activation. Trends Biochem Sci 30(5):235–239

Laphanuwat P et al (2018) Cyclin D1 depletion interferes with oxidative balance and promotes cancer cell senescence. J Cell Sci 131(12):jcs214726

Larochelle S et al (2001) T-loop phosphorylation stabilizes the CDK7–cyclin H–MAT1 complex in vivo and regulates its CTD kinase activity. EMBO J 20(14):3749–3759

Li J-Q et al (2002) Cyclin a correlates with carcinogenesis and metastasis, and p27kip1 correlates with lymphatic invasion, in colorectal neoplasms. Hum Pathol 33(10):1006–1015

Li Z et al (2010) Alternative cyclin D1 splice forms differentially regulate the DNA damage response cyclin D1 on DNA damage response. Cancer Res 70(21):8802–8811

Li Z et al (2008) Alternate cyclin D1 mRNA splicing modulates p27KIP1 binding and cell migration. J Biol Chem 283(11):7007–7015

Löbrich M, Jeggo PA (2007) The impact of a negligent G2/M checkpoint on genomic instability and cancer induction. Nat Rev Cancer 7(11):861–869

Lohka MJ et al (1988) Purification of maturation-promoting factor, an intracellular regulator of early mitotic events. Proc Natl Acad Sci 85(9):3009–3013

Ma Y et al (2007) Transgenic cyclin E triggers dysplasia and multiple pulmonary adenocarcinomas. Proc Natl Acad Sci 104(10):4089–4094

Malik S, Roeder RG (2010) The metazoan Mediator co-activator complex as an integrative hub for transcriptional regulation. Nat Rev Genet 11(11):761–772

Malumbres M (2014) Cyclin-dependent kinases. Genome Biol 15(6):1–10

Malumbres M, Barbacid M (2005) Mammalian cyclin-dependent kinases. Trends Biochem Sci 30(11):630–641

Malumbres M, Barbacid M (2009) Cell cycle, CDKs and cancer: a changing paradigm. Nat Rev Cancer 9(3):153–166

Marshall RM et al (2005) Cyclin T1 expression is regulated by multiple signaling pathways and mechanisms during activation of human peripheral blood lymphocytes. J Immunol 175(10): 6402–6411

Mavrommati I et al (2018) β-TrCP-and casein kinase II-mediated degradation of cyclin F controls timely mitotic progression. Cell Rep 24(13):3404–3412

Mehraj U et al (2022a) Expression pattern and prognostic significance of baculoviral inhibitor of apoptosis repeat-containing 5 (BIRC5) in breast cancer: a comprehensive analysis. Adv Cancer Biol Metastasis 4:100037

Mehraj U et al (2021a) Tumor microenvironment promotes breast cancer chemoresistance. Cancer Chemother Pharmacol 87(2):147–158

Mehraj U et al (2021b) Prognostic significance and targeting tumor-associated macrophages in cancer: new insights and future perspectives. Breast Cancer 28(3):539–555

Mehraj U et al (2022b) Expression pattern and prognostic significance of CDKs in breast cancer: an integrated bioinformatic study. Cancer Biomark 34:1–15

Mir MA (2015) Developing costimulatory molecules for immunotherapy of diseases. Academic Press. https://doi.org/10.1016/C2014-0-02898-5

Mir MA, Agrewala JN (2008) Signaling through CD80: an approach for treating lymphomas. Expert Opin Ther Targets 12(8):969–979

Mir MA, Mehraj U (2019) Double-crosser of the immune system: macrophages in tumor progression and metastasis. Curr Immunol Rev 15(2):172–184

Mir MA et al (2020a) Nanobodies: the "magic bullets" in therapeutics, drug delivery and diagnostics. Hum Antibodies 28(1):29–51

Mir MA et al (2020b) Targeting different pathways using novel combination therapy in triple negative breast cancer. Curr Cancer Drug Targets 20(8):586–602

Mir MA, Sofi S, Qayoom H (2022) Conventional adjuvant chemotherapy in combination with surgery, radiotherapy, and other specific targets. In: Combinational therapy in triple negative breast cancer. Elsevier, San Diego, pp 95–120. https://doi.org/10.1016/B978-0-323-96136-3.00008-X

Moiola C et al (2010) Cyclin T1 overexpression induces malignant transformation and tumor growth. Cell Cycle 9(15):3191–3198

Murakami H et al (1999) Determination of the prognostic significance of cyclin B1 overexpression in patients with esophageal squamous cell carcinoma. Virchows Arch 434(2):153–158

Murali A et al (2014) Association of single nucleotide polymorphisms in cell cycle regulatory genes with oral cancer susceptibility. Br J Oral Maxillofac Surg 52(7):652–658

Nelsen CJ et al (2005) Short term cyclin D1 overexpression induces centrosome amplification, mitotic spindle abnormalities, and aneuploidy. J Biol Chem 280(1):768–776

Neumeister P et al (2003) Cyclin D1 governs adhesion and motility of macrophages. Mol Biol Cell 14(5):2005–2015

Okamoto K, Beach D (1994) Cyclin G is a transcriptional target of the p53 tumor suppressor protein. EMBO J 13(20):4816–4822

Okamoto K et al (1996) p53-dependent association between cyclin G and the B' subunit of protein phosphatase 2A. Mol Cell Biol 16(11):6593–6602

Pagano M, Draetta G (1991) Cyclin A, cell cycle control and oncogenesis. Prog Growth Factor Res 3(4):267–277

Pagano M et al (1992) Cyclin A is required at two points in the human cell cycle. EMBO J 11(3):961–971

Palugulla S et al (2018) Genetic polymorphisms in cyclin H gene are associated with oxaliplatin-induced acute peripheral neuropathy in South Indian digestive tract cancer patients. Cancer Chemother Pharmacol 82(3):421–428

Patel H et al (2016) Expression of CDK7, Cyclin H, and MAT1 is elevated in breast cancer and is prognostic in estrogen receptor–positive breast cancer. Clin Cancer Res 22(23):5929–5938

Peng J et al (1998) Identification of multiple cyclin subunits of human P-TEFb. Genes Dev 12(5):755–762

Pfau SJ, Amon A (2012) Chromosomal instability and aneuploidy in cancer: from yeast to man: 'exploring aneuploidy: the significance of chromosomal imbalance' review series. EMBO Rep 13(6):515–527

Phatnani HP, Greenleaf AL (2006) Phosphorylation and functions of the RNA polymerase II CTD. Genes Dev 20(21):2922–2936

Pines J, Hunter T (1990) Human cyclin A is adenovirus E1A-associated protein p60 and behaves differently from cyclin B. Nature 346(6286):760–763
Pines J, Hunter T (1991) Human cyclins A and B1 are differentially located in the cell and undergo cell cycle-dependent nuclear transport. J Cell Biol 115(1):1–17
Poon RY et al (1994) Cell cycle regulation of the p34cdc2/p33cdk2-activating kinase p40MO15. J Cell Sci 107(10):2789–2799
Porter DC, Keyomarsi K (2000) Novel splice variants of cyclin E with altered substrate specificity. Nucleic Acids Res 28(23):e101–e101
Porter DC et al (2001) Tumor-specific proteolytic processing of cyclin E generates hyperactive lower-molecular-weight forms. Mol Cell Biol 21(18):6254–6269
Price DH (2000) P-TEFb, a cyclin-dependent kinase controlling elongation by RNA polymerase II. Mol Cell Biol 20(8):2629–2634
Qayoom H et al (2022) Expression patterns and therapeutic implications of CDK4 across multiple carcinomas: a molecular docking and MD simulation study. Med Oncol 39(10):1–13
Qayoom H et al (2021) An insight into the cancer stem cell survival pathways involved in chemoresistance in triple-negative breast cancer. Future Oncol 17(31):4185–4206
Redon R et al (2002) Amplicon mapping and transcriptional analysis pinpoint cyclin L as a candidate oncogene in head and neck cancer. Cancer Res 62(21):6211–6217
Reutens AT et al (2001) Cyclin D1 binds the androgen receptor and regulates hormone-dependent signaling in a p300/CBP-associated factor (P/CAF)-dependent manner. Mol Endocrinol 15(5): 797–811
Roy R et al (1994) The MO15 cell cycle kinase is associated with the TFIIH transcription-DNA repair factor. Cell 79(6):1093–1101
Rudolph P et al (2003) Differential prognostic impact of the cyclins E and B in premenopausal and postmenopausal women with lymph node-negative breast cancer. Int J Cancer 105(5):674–680
Sakamaki T et al (2006) Cyclin D1 determines mitochondrial function in vivo. Mol Cell Biol 26(14):5449–5469
Sala A et al (1997) Activation of human B-MYB by cyclins. Proc Natl Acad Sci 94(2):532–536
Santos SDM et al (2012) Spatial positive feedback at the onset of mitosis. Cell 149(7):1500–1513
Sherr CJ et al (2016) Targeting CDK4 and CDK6: from discovery to therapy. Cancer Discov 6:353–367. https://doi.org/10.1158/2159-8290.CD-15-0894
Shiekhattar R et al (1995) Cdk-activating kinase complex is a component of human transcription factor TFIIH. Nature 374(6519):283–287
Shimizu A et al (1998) Cyclin G contributes to G2/M arrest of cells in response to DNA damage. Biochem Biophys Res Commun 242(3):529–533
Shoker BS et al (2001) Immunodetectable cyclin D1is associated with oestrogen receptor but not Ki67 in normal, cancerous and precancerous breast lesions. Br J Cancer 84(8):1064–1069
Sikorski TW, Buratowski S (2009) The basal initiation machinery: beyond the general transcription factors. Curr Opin Cell Biol 21(3):344–351
Sivakumar S, Gorbsky GJ (2015) Spatiotemporal regulation of the anaphase-promoting complex in mitosis. Nat Rev Mol Cell Biol 16(2):82–94
Skotzko M et al (1995) Retroviral vector-mediated gene transfer of antisense cyclin G1 (CYCG1) inhibits proliferation of human osteogenic sarcoma cells. Cancer Res 55(23):5493–5498
Smith ML et al (1997) The p53-regulated cyclin G gene promotes cell growth: p53 downstream effectors cyclin G and Gadd45 exert different effects on cisplatin chemosensitivity. Exp Cell Res 230(1):61–68
Smits VAJ, Medema RH (2001) Checking out the G2/M transition. Biochim Biophy Acta 1519(1-2):1–12
Sofi S et al (2022a) Targeting cyclin-dependent kinase 1 (CDK1) in cancer: molecular docking and dynamic simulations of potential CDK1 inhibitors. Med Oncol 39(9):1–15
Sofi S et al (2022b) Cyclin-dependent kinases in breast cancer: expression pattern and therapeutic implications. Med Oncol 39(6):1–16

Sofi S, Mir MA (2021) Novel biomarkers in breast cancer. In: Combination therapies and their effectiveness in breast cancer treatment. Elsevier, San Diego, pp 47–92
Soria J-C et al (2000) Overexpression of cyclin B1 in early-stage non-small cell lung cancer and its clinical implication. Cancer Res 60(15):4000–4004
Spruck CH et al (1999) Deregulated cyclin E induces chromosome instability. Nature 401(6750): 297–300
Suzuki T et al (2007) Nuclear cyclin B1 in human breast carcinoma as a potent prognostic factor. Cancer Sci 98(5):644–651
Tassan J-P et al (1994) Cell cycle analysis of the activity, subcellular localization, and subunit composition of human CAK (CDK-activating kinase). J Cell Biol 127(2):467–478
Tchakarska G et al (2011) Cyclin D1 inhibits mitochondrial activity in B cells. Cancer Res 71(5): 1690–1699
Van Diest PJ et al (1997) Cyclin D1 expression in invasive breast cancer. Correlations and prognostic value. Am J Pathol 150(2):705
Wang A et al (1997) Overexpression of cyclin B1 in human colorectal cancers. J Cancer Res Clin Oncol 123(2):124–127
Wang G et al (2018) Cyclin D1-mediated microRNA expression signature predicts breast cancer outcome. Theranostics 8(8):2251
Wegiel B et al (2008) Multiple cellular mechanisms related to cyclin A1 in prostate cancer invasion and metastasis. J Natl Cancer Inst 100(14):1022–1036
Wingate H et al (2009) Low molecular weight cyclin E is specific in breast cancer and is associated with mechanisms of tumor progression. Cell Cycle 8(7):1062–1068
Winters ZE et al (2001) Subcellular localisation of cyclin B, Cdc2 and p21WAF1/CIP1 in breast cancer: association with prognosis. Eur J Cancer 37(18):2405–2412
Wu LT et al (1994) Molecular-cloning of the human CYCG1 gene encoding a G-type cyclin-overexpression in human osteosarcoma cells. Oncol Rep 1(4):705–711
Xiang X et al (2014) A distinct expression pattern of cyclin K in mammalian testes suggests a functional role in spermatogenesis. PLoS One 9(7):e101539
Yankulov KY, Bentley DL (1997) Regulation of CDK7 substrate specificity by MAT1 and TFIIH. EMBO J 16(7):1638–1646
Zauberman A et al (1995) Identification of p53 target genes through immune selection of genomic DNA: the cyclin G gene contains two distinct p53 binding sites. Oncogene 10(12): 2361–2366
Zeng X et al (2010) The Ras oncogene signals centrosome amplification in mammary epithelial cells through cyclin D1/Cdk4 and Nek2. Oncogene 29(36):5103–5112
Zhang J et al (2018) Cyclin D–CDK4 kinase destabilizes PD-L1 via cullin 3–SPOP to control cancer immune surveillance. Nature 553(7686):91–95
Zhong Z et al (2010) Cyclin D1/cyclin-dependent kinase 4 interacts with filamin A and affects the migration and invasion potential of breast cancer cells cyclin D1 and filamin A in breast cancer cell migration. Cancer Res 70(5):2105–2114
Zwijsen RML et al (1998) Ligand-independent recruitment of steroid receptor coactivators to estrogen receptor by cyclin D1. Genes Dev 12(22):3488–3498

The manufacturer's authorised representative in the EU is Springer Nature Customer Service Centre GmbH, Europaplatz 3, 69115 Heidelberg, Germany. If you have any concerns regarding our products, please contact ProductSafety@springernature.com

Printed and bound by CPI Group (UK) Ltd, Croydon, CR0 4YY
30/09/2026
02225244-0011